OXFORD MONOGRAPHS ON GEOLOGY AND GE

OXFORD MONOGRAPHS ON GEOLOGY AND GEOPHYSICS

Paleoclimatology

THOMAS J. CROWLEY
GERALD R. NORTH

OXFORD UNIVERSITY PRESS · New York
CLARENDON PRESS · Oxford

Oxford University Press

Oxford   New York
Athens   Auckland   Bangkok   Bombay
Calcutta   Cape Town   Dar es Salaam   Delhi
Florence   Hong Kong   Istanbul   Karachi
Kuala Lumpur   Madras   Madrid   Melbourne
Mexico City   Nairobi   Paris   Singapore
Taipei   Tokyo   Toronto

and associated companies in
Berlin   Ibadan

Library of Congress Cataloging-in-Publication Data
Crowley, Thomas J., 1948–
Paleoclimatology / Thomas J. Crowley and Gerald R. North,
p. cm.—(Oxford monographs on geology and geophysics; no. 18)
Includes bibliographical references.
ISBN 0-19-503963-7
ISBN 0-19-510533-8 (PBK.)
1. Paleoclimatology.   I. North, Gerald R.   II. Title.   III. Series.
QC884.C76   1990   551.6—dc20   89-26604

1 3 5 7 9 8 6 4 2

Printed in the United States of America
on acid-free paper

PREFACE

The last fifteen years have witnessed not only considerable observational advances in paleoclimatology, but also the growth of a new field—theoretical paleoclimatology. For the first time, physicists and atmosphere and ocean scientists are applying quantitative climate models to interpret the many fascinating observations uncovered by geologists. In some cases we have a good understanding of the observed changes. In other cases there is a considerable gap between models and data.

The purpose of this book is to summarize results from both observational and modeling studies in order to assess our understanding of past climate change. We discuss results from both Quaternary and pre-Quaternary studies and review simulations covering a large number of different models. For the most part we have kept the discussion of observations and models separate, so that those more interested in one will not be too distracted by the other.

Although the book is part of a monograph series, we include a fair amount of background material so that the book may be used as a graduate level text. For the same reason we have added an introductory section on climate models that addresses their structure,

power, and limitations. We conclude by summarizing the present state of our knowledge, identifying areas that require more research, and applying our knowledge of the past to interpreting the consequences of a future greenhouse warming.

A number of acknowledgments are due. We thank the following for comments on individual chapters: Walter Alvarez, Raymond Bradley, Andrew McIntyre, Robert Oglesby, Judith Parrish, William Ruddiman, and Thompson Webb III. We also thank Eric Barron and William Hyde for comments on the entire manuscript. We are grateful to our editor Joyce Berry for her encouragement and patience in the completion of this project. We also wish to thank our employers, Applied Research Corporation and Texas A&M University, for their support. Last, we owe a special debt to Ann Galloway for the immense amount of work done in producing a readable manuscript and legible figures. This work was supported in part by National Science Foundation grants to T.J.C. and G.R.N.

College Station, Tex. T.J.C.
G.R.N.

CONTENTS

PART I
INTRODUCTION TO CLIMATE MODELS

1. ELEMENTARY MODELS OF THE CLIMATE SYSTEM

1.1 Introduction

1.1.1 Goals and Elementary Concepts

The goal of this book is to link recent paleoclimate data analyses with progress in climate modeling. In the process, the advances in these disciplines will be brought into better perspective alongside one other, and in principle the significance of each can then be understood more clearly.

The first two chapters of the book represent an introduction to climate models. Before proceeding to the discussion of models, it is useful to briefly discuss the nature of the climate system and climate variability through earth history. The climate system is composed of many subsystems (Fig. 1.1): the atmosphere, cryosphere, biosphere, and hydrosphere. Each component of the climate system has its own characteristic physics and time scales of response. For example, both the atmosphere and the ocean can be considered fluids on a rotating sphere, but with different time scales of response (weeks to months for the atmosphere, years to centuries for the ocean). The equations of motion are similar for both the atmosphere and ocean, but there are also some significant differences in other features of the system (e.g., clouds and radiative processes for the atmosphere and solid lateral boundaries for the ocean). When we consider biosphere processes (e.g., vegetation and soil moisture feedbacks), the time scales are closer to those of the atmosphere than the ocean, but the physics of the processes is completely different. Similarly, numerical models of the cryosphere differ from any of the above subsystems, and fluctuations (except for sea ice) are generally on a much longer time scale than for the atmosphere and ocean.

The different components of the climate system interact with each other in a complex manner, with the combination of different physics and different time scales of response resulting in

a relatively complicated pattern to the nature of climate change. A convenient way of illustrating the net result of these interactions is to graph variability on a variance spectrum (a brief summary of time series terminology is given in Appendix C). A sketch of the surface temperature spectrum originally developed by Mitchell (1968) is shown in Fig. 1.2. This particular diagram is qualitative, but it conveniently summarizes and classifies many different response characteristics of the system.

In examining the climate record, we find that variability exists over a great continuum of temporal scales ranging from billions of years to seconds. Note that the spectrum has a number of bumps and spikes superimposed on smoother features. There is also a general decline from left to right. The very narrow band responses at the annual cycle and its harmonics and those at the diurnal cycle are essentially linear system responses to periodic forcing by the solar heating cycles. Other sharp peaks occur at nearly discrete periods in the range from 100,000 down to 20,000 years. As we will see later, these have been identified with external periodic forcings due to the earth's orbital elements being changed (Chap. 7).

In addition to the "spikes," the climate spectrum is characterized by a "background" record due to nonperiodic phenomena. This background record has some plateaus, and there is considerably more variance in the lower frequency end of the spectrum. Much of this background variance is thought to originate from sluggish parts of the system accumulating perturbations due to shorter time scale forcing phenomena. In the long run these anomalies would average to zero but can be of quite long duration. For example, weather events are very short lived (few days) and the upper ocean will add up these essentially random perturbations over times comparable to the upper ocean adjustment time. By chance the sum of these random

3

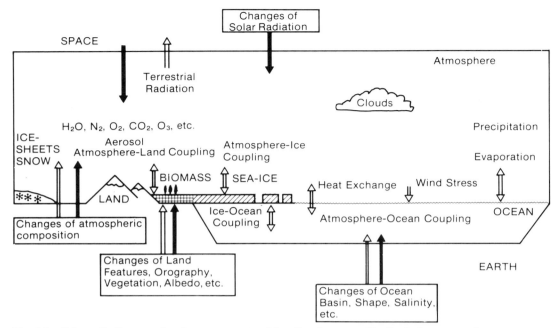

Fig. 1.1. Schematic diagram of major components of the climatic system. Feedbacks between various components play an important role in climate variations. [Adapted from National Research Council, 1975] *Reproduced with permission of the National Academy of Sciences.*

Fig. 1.2. Schematic variance spectrum of the surface temperature versus frequency [actually labeled in terms of the period = (frequency)$^{-1}$ in years], illustrating time scale for different components of the climate system.

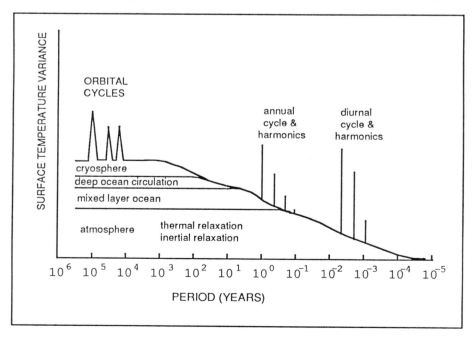

inputs over a finite period of time will not always be zero. This will occasionally lead to the buildup of a thermal anomaly of upper ocean temperatures lasting months or even a few years.

The spectral representation of such a response of a slow component is the plateau-like feature in the spectrum—flat out to about the period of a few years, then tapering off (e.g., Hasselmann, 1976). The tapering off of the spectrum from low to high frequencies is called "red noise" and is very common in geophysical data. Similar reasoning holds for the other slowly adjusting components. On the other hand, if the relaxation time is short, as in the case of weather, the spectrum is flat all the way out to periods of a few days. The latter case is called "white noise" (see Appendix C).

Even though the climate system is very complex, it is possible to make some simplifying assumptions when attempting to model it. In some cases we can take one subsystem to be quasi-static with respect to another. For example, it is often convenient to set the continental glaciers in place (mathematically) and then find the equilibrium climate of the atmospheric subsystem, taking the glaciers as part of the given boundary conditions that force or constrain the atmosphere. In another example, some of the subsystems are linear in their response to interactions with other components. Additionally, some nonlinear systems can be modeled quite satisfactorily; progress is advancing on other systems. Climate modeling becomes essentially the sorting out of isolatable, relevant subsystems that can be approximated well by limited mathematical models. Through such an approach we can learn about many features of the climate system, even though we cannot yet model the whole system. On this optimistic note we take our next step in examining the nature of climate and climate models.

1.1.2 Note on Climate Modeling Chapters

The first two chapters contain some elementary aspects of climate theory and climate models which will be applied in subsequent chapters. These chapters are primarily aimed at those who have not had much exposure to climate models but would like to learn more in order to better understand the dynamics of the system whose history they are reconstructing. Our purpose is hardly a complete development of the theory of climate modeling. That would be too ambitious and belongs in a book by itself (e.g., Washington and Parkinson, 1986; Schlesinger, 1988). Rather, we hope to bring readers to a point where they will have learned enough to exercise sound judgments about the uses of these tools, and perhaps embark on their own independent reading program in the climate modeling literature. Those who are familiar with climate models may wish to skip directly to the chapters that discuss data and simulations.

The first chapter introduces the earth as a planet in radiative balance. This balance has implications for the large-scale surface temperature field. Simple models are used to illustrate factors responsible for controlling planetary temperatures, heat transport by the atmosphere–ocean system, and seasonal cycle variations. The significance of terms such as climate sensitivity, stability, and ice-albedo feedback are discussed. The second chapter discusses the general circulation of the atmosphere and ocean.

Before proceeding further it is important to introduce the concept of a hierarchy of climate models. This concept dates to one of the earliest surveys of climate modeling (Schneider and Dickinson, 1974). The basic notion is that in a complex system such as ours, there exists a sequence of ever more complicated models, each incorporating more detailed treatments of climate processes, and that the sequence converges to a very accurate simulation as the level of complexity increases. Experience has shown this approach to be very useful, not only due to the insight gained from looking at simple bulk considerations first, but also due to the economic constraint imposed by lack of computer time for the more complex models. We shall begin our study with energy balance models (EBMs), which are based on energy conservation and simple formulas for the various flux terms. The second chapter introduces general circulation models (GCMs), which explicitly calculate fields such as temperature, winds, and precipitation.

The reasons for starting with the simpler models is that they afford several advantages, especially to the beginner. By isolating only the large-scale surface temperature field, we select the single most constrained component of the system due to the necessity for near-radiative balance of the planet as a whole. For this reason even fairly simple models calculate approximately the right answers to a number of questions of interest to paleoclimatologists. In addition, of course, EBMs are easy to understand and can be solved at almost no cost compared to the more realistic members of the model hierarchy.

1.2 Energy Balance Models of the Present Climate

1.2.1 Radiation and Climate

The simplest class of climate models is the energy balance model. As the name implies, the models focus on the required balance between incoming and outgoing radiation at the top of the atmosphere. Even though they consider only temperature, a number of significant insights are possible. Before addressing the details of these models, it is useful to briefly review some basic facts about the planetary radiation balance.

Radiant energy emitted by the sun occupies a broad band of wavelengths from gamma rays to radio waves (Fig. 1.3), but is peaked at a wavelength of 0.5 μm (visible part of the spectrum). This radiation is significantly affected by its passage through the earth's atmosphere (Fig. 1.4). Some of it is absorbed (19%) by gases and clouds. About 30% is reflected directly back to space by the earth–atmosphere system. The earth's surface absorbs slightly more than 50% of the total incoming energy at the top of the atmosphere. About 60% of this net amount warms the air by sensible and latent heat exchange (Fig. 1.4). The remaining 40% is emitted as radiation, primarily in the infrared portion of the electromagnetic spectrum (see Fig. 1.3).

Part of the upwelling radiation is intercepted

Fig. 1.3. (a) Normalized black body curves for sun (5780°K) and earth (255°K), plotted so that irradiance is proportional to the areas under the curves. (c) Atmospheric absorption in clear air for solar radiation with a zenith angle of 50° and for diffuse terrestrial radiation. (b) Same as (c) but for the portion of the atmosphere lying above the 11-km level, near the midlatitude tropopause. [Adapted from Goody, 1964 by Wallace and Hobbs, 1977]

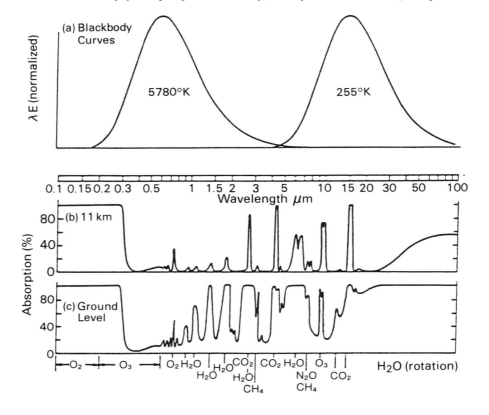

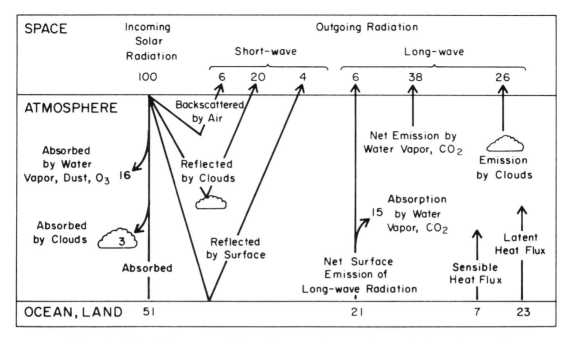

Fig. 1.4. Schematic of the relative amounts of the various energy inputs and outputs in the earth–atmosphere system [After National Research Council, 1975] *Reproduced with permission of the National Academy of Sciences.*

and absorbed by layers of the stratified atmosphere primarily through triatomic trace gases such as H_2O, CO_2, and O_3, as revealed in the absorption spectrum (Fig. 1.3). Carbon dioxide absorbs at about 2.7, 4, 10, and 14 μm; a water vapor continuum exists from ~ 12 to 18 μm; and so on. Water vapor is the most important of these infrared absorbers. Its concentration in the atmosphere is highly variable even on short (weather) time scales. Its saturation vapor pressure also increases approximately exponentially as temperatures increase, with the vapor pressure of water roughly doubling for each increment of 10°C in the range of interest. This is called the Clausius–Clapeyron relationship, and it is an important feature of the earth's climate (and climate models). It is an interesting fact that the relative humidity (ratio of actual concentration to the saturation value) appears to stay at an approximately constant value near 50% as the climate changes.

At some wavelengths there is very little absorption of the upwelling infrared; radiation in these bands is lost directly from the surface to space except when clouds intervene. The amount of radiation absorbed by atmospheric gases is very important for maintaining habitable temperatures on earth. The absorbing layers radiate as warm bodies themselves, thus producing a flux of downwelling radiation which heats the surface to temperatures far above the values that would otherwise have occurred.

Climate modelers have learned to model the radiation fluxes and their interaction with the layers of matter in a vertical column of air above the earth's surface. Not only must the details of the wavelength dependencies of the various absorptivities, emissivities, and scattering properties be included, but allowance for vertical convective motions must also be included. Such convection will occur when the atmospheric thermal layering is buoyantly unstable. The net effect of the convection is to remove heat from the surface and carry it higher in the atmosphere mechanically. Even without including horizontal motions of the atmosphere, these so-called radiative convective models (RCMs) are quite successful in describing the vertical temperature structure of the atmosphere and in leading the way for more accurate and efficient radiation calculations in general circulation models (Ramanathan and Coakley, 1978).

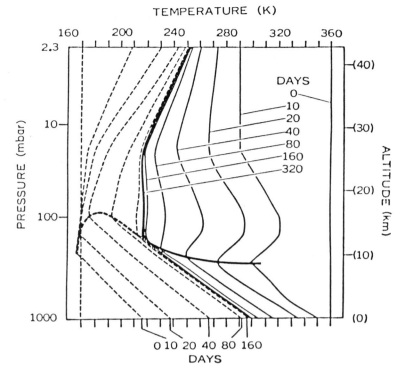

Fig. 1.5. Simulation of vertical atmospheric temperature structure as it equilibrates from two originally isothermal conditions. Temperature (abscissa) versus altitude (shown as pressure on the left and height in km on the right) as a function of time for two cases in a radiative convective model simulation. The rightmost sequence is started from a vertically isothermal condition of 360°K with profiles shown for different times (days) after the model starts to integrate. Dashed profiles on the left are the same type of sequence for a very cold isothermal initial condition. [From Manabe and Strickler, 1964] *Reprinted from Journal of the Atmospheric Sciences with permission of the American Meteorological Society.*

An example of an early RCM computation is shown in Fig. 1.5 (Manabe and Wetherald, 1967). In this computation the planet is started in two different vertically uniform initial states, 170°K and 360°K. The model is allowed to relax in time toward equilibrium. First we note that each initial state leads to the same final equilibrium profile and that the relaxation time (exponential decay constant) is about 40 or 50 days. The nearly discontinuous derivative at about 10 km in the equilibrium solution is called the tropopause. That this very simple model of the vertical structure of the atmosphere is very close to observations gives us confidence that at least this aspect of the climate system is approachable by models. Discussions of the vertical profile based on even simpler radiative equilibrium models without convection can be found in the literature (Goody and Walker, 1972; Houghton, 1977).

1.2.2 Planetary Energy Balance Models

We consider next the heat balance of the globe with simple rules for the radiation terms so that we can make elementary computations of the balance. We select the simplest class from the climate model hierarchy—the zero-dimensional energy balance models (cf. North et al., 1981; North, 1988). [By zero-dimensional we mean there are no spatial dimensions (latitude, longitude, height) and calculations are for global quantities (e.g., global average temperature).]

The simplest EBMs are based on the assumption that the planet is in a steady-state balance of the rate of incoming absorbed radiation energy from the sun and the outgoing terrestrial radiation rate to space. If the incoming absorbed radiation is independent of planetary temperature, and the emitted terrestrial radiation increases as the average temperature increases, there will be a single temperature for which the balance can be achieved. If the planet's outgoing radiation is too low, the temperature will rise until the outgoing rate exactly matches the incoming. Similarly, if the planet is too warm for a match, the outgoing excess will cause a planetary cooling until a balance is achieved.

In equation form, we can write the absorbed solar radiation per unit time as $\sigma_0(1 - \alpha_p)\pi a^2$, where a is the radius of the earth, σ_0 is the solar constant (amount of energy per unit time passing through a unit area perpendicular to the beam; for the earth's mean annual distance from the sun, this is about 1370 W/m^2), and the fraction of incident radiation averaged over the globe that is reflected back to space is the planetary albedo α_p (for the earth at present $\alpha_p = 0.30$). The outgoing radiation per unit area per unit time to space can be calculated using the Stefan–Boltzmann law, which states that $IR = \sigma T^4$, where σ is the Stefan–Boltzmann constant (5.67×10^{-8} W/(m^2 °K^4)). However, for the temperature range at which the earth radiates, this fourth-power expression can be approximated by a straight line, $A + BT$, where T is the planetary average temperature (°C) and the values of A and B are to be determined from measurements to be discussed later. Given the values of A and B, we have the equality:

$$4\pi a^2(A + BT) = \pi a^2 \sigma_0(1 - \alpha_p)$$

(total outgoing) = (total incoming absorbed)

(1.1)

or

$$T = \frac{\sigma_0}{4B}(1 - \alpha_p) - \frac{A}{B} \qquad (1.2)$$

If the earth were a perfect black body we could calculate A and B from the Stefan–Boltzmann law, leading to values of 315 W/m^2 and 4.6 W/(m^2 °C). Insertion of these values leads to an estimate of about -16°C for the earth's mean radiating temperature, a value about 31° too low. The reason for the discrepancy is that the earth is far from a perfect black body radiator. The effective radiating temperature of the planet is about -16°C, but most of the radiation to space is leaving the planet from sources high in the atmosphere, not from the earth's surface. This is illustrated by the equilibrium pro-

file in Fig. 1.5 where the temperature at 10 km is seen to be much lower than at the surface (the exponential decrease of density versus altitude makes the contributions above the tropopause negligible for the present purposes).

Satellite data have been used to estimate the values of A and B (Short et al., 1984). This method and the calculations from detailed radiative convective models (Ramanathan and Coakley, 1978) lead consistently to values of A and B in the neighborhood of 210 W/m^2 and 2.1 (W/m^2 °C), provided certain assumptions about the constancy of cloudiness parameters hold. Once these values are inserted we obtain the approximately correct value of 15°C. The substitution of the empirical values A and B effectively has taken into account the modifying effect of the atmosphere (greenhouse effect).

The above example shows how one blends theory and empirical information to form a simple model of the earth's steady-state temperature. A similar path of reasoning at this very primitive level leads to reasonable results for the planets Venus and Mars once account is taken of their atmospheric compositions (e.g., Goody and Walker, 1972). These facts give us confidence that the energy balance approach is a useful way to introduce ourselves to the climate system. Let us now extend our range of questions to include the effect of heat transport parallel to the earth's surface.

1.2.3 One-Dimensional EBMs

Observations of latitudinally averaged outgoing radiation indicate that more radiation is emitted to space in high latitudes than is absorbed at the surface. This pattern implies that heat is being exported from low latitudes to high latitudes across latitude circles by atmospheric and oceanic currents. Models with latitude dependence must take this transport process into account by including a flux term for it. We describe the climate of the planet by the zonally averaged (averages around latitude belts) temperature $T(x)$, where for convenience we have used $x = \sin(latitude)$ to denote the equator-to-pole coordinate. It has the value 0 at the equator and 1 at the pole. This turns out to be a convenient choice of variable since Δx, the width of a zonal strip, is proportional to the area of the strip and for energy budget studies this is a more physical quantity than true distance or latitude on a meridian.

We start by ignoring the seasons (use mean annual solar input) and consider each strip to be in steady state with respect to its energy budget. For each strip there will be an infrared radiation energy rate escaping to space per unit area, $A + BT$, with the same constants A and B as before; an energy rate absorbed from solar radiation, $QS(x)a(x)$, where $S(x)$ is the distribution of radiation reaching the top of the atmosphere averaged through the length of the average day (it is normalized so that its integral over the hemisphere with respect to x is unity), $a(x)$ is the coalbedo (1 − albedo) at latitude x and $Q = \sigma_0/4$ (reduction results from earth's cross-section divided by its surface area). To a good approximation $S(x)a(x)$ can be written as a parabola in x with symmetric maximum at $x = 0$, the equator. Now we must account for the energy that flows into (or out of) a strip from the adjacent strips poleward and equatorward of it. We consider next a simple form for the heat flow.

In a detailed model including atmospheric (and oceanic) dynamics, the net flow of heat into a latitude belt is obtained by integrating through time the complete set of equations of motion for the geophysical fluid components (see Chap. 2). These motions are very complicated and fully three-dimensional, as can be seen from any weather map. The circulations can be usefully separated into two parts: the mean flow components and the eddy or random components. In the earth's atmosphere the eddy components dominate outside the tropics. The reason for the existence of eddies is the constant presence of unstable conditions in the steady atmospheric flow due to thermal contrasts between different regions.

Atmospheric eddies have a characteristic time scale of a few days. By contrast, the radiative relaxation time for a column of air as seen from the last section is about a month. The scale separation between the fast-eddy motions and the slower forcing (e.g., seasonal heating cycle) and radiative relaxation time scales suggests that we may model the heat flowing into and out of a strip as down-gradient diffusion, which we adopt for the present illustrations. Here is a good example of where a slower component (large-scale

thermal field) treats a faster component (atmospheric eddies) only on an average basis.

The flux of heat across an imaginary border (latitude circle) is taken to be proportional to the gradient of the temperature field and the length of the border (circumference of the latitude circle). The difference between what flows into the strip (from the equatorward edge) and what flows out (poleward) per unit of area is expressed mathematically as the divergence of the flux. This difference goes into heating the strip and it must be balanced by the radiative terms. The latitude-dependent energy balance equation can be written:

$$-\frac{d}{dx} D(1 - x^2) \frac{d}{dx} T + A + BT$$
$$= QS(x)a(x) \quad (1.3)$$

where the first term represents the divergence of heat flux in terms of the latitude variable x (North et al., 1981). This expression is a second-order differential equation in x and requires end-point boundary conditions ($x = \pm 1$) to be given in order that the solution be uniquely determined (the reader may recall that two arbitrary constants must be fixed in solving a second-order differential equation). These conditions are that no heat is to flow per unit area into the poles.

The above conditions determine the solution to the linear problem (no ice-albedo feedback). The model can be solved analytically for the case that D is a constant independent of x (North, 1975a,b). The solution is especially simple when the right-hand side of (1.3) is an even second-degree polynomial in x. It is readily found that a solution for $T(x)$ is also a parabola, which is an even function of x. Some instructive examples for three values of D are shown in Fig. 1.6. In the case of infinite D the planet is isothermal (analogous to the equipotential surface formed by an electrical conductor). For zero D the temperature has the same shape as the absorbed heating $QS(x)a(x)$, which leads to very steep temperature gradients. The real earth is in between the two cases. This suggests that for many purposes a second-degree polynomial is a sufficiently good approximation (see North, 1975b for the solution in terms of Legendre polynomials). The simplest approach is to adjust

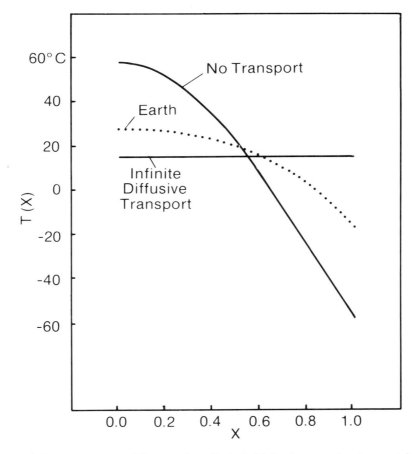

Fig. 1.6. Energy balance temperature (°C) versus sine of latitude (x) for the cases of no transport, infinite transport, and present case (schematic). [From North et al., 1981] *Courtesy of American Geophysical Union.*

the thermal diffusion coefficient D (a measure of the curvature of T versus x) until a good fit to the observations is achieved (Fig. 1.7). Now the model is set for sensitivity studies. This is a good example of a linear system in which the spatial scale of the input (parabola in x) is reproduced in the scale of the response of the system (also parabola in x). The parabolic forms in x mean that the scales are essentially global and semihemispheric.

1.2.4. Two-Dimensional EBMs with Realistic Geography

We will now skip over several intermediate categories of EBMs (cf. North et al., 1981) to one that is particularly useful for certain paleoclimate studies. In the previous section, we discussed how horizontal heat transport modifies

the local climate on a planet. That discussion applied to mean annual conditions, i.e., no seasons. However, the seasonal cycle is the largest climate change we know of—the change in temperatures over North America from winter to summer is far greater than glacial-interglacial changes in mean annual temperatures of the Pleistocene. There are really two separate features to consider for the seasonal cycle: the effects of geography and those of seasonal forcing. The geographic term is very important because it includes the different effective heat capacities of land versus ocean. These are critical because the differing heat capacities of land and sea areas almost completely dominate the seasonal temperature changes on the earth's surface (Fig. 1.8). In this section we will examine the more complicated situation of differential heat capacity with realistic land–sea geography (North et

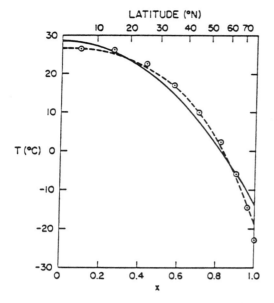

Fig. 1.7. Mean annual Northern Hemisphere temperature versus sine of latitude for a two-mode (parabolic solution) EBM model with diffusive transport (solid curve), the observations (circled dots), and a three-mode (quartic in x) model (dashed curve). [From North, 1975b] *Reprinted from Journal of the Atmospheric Sciences with permission of the American Meteorological Society.*

al., 1983; Hyde et al., 1989). The interested reader is referred to North and Coakley (1979) and North et al. (1981) for a discussion of the intermediate cases.

We construct a seasonal model along the lines of previous sections. In the seasonal model,

however, a storage term $C(\vec{r})\,\partial T/\partial t$ must be added, where $\vec{r}$ designates a point on the sphere, and $C(\vec{r})$ is the latitude- and longitude-dependent heat capacity per unit area. The model forcing is given by an expression for the distribution of heat reaching the top of the atmosphere $S(x, t)$, which is a function of latitude and time of the year t. We first note that the solution is cyclical with a period of 1 year. This allows us to express the time variation of $T(\vec{r}, t)$ in a Fourier series consisting of the mean annual, seasonal harmonic, and semiannual harmonics for $T(\vec{r}, t)$. Examination of seasonal data fields ensures that this truncation in two harmonics is close to nature. Linearity of the system is strongly suggested by the absence of appreciable higher harmonics in both the forcing and the (observed) response fields (see North and Coakley, 1979). The temperature at any given point on the earth's surface satisfies

$$C(\vec{r})\,\frac{\partial T}{\partial t} - \nabla \cdot (D(x)\nabla T) + A + BT$$
$$= QS(x, t)a(\vec{r}) \quad (1.4)$$

where $C(\vec{r})$ is the local heat capacity per unit area, $D(x)$ is the local diffusion coefficient representative of heat transport, Q is the solar constant divided by 4 (see Section 1.2.3), $S(x, t)$ is the distribution function for insolation, and $a(\vec{r})$ is the local coalbedo.

The model heat capacity is a step function which can take on any of three values: large over

Fig. 1.8. Annual range in temperature at the earth's surface (°C). Topography is not plotted in order to emphasize the dominant control of the land–sea distribution on the seasonal cycle. [After Monin, 1975] *Reproduced with permission of the World Meteorological Organization.*

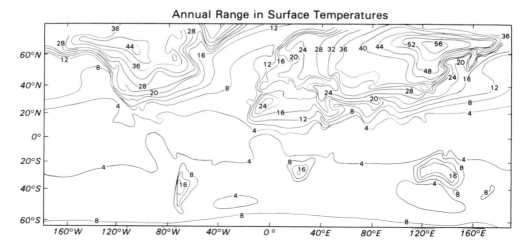

ocean, much smaller over sea ice, and very small over land (ratio of 60:5:1). In the first case the heat capacity is taken to represent the heat capacity of a 75-m mixed-layer ocean, while in the third case it represents roughly half the heat capacity of an atmospheric column. The 60:1 ratio approximately reflects the ratios of the relaxation time for a thermal perturbation of the two media (5 years for mixed-layer ocean versus 1 month for an atmospheric column). The situation is ambiguous over sea ice, where in winter the ice has a low effective heat capacity because it cuts off the warmer ocean from the atmosphere and hence allows large air temperature decreases for small changes in forcing analogous to land surfaces. In summer sea ice behaves as if it had a very high heat capacity, owing to the heat required to melt the solid ice (latent heat of fusion).

The diffusion approximation is the same as used earlier. It is now latitudinally dependent in order to imitate the greater atmospheric heat transport in low latitudes (cf. Lindzen and Farrell, 1977). The outgoing infrared radiation is as before. The albedo terms are also consistent with satellite observations (Stephens et al., 1981). The coalbedo varies with latitude to account for the "solar zenith angle effect," the fact that reflection increases when the sun is closer to the horizon, especially over clouds and open water. It can be particularly important in high latitudes where the incident radiation is at a low angle (Lian and Cess, 1977). For example, the normal coalbedo of any high-latitude surface is about 0.6, significantly lower than tropical land or water (about 0.9). The albedo term also allows for regionally varying snow or ice fields. Where they occur, a higher albedo than the "zonal average coalbedo" is inserted. From the above discussion, the reader can see that although this equation involves some adjustable parameters, they are constrained by data.

The method of solving the model is discussed in North et al. (1983). The relevant fields are expressed as spherical harmonics in space and complex exponentials in time, the latter being just a compact way of expressing the truncated Fourier series:

$$T(\vec{r}, t) = \sum_{l=0}^{L} \sum_{m=-l}^{l} \sum_{n=0}^{2} T_{lm}^{n} Y_{lm}(\vec{r}) e^{2\pi i n t} \quad (1.5)$$

where the Y_{lm} are the complex spherical harmonics with meridional wavenumber l and zonal wavenumber m, and n indexes the time harmonics ($n = 0$ is the annual average, $n = 1$ is the annual cycle, etc.). The degree of truncation L is 11 for the temperature and solar input, in which case 22 polynomials are required for C and a. The time series is truncated at $n = 2$ (the semiannual cycle). This method converts the partial differential equation into 432 simultaneous linear complex algebraic equations which must be solved for the components T_{lm}^{n}; in so doing we eliminate the problem of integrating in time until a repeating seasonal cycle is achieved. Given a complete set of boundary conditions, the model can be solved in under 3 min of CPU time on a VAX 8300. A time-marching version of the above linear EBM requires two orders of magnitude more CPU time to produce an equilibrium annual cycle, primarily because the presence of the 5-year oceanic time constant requires that at least 15 years must be simulated to damp out the transient part of the solution which depends on the initial conditions. For comparison, a full seasonal cycle simulation on the NCAR GCM (known as the Community Climate Model, CCM) requires about 70 hr of time on a Cray-1 equivalent supercomputer in order for the model to reach equilibrium.

Aside from computational efficiency the spectral decomposition method (spherical harmonics in space and Fourier series in time) facilitates physical interpretation in many cases. For example, in the linear model the second harmonic response is a direct consequence of the second harmonic forcing after passing through the filter induced by geography. Similarly, the seasonal phase lags of the harmonics are easily interpreted as consequences of geography (about 1 month over land, 3 months over ocean). The degree of truncation ($L = 11$) acts as a smoothing on the output field; no details finer than about 2000 km can be captured by the model.

The model simulates the present seasonal cycle remarkably well (Fig. 1.9), a result compatible with the dominance of the land–sea distribution on the seasonal cycle. The semiannual cycle is also in qualitative agreement with the data (not shown; see North et al., 1983; Hyde et al., 1989). Phase lags over the ocean are virtually a quarter cycle (3 months) in the model but are

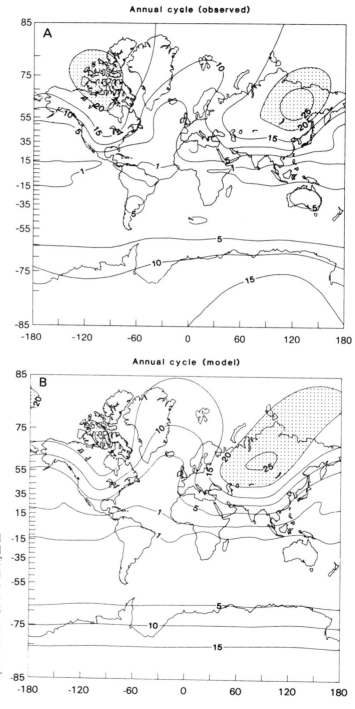

Fig. 1.9. Comparison of the modeled present seasonal cycle (two-dimensional EBM) with data. (A) The amplitude of the annual harmonic (°C) in the observed annual cycle of surface temperature. Climatological data were smoothed by a spatial filter which preserves features having a length scale larger than about 1600 km. For further comparisons including phase lags and higher harmonics, see North et al., 1983 and Hyde et al., 1989. (B) As in (A), but for the two-dimensional EBM. [From North et al., 1983] *Courtesy of American Geophysical Union.*

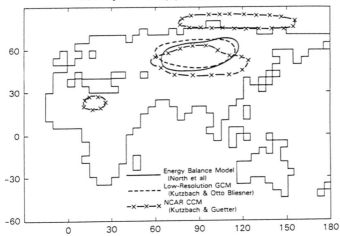

JULY [ΔT = 4°C] [9KYBP–PRESENT]

Fig. 1.10. Comparison of energy balance and general circulation model responses to increased summer insolation at 9000 BP. The 4°C contours indicate the difference between the model response for 9000 years ago and the present (the earlier time period is warmer). The energy balance model is from North et al. (1983); the low-resolution model and the NCAR CCM are from Kutzbach and Otto-Bliesner (1982) and Kutzbach and Guetter (1984), respectively. Note that all models have comparable sensitivity. [From Crowley et al., 1986] *Reproduced from Science 231:579–84, Copyright 1986 by the AAAS.*

usually closer to 2 months in the data (North et al., 1983). One feature that the model can not capture are areas strongly affected by ocean heat transport, e.g., the Gulf Stream. The advection of such warm water has a relatively small effect on the seasonal cycle amplitude and phase but a very significant effect on the mean annual temperatures. We therefore have to be careful about interpretations when we consider regions that may have been strongly affected by changes in such features.

Any climate model "tuned" to the present climate will give a different seasonal cycle solution for altered forcing or boundary conditions. A key question for any climate model is whether the altered response is reasonable and, for EBMs, has sensitivity characteristics comparable to GCMs. For the two-dimensional EBM just described, several lines of evidence indicate that for some problems the EBM compares quite favorably with GCMs. For example, 9000 years ago the seasonal cycle of insolation was different from the present (see Fig. 4.15)—more summer insolation and less winter insolation. Based on the relationship between the observed seasonal cycle and present insolation, we would predict the greatest warming to occur in the interior of the largest landmass. Figure 1.10 shows the results of the summer warming for two GCMs (a low-resolution model by Kutzbach and Otto-Bliesner, 1982; the NCAR GCM, Kutzbach and Guetter, 1984), and a two-dimensional EBM (solid contour). Shown is the isotherm for a 4°C warming. The magnitude of

summer warming at 9000 BP appears to be a very robust response of climate models.

Since the EBM is linear, the above comparison further implies that the average GCM temperature response to changes in seasonal forcing in continental interiors is linear (Crowley, 1988). Given the clearly nonlinear nature of many climate processes, the reader may wonder how such linearity can be obtained. A simple example might clarify the matter. Consider how pressure may change in a cylinder as a piston moves up or down. There are two ways of addressing this problem. One would be to integrate the effect of numberless nonlinear interactions of individual molecules in the cylinder. Another would be to utilize the ideal gas law ($PV = nRT$), which is both simple and linear! The energy balance model functions like the ideal gas law. By taking the ensemble average of many complex interactions, some of the processes can be reduced to more simplified expressions. This is not to imply that all meteorological processes can be so simplified, and there are some limitations even for temperature, but the redeeming feature of the above comparison is that it demonstrates that EBMs can be employed with some usefulness to attack a number of problems not yet addressed with GCMs.

1.3 Perturbing the Energy Balance

A considerable degree of insight can be derived from changing the forcing in energy balance models. The following sections discuss some useful concepts and an intriguing result that occurs when forcing varies in a nonlinear model.

1.3.1 Climate Sensitivity

A basic property of the climate system is its sensitivity to changes in external parameters such as the solar constant. As an example, consider the globally averaged mean annual temperature estimated from energy balance as was just considered above. We may form a sensitivity parameter by computing the derivative:

$$\beta = \frac{Q}{100}\frac{dT}{dQ} \qquad (1.6)$$

The sensitivity parameter is the change in globally averaged temperature for a 1% change in solar constant. For the simple (linear) model above we can show by elementary manipulations that

$$\beta_{\text{Lin}} = \frac{Qa_P}{100B} \qquad (1.7)$$

Note that in going from the black body to the real-earth values of A and B, the coefficient B decreases significantly. The decrease of B comes about because the amount of water vapor in the atmosphere is an increasing function of surface temperature (the relative humidity tends to remain roughly constant in the atmosphere as other conditions change).

Another way of looking at the water vapor feedback effect is to think of A as a function of W, the amount of water vapor in the column of air. Since W is actually a function of the temperature $W(T)$, we must allow for this effect in the derivation of (1.7). The result is that B is to be replaced by $B' = B - (dA/dW)(dW/dT)$, which turns out to be about half the size of the original (black body) B. The presence of water vapor feedback leads to an increase in the sensitivity by about a factor of 2 (cf. Raval and Ramanathan, 1989). Climate modelers refer to this multiplicative increase as the "gain" due to the water vapor feedback in the system. Including the water vapor feedback in (1.7) leads to a sensitivity parameter of about 1.12°C per 1% change in solar constant.

Climate sensitivity also varies by the frequency of forcing (North et al., 1984). Consider

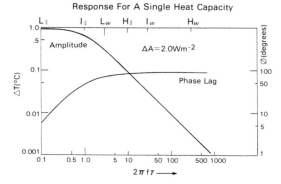

Fig. 1.11. Example of the sensitivity of system response to forcing, with sensitivity varying according to the frequency of forcing. Illustration is for the amplitude (°C) and phase lag (degrees) for a simple linear global model with time characteristic τ forced at frequency f. Note that the amplitude of response (and therefore sensitivity) increases as frequency of forcing decreases and the effect is larger for land than for water. The labels L_ℓ, I_ℓ, H_ℓ refer to low, intermediate, and high frequencies over land (e.g., once per decade, once per year, and once per month). L_w, I_w, H_w refer to these same frequencies but over water. [From North et al., 1984] *Courtesy of American Geophysical Union.*

Fig. 1.11, which illustrates that the amplitude of the temperature response of an EBM is dependent on the frequency of forcing. The temperature response to a 2 W/m² forcing (slightly larger than the ice-age CO_2 radiative forcing; see Chap. 3) varies from ~0.01 to 1.0°C. The smaller response at high frequencies reflects the fact that the system has not reached equilibrium due to an increase in heating before the change of forcing is reversed in sign (Hansen et al., 1981; Cess and Potter, 1984).

1.3.2 Ice-Albedo Feedback

The present temperature field of the earth is such that permanent ice exists year-round at both poles. These ice caps have a much lower local coalbedo than surrounding bare surfaces. If the planet is cooled, we expect the ice caps to grow in area, leading to a lower planetary coalbedo a_P. Since a lower rate of absorption of solar radiation will lead to a lower global temperature, we have the potential for a strong positive feedback mechanism in the global climate system. Some of the earliest modeling studies of climate (Budyko, 1968, 1969; Sellers, 1969) incorporated this feedback and the findings helped stimulate the great interest in climate modeling in the last two decades. In the global average model of the last section we can incorporate ice-albedo feedback in an approximate form and try to investigate its consequences.

Equation (1.1) must now allow for a temperature dependence of coalbedo $a_P(T)$. The easy way to solve the now nonlinear form of (1.1) is to graph the left-hand side against the right (Craford and Källén, 1978; North et al., 1981). Figure 1.12 shows the T versus Q curve, which we refer to as the operating curve. It depicts the steady-state solutions of climate for various values of externally imposed conditions. For a large value of Q, there will be a solution with the earth ice-free; this solution is similar to the linear solution found in a previous section since in this range of T, $a_P(T)$ is constant (denoted by the indicator I on the graph). Si⌐ for small Q there is an ice-covered e⌐ with again $a_P(T)$ constant but r⌐ ⌐noted as III on the graph). F⌐ ⌐ of Q there is a

Fig. 1.12 ⌐peratures corresponding to the cli⌐ ⌐r a zero-dimensional climate mo⌐ ⌐e cap as a function of solar con-sta⌐ ⌐ present value. The roots I, II, III cor-respo⌐ ⌐olutions for the present level of solar forc-ing. [A⌐ ⌐orth et al., 1981] *Courtesy of American Geophysi⌐al Union.*

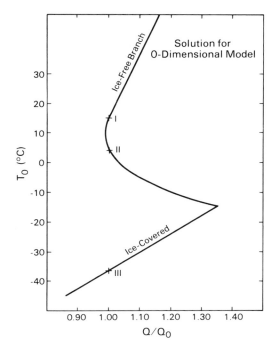

possibility for three solutions, one with a small ice cap (I), one with a large ice cap (II), and one completely ice-covered (III).

Several profound aspects of this climate system are readily apparent in Fig. 1.12. First, note that if the system is in equilibrium on the upper branch (which seems to correspond roughly to the present climate), and if the solar constant is lowered by just a few percent from its present value (about 340 W/m²), the system plunges to the only other steady-state climate available, namely, the ice-covered earth. On the other hand, if we were on the ice-covered earth branch and increased the solar constant, we would not jump to the ice-free branch until the solar constant was raised to about 1.4 times its present value. This last property is the origin of the so-called Faint Young Sun Paradox. Since the sun is thought to have steadily increased its luminosity over the last few 10^9 years (e.g., Ulrich, 1975), the question of why the earth is not ice-covered must be addressed. The most likely way out of the paradox is that the concentration of CO_2 on the early earth might have been much greater (like Venus) and therefore prevented the initial ice-over (Owen et al., 1979; see Chap. 12).

Another interesting feature of the operating curve for the zero-dimensional model is that the intermediate branch cannot be accessed by raising or lowering the solar constant from the upper or lower branches. It is curious that the intermediate branch also has the peculiar property that as the solar constant is increased the ice cap grows. This is especially disturbing since no moisture budgets are included in the model. The unphysical nature of this branch will be resolved in the next section.

The example just given illustrates the complex structure of even the simplest ice feedback model. The multiple-branch structure shown in Fig. 1.12 persists in models with latitude dependence and even exists in GCMs (Wetherald and Manabe, 1975). Further insight is derived from a comparison of a two-dimensional EBM's (North et al., 1983) global annual temperature as a function of the solar constant (Fig. 1.13). The linear model (straight line) has a sensitivity $\beta_0 = 1.14°C$. For a nonlinear ice feedback version of the two-dimensional seasonal model (North et al., 1983), the lower (ice age?) branch has a sensitivity twice as large (2.40°C). Thus,

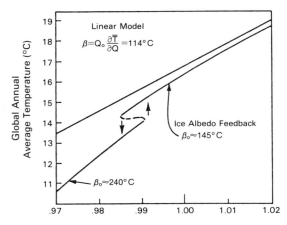

Fig. 1.13. Operating curve for the two-dimensional EBM described in text and in Fig. 1.9. The upper operating curve is the case for no snow feedback. The lower curve, which includes a simple nonlinear snow feedback mechanism, shows a larger sensitivity and a discontinuity at about $Q/Q_0 = 0.985$ (ice age analog?). Note the increase in sensitivity as the ice cap grows. [From North et al., 1983] *Courtesy of American Geophysical Union.*

ice-albedo feedback affects sensitivity, and the effect may vary with the forcing.

1.3.3 Stability

When a physical system has more than one admissible solution for a given set of conditions, we must examine another property of the solutions to see if they can all be realized in nature. We now ask the question about whether an individual solution is stable. A stable solution is one that will return to its equilibrium value if it is perturbed slightly. An unstable solution is one in which a small perturbation leads to an excursion of the system to some other branch. An unstable solution is rarely realized in nature, since there are inevitably small perturbations present.

Most stability analyses for physical systems are very difficult. However, the simple climate model illustrated in Fig. 1.12 provides one case in which the analysis requires only a few simple manipulations. By following this simple sequence, the reader can better understand the stability concept and actually follow one case from beginning to end (those not interested in this level of detail might want to skip to the next section).

Consider mean annual models but with departures from steady state, so that the left-hand side of (1.1) no longer equals the right. Any excess of heating rate compared to cooling to space will result in a change in temperature. The coefficient of proportionality is the effective heat capacity per unit surface area C, which for an all-land planet might be considered to be the heat capacity at constant pressure for a column of air with unit cross-sectional area. For large departures of temperature (or over ocean) larger values of C must be used, but for our present purposes consider it to be constant and having the atmospheric column value. We may write the new energy balance as

$$C\frac{dT}{dt} = -A - BT + Qa_P(T) \qquad (1.8)$$

This is a nonlinear first-order differential equation for the evolution of $T(t)$. The standard method of conducting a linear stability analysis is to substitute the equilibrium global temperature with a small departure added:

$$T(t) = T_0 + T_1(t) \qquad (1.9)$$

where T_0 is the solution of (1.1) of the equilibrium problem including ice-albedo feedback and $T_1(t)$ is an infinitesimal time-dependent departure. We may expand $a_P(T)$ in a Taylor series since $T_1(t)$ is so small:

$$a_p(T) = a_p(T_0) + a'_p(T_0)T_1 + \cdots \qquad (1.10)$$

Inserting (1.9) and (1.10) into (1.8) and using (1.1), we obtain a linear differential equation for $T_1(t)$:

$$C\frac{dT_1}{dt} = [-B + Qa'_P(T_0)]T_1 \qquad (1.11)$$

A simpler expression for the bracketed quantity can be obtained if we differentiate (1.1) with respect to Q, noting that Q is a function of T_0 and vice versa as in Fig. 1.12.

$$-a_P\frac{dQ}{dT_0} = [-B + Qa'_P(T_0)] \qquad (1.12)$$

We finally arrive at the result:

$$\frac{dT_1}{dt} = -\lambda T_1 \qquad (1.13)$$

with

$$\lambda = \frac{a_P}{C}\frac{dQ}{dT_0}$$

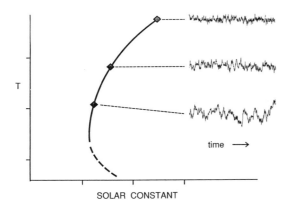

Fig. 1.14. Schematic depiction of the time series of temperature fluctuations about three different equilibrium climates near a bifurcation point. Operating curve is for an energy balance model of global average temperature. Different time series represent solutions to stochastic forcing of a linearized version of the model. Time and temperature scales are arbitrary, but the relative magnitude of different temperature amplitudes represents actual model output. For reference, the smallest temperature fluctuations may be considered representative of a system in equilibrium. Analysis of the instrumental record of surface temperature fluctuations over the last 120 years suggests that the natural variability of such a system is about ±0.5°C. [From Crowley and North, 1988] *Reproduced from Science 240:996–1002, Copyright 1988 by the AAAS.*

forcing increase near "bifurcation points" (Fig. 1.14). The steeper the slope, the longer the return time, and the larger the fluctuation amplitude for a given random disturbance. In the case of a simple linear model with no ice feedback, the e-folding time (time to decay to $1/e$ of the original amplitude) for return is given by C/B, which for a column of air is about 1 month (North and Coakley, 1979; North and Cahalan, 1981).

1.3.4 Small Ice Cap Instability

The behavior of climate models incorporating ice-albedo feedback is especially interesting for parameter values near the first formation of an ice patch. EBM studies suggest that an operating curve like that in Fig. 1.15 results when we include ice-albedo feedback. Note that more than one equilibrium solution (e.g., A, A′) may exist for the same external boundary condition; inter-

Fig. 1.15. Schematic graph of equilibrium solutions of an energy balance model with ice-albedo feedback. The dependent variable is the ice cap radius and the independent variable is the solar constant (or carbon dioxide) increasing to the right. See text for further explanations. [From North and Crowley, 1985] *Reproduced by permission of the Geological Society from "Application of a seasonal climate model to Cenozoic glaciation," by G. North and T. Crowley, in Journal of the Geological Society (London) 142, 1985.*

The solution to this elementary differential equation is $T_1(t) = T_1(0) \exp(-\lambda t)$, which decays back to the nearby steady-state solution if $\lambda > 0$ or grows exponentially away from the nearby solution if $\lambda < 0$. The sign of λ is determined by the sign of the local slope at I, II, or III in Fig. 1.12. One can now see that the negatively sloped intermediate branch of Fig. 1.12 is unstable to infinitesimal perturbations and is therefore of no physical consequence, since even if the climate should find itself in such a state, it would soon wander away because of inevitable small disturbances. The result holds in slightly different forms in more complicated climate models than those discussed above. It is known as the Slope Stability Theorem (Schneider and Gal-Chen, 1973; North, 1975a; Cahalan and North, 1979; Lin and North, 1990).

Aside from the stability characteristics being determined by the sign of the slope of the operating curve, it is interesting that the time to return to the steady-state operating curve in a stable case depends on the magnitude of the slope, and that the variations for a given level of noise

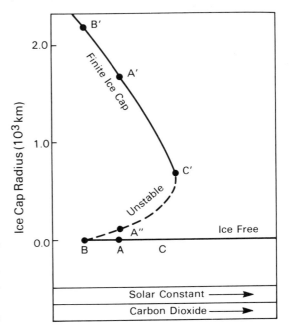

mediate solutions (A″) are unstable. As boundary conditions are slowly changed the possibility of a rapid transition between climate states results. This feature has sometimes been called "the small ice cap instability" (North, 1984), since ice caps smaller than C′ are unstable. Geologists have postulated on intuitive grounds that this mechanism may be important for glacial inception during the Pleistocene (Brooks, 1949; Ives et al., 1975). One equilibrium solution in the model (A) may correspond to an ice-free earth; the other equilibrium solution (A′) may correspond to a glaciated earth (finite ice cap). Although the real climate system may contain more than two branches, e.g., polar ice caps and midlatitude ice sheets, the basic concept of discontinuous steps remains the same.

Another feature of the "simple" model is that system sensitivity depends on its history. For example, suppose the initial climate state in Fig. 1.15 was at A, a solution representing an ice-free state. Now let the solar constant (or CO_2) decrease. Lowering the solar constant moves the equilibrium solution for the earth along the lower line to point B. The earth is still ice-free

up to this stage. However, decreasing the solar constant by a further small increment forces the equilibrium point to jump to the upper solution curve at point B′, representing a glaciated state. Reversing the forcing may not necessarily reverse the results. To return to the ice-free state, the solar constant (or CO_2) may have to increase greatly until another critical point (C′) is reached. The solution state might then dramatically return to the original ice-free branch at point C.

Experiments (Fig. 1.16) with a zonally symmetric (polar land cap) seasonal energy balance climate model (Mengel et al., 1988) suggest that a discontinuous transition from a polar glaciated to a polar ice-free state could occur for infinitesimal perturbations in forcing when the system is poised at a bifurcation on the operating curve. For example, the transition from ice-free to ice-covered state in Fig. 1.16 is associated with an insolation perturbation (.07 W/m²) almost three orders of magnitude smaller than orbital variations of the Pleistocene (see Chap. 7)! Dramatically enhanced sensitivities near bifurcations may therefore play some role in ice sheet

Fig. 1.16. Calculation illustrating that the transition between a seasonally ice-free state and permanent snow cover can occur for a very small change in forcing. Example based on results from a highly idealized seasonal climate model with land poleward of 70°. Ordinate is latitude; forcing represents changes in the solar constant (Q), normalized to present values. [From Mengel et al., 1988] *Reproduced with permission of Springer-Verlag.*

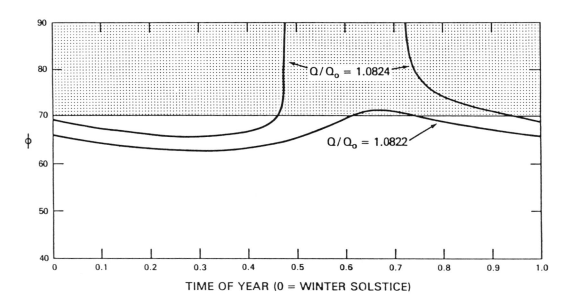

SEASONAL VARIATION OF SNOWLINE

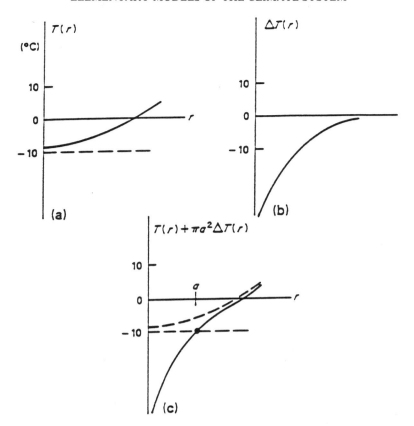

Fig. 1.17. Schematic explanation for the small ice cap instability. Figure illustrates temperature solutions versus distance from the pole for (a) the temperature for a climate state corresponding to point A in Fig. 1.15; (b) the depression of temperature due to a point sink of heat at the pole corresponding to a small patch of highly reflective ice. The depression is everywhere approximately proportional to the ice disk area; (c) the sum of curves in (a) and (b), where the size of the ice disk has been chosen to be of radius a; this value is also the correct equilibrium value for an ice cap corresponding to point A' in Fig. 1.15. [From North and Crowley, 1985] *Reproduced by permission of the Geological Society from "Application of a seasonal climate model to Cenozoic glaciation," by G. North and T. Crowley, in Journal of the Geological Society (London) 142, 1985.*

formation or decay. Of course, the rate of growth of ice volume depends on the mass balance of a glacier and its computation is well outside the scope of these simple EBMs. On the other hand, the eligibility for growth may be governed by small ice cap instability considerations.

Figure 1.17 illustrates schematically the underlying reasons for the small-ice-cap critical point phenomenon in diffusive energy balance climate models. The interested reader is referred to North (1984) for a more rigorous examination of the problem. First, consider a situation analogous to conditions on an ice-free earth, e.g., point A of Fig. 1.15. For a given level of forcing, temperatures will increase with distance from the pole (Fig. 1.17a). If the heating decreases, the shape of the temperature curve will remain the same but will shift downward along the y axis until some critical temperature is reached that might represent, for example, the mean annual temperature below which ice is permanently present. By convention, we choose $-10°C$ as this critical temperature (Budyko, 1969).

The presence of a small patch of ice at the pole effectively produces a steady heat sink because of its higher albedo. Temperatures now decrease below $-10°C$ over the ice. The linear response of the climate model to this heat sink can be found. The result is a sizable thermal depression over a large distance from the sink (Fig.

1.17b), with the length scale dependent on a balance between transport and the adjustment time scale for the radiative perturbation. Adding the curves in 1.17a and b leads to the solid temperature curve in Fig. 1.17c. For parameter values representative of the earth's transport and radiation, the effective range of the influence function is about 20° on a great circle. Thus a strong localized heat sink will have a nonnegligible effect as far as 2000 km away, potentially depressing temperatures below $-10°C$ over a wide area. A second equilibrium solution with glaciation is implied. Ice-age experiments with general circulation models (Manabe and Broccoli, 1985b) and glacial-maximum ice positions suggest approximately the same horizontal range to the influence function as the simple climate model (Chap. 4).

There is some support for the existence of abrupt transitions in more realistic climate models. Threshold phenomena occur in an atmospheric general circulation model (Held et al., 1981), an ocean general circulation model (F. Bryan, 1986), a coupled ocean–atmosphere model (Manabe and Stouffer, 1988), and a thermohaline model of deep–water circulation (Peterson, 1979). Abrupt transitions can occur in other types of climate models (e.g., Lorenz, 1968; Saltzman and Sutera, 1987) and are a fairly common feature in complex physical systems. In later chapters we will apply this concept

to two significant features of earth history—the long-term evolution of climate and extinction events. But for now we end the discussion on "simple" climate models and move on to complex models of the atmospheric and oceanic circulation (Chap. 2).

1.4 Summary

The climate system is made up of many components which have significantly different physics and time scales of response. It is nevertheless sometimes possible to examine the system with relatively simple models which provide considerable insight into processes operating in more complicated models. For example, energy balance models yield very useful information about sensitivity and stability of climate. Seasonal two-dimensional EBMs are particularly useful for a number of paleoclimate applications, as they allow economic testing of changes in the seasonal cycle due to orbital forcing (Chap. 7) or land–sea distribution (Chaps. 10 and 11), and appear to have a sensitivity comparable to that of GCMs for changes in seasonal forcing.

Although EBMs are very useful, they also clearly have their limits. In order to get a fuller understanding of present and past climates, it is necessary to consider general circulation models of the atmosphere and ocean. We will discuss these topics in the next chapter.

2. GENERAL CIRCULATION OF THE ATMOSPHERE AND OCEANS

In this chapter we introduce general circulation models (GCMs) of the atmosphere; we will also consider the circulation of the oceans. We go beyond the phenomenological models of the last chapter, adopting a more deductive approach to the problem. The basic idea is that the motions of the atmosphere and ocean are governed by the laws of classical physics and these can be cast into the form of partial differential equations, which can be solved approximately on modern computers. Our goal is to introduce these concepts to the earth scientist who may have little interest in actually constructing such a model but who wants to understand how the models are constructed and how believable are their results. There are several excellent sources of more detailed information on climate models (e.g., Houghton, 1984; Washington and Parkinson, 1986; Schlesinger, 1988). Our purpose here is to introduce the subject in sufficient depth as to make nonspecialists aware of the power and limitations of GCMs in order to appreciate the later chapters in this book and also be equipped for more detailed independent reading.

2.1 Principal Features of the Atmosphere and Ocean Circulation

2.1.1 Atmospheric Circulation

Let us begin our discussion of the general circulation by examining some basic features of the atmosphere and ocean. One of the most prominent features of the general circulation (Fig. 2.1) involves the subtropical high-pressure systems, midlatitude zones of descending, dry air that rotate clockwise in the Northern Hemisphere and counterclockwise in the Southern Hemisphere. The subtropical highs are permanent features over the ocean and seasonally present over the continents. In lower latitudes, the outflowing air from these systems leads to easterly or "trade" winds that converge in the

Intertropical Convergence Zone (ITCZ), the regions of most intense tropical convection. Ideally this is a thin band of deep clouds encircling the earth near the equator. The ITCZ moves north and south with the seasons, following the solar heating, but with about a 1- to 2-month lag (due to the thermal inertia of the land and ocean).

On the poleward side of the subtropical highs, the outflowing air is westerly (i.e., from the west). It forms a discontinous boundary with denser air that has been cooled at high latitudes and is moving equatorward. This transition zone (polar front) is the region of the midlatitude low-pressure systems, the source of much of our midlatitude "weather." The low-pressure systems, which have inflowing air at lower elevations, develop through a process known as baroclinic instability. This process has been much studied by dynamical meteorologists, but we will take leave of it, except to note that even small perturbations along a "frontal zone" between mid- and high-latitude air can, under certain conditions, lead to a rapid growth with the resultant formation of a traveling low-pressure system.

A longitudinal cross-section of the atmosphere (Fig. 2.2) provides further information as to the factors responsible for the general circulation. The descending air in the subtropical highs is part of a three-dimensional circulation, with warm air rising in the tropics (in the ITCZ), spreading poleward aloft, and then descending in midlatitudes. The meridional circulation in low latitudes is known as the Hadley cell. Another region of ascending air occurs along the tilted axis of the polar front, with the jet stream being the physical manifestation of the large temperature (and pressure) gradient at lower elevations. Note that there are actually two jet streams—a polar jet and a subtropical jet. The polar jet is a more irregular but often stronger feature, and fluctuates with the vagaries of the

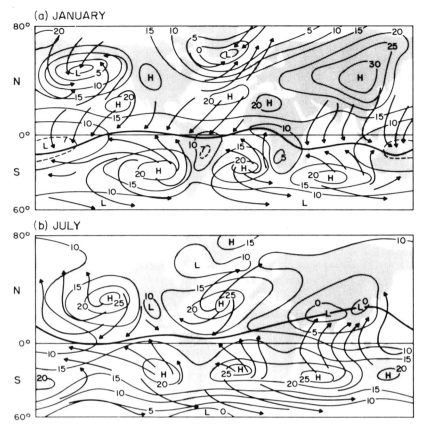

Fig. 2.1. Average surface winds and sea level pressures (in mb above 1000 mb) for (a) January and (b) July. [After Riehl, 1979; redrawn by Washington and Parkinson, 1986] *Reproduced with permission of Academic Press.*

Fig. 2.2. Meridional cross-section of the atmospheric circulation for the Northern Hemisphere in winter. [From Barry, 1967] *Reproduced with permission of Methuen and Co.*

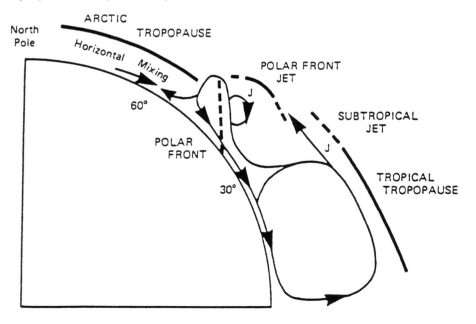

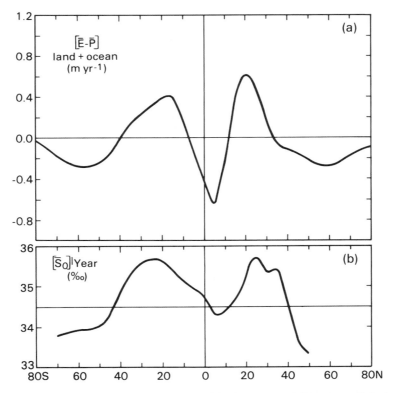

Fig. 2.3. Zonal mean profiles of (a) evaporation-minus-precipitation (E-P) and (b) surface salinity in the world ocean (S_o). [Modified from Peixoto and Oort, 1983] *Reproduced with permission of American Physical Society.*

weather. The subtropical jet reflects the poleward limit of the Hadley cell.

The circulation in the Hadley cell is due to the motion induced by differential solar heating between the equator and the pole; air rises at the equator and sinks at subtropical latitudes. The process is complicated by the earth's rotation which causes the upper air parcels in the northern hemisphere to be deflected to the right (eastward) along their trajectories. This eastward deflection of upper air paths, along with the conservation of angular momentum (westerly air drawn closer to the polar axis must speed up), causes an intense band of westerly flow at upper levels. A similar "deflection to the right" (in the Northern Hemisphere) occurs for air parcels moving from high to low pressures. In the Southern Hemisphere the same reasoning applies, with the deflection being to the left instead of the right.

Outflowing air from the subsiding branch of the Hadley cell travels along the surface of the earth accumulating moisture from the ocean surfaces, drying out land surfaces (note the net evaporation peak in Fig. 2.3a), and strongly influencing ocean surface salinity (Fig. 2.3b). The Coriolis Effect guides this surface "return air" along lines running from northeast to southwest. This air laden with water vapor is then convected vertically at the ITCZ where large amounts of rain fall (note the net precipitation peak in Fig. 2.3a). In the process enormous amounts of latent heat are released resulting in a buoyant force, further driving the convection cycle. Likewise, outflow on the poleward branch of the Hadley cell occurs in the region of strong westerlies (note the net precipitation peak in midlatitudes in Fig. 2.3a). Interaction with cooler air from higher latitudes results in instabilities that lead to the development of midlatitude low-pressure systems.

2.1.2 Oceanic Circulation

The atmosphere drives the oceanic circulation in two ways. The winds drive the surface cur-

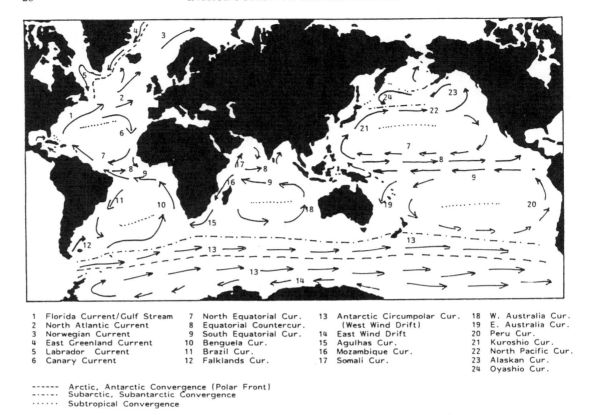

1	Florida Current/Gulf Stream	7	North Equatorial Cur.	13	Antarctic Circumpolar Cur.	18	W. Australia Cur.
2	North Atlantic Current	8	Equatorial Countercur.		(West Wind Drift)	19	E. Australia Cur.
3	Norwegian Current	9	South Equatorial Cur.	14	East Wind Drift	20	Peru Cur.
4	East Greenland Current	10	Benguela Cur.	15	Agulhas Cur.	21	Kuroshio Cur.
5	Labrador Current	11	Brazil Cur.	16	Mozambique Cur.	22	North Pacific Cur.
6	Canary Current	12	Falklands Cur.	17	Somali Cur.	23	Alaskan Cur.
						24	Oyashio Cur.

------ Arctic, Antarctic Convergence (Polar Front)
-·-·- Subarctic, Subantarctic Convergence
······ Subtropical Convergence

Fig. 2.4. Schematic of the earth's major surface ocean currents.

rents of the ocean (Fig. 2.4), which basically reflect the structure of the overlying atmosphere except that the presence of solid boundaries causes the development of rapid, narrow poleward moving western boundary currents (e.g., the Gulf Stream) or the broad, more sluggish eastern boundary currents (e.g., the Peru Current). The centers of the subtropical gyres are situated beneath the Hadley cells and are zones of surface convergence and downwelling. In the Southern Ocean, where there are no lateral boundaries, the Antarctic Circumpolar Current circumscribes the continent. The currents in this system are very strong and penetrate almost the entire water column.

The wind-driven circulation modifies the sea-surface temperature field (Fig. 2.5). The western boundary currents cause poleward penetrations of warm water in some regions, while the eastern boundary currents lead to the opposite effect. In addition, there is significant equatorial upwelling associated with a changeover in sign

of the Coriolis Effect. Away from the upwelling areas, the tropics have very stable temperatures, peaking near 30°C in the western Pacific. Sea surface temperatures (SSTs) above 27.5°C appear to be critical for unleashing convective instability associated with tropical precipitation (Graham and Barnett, 1987). Apparently, this figure represents a threshold value for saturation vapor pressure; once it is reached there is enough latent heat to drive the system toward unstable convection.

An examination of poleward ocean heat transport (Fig. 2.6) reveals some dramatic differences between the ocean basins. Whereas the Pacific transport is symmetric about the equator (Hastenrath, 1980), Atlantic transport is northward-directed, even at low latitudes in the Southern Hemisphere. This figure cannot be understood unless we consider the second factor that affects the surface circulation (salinity) and which in turn affects the deep-water circulation. The surface salinity of the ocean (Fig. 2.7) re-

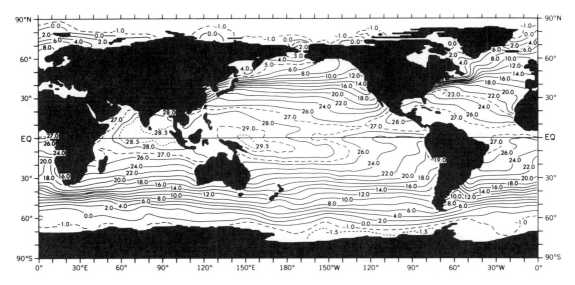

Fig. 2.5. Observed surface temperature of the oceans averaged over all seasons. [From Levitus, 1982]

flects the precipitation-minus-evaporation pattern in the overlying atmosphere (see Fig. 2.3), with the centers of the subtropical gyres, situated beneath the descending branch of the Hadley cell, having the highest salinities. There is also a very significant difference in salinity (about 1.5‰) between the North Atlantic and North Pacific, a feature attributable mainly to water vapor transport from the Atlantic to the Pacific basins (cf. Weyl, 1968; Bryan and Oort, 1984).

The relatively high surface salinity of North Atlantic water is very important for deep-water

Fig. 2.6. Map scheme of annual mean meridional heat transport within the oceans. Heavy cross-bar denotes latitude of zero, and "max" that of maximum meridional transport, with numbers indicating amounts in units of 10^{13} W. Broken lines show the meridians used as boundaries between oceans in the high southern latitudes. [From Hastenrath, 1980] *Reprinted from Journal of Physical Oceanography with permission of the American Meteorological Society.*

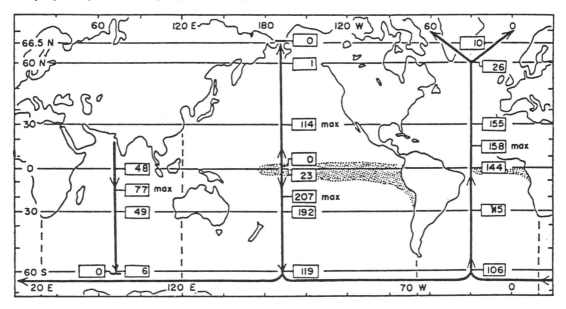

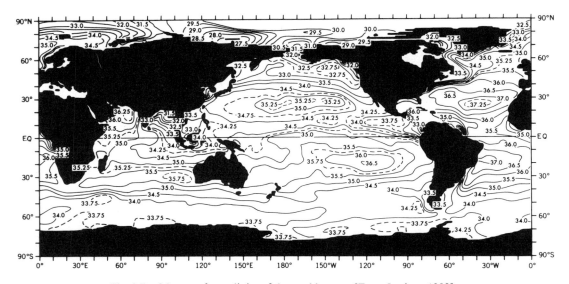

Fig. 2.7. Mean surface salinity of the world ocean. [From Levitus, 1982]

formation (Fig. 2.8). Salinity is the controlling factor on density at low temperatures. Cooling of North Atlantic water in winter leads to formation of North Atlantic Deep Water (NADW), which is both warm and salty (compared to other deep water). About 15 Sv (Sverdrups; 1 Sv $= 1 \times 10^6$ m^3/sec) of this water mass is exported across the equator to the rest of the world ocean. To conserve volume, it is replaced by warm near surface water from the South Atlantic crossing the equator; thus the equatorward-directed heat flow. Whereas the age of the deep water in the North Atlantic basin is less than 500 years (Stuiver et al., 1983), deep water in the Pacific

basis is as much as 1500 years old (Broecker, 1963). The age difference between the water masses in the two basins significantly affects the distribution of chemical properties in the deep ocean (Berger, 1970).

The other main source for Southern Hemisphere bottom water formation is around Antarctica (e.g., Foster and Carmack, 1976). Formation of sea ice on continental shelves causes salt ejection and development of dense brines, which periodically move as density flows down the continental shelves, mixing with upwelling Circumpolar Deep Water (itself a mixture of NADW and Pacific water). This new water

Fig. 2.8. Schematic diagram of deep-water circulation in the western Atlantic basins. NADW = North Atlantic Deep Water; AABW = Antarctic Bottom Water; AAIW = Antarctic Intermediate Water; CPDW = Circumpolar Deep Water.

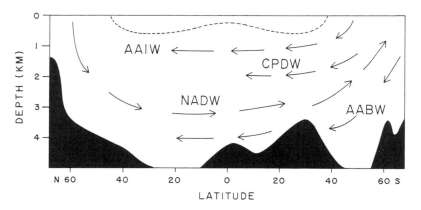

mass, the Antarctic Bottom Water (AABW), flows into all the major basins of the world ocean. Some of the AABW may result from open-ocean convection in sea ice-free regions known as polynyas (e.g., Martinson et al., 1981).

In addition to deep/bottom-water formation, another water mass forms at intermediate depth levels (~ 1500 m) in the Antarctic. This water mass (Antarctic Intermediate Water, AAIW) forms along the Antarctic polar front (Fig. 2.4), a surface convergence zone in the easterly flowing Antarctic Circumpolar Current. The water mass is composed of low-temperature, low-salinity water made up in part of melted Antarctic sea ice south of the polar front. AAIW spreads northward at subthermocline levels through the Pacific, Indian, and Atlantic basins (Fig. 2.8).

2.2 Modeling the General Circulation of the Atmosphere

2.2.1 Governing Equations

The atmosphere is a continuum of matter, each infinitesimal element of which is subject to the law of gravitation, the conservation of mass, energy, and momentum as well as the laws of thermodynamics. Detailed rules about how radiation is absorbed and energy is extracted from the surfaces must also be included. Of singular importance is the treatment of moisture in its role both as a substance relevant to surface climatology (precipitation and evaporation) and as a conveyor of stored energy via latent heat to sensible heat conversion. Finally, phenomenological equations of state must be included, such as the ideal gas law, connecting local pressure, temperature, and density. A complete model of the atmosphere will therefore involve as dependent field variables (functions of position and time): the temperature, three components of air velocity, local density or pressure, and the concentration of water vapor.

GCMs explicitly treat the different variables by dividing the earth into a number of grid boxes and defining the forces on each box. Since we have already introduced the concept of heat balance of an infinitesimal area element in the last chapter, we start with the First Law of Thermodynamics, which refers now to a small element of the atmospheric material treated as a thermodynamic system. The so-called enthalpy form of this equation may be stated as follows (Holton, 1979):

$$C_P \left(\frac{\partial T}{\partial t} + \vec{V} \cdot \nabla T \right) = \dot{Q} + \frac{1}{\rho} \frac{dP}{dt} \quad (2.1)$$

where C_P is the specific heat capacity for air at constant pressure; $\vec{V}$ is the air velocity field which is a function of position and time; $\dot{Q}$ is the heating rate per unit mass of the infinitesimal volume element by influences external to the element, such as radiation emission and absorption, evaporation cooling or condensation heating, or conduction of heat across a face of the element; ρ is the local density of the atmospheric matter; and P is the local pressure inside the volume element, which is instantaneously the same as the pressure of the environment surrounding the element to a good approximation.

The aforementioned terms have the following interpretation: the first term is proportional to the rate of change of local temperature in response to an imbalance of heating to the parcel; the second term involving velocity is the heat advected into a local fixed-volume element due to fluid bringing in neighboring air of a different temperature (there must be a nonvanishing local gradient); the $\dot{Q}$ term is the main source of heating due to outside influences; the last term represents the change in energy per unit time due to compression or expansion work being performed on a fluid element as it moves into the fixed-volume element from regions of different pressure.

Equation (2.1) is essentially the same as the one used in Chap. 1 in connection with EBMs. The main differences are that the EBM integrates (2.1) vertically so that a single value characterizes the entire column; in the vertical integration the time derivative of pressure can be dropped. The EBMs also do a type of ensemble averaging; that is to say, they determine a mean population temperature $\langle T(\vec{r}, t) \rangle$, where t might be time of year, and the brackets indicate averaging over an ensemble of seasonal realizations. It is in this sense that the approximation

$$\langle C_P \vec{V} \cdot \nabla T \rangle \approx -D\nabla^2 \langle T \rangle \quad (2.2)$$

was used to "close" the system down to one variable, $\langle T(\vec{r}, t) \rangle$. In GCMs we do not adopt such a simple approximation but rather attempt to calculate explicitly, in a time-marching manner, the time- and space-dependent wind vectors so that they can be inserted into (2.1) at each time step. The approximation (2.2) is equivalent to assuming the atmosphere is composed mainly of random eddies. That idealization may now be removed.

The velocity field evolves due to forces on the fluid element. The Navier–Stokes equation is an expression of Newton's second law of motion applied for fluids on a rotating sphere:

$$\frac{\partial \vec{V}}{\partial t} + \vec{V} \cdot \nabla \vec{V} - 2\vec{\Omega} \times \vec{V}$$

$$= -\frac{1}{\rho} \nabla P + \vec{B} + \vec{F} \quad (2.3)$$

where $\vec{\Omega}$ is the angular rotation rate vector pointing along the earth's axis out of the North Pole (radians/sec); $\vec{B}$ is a buoyancy force approximately proportional to the difference in density inside and outside the parcel; and $\vec{F}$ is a friction force that might arise directly from viscosity (negligible) or indirectly due to small eddies not resolved by the numerical grid. The term involving the earth's rotation comes about from our use of a coordinate system rotating with the earth; it is the so-called Coriolis term and it is an essential ingredient needed to understand most of the familiar meteorological cyclone system behavior in the midlatitudes. The pressure gradient term in (2.3) causes an acceleration of a parcel because of an imbalance of force on one face of the volume element to that of an opposite face (force on parcel = pressure times area of face vectorially added over all faces of the element).

The equation of continuity,

$$\frac{\partial \rho}{\partial t} + \nabla \cdot (\rho \vec{V}) = 0 \quad (2.4)$$

ensures that mass is not created or destroyed along the motion of the fluid parcels. A similar equation holds for water:

$$\frac{\partial q}{\partial t} + \vec{V} \cdot \nabla q = E - p \quad (2.5)$$

where q is the concentration of water vapor, E is the evaporation rate, and p is the precipitation or condensation rate. The equations may be partially joined by the use of the equation of state for air:

$$P = \rho R_a T \quad (2.6)$$

where R_a is the gas constant for air (possibly modified to include effects of moisture).

In addition to the above equations, a number of parameters must be specified (Table 2.1), and the whole set must be solved simultanously and evolving in time. A major part of the solution algorithm must be devoted to "radiative transfer," i.e., computing the detailed radiative absorption and emission from each element, the construction of clouds from the moisture field, and the subsequent interaction of these features through buoyancy and radiation imbalance. There will be additional changes in external boundary conditions (Fig. 1.1), such as snow cover and sea ice. These latter evolving features constitute what is called the "physics" of the model, in contrast to solving the pure fluid evolution ("dynamics") manifest in (2.3) and (2.4). A particularly difficult field to model, but one which is very important, involves clouds, which can significantly affect the earth's radiation budget (e.g., Ramanathan et al., 1989).

Typical GCMs devote about an equal amount of computation time to the physics and the fluid dynamics in climate simulations. In weather forecasting, one devotes a much greater fraction of computing resources to the fluid dynamics and employs much more approximate forms for the physics. The reason for this is that short-term weather is largely governed by advection of neighboring air and its complexes, whereas in climate simulations the physics dominates more and more as the details of individual weather events are smoothed over. Furthermore, the extension to time averages necessarily causes the longer time scale phenomena to be more impor-

TABLE 2.1. *Prescribed parameters and boundary conditions in atmospheric general circulation models*

Radius, surface gravity, and rotation speed of the planet
Solar constant and orbital parameters of the planet
Total atmospheric mass and composition
Thermodynamic and radiation constants of the
    atmospheric gases and clouds
Surface albedo
Surface elevation

Source: Schlesinger, 1984.

tant (i.e., the "slow physics"); an extreme example would be the movement of glaciers.

2.2.2 Numerical Solutions of the Models

Users of GCM model output need not know all the numerical procedures that go into solving the models. However, some basic design concepts should be known to understand why some techniques are more appropriate to certain experiments. Virtually all GCMs are solved for the whole spherical earth. The numerical procedure for solving such models falls into two alternative methods: grid point and spectral. The two alternative approaches each have certain advantages, which mainly depend on the desired resolution of the simulation.

Since the grid point method is easiest to visualize we begin with it. The sphere is partitioned into a grid according to equal intervals in the spherical coordinates (latitude, longitude, height above earth's surface). This scheme must be modified near the poles (e.g., Hansen et al., 1983) since the lines of longitude converge to very short spatial distances, which eventually causes complications in the numerical integration in time (when the spatial grid is dense, the time stepping must be correspondingly short to prevent instability). Some very sophisticated schemes, mostly developed by Arakawa (e.g., 1988), use octagonal grids on the sphere, staggered in height. These have the special property that at each numerical time step certain quantities that should be conserved in a true continuum formulation (such as energy and angular momentum) are exactly preserved. Values of all the fields on this spatial grid are then initialized and advanced to the next time step with each variable updated by its equation of motion.

For illustrative purposes consider a one-variable climate system which might be represented as

$$\frac{dT}{dt} = f(T) \qquad (2.7)$$

Here T might represent the globally averaged temperature. Its numerical finite difference form can be written as

$$T_{t+\Delta t} = T_t + f(T_t) \cdot \Delta t \qquad (2.8)$$

The right-hand side can be evaluated from knowledge of T_t. We advance the calculation step by step toward equilibrium. On the spherical grid we have $T_t(\theta_i, \phi_j)$, where the latitude is evaluated at the ith gridpoint θ_i and the longitude is evaluated at the jth gridpoint ϕ_j. There is in principle an equation like (2.7) for each grid point (i, j) on the sphere and the equations are coupled together. In other words, the right-hand side of the equations like (2.8) will have the temperature at other points included in the functional dependence. The equations must then be solved simultaneously and advanced in time at each time step. In reality this must be done for all the fields of which temperature is only one. The time-stepping scheme in (2.8) is a so-called first-order scheme; in practice, modelers use higher order schemes that reduce the so-called truncation error allowing larger time steps for a given level of error tolerance.

The spectral method (e.g., Bourke, 1988) utilizes expansions of the field variables into spherical harmonics (something like a Fourier series). For example, with the seasonal EBM discussed in Section 1.2.4, Eq. (1.4) can be written as

$$T(\theta, \phi, t) = \sum_{l=0}^{L} \sum_{m=-l}^{l} T_{lm}(t) Y_{lm}(\theta, \phi) \qquad (2.9)$$

and the set of $T_{lm}(t)$ plays the role of the aforementioned grid point quantities $T_t(\theta_i, \phi_j)$.

One truncates the series at level L and that truncation level determines essentially the spatial resolution of the model. The time-dependent spherical harmonic coefficients take the place of grid point evaluated fields. Spherical harmonic degree 10 corresponds to a polynomial in the sine of latitude of degree 10. Such a polynomial has 10 zeros from pole to pole and hence we might expect a resolution of about $180/11 = 16°$, corresponding to about 1600 km on any great circle. The resolution in spatial distance is inversely proportional to the spherical harmonic degree of truncation. The spherical harmonics have a special relationship to the gradient operator (they are eigenfunctions of the Laplacian operator) and this allows certain terms in the hydrodynamic equations to be evaluated very quickly, especially on computers that have array processors. The spectral form of the equations is well tailored to the hydrodynamic and thermodynamic equations but is very cumbersome in dealing with the physics processes, which must invariably be calculated in grid point representation. This necessitates the rapid transformation from one form to the other and this is facilitated by such modern algorithms as the Fast Fourier Transform.

Most intermediate resolution climate models make use of the spectral formulation. As resolution is increased, it has been shown that the time-saving advantages of the spectral method are eventually lost and at a resolution of about 100 km it appears that the grid point method reemerges as the formulation running fastest on modern computers (e.g., Girard and Jarraud, 1982).

Typical GCMs used in climate simulations in the late 1980s have an equivalent spatial resolution of about 4–6° of latitude (400–600 km) in the horizontal directions and 9–12 levels in the vertical (usually equidistant in atmospheric mass rather than actual distance). This coarse resolution raises an important issue that is still unresolved, namely, the accounting for important phenomena that occur at smaller scales than the model's grid spacings. Examples include clouds and their impact on radiation, precipitation and the associated latent heat releases, and turbulent phenomena in the so-called planetary boundary layer, which is in the lowest kilometer or two of the atmosphere. The models take these processes into account at the grid scale by using simplified formulas that are supposed to include the subgrid scale physics. A significant problem in climate modeling involves difficulties encountered in validating the formulations of subgrid scale parameterizations, particularly aspects related to turbulence. Considering the uncertainties, it is remarkable that GCMs perform as well as they do. The reader is referred to Schlesinger (1984) for a discussion of GCM parameterization. Table 2.2 is a list of many of the important parameterizations used in GCMs.

It is generally believed that increasing the resolution will increase the model's ability to simulate the true climate. Unfortunately, increasing the resolution incurs considerable cost, since an increase in spatial resolution forces an accompanying increase in temporal resolution to preserve numerical stability. Moreover, when the resolution is increased one must necessarily increase the burden of computing the physics, since it would only be consistent to improve the formulation of these phenomena before increasing computer time for each time step. Hansen et al. (1983) presented a rule of thumb for estimating computational time:

$$\text{Time} \propto n_z \, [0.4(n_x n_y)^{3/2} + 0.6 n_x n_y] \quad (2.10)$$

where n_x, n_y, and n_z are the number of latitudes, longitudes, and layers relative to the baseline number, respectively.

2.2.3 Some Examples of GCM Output

There are obviously many more details of GCMs that could be discussed, but we will have to move beyond this brief sketch toward examining aspects of GCMs of more relevance to paleoclimate studies. There are many GCMs in operation today, and some of the more important ones that have been used for paleoclimate application are listed in Table 2.3. We will show here a few typical results, especially of surface predictions, since these are of greatest interest to paleoclimatology. It must be kept in mind that surface quantities may be affected by the performance of a GCM at higher altitudes and for variables not necessarily of direct interest to geologists. Specialists will also be interested in the performance of, for example, high-altitude polar temperatures, since these will reflect the validity of the simulation of such processes as radiation transfer through the atmospheric gases.

Figure 2.9 shows the map of January surface temperature for the NCAR model with a slab mixed-layer ocean (see text following). Along with it are shown the corresponding observations. Note that the model qualitatively repro-

TABLE 2.2. *Subgrid scale processes that are parameterized in atmospheric general circulation models*

Turbulent transfer of heat, moisture, and momentum between the earth's surface and the atmosphere
Turbulent transfer of heat, moisture, and momentum within the atmosphere by dry and moist (cumulus) convection
Condensation of water vapor
Transfer of solar and terrestrial radiation
Formation of clouds and their radiative interaction
Formation and dissipation of snow
Soil heat and moisture physics

Source: From Schlesinger, 1984.

TABLE 2.3. *Some common GCMs used in paleoclimate applications*

Model	Descriptions	Sample applications
NOAA/GFDL	Manabe and Stouffer (1980)	Manabe and Broccoli (1985a)
		Manabe and Bryan (1985)
NCAR	Washington and Meehl (1984)	Kutzbach and Guetter (1986)
		Barron and Washington (1985)
NASA/GISS	Hansen et al. (1983)	Hansen et al. (1984)
		Rind et al. (1989)
OSU	Schlesinger and Gates (1980)	Gates (1976a,b)
MPI	von Storch (1988)	Lautenschlager and Herterich, 1990a
LMD	Sadourney and Laval (1984)	Joussame and Jouzel (1987)
UKMO	Wilson and Mitchell (1987)	Mitchell et al. (1988)
LR-GCM	Otto-Bliesner et al. (1982)	Kutzbach and Otto-Bliesner (1982)
		Kutzbach and Gallimore (1989)

Abbreviations:
NOAA/GFDL is the Geophysical Fluid Dynamics Laboratory at the National Oceanographic and Atmospheric Administration.
NCAR is the National Center for Atmospheric Research.
NASA/GISS is the Goddard Institute for Space Sciences at the National Aeronautics and Space Administration.
OSU is Oregon State University.
MPI is the Max Planck Institute für Meteorologie.
LMD is the Laboratoire Météorologie Dynamique.
UKMO is the United Kingdom Meteorological Office.
LR-GCM is an abbreviation for a low-resolution atmospheric GCM useful in pilot studies.

duces virtually all of the features of this field. Figure 2.10 shows the comparison for precipitation (mm/day) for the same month and model. Here the agreement, although reasonable, is quantitatively less satisfactory. Experience has shown that GCM "performance" varies by model, field, and spatial scale (e.g., Schlesinger and Mitchell, 1985; Wigley and Santer, 1988).

Precipitation is hard to simulate in any model. Among other things monthly averaged precipitation is quite variable from year to year in nature (the standard deviation is about equal to the mean) and even the observations of precipitation are very poor, especially over the oceans (e.g., Mintz, 1981). The main modeling difficulty is related to the fact that precipitation processes occur at the subgrid scale, and models with coarse resolution simply cannot be expected to faithfully reproduce them at this time. We will have occasion to discuss the problem of precipitation simulation many times in the course of this book. Figure 2.11 shows a summary of some recent GCM simulations of zonally averaged precipitation for the present mean annual climate. Note that there is still considerable bias surviving even after zonal averaging, which smooths out many local differences.

Another way to evaluate GCM performance

is to examine different GCM responses to altered boundary conditions. Consider Fig. 2.12, which shows the surface temperature field as a function of latitude and time of year for five different model runs of doubled CO_2. Certainly the common agreement is much greater than the intermodel differences. All models show greater warming at high latitudes, with the warming greatest in the winter months. However, the absolute value of warming varies between models. A similar comparison for net precipitation minus evaporation (Fig. 2.13) shows much less agreement because precipitation is a harder field to model (Kellogg and Zhou, 1988). A comparison of 14 different GCMs indicates that one of the prime factors responsible for model-model differences involves the way clouds are depicted in the models (Cess et al., 1989).

2.3 Designing Model Experiments

The present state of progress in model development (as well as economic considerations) dictates that careful planning should precede model experimentation. Certain time scales in climate determine the length of a run necessary for simulation of different climatic phenomena. As a crude rule one may think of the relaxation time of the atmosphere alone as being of the order of a month. If the mixed layer of the ocean is cou-

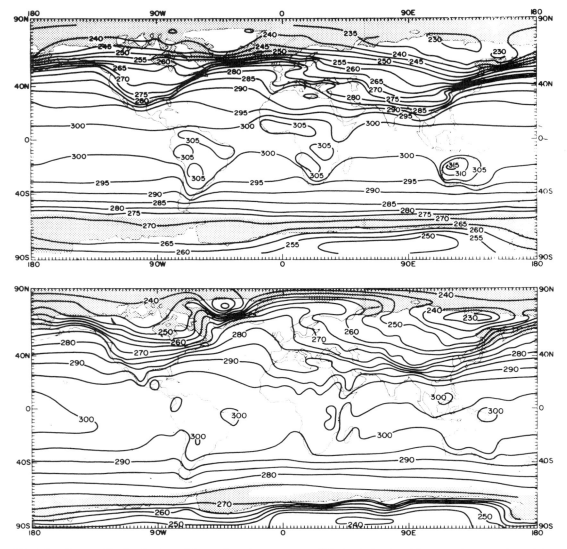

Fig. 2.9.   Geographic distribution of the surface air temperature (°C): (top) simulated for DJF with the NCAR GCM by Washington and Meehl (1984); (bottom) observed for January from Schutz and Gates (1972) based on Crutcher and Meserve (1970) and Taljaard et al., (1969). Simulated sea ice limits show the edge of sea ice 0.2 m thick. Observed sea ice limits from Alexander and Mobley (1976). [From Washington and Meehl, 1984] *Courtesy of American Geophysical Union.*

pled to the atmosphere as a slab of matter capable of storing heat but usually not transporting it horizontally, the time for relaxation is about 5 years. If the deeper layers of the ocean are coupled into the system dynamically, there may be several hundred year time scales involved. Finally, if continental ice sheets are taken into account, the time scale could be thousands of years—no GCM runs have attempted to include the dynamics of ice sheets to date.

Since the atmosphere and the ocean's mixed layer are regarded as fluctuating media whose long-term statistics can be estimated by making long runs (several characteristic time scales), modelers have singled out these components of the system for intensive study. Hence, much of the strategy in GCM modeling takes advantage of the short time constants of these individual components. In the following section we list various types of GCM runs.

2.3.1 Simulations versus Sensitivity Studies

Before discussing some specific types of model runs, it is necessary to outline the two basic different types of runs which apply to all the categories listed below. Model runs are either simulations or sensitivity studies. In the first case, we are actually trying to simulate a climate for a past time. Results are most believable when they can be compared with independent types of data. Such simulations require realistic specification of different boundary conditions (e.g., ice sheets, land albedo, solar insolation, CO_2, sea surface temperature (SST)), with the validation data kept independent from the altered boundary conditions (e.g., lake levels as a proxy for precipitation). In most cases, we do not have enough information about all the relevant past

Fig. 2.10. Geographic distribution of the precipitation rate (mm/day) for DJF: (top) simulated with the NCAR GCM by Washington and Meehl (1984); (bottom) observed based on Jaeger (1976). [From Washington and Meehl, 1984] *Courtesy of American Geophysical Union.*

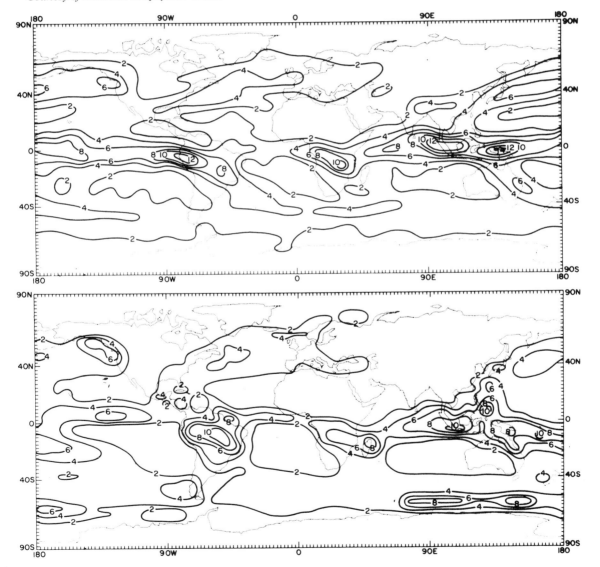

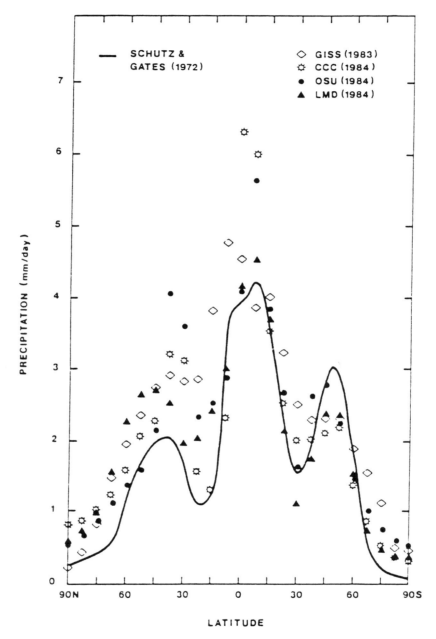

Fig. 2.11. Zonally averaged January precipitation rate simulated by four different GCMs in recently performed control integrations compared with observations. GISS is the NASA/Goddard Institute for Space Sciences; CCC is the Canadian Climate Centre; OSU is Oregon State University; and LMD is the Laboratoire Météorologie Dynamique. Observations (solid line) are from Schutz and Gates (1972). [From Wigley and Santer, 1988] *Reprinted by permission of Kluwer Academic Publishers.*

ZONAL MEAN TEMPERATURE DIFFERENCES

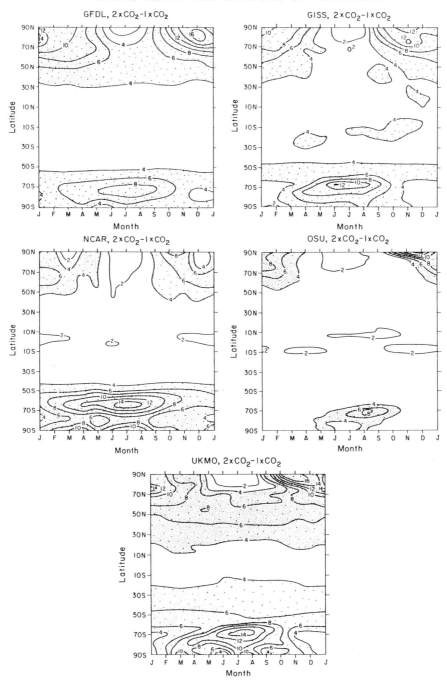

Fig. 2.12. Comparison of different GCM temperature response to the same forcing (doubled CO_2). Figure illustrates time–latitude distribution of the zonal mean surface air temperature change (°C), $2 \times CO_2 - 1 \times CO_2$, simulated by (a) the NOAA/GDFL (Geophysical Fluid Dynamics Laboratory) model of Wetherald and Manabe (1986); (b) the NASA/GISS (Goddard Institute for Space Science) model of Hansen et al. (1984); (c) the NCAR (National Center for Atmospheric Research) model of Washington and Meehl (1984); (d) the OSU (Oregon State University) model of Schlesinger and Zhao (1989); and (e) the UKMO (United Kingdom Meteorological Office) model of Wilson and Mitchell (1987). Stipple indicates temperature increases larger than 4°C. [From Schlesinger, 1989] *Reprinted by permission of Kluwer Academic Publishers.*

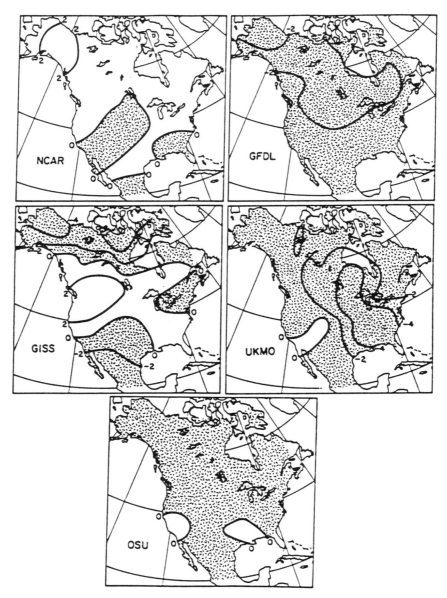

Fig. 2.13. Comparison of different GCM P-E response over North America to the same forcing (doubled CO_2). Increases and decreases of soil moisture in summer relative to the control when carbon dioxide is doubled ($2 \times CO_2$). The areas shaded with hen scratches are those where there is a decrease or change to a dryer condition; clear areas show an increase. Units are centimeters of water. [From Kellogg and Zhao, 1988] *Reprinted from Journal of Climate with permission of the American Meteorological Society.*

boundary conditions to conduct paleoclimate simulations. An outstanding exception involves climate model simulations for the last 18,000 years (Chap. 4).

"Realistic" simulations can provide a great deal of valuable insight into past climates, and some specific examples are discussed in Chap. 4. In many cases, however, we do not have enough information about some of the required boundary conditions. For example, Cretaceous simulations cannot be conducted with realistic SSTs as an altered boundary condition because we do not have good estimates of Cretaceous SSTs.

Even if we do not have enough information for simulations, we can learn a great deal about past climates with sensitivity studies. In these experiments, we change only one or a few boundary conditions and evaluate how models respond to the altered conditions. For example, since we know that the land–sea distribution exerts the dominant influence on the annual cycle, we might want to examine the effects of a supercontinent on the annual cycle. In this case, we would change only the land–sea distribution (see Chap. 11). Or we might want to know the effect of topography on model climates. In this case, we might conduct two experiments—one with mountains, one without (see Section 8.2.2). For the oceans, we might want to know the effect of oceanic gateways on poleward heat transport. Thus, we might conduct a control/perturbed case with closed/open gateways (see Fig. 10.25). In all these cases we learn something valuable about the relative importance of different mechanisms which have been proposed, by either geologists or climate modelers, as important factors contributing to past climate change.

2.3.2 Types of Experiments

There are several types of sensitivity experiments that can be conducted.

2.3.2.1 *"Swamp" Models* Swamp models are for mean annual conditions only. By ignoring the seasonal cycle, one can eliminate the participation of the oceanic mixed layer in storing seasonal heat imbalances and proceed to find some aspects of the mean annual circulation of the atmosphere. The ocean is treated as having no heat capacity (heat is not conducted down into its mixed layer); however, it is considered to be an infinite source of moisture for evaporation into the atmosphere (hence the term "swamp"). Models of this type are useful for experiments where it might be suspected that seasonal effects can be ignored, such as response to an overall change in solar luminosity or a doubling of CO_2. This type of run makes the assumption that the seasonal cycle is essentially a linear phenomenon (supported, for example, by

the success of the linear EBM in Section 1.2.4). In reality we know that there are some limitations to this assumption; for example, we know that changes in seasonal sea ice variations will cause a "rectifier" effect—winter conditions might be affected more than summer and this would lead to changes in the mean annual temperature that would be missed in a linear formulation. Nevertheless, a number of useful results can be obtained with swamp models, especially since a typical run need only be a few hundred model days before very stable equilibrium fields can be obtained.

2.3.2.2 *"Perpetual" Runs* The perpetual run experiment takes advantage of the short time scale of the atmosphere (1 month), which allows simulation for a particular month of the year. One fixes the solar incidence angle for that month and, to avoid the longer time scale of the mixed layer (5 years), prescribes the position of sea ice and the entire field of SSTs. Because the SSTs are prescribed, heat is allowed to flow from the ocean surface to the atmosphere and vice versa. The model is now run subject to these boundary conditions for many months (hence the term "perpetual"), essentially to statistical equilibrium, and averages are taken over the last few months.

Perpetual runs are especially useful in testing parameterizations and comparing with data for a particular month, usually February or August. Several problems prevent this type of experiment from being adequate for many purposes; in fact, the results of such experiments can be misleading in some instances. For example, since 70% of the earth is covered with water and these temperatures are prescribed, there is always the possibility that the simulations are so constrained by these boundary conditions that sensitivities to parameter changes are constrained as compared to their potential response in a "free" system. Furthermore, running GCMs in the perpetual mode enhances the model sensitivity (see following) because there is a longer time available to reach thermodynamic equilibrium than exists in the real world. For example, temperature sensitivity to seasonal forcing over land is considerably less than for the same forcing annually averaged (see Fig. 1.11 and discussion therein; also see Zwiers and

Boer, 1987 and Hyde et al., 1989). It is obvious that such experiments as CO_2 doubling experiments cannot be performed in the perpetual mode, since the ocean would be locked to the same SST field as at present. In most paleoclimate experiments, the SST is not known as an input field, except in some notable cases such as provided by CLIMAP (Chap. 4).

2.3.2.3 *Mixed-Layer Models* The mixed-layer model configuration is the same as that employed in the seasonal EBM of the last chapter (Section 1.2.4). The mixed layer is approximated as a perfect heat-conducting medium in the vertical direction throughout its depth; no heat is usually transported in the horizontal directions. This model reproduces the seasonal cycle best, since it actually includes the transient effects of storage of heat in the mixed layer. The model still is not perfect, since it ignores the seasonal variations in mixed-layer depth (including upwelling) and oversimplifies the vertical heat conduction physics. The latter error often leads to a 90-day lag in maximum SST after heating maximum, but in reality the lag is usually closer to 60 days. The slab ocean also does not include the effects of currents in the ocean basins, such as the Gulf Stream. These currents have little direct effect on seasonal cycling of surface air temperatures, but have a strong influence on the mean annual thermal distribution (see Fig. 1.9b). Some modelers have modified their mixed-layer formulations to include some of these effects, but it is not a uniformly established practice. Mixed-layer model runs usually cover about 15 model years, after which an essentially repeating seasonal cycle establishes itself. Such runs can require more than 20 times more computer time than perpetual model runs. Thus we see how modelers are forced to consider computer expenses when designing climate experiments.

2.4 Ocean Models

In a somewhat parallel form to atmospheric GCMs, ocean GCMs have been developed to examine the three-dimensional circulation of the ocean. Many of the equations are the same for the two media and several models are now running (e.g., Maier-Reimer et al., 1982; Meehl et al., 1982; Cox, 1985; Han, 1988; Semtner and Chervin, 1988). The models do a reasonable job of simulating observed fields (e.g., Fig. 2.14), and one has even been coupled to a carbon cycle model (Maier-Reimer and Hasselmann, 1987).

2.4.1 Ocean GCMs

We mention here a few distinctions between the ocean and atmospheric dynamics that influence what can and cannot be done with ocean models. First of all, there is a crucial difference in length scale between the two media. The so-called Rossby radius of deformation is a relevant length scale for dynamic flows (Washington and Parkinson, 1986). Roughly speaking, it is the distance above which earth rotation effects become important. It is different for the two media because natural gravitational adjustments propagate at different speeds in the two fluids (factor of 10 slower in the oceans). This length scale is about 1000 km in midlatitudes for the atmosphere and only about 100 km for the oceans. The significance is that in numerical models, all scales greater than these must be resolved by the numerical grid or at least accounted for by some fairly explicit parameterization scheme. From the outset fully satisfactory ocean models will need to have a grid resolution about a factor of 10 finer in order to achieve comparable accuracies.

In the ocean's favor are the absence of complicating radiation processes such as clouds and water vapor changes that tend to slow down the atmospheric models. On the other hand, ocean basins are topographically complicated and preclude the use of spectral (e.g., spherical harmonic) methods in almost all cases (see Section 2.2.2). Some other differences are as follows: (1) Water can be considered incompressible to simplify the continuity equation. (2) Buoyancy differences from place to place are included as a force in the momentum equations of ocean models. Such small but persistent forces arise from small thermal or salinity contrasts. (3) The friction terms in the momentum equations are generally much larger for oceanic flows because of the need to account for strong eddy motions at the 100-km scale.

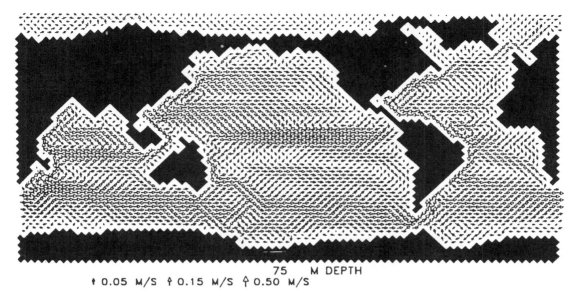

Fig. 2.14. Horizontal ocean current velocity field (75 m) in an ocean GCM. Current velocities are proportional to the size of the arrows. [From Maier-Reimer and Hasselmann, 1987] *Reproduced with permission of Springer-Verlag.*

A particularly useful result (K. Bryan, 1986) from one ocean GCM study involves comparison of a high-resolution (eddy-resolving) and low-resolution ocean model (Fig. 2.15). Although higher resolution is needed for accurate treatment of tracer transport in the sea, it is apparently not needed to satisfactorily simulate poleward ocean heat transport. Since heat transport is one of the variables of greatest interest to geologists, the results suggest that inadequate

Fig. 2.15. Comparison of total time-averaged transport from an eddy-resolving ocean model (solid line; from Cox, 1985) and a coarse resolution, non-eddy-resolving model (dashed line). 1 PW = 10^{15} W. [From K. Bryan, 1986] *Reprinted from Journal of Physical Oceanography with permission of the American Meteorological Society.*

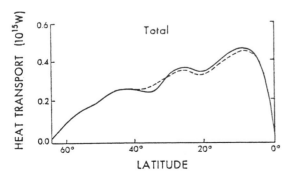

representation of detailed ocean eddy dynamics may not be a critical hindrance in ocean GCM use for paleoclimate simulation.

The time constant of the deeper ocean is hundreds of years; hence, one must estimate the appropriate run time for an ocean GCM simulation. If we are concerned about the problem of doubling CO_2 in the next 50 years, we may not need to know details of the deep-ocean response. On the other hand, we might be interested in paleoclimate questions such as the positioning of the Gulf Stream or changes in NADW formation during the last major glaciation. This latter application might require running the dynamic ocean model all the way to statistical equilibrium.

2.4.2 Coupled Ocean–Atmosphere Models

A final type of model run involves fully interactive ocean–atmosphere models. In this type of configuration, neither the atmosphere nor the ocean is prescribed; each interacts with the other in a time-dependent manner. In theory, such models should give the results that are least constrained by assumptions about how the system should work, and such models are required for fully satisfactory treatment of future environmental consequences of a greenhouse warming. However, in practice coupled ocean–atmosphere models are just reaching the stage where they yield moderately satisfactory simulations even for the present climate (e.g., Washington and Meehl, 1989). The basic problem with fully satisfactory simulations involves coupling between the two media (e.g., Bryan, 1988; Hasselmann, 1988). For example, the response time of the atmosphere is fast compared to the ocean, especially the deep ocean. Vast amounts of computer time would be required to run the coupling in a real-time "synchronous" mode, so various types of "asynchronous" coupling routines have been developed to shorten the spin-up time of the system.

A further problem is posed by our inadequate treatment of the fluxes between the ocean and the atmosphere. When fluxes are not prescribed, they often rapidly and radically diverge from (poorly known) observed fluxes, resulting in model output that is not very realistic. For example, an early version of a NOAA/GFDL coupled ocean–atmosphere model (Bryan et al., 1975) produced a North Atlantic temperature and salinity field for the present that is somewhat like the ice-age configuration (Chap. 3)—the Gulf Stream does not penetrate into higher latitudes of the North Atlantic. This results from too much model-generated precipitation minus evaporation, which causes a low salinity lens in the subpolar North Atlantic that does not exist today. The result is that the Gulf Stream flows along the new density gradient farther southward than it does at present.

There have been various methods proposed to solve some of the previously discussed problems (e.g., Sausen et al., 1988), but to date modelers are still in an experimental phase for fully coupled runs. Ultimately, we have to turn to these models, especially for greenhouse studies. But we may be several years away from satisfactory coupled experiments.

2.5 Cautions, Caveats, and Conclusions

We may now try to summarize some of what we have learned about climate models, particularly with respect to their paleoclimate utility. It is important for geologists to understand the con-

cept of the hierarchy of climate models, which we discussed in Chap. 1. It is also useful to understand the basic elements of GCMs, so that a better feeling can be gained for how the models work, and which models are appropriate for particular paleoclimate applications. We also believe it is important for geologists to understand what we call the "hierarchy of climate model believability," in which we distinguish between model complexity and adequate simulation of the field under special consideration. For example, we demonstrate in Chap. 1 that even EBMs do a satisfactory job of simulating the seasonal cycle of surface temperatures. The sensitivities are comparable to those of GCMs.

GCMs are essential for understanding past climates. But it is important to understand that not every field produced by a GCM is done with the same level of fidelity to nature. Models simulate temperature better than precipitation, and some types of precipitation (large-scale monsoonal flows) better than others (e.g., midcontinent summer convective precipitation). If the deep ocean is involved, there may be further serious errors in a local surface field. In order to evaluate the level of believability of model results it is useful to compare the simulated field with the observed field. If they are in reasonably satisfactory agreement, then that particular field may be a good target for a paleoclimate study (assuming the boundary conditions can be satisfactorily specified). Another way to evaluate model behavior is to compare one GCM product with a different GCM product of the same field. If they agree well, believability increases. As demonstrated in this chapter, adequate simulation varies by GCM field. Finally, it is very desirable to develop a solid physical understanding for the reasons why models produce the changes they simulate. If we can link a perturbed response field to a well-understood altered boundary condition, we have a greater chance of achieving real physical insight from a model simulation.

Geologists must therefore learn to pick their way through the maze of GCM output to focus on the results most likely to be believable. This selection process can only come from experience. It is not enough to accept everything a model produces or reject everything a model produces because "it's just a model." We have to learn how to carve out the believable part from the more questionable results. That ability comes only from experience and a willingness to understand some basics of how the models are put together. But the benefits of understanding the models are significant. Responsible use of climate models can help in the formulation of theories of climate change and in some cases such models can lead the geologist to collect and analyze data in new ways.

PART II
QUATERNARY CLIMATES

3. RECONSTRUCTING CLIMATE OF THE LAST 20,000 YEARS

Various kinds of geologic evidence indicate that past climates of the earth have fluctuated between ice-free and glaciated states. Although there have been numerous changes of climate in earth history, we will begin our summary of past climates by focusing on the most recent case of a large climate change (the last glacial maximum). There is a wealth of data (Fig. 3.1) covering this time period (14,000–22,000 BP), the deglaciation (approximately 10,000–14,000 BP), and the present Holocene interglacial (0–10,000 BP). To a first approximation we can use these time intervals to develop a picture of what the earth may have been like during other comparable periods for which we have a less complete record. This information is essential for detailed testing of climate models—a topic we will address in Chap. 4.

Although work on ice-age climates has been under way for more than a century, a more concentrated effort began in the early 1970s to synthesize the work of past generations of geologists, add new information from the rapidly expanding area of Pleistocene oceanography, and reconstruct the climate of the earth's surface at the midpoint of the last glacial maximum (18,000 BP). This group was called CLIMAP (Climate/Long-Range Investigation, Mapping and Prediction), and several important publications summarize their findings (CLIMAP, 1976, 1981, 1984; Cline and Hays, 1976). The success of the CLIMAP project inspired a similar effort to expand the time slice approach into the Holocene and reconstruct the climates of the last 18,000 years at 3000-year intervals. This latter group is called COHMAP (Cooperative Holocene Mapping Project; COHMAP, 1988). Results presented below summarize the work of these cooperative research projects and that of numerous other scientists who have made important contributions to understanding this period in the earth's climate history.

Before embarking into the nature of the records, it is necessary to briefly comment on techniques and methodologies. A vast array of techniques have been used to reconstruct past climates. For example, work on land involves examination of glacial moraines, pollen, and lake levels. Oceanic records involve examination of both the biogenic and nonbiogenic components of deep-sea sediments. A wealth of information is also available from ice cores. Although we elaborate in Appendix B on methodologies used to make quantitative estimates from paleoecological data, in general it is beyond the scope of this book to discuss in detail the various methodologies used to produce the climate reconstructions we discuss in this chapter. We refer the readers to two books which successfully cover many aspects of this problem (Bradley, 1985; Hecht, 1985).

3.1 Last Glacial Maximum

The last glacial maximum extended from about 22,000 to 14,000 BP (e.g., Dreimanis and Goldthwait, 1973). The glaciation has been variously called the Wisconsin, Weichselian, or Würm, depending on whether eastern North America, western Europe, or the Alps is referred to. Scientists have long been intrigued by the details of climate from a time interval so different from the present. The reconstruction of events has been greatly aided by radiocarbon age dating, which has a precision of 1000–2000 years over much of the time interval studied. The advent of accelerator mass spectrometry (AMS) ^{14}C dating has reduced this number to well under 1000 years in many cases (e.g., Bard et al., 1987).[1]

---

1. [Note added in proof] Recent comparisons of ^{14}C and U/Th dates indicate that ice-age ^{14}C dates may be too young by as much as 3,500 years (Bard et al., 1990).

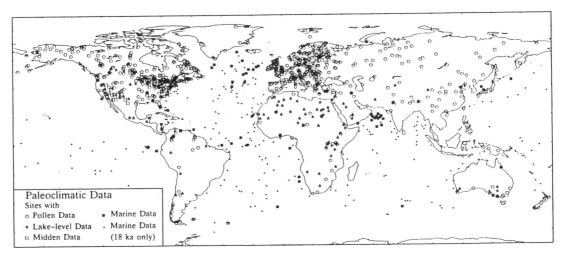

Fig. 3.1. Location of sites in a global paleoclimate database 0–18,000 BP. [From COHMAP, 1988] *Reproduced by permission from Science 241:1043–52. Copyright 1988 by the AAAS.*

3.1.1   Changes in Snow and Ice Cover

The most dramatic features of the last glacial maximum were the great ice sheets (Fig. 3.2). Although ice cover increased in many areas, the largest accumulations were in eastern North America (Laurentide Ice Sheet) and northwestern Europe (Fennoscandian Ice Sheet), with smaller increases in western North America (Cordilleran Ice Sheet), European Russia, and West Antarctica. Some authors have suggested that the ice sheets may have been linked via a thicker Arctic Ocean ice shelf (Broecker, 1975; Hughes et al., 1977; Denton and Hughes, 1981a). This conjecture remains a topic of debate. Although Kuhle (e.g., 1987) suggested there was an ice sheet on the Tibetan Plateau, this conclusion has also been disputed (Zheng, 1989).

Estimated thicknesses of the Laurentide and Fennoscandian ice sheets are on the order of 3500–4000 m thick (e.g., CLIMAP, 1981). Their formation required evaporation of 50–60×10^6 km^3 of water from the oceans (e.g., Hughes et al., 1981). The best estimate for the consequent sea level lowering is 121 ± 5 m (Fairbanks, 1989; see below, Fig. 3.14). Approximately one-half to two-thirds of global ice volume was used to construct the huge Laurentide Ice Sheet (Hughes et al., 1981), which had the

dimensions of a large mountain range. It extended from the Rocky Mountains to the Atlantic shore and from the Arctic Ocean southward to about the present positions of the Missouri and Ohio rivers. In Europe the Fennoscandian Ice Sheet reached northern Germany and the Netherlands. The weight of the massive ice sheets depressed the crust by as much as 700–800 m (e.g., Flint, 1971) resulting in gravity anomalies that are still detectable (e.g., Peltier, 1987).

The large changes in Northern Hemisphere ice cover on land were paralleled by large changes over the ocean. Oceanic polar fronts and sea ice migrated equatorward in both hemispheres (CLIMAP, 1976, 1981). Some of the most dramatic changes occurred in the North Atlantic (Fig. 3.2). At present, the northeastern Atlantic is at least seasonally ice-free as far north as 78°N in the Norwegian Sea. This condition reflects the advection of warm North Atlantic Current (i.e., Gulf Stream) water into this region (Fig. 2.4). During the last ice age, the oceanic polar front migrated to about 45°N (McIntyre et al., 1976). North of this latitude, the ocean was mainly covered by sea ice during the winter. This configuration resulted in a substantial region of the earth's surface (120°W to 90°E, north of 45–50°N) covered either by ice or tundra (see below). The expanded sea ice apron probably contributed to the large area of polar desert in Europe "downwind" from the sea ice zone (cf. Kutzbach and Wright, 1985).

Virtually all regions of the world capable of supporting snowcaps witnessed expanded ice cover at 18,000 BP. In the tropics, mountain glaciers formed or expanded in New Guinea, Hawaii, eastern Africa (e.g., Kilimanjaro), and the Andes (e.g., Hastenrath, 1971, 1985; Bowler et al., 1976; Porter, 1979). These changes, together with snowline depressions of about 1000 m (Fig. 3.3) and variations in pollen, are consistent with a temperature depression of 5–6°C at elevations greater than 2000 m (Rind and Peteet, 1985; Bonnefille and Riollet, 1988).

In the Southern Hemisphere glaciation was approximately synchronous with that of the Northern Hemisphere (e.g., Mercer, 1984). Glaciers expanded in the southern Andes (Porter, 1981) and New Zealand (Porter, 1975). Periglacial features (environments typical of glacial margins) occur in lower elevations of southeast Australia (e.g., Galloway, 1965; Costin, 1972).

In Antarctica the West Antarctic Ice Sheet (cf. Fig. 10.15) may have increased in volume (Stuiver et al., 1981). At present this ice sheet is constrained by proximity of the ocean; lower sea level would have allowed a significant expansion (e.g., Denton et al., 1986).

Initially, CLIMAP (1981) suggested that the interior East Antarctic Ice Sheet elevations were about 400–500 m greater than the present at 18,000 BP. This conclusion was based on the assumption that the ice sheets had reached an equilibrium profile. A reevaluation concluded that interior elevations were not significantly different from those of the present (Denton et al., 1989; Jouzel et al., 1989).

There were also large changes in Antarctic sea ice cover. At present winter sea ice cover in the Southern Ocean is approximately equal to the amount of ice on Antarctica ($\sim 15 \times 10^6$ km^2; Zwally et al., 1983). At 18,000 BP (Fig. 3.4) sea ice was approximately twice the area of Antarctic ice cover (Cooke and Hays, 1982). Summer meltback may also have been reduced (Cooke

Fig. 3.2. Northern Hemisphere ice distribution (land and sea) for January 18,000 BP. [Modified from Denton and Hughes, 1981b] *Reproduced with permission of G. H. Denton and T. J. Hughes (Eds.), "The Last Great Ice Sheets." Copyright 1981, John Wiley and Sons.*

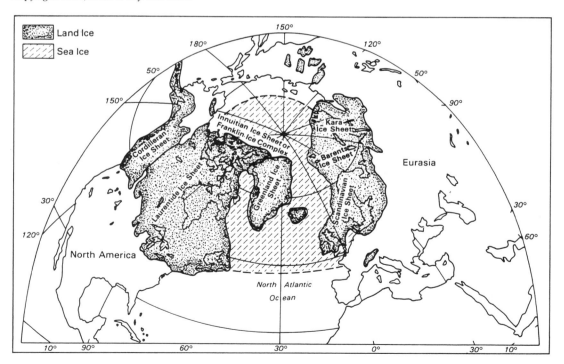

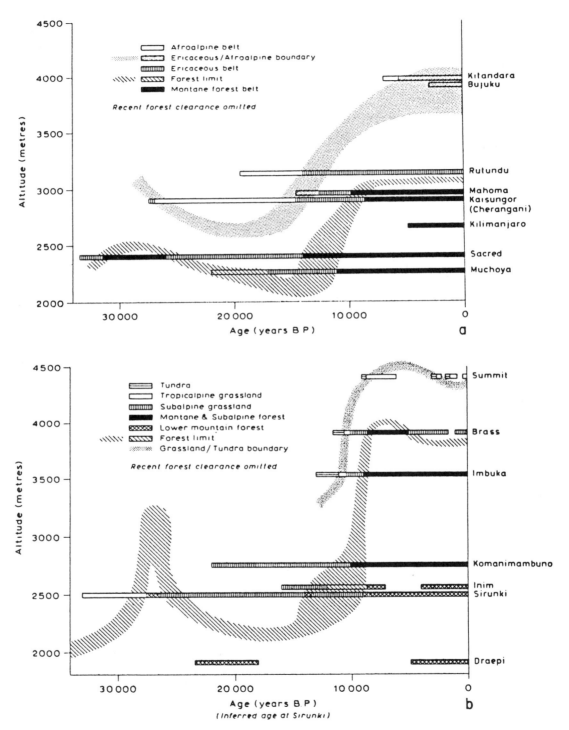

Fig. 3.3. Summary diagram of late Quaternary vegetational changes in (a) the eastern African mountains (Flenley, 1979b) and (b) New Guinea highlands (Hope et al., 1976; Flenley, 1979b). Note the ~1-km lowering of snowlines and vegetation zones. [From Rind and Peteet, 1985] *Reproduced by permission of Quaternary Research.*

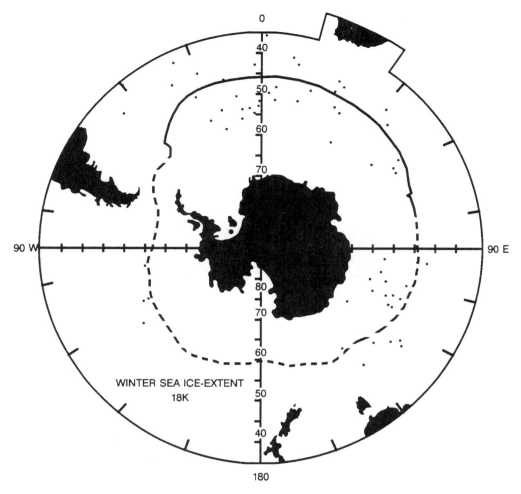

Fig. 3.4. Estimated July Antarctic sea ice extent at 18,000 BP. [After Cooke and Hays, 1982] *Reproduced by permission of the University of Wisconsin Press.*

and Hays, 1982). Interpretation of Antarctic sea ice history is subject to some uncertainties (Burckle et al., 1982; see also discussion in Crowley and Parkinson, 1988a), so these conclusions may be altered in the future.

3.1.2   Additional Mid- and High-Latitude Temperature Changes

Numerous records (e.g., Fig. 3.1) indicate large changes of climate in areas bordering the ice sheets (e.g., Watts, 1983; Wright, 1987). In North America and Europe (Fig. 3.5) tundra extended southward from the ice margins, with the geographic extent much greater in western Eu-

rope than in North America (some of the tundra deposits are not well dated but it is reasonable to assume they reflect full-glacial conditions). In North America, the thin strip of tundra (Péwé, 1983) was replaced southward by a spruce-pine boreal forest (e.g., Delcourt and Delcourt, 1985). South of 35°N, oak-hickory forests, local prairie vegetation, and low lake levels (in Florida) support a picture of overall ice-age aridity. The 34°N boundary has been interpreted as the mean position of the polar front (Delcourt, 1979; Delcourt and Delcourt, 1983), which today is located about 1200 km to the north in southern Canada. In the regions south of the Laurentide Ice Sheet, there are combinations of

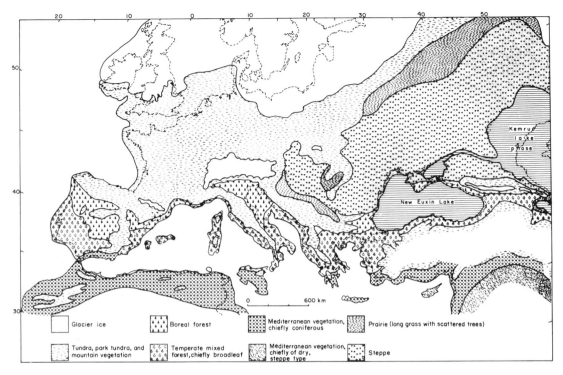

Fig. 3.5. Sketch map of inferred European vegetative patterns during the last glacial maximum. [From Flint, 1971; adapted from Büdel, 1949] *Reproduced with permission from "Glacial and Quaternary Geology," R. F. Flint. Copyright 1971, John Wiley and Sons.*

mammals (cold and warm elements) which do not occur at present (Lundelius et al., 1983; cf. Wright, 1987).

In Europe, polar deserts may have covered most of the area between the southern margin of the Fennoscandian Ice Sheet and the northern margin of expanded Alpine glaciers (Fig. 3.5). This region was populated by typical Arctic vertebrates such as reindeer, mammoth, Arctic fox, and lemming (e.g., Flint, 1971; Kukla, 1977). Farther east, windblown eolian deposits (loess), typical of cold and dry steppe conditions, extended from the eastern European plains across Eurasia to central China (Kukla, 1977, 1987). Since sea level changes had exposed the Bering Strait, the steppe belt may have continued through Siberia all the way to Alaska (Guthrie, 1982). Despite the harsh conditions, the Siberia–Alaska corridor (Beringia) is notable for its large numbers of fossils of large vertebrates (mammoth, bison, elk, horses; Guthrie, 1982). There is evidence that early man hunted these populations during his migration to

North America (Morlan and Clinq-Mars, 1982).

A variety of paleoclimate indices have been used to estimate temperature changes over land. Although there are some differences in the results, estimates generally suggest that midlatitude surface temperatures decreased about 10°C (annual average) in regions near the Laurentide and Fennoscandian ice sheets (e.g., Flint, 1971; Barry, 1983; Guiot et al., 1989). Temperature changes south of the Laurentide Ice Sheet may have been more severe in winter than in summer, a feature perhaps reflecting outbreaks of polar air from the ice sheet (see Chap. 4). Delcourt (1979) and Watts (1980) estimated winter temperatures perhaps 15–20°C below present for Tennessee and South Carolina, changes consistent with freeze–thaw churning of the soil (Clark, 1968; Péwé, 1983). Estimated winter temperature decreases for northwestern Europe are also 15–20°C (Kutzbach and Wright, 1985).

Except for central China (10°C), estimates for temperature decrease are in the range of 5–8°C for a large number of midlatitude land areas, e.g., the western United States, southeast Australia, New Zealand, and Chile (e.g., Bowler et al., 1976; Heusser and Streeter, 1980; Liu et al., 1985; Spaulding and Graumlich, 1986; Markgraf, 1989). Temperature estimates for western North America regions affected by maritime air masses are 4–5°C (Heusser et al., 1985).

3.1.3 Precipitation Changes

With a few exceptions much of the planet seems to have been drier during the last ice age. Precipitation patterns varied by latitude belt. In high latitudes, accumulation rate changes in Greenland and Antarctic ice cores suggest that precipitation decreased about 50% in polar regions (Beer et al., 1985; Herron and Langway, 1985; Lorius et al., 1985).

Conditions were moist in some midlatitude land regions affected by equatorially displaced westerlies and the low-pressure systems that form in this belt. Extensive lakes (Fig. 3.6) developed in the Great Basin of western North America (e.g., Smith and Street-Perrott, 1983; Benson and Thompson, 1987) and European Russia (Grosswald, 1980). The proximity of ice

sheets may have contributed to the formation of these lakes by a combination of lower air temperatures (implying less evaporation), some increased precipitation, and/or meltwater runoff/ physical blockage of drainage systems (e.g., Brakenridge, 1978; Grosswald, 1980; Benson and Thompson, 1987). A combination of a 2.4 times precipitation increase and 5–7°C temperature decrease may explain some Pleistocene lake level changes in the western United States (Benson and Thompson, 1987). However, there are a number of uncertainties in these calculations (Smith and Street-Perrott, 1983; Benson and Thompson, 1987), so estimates should be treated with some caution.

Conditions were also relatively moist in northwestern Africa, the Middle East, southern Australia, and southern South America (Fig. 3.6; Street-Perrott and Harrison, 1985; Markgraf, 1989). These regions, presently on the poleward margins of subtropical arid belts, benefitted from the equatorward displacements of midlatitude low-pressure systems, which would presumably track along the boundary of maximum temperature gradient (sea ice margins) in both hemispheres (e.g., Figs. 3.2 and 4.1).

The overall increase in surface aridity at

Fig. 3.6. Global map of lake level status at 18,000 BP. [From Street-Perrott and Harrison, 1985] *Reproduced with permission of F. A. Street-Perrott and S. P. Harrison, Lake levels and climate reconstruction, in "Paleoclimate Analysis and Modeling." A. D. Hecht (Ed.). Copyright 1985, John Wiley and Sons.*

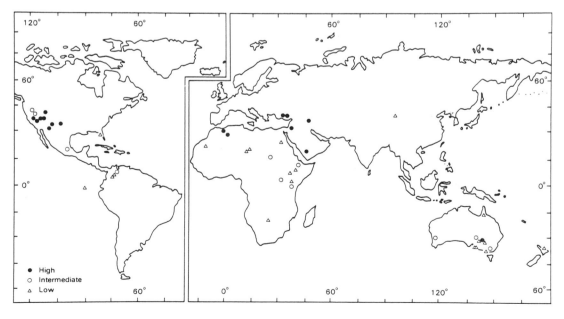

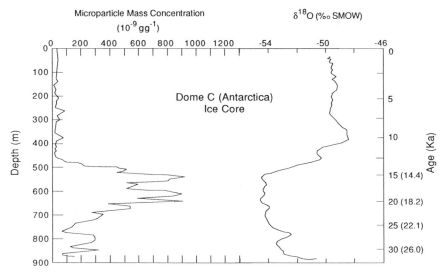

Figure 3.7. Dust and isotope records from the Dome C ice core on the East Antarctic plateau (see Fig. 10.15). The differences in chronology (parentheses) reflect assumptions about variable or constant accumulation. [After A. Royer et al., 1983] *Reprinted by permission of Kluwer Academic Publishers.*

18,000 BP is consistent with an increase in atmospheric dust concentrations as recorded in the Greenland and Antarctic ice cores (Petit et al., 1981, 1990; Hammer et al., 1985; see Fig. 3.7) and increases in wind-blown eolian sediments in equatorial Atlantic deep-sea cores (Kolla et al., 1979; Sarnthein et al., 1981; Pokras and Mix, 1985). Ice-age aridity is also consistent with a massive transfer of carbon from terrestrial to marine reservoirs (Shackleton, 1977; Keigwin and Boyle, 1985; Shackleton and Pisias, 1985; cf. Fig. 6.9). Glacial-interglacial carbon transfer (Curry et al., 1988; Duplessy et al., 1988) is approximately equivalent to about one-fourth to one-third of the total amount of carbon stored in plant, soil, and continental shelf reservoirs, i.e., about $0.5–0.7 \times 10^{18}$ g of carbon (Duplessy et al., 1988). As a result of increased aridity and ice extent, surface albedo (summer) increased from 0.14 to 0.22 (Gates, 1976a).

3.1.4 Ice-Age Tropics

In addition to the good evidence for snowline lowering in tropical mountains, lowland climates changed. Our knowledge of precipitation changes is better than temperature changes. Sparse estimates for the latter are ~4–5°C (Liu and Colinvaux, 1985; Markgraf, 1989), and

time control for some of these estimates could be better.

There is moderately good evidence that tropical lowlands were drier during the last glacial maximum. Lake levels in tropical Africa and Central America were very low (Street-Perrott and Harrison, 1985; see Fig. 3.6), with water levels in some of the larger eastern African lakes 250–500 m below present (Scholz and Rosendahl, 1988; cf. Gasse et al., 1989b). Sand dunes expanded in the sub-Sahara and Central America (Ward, 1973; Sarnthein, 1978). It has been suggested that the Amazon rainforest may have been reduced to a few "refugia" (localized regions of higher precipitation; Haffer, 1969; Prance, 1982). This hypothesis is based on biological observations of present geographic variations in the diversity of various types of Amazon Basin biota, e.g., woody angiosperms (flowering plants) and butterflies (Fig. 3.8). The "islands" of high diversity may reflect moist regions, whereas the lower diversity regions have been interpreted as more savannah-like. [An alternate interpretation to heterogeneous diversity patterns is that they reflect non-equilibrium interactions within the biosphere and between the biosphere and the environment (cf. Connell, 1978; Colinvaux et al., 1988).]

Inferences about tropical rainfall history

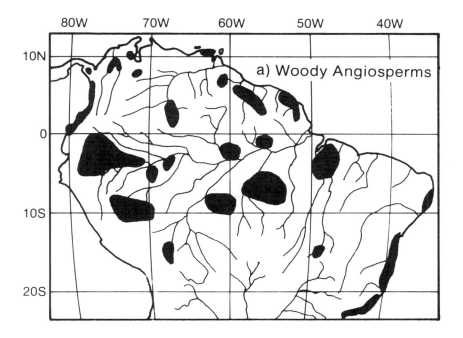

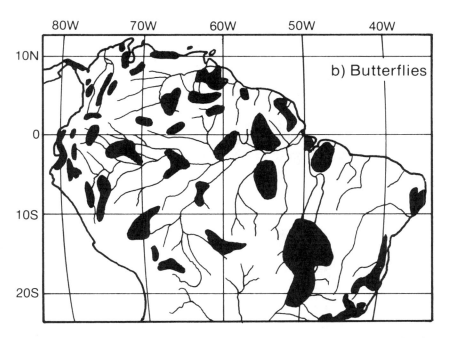

Figure 3.8. Proposed refuge areas for certain species of (a) woody angiosperms and (b) butterflies in the Amazon Basin during postulated dry climatic phases of the Pleistocene. [From Bradley, 1985; based on results in Prance, 1982; and Brown, 1982] *Extract taken from "Quarternary Paleoclimatology" by R. S. Bradley. Reproduced by kind permission of Unwin Hyman Ltd.*

could be considered suspect because of the general scarcity of good sites from lowland regions (cf. Colinvaux, 1989). Yet verification of this conclusion is very important both for validating climate models (Chap. 4) and because of the significant implications for tropical rainforest history. We will therefore explore in more detail other types of evidence for lowland tropical aridity.

Given the sampling limitations, there is a fair amount of supporting evidence for tropical precipitation decreases. For example, there are increased ice-age concentrations of feldspar and the clay mineral illite in deep-sea sediments originating from the Amazon and Zaire (Congo) Rivers (Damuth and Fairbridge, 1970; Jansen et al., 1984). Feldspar and illite are normally altered under moist conditions; thus their preservation suggests drier conditions. Pollen, spores, and freshwater diatom fluxes in Zaire River sediments also indicate drier conditions (Jansen et al., 1984; Gasse et al., 1989a).

Similarly, pollen records from (a few) lowland sites in Central and South America support interpretations of drier conditions (e.g., Colinvaux, 1972; van der Hammen, 1974; Flenley, 1979a; Bradbury et al., 1981; Leyden, 1984; Absy, 1985). An overall increase in net evaporation (plus lowered sea level) has been linked to significantly higher salinities in some marginal seas (Mediterranean and Red seas). Estimated salinity increases for the Mediterranean and Red seas are 1–3 and 10‰, respectively (Thunell et al., 1987, 1988; Thunell and Williams, 1989). The hypersaline conditions in the Red Sea probably approached the tolerance limit of planktonic organisms.

There are a few caveats to the conclusion of enhanced tropical aridity at the glacial maximum. An eolian record from the central equatorial Pacific suggests wetter conditions in its source area (South America?) during the last glacial maximum (Rea et al., 1990). Additionally, time control is not always as good as desired and some evidence may actually come from older periods. As discussed in Section 6.4.6, there is some indication that in certain regions tropical SSTs were colder prior to 18,000 BP (cooler and drier conditions should accompany lower SSTs; see Section 4.1.2).

Evidence for enhanced lowland tropical arid-ity is also more variable in the Indonesian sector. Pollen records from northern Queensland (about 17°S) and New South Wales (31°S) suggest significant drying at ~18,000 BP (Kershaw, 1986; Dodson and Wright, 1989). However, (sparse) lowland records from Sumatra, Java, and Kalimantan (Borneo) indicate at best only minor drying over the same time interval (Caratini and Tissot, 1988; Stuijts et al., 1988). It would be desirable to acquire more lowland data from this important region.

It is difficult to estimate precipitation reductions for lowland tropical areas. If tropical rainforests were reduced as much as they appear to have been, it would seem that a 30–50% decrease in net precipitation minus evaporation would be required. As we will see in Section 4.1.2, climate models in general do not produce such a decrease. However, it is necessary to emphasize that there is an important biological feedback which may not be satisfactorily simulated in GCMs. Observations in the Amazon basin (e.g., Salati, 1985) indicate that as much as one-half of Amazon rainfall is "recycled" due to plant transpiration. Therefore, a 30–50% decrease in net precipitation minus evaporation may require perhaps only an initial 15–25% decrease in precipitation, with decreased plant cover providing the feedback that amplifies the perturbation into a response such as that inferred for the glacial maximum. Since present interannual fluctuations in Asian monsoon rainfall are in the range of 10–20% (e.g., Rasmusson and Carpenter, 1983), and such changes are known to cause serious drought, it is possible than an initial reduction in rainfall by 15–25% coupled with biofeedback could produce the changes inferred from Pleistocene observations.

3.1.5 Atmospheric Circulation Changes

There were also changes in ice-age wind directions and speeds. Wells (1983) examined late-glacial distribution of eolian features on North America (Fig. 3.9) and concluded that average wind directions changed from present southwesterlies to ice-age northwesterlies. The advection of very cold air by the strong northwesterlies might have had a significant effect on evaporation rates in the Gulf Stream. The largest amount of latent heat transfer from the ocean to

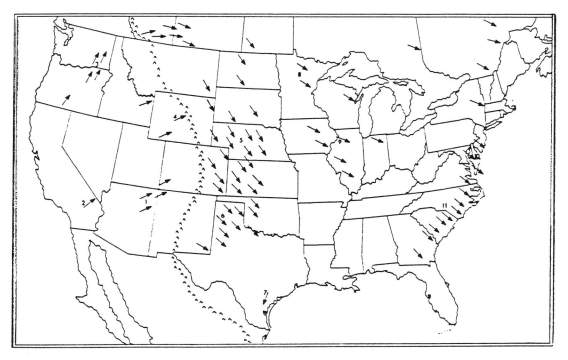

Fig. 3.9. Trends of late glacial features in prevailing surface winds on North America, based on studies of eolian features (dune orientation, etc.) [From Wells, 1983] *Reprinted by permission of Kluwer Academic Publishers. Copyright © 1983 D. Reidel Publishing Co.*

the atmosphere presently occurs in such regions (e.g., Budyko, 1978), with synoptic scale events associated with enormous fluxes of as much as 800–1000 W/m² (SethuRaman et al., 1986; SethuRaman and Riordan, 1988). Such values may have been exceeded during the last glacial maximum.

Studies of upwelling indices and windblown (eolian) material in deep-sea cores suggest glacial age increases of about 20% for the North Pacific westerlies, about 30% for the North Pacific trades, and about 50% for the North Atlantic trades (e.g., Sarnthein et al., 1981; Pedersen, 1983; Janecek and Rea, 1985; Hooghiemstra et al., 1987). Upwelling variations along the Peru Current are consistent with a 30–50% increase in the South Pacific trades (Molina-Cruz, 1977). Despite these changes, records from the eastern equatorial North Atlantic indicate that wind belts did not shift latitudinally with enhanced velocities (Sarnthein et al., 1981; Hooghiemstra et al., 1987).

Ice cores also record information about changes in ice-age winds (Petit et al., 1981; Her-

ron and Langway, 1985; DeAngelis et al., 1987). Estimates are based on measurements of chloride concentrations in the cores. The chloride originates as atmospheric sea salt. Concentration of sea salt over the ocean is a function of height above the sea surface and wind speed (the stronger the winds, the greater the concentration of sea salt). Increased ice-age aerosol concentrations (cf. Fig. 6.22) can be used to infer increased speeds of westerlies in the North Atlantic and Southern oceans of about 5–8 m/sec— 50–80% above the present average speed of about 10 m/sec for the Southern Ocean.

Thus, the small and varied proxy data sets consistently indicate that global surface wind speeds may have increased anywhere from 20 to 50% and perhaps more. Such large changes in forcing might be expected to exert a strong effect on the ocean circulation. The stronger winds would also affect sea ice formation. For example, stronger winds in the Southern Ocean may have caused a 100 W/m² increase in ocean heat loss to the atmosphere (Crowley and Parkinson, 1988b).

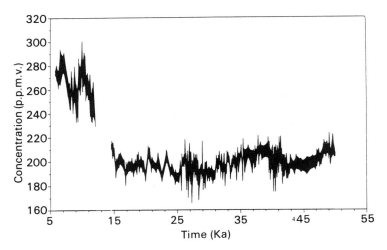

Fig. 3.10. Atmospheric CO_2 concentration in Byrd (Antarctica) ice core. [From Neftel et al., 1988] *Reprinted by permission from Nature 331, 609–611, copyright © 1988, Macmillan Magazines Ltd.*

3.1.6   Atmospheric Composition

In addition to higher dust levels (Section 3.1.4), ice cores record another remarkable feature of the ice-age earth. Studies of gas inclusions in Greenland and Antarctic ice cores indicate CO_2 glacial concentrations (Fig. 3.10) of about 200 ppm (Berner et al., 1980; Delmas et al., 1980; Barnola et al., 1987), about 75–80 ppm less than the estimated preindustrial value of about 280 ppm (Neftel et al., 1985; cf. Fig. 14.3). Methane concentrations also decreased by one-half (Stauffer et al., 1988). The fluctuations in CO_2 may be due to variations in marine biological productivity (e.g., Broecker, 1982a,b) and will be discussed more in Chap. 6. Causes for atmospheric methane variations have been less well studied, but may reflect increased aridity, as wetlands are an important source of methane (e.g., Matthews and Fung, 1987).

Since both CO_2 and CH_4 absorb outgoing infrared radiation from the earth (cf. Fig. 1.3), changes in their concentration will cause a climate feedback. The perturbation in CO_2 radiative forcing is about ~1.7 W/m^2 and that for methane is about 0.1–0.2 W/m^2 (for reference, a CO_2 doubling causes a 4.3 W/m^2 change). With feedbacks (see Sect. 1.3.2), the CO_2 changes translate to ~1.5°C in global average temperatures—about 40% of the glacial-interglacial signal (see Chap. 4). Thus, the CO_2 changes represent a very significant amplifica-

tion of glacial-interglacial climate change. However, the methane variations appear to be too small to have a significant climate effect (~0.1–0.2°C).

Other changes in atmospheric "composition" during the last ice age involve a possible increase in cosmic dust flux (LaViolette, 1985) and perhaps a 20% increase in the production rate of the cosmogenic isotopes ^{10}Be and ^{14}C (Beer et al., 1988). Changes in the latter may reflect in part weakening of the magnetic dipole field between 50,000 and 10,000 BP (Barbetti, 1980).

3.1.7   Surface Ocean Temperature Changes

Quantitative studies of microfossil assemblages have led to a great increase of our knowledge of paleoceanographic changes (e.g., Ruddiman, 1985). Imbrie and Kipp (1971) first demonstrated that the various assemblages of planktonic foraminifera (e.g., tropical, subtropical, polar, etc.) could be economically mapped using now standard techniques of factor analysis (also known as principal component analysis or empirical orthogonal functions). The reader is referred to Appendix B for a discussion of how faunal and floral abundances are converted into temperature estimates.

The publication of the historic Imbrie–Kipp paper in 1971 opened the door to a new age in the science of paleoclimatology because it enabled geologists to derive estimates of a physical variable (sea surface temperature, SST) that is a critical boundary condition for the atmosphere and atmospheric models. The study led directly to the development of the CLIMAP project and eventually stimulated a widespread interest in modeling past climates.

In regions affected by the migrations of sea ice and oceanic polar fronts (Section 3.1.1), SST decreased by as much as 6–10°C (Fig. 3.11). Along eastern boundary currents (cf. Section 2.1.2), present sites of equatorward penetrations of cool waters, SST decreased by about 6°C. Cool conditions sometimes extended into eastern regions of equatorial oceans and probably reflect in part increased upwelling along the equator.

Over large parts of the remaining tropical oceans, SST changes were apparently much smaller—only 1–2°C. In most areas, this zone of stable SST extended poleward to about 40° latitude. Since tropical conditions did not change much and polar conditions changed greatly, there was a compaction of "living space" in the transitional regions in between (cf. Fig. 6.20).

The globally averaged decrease in SST was about 1.6°C (CLIMAP, 1981). This result represents one of the most significant contributions of the CLIMAP project. It is also perhaps the most difficult result to reconcile with some climate modeling results (e.g., Webster and Streeten, 1978; Rind and Peteet, 1985; cf. Section 4.1.4). These studies suggest that high tropical SSTs are incompatible with the magnitude of temperature depressions in tropical highlands and much enhanced aridity of tropical lowlands. The CLIMAP conclusion has been tested by several approaches and still appears valid (Prell, 1985; Broecker, 1986; Brassell et al., 1986; Anderson et al., 1989). It would be desirable to further check the conclusion by examining additional lowland tropical sites, further examining the ocean sediment record, and further testing model calculations.

Much less is known about salinity variations during the last glacial maximum. Surface salinities were probably lower in the area of displaced polar fronts (e.g., McIntyre et al., 1976). As discussed in Section 3.1.4, salinities increased in the marginal basins of the Mediterranean and Red seas. Based on Atlantic–Pacific differences

Fig. 3.11. Difference between modern August SST and estimated SST at the last glacial maximum 18,000 BP. Negative values (in °C) mean that the ice-age ocean was colder. [From CLIMAP, 1981] *Reprinted by permission of The Geological Society of America. From Geological Society of America Map Chart Series MC-36.*

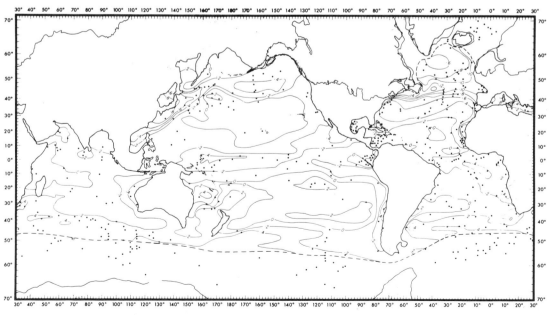

18K   AUGUST   ΔT

in $\delta^{18}O$ composition of mixed-layer dwelling foraminifera, Broecker (1989) suggested that the present differences in surface salinity between the North Atlantic and North Pacific (cf. Fig. 2.7) were enhanced by about 50%, with low-latitude Atlantic salinities increasing by perhaps 1.0‰ (cf. Keigwin and Boyle, 1989).

3.1.8 Deep-Ocean Changes

Several lines of geochemical evidence suggest that North Atlantic Deep Water (NADW; cf. Chap. 2) production rates may also have decreased at 18,000 BP (Boyle and Keigwin, 1982; Crowley, 1983a; Curry et al., 1988). Some of the evidence for deep-ocean change is based on variations in the cadmium/calcium (Cd/Ca) ratios of bottom-dwelling (benthic) forams (Fig. 3.12). In the present ocean Cd concentrations are similar to those of phosphorus, which increases with water-mass age, because of the steady rain of organic matter to the deep sea and its subsequent oxidation. Changes in phosphorus (and cadmium) concentration therefore reflect changes in the age of the water mass (Hester and Boyle, 1982).

Observations indicate that production rates of NADW decreased by perhaps one-third to one-half during the last glacial maximum (Boyle and Keigwin, 1985/1986). Deep water still formed in the northeastern Atlantic, possibly on the European continental shelf as a result of brine formation and sea ice freezing (Duplessy et al., 1980). However, intermediate water production increased in both the North Atlantic and North Pacific (Boyle and Keigwin, 1987; Oppo and Fairbanks, 1987; Zahn et al., 1987; Boyle, 1988b; Duplessy et al., 1988). Decreased oxygen levels in deep waters of some isolated eastern Atlantic basins may also have caused conditions conducive to the formation of organic-rich sediments (e.g., Curry and Lohmann, 1983).

Whereas there is consistent evidence for decreased NADW production, Antarctic Bottom Water estimates are more variable. Some studies suggest that it decreased (Ledbetter, 1984), stayed the same (Duplessy et al., 1988), increased (Johnson et al., 1977), or varied in a manner not related to simple glacial-interglacial fluctuations (Corliss et al., 1986). The evidence for possible North Pacific deep water formation is somewhere between the AABW and NADW

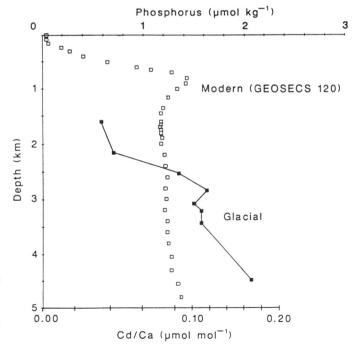

Fig. 3.12. Depth profile of phosphorus in the modern western North Atlantic compared to glacial profile. Glacial changes, reconstructed from Cd/Ca ratios, are inferred to reflect changing phosphate content (and therefore age) of intermediate and deep waters (increasing phosphate content implies older waters). [From Boyle and Keigwin, 1987] *Reprinted by permission from Nature 330, 35–40, copyright © 1987, Macmillan Magazines Ltd.*

stories. One analysis (Keigwin, 1987) suggested no North Pacific deep water formation at 18,000 BP, while a more complete assessment of $\delta^{13}C$ records provided some support for the concept (Curry et al., 1988).

Reduced deep-ocean overturn is consistent with accelerator mass spectrometry ^{14}C dating of the difference in ages between glacial maximum ages of surface and deep waters. Whereas the present difference is about 1500 years in the North Pacific (Broecker, 1963), glacial maximum differences are ~2000 years (Shackleton et al., 1988; Broecker et al., 1988c). However, there are still some vexing problems associated with dating foraminifera (Broecker et al., 1988b), so these results should be considered preliminary. The greater age of deep waters enhanced carbon storage in the deep sea, with the net effect on atmospheric pCO_2 levels being about 10 ppm (Keir, 1983). It has also been suggested that deep-water temperatures decreased by 1.0–2.0°C (Chappell and Shackleton, 1986; Labeyrie et al., 1987). This suggestion has been challenged and remains a matter of some uncertainty (Mix and Pisias, 1988).

3.1.9 Ocean Productivity Changes

Interest in ocean productivity changes was greatly heightened after Broecker (1982a,b) demonstrated that such changes were linked to atmospheric CO_2 variations. The marine productivity picture is quite complex, with different regions recording different trends. Advances of sea ice cover were probably associated with productivity decreases in high latitudes (Luz and Shackleton, 1975). There were significant increases in eastern boundary current upwelling along the Benguela, Peru, and Canary currents (cf. Fig. 2.4). Upwelling increased in the eastern equatorial Pacific and Atlantic (e.g., CLIMAP, 1981; Pedersen, 1983; Lyle, 1988; Sarnthein et al., 1988; Mix, 1989). The upwelling increases are consistent with inferred changes in tradewind strength based on analysis of grain size variations in deep-sea cores (Parkin and Shackleton, 1973; Janecek and Rea, 1985).

Ice cores also contain information about past changes in oceanic productivity. Inferences are based on observed variations in methanesulfonic acid in the cores. This compound is derived from dimethyl sulfide (DMS), which is emitted from marine plankton. Observations indicate increased concentrations in glacial age samples of Antarctic ice (Saigne and Legrand, 1987; cf. Fig. 6.17).

Ecological changes associated with productivity variations may have affected the organic carbon/carbonate rain ratio ($C_{org}/CaCO_3$) of detritus settling out of the mixed layer. At present this ratio averages 4:1. Broecker and Peng (1982) calculated that if this ratio were 8:1, it would result in about a 30 ppm drop in pCO_2. Some evidence exists for past changes in the flux of the components of this ratio, but changes in the ratio itself are more uncertain. For example, glacial carbonate accumulation rates decreased in high latitudes and in the equatorial Atlantic and Indian oceans but increased in the equatorial Pacific (Luz and Shackleton, 1975; Broecker and Peng, 1982; Curry and Lohmann, 1986; Peterson, 1986; Rea et al., 1986). Siliceous diatom productivity increased in the equatorial Atlantic (Stabell, 1986; Pokras, 1987).

In addition to the productivity-CO_2 connection, there may be another climate feedback to marine productivity variations (Charlson et al., 1987). DMS represents the primary source of cloud condensation nuclei (CCN) in the marine atmosphere (Bates et al., 1987). Changes in marine productivity and the atmospheric CCN could affect cloud populations and cloud albedo in the equatorial oceans (DMS is primarily produced in these regions). Calculations indicate that the effect on global average temperature could be on the order of 1°C (Charlson et al., 1987), about one-fourth of the ice-age signal (Section 4.2.1) and comparable in magnitude to the CO_2 radiative perturbation. However, the magnitude of this feedback has been questioned (Schwartz, 1988), so its ice-age significance requires further evaluation.

3.2 Deglaciation

The melting of the great ice sheets represents one of the most rapid and extreme examples of climate change recorded in the geologic record. Although residual remnants of the Laurentide Ice Sheet lasted until 6000–7000 BP, most of the

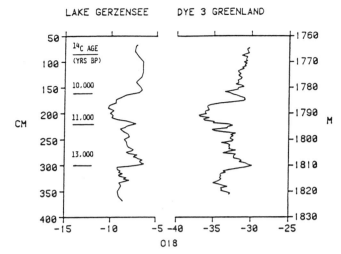

LAKE GERZENSEE   DYE 3 GREENLAND

Fig. 3.13. Comparison of a section of the $\delta^{18}O$ profile from the Dye 3 ice core (right) with the $\delta^{18}O$ record in carbonate from Lake Gerzensee, Switzerland (left), indicating two stages to the deglaciation. The strong similarities suggest that both records represent the same sequence of climatic events and thus the same time period. [From Oeschger, 1985] *Courtesy of American Geophysical Union.*

ice disappeared within an interval of about 5000 years (14,000–9000 BP).

3.2.1 Temporal Structure of Deglaciation

Although the major phase of deglaciation was initiated about 13,000–14,000 BP (Bryson et al., 1969; Berger et al., 1985; Mix and Ruddiman, 1985), there is some evidence for earlier significant melting on Antarctica (16,000–17,000 BP) and the northern Norwegian Sea (15,000 BP; Labeyrie et al., 1986; Jones and Keigwin, 1988). The time interval 16,000–13,000 BP is sometimes known as the "Late Glacial" and is the time of maximum aridity in African lakes (Street and Grove, 1979). Some ¹⁴C-dated Pacific records suggest sporadic production of North Pacific deep water during the early transition period (Berger, 1987; Shackleton et al., 1988; cf. Dean et al., 1989).

There is considerable structure to the pattern of deglaciation after ∼13,000 BP (e.g., Ruddiman and McIntyre, 1981b; Berger et al., 1985; Ruddiman and Duplessy, 1985; Mix and Ruddiman, 1985; Fairbanks, 1989). It occurred in two main steps (Fig. 3.13): an abrupt warming (∼13,000 BP), followed by a climate reversal at about 11,000 BP (termed the "Younger Dryas" in Europe), and then another abrupt warming (∼10,000 BP). A third step may have occurred about 8000 BP, marking the final outflow of Laurentide ice from Hudson Strait and Hudson Bay (Mix and Ruddiman, 1985; Paterson and

Hammer, 1987). The first two warming trends are also manifested as rapid changes in sea level (Fig. 3.14; Fairbanks, 1989).

3.2.2 First Stage of Warming

A key question in paleoclimatology involves the rate, areal extent, and magnitude of warming at

Fig. 3.14. Sea level record for the last 17,000 years, based on AMS ¹⁴C dating of corals growing in the surf zone offshore from Barbados. $\Delta\delta^{18}O$ values are estimated $\delta^{18}O$ changes due to ice volume effect. [From Fairbanks, 1989] *Reprinted by permission from Nature 342:637–642, copyright © 1989, Macmillan Magazines, Ltd.*

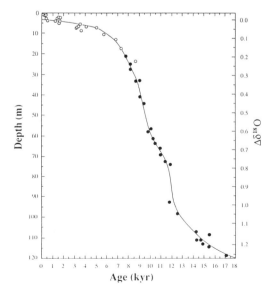

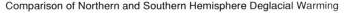

Comparison of Northern and Southern Hemisphere Deglacial Warming

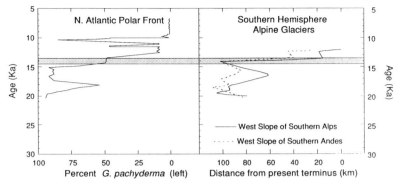

Fig. 3.15. Comparison of deglacial warming in the Northern and Southern Hemispheres. North Atlantic record from Broecker et al. (1988d); Southern Hemisphere records modified from many sources. [After Broecker and Denton, 1989] *Reprinted with permission from Geochimica et Cosmochimica Acta 53:2465–2501, W. S. Broecker and G. H. Denton, "The role of ocean-atmosphere reorganizations in glacial cycles." Copyright 1989, Pergamon Press.*

the end of the last ice age (Broecker and Denton, 1989). In many areas, the first stage of the deglaciation was both abrupt and very large. AMS ^{14}C dates of the North Atlantic polar front retreat indicate that the warm step may have occurred in as short a time as 200–300 years (Bard et al., 1987). Fossil Coleoptera (beetles) from England, which are very "fastidious recorders of the environment," suggest summer temperatures as warm as the present by about 12,000 BP (Coope, 1975; Atkinson et al., 1987). δ^{18}O records from southern Greenland indicate near-interglacial level values (Fig. 3.13).

A compilation of records of alpine glacial retreat along a transect from Alaska to Chile indicates that similar warming had occurred in many areas by ∼ 13,500 BP (Broecker and Denton, 1989). Temperatures increased in many mid- and high-latitude Southern Hemisphere regions (Fig. 3.15; cf. Fig. 3.7) [e.g., New Zealand, Tasmania, Antarctica, and surrounding islands (Broecker and Denton, 1989; Heusser, 1989; Jouzel et al., 1989)]. As noted earlier, Antarctic sea ice retreat began about 15,000 BP (Labeyrie et al., 1986), with SSTs in the Southern Ocean reaching present levels by ∼ 13,000 BP (Labracherie et al., 1989).

The large scale to the extent of the warming places severe constraints on mechanisms for deglaciation, for much of the Laurentide and Fennoscandian Ice Sheets were still present, so reduction in albedo and higher planetary temperatures could not result from this source.

CO_2 apparently reached interglacial levels after 12,000 BP (see below), so that a different mechanism would have to be invoked for the Southern Hemisphere warming. Broecker (1989; e.g., Broecker et al., 1985, 1988a) has argued that such rapid changes can best be explained by an abrupt shift in the ocean–atmosphere circulation (cf. Section 4.4.1).

3.2.3 Younger Dryas Cooling

The climate deteriorated again during an interval known as the Younger Dryas (∼ 10,000–11,000 BP). The North Atlantic polar front readvanced far southward (Fig. 3.16) and cool-

Fig. 3.16. Map of position of North Atlantic polar front over the last 20,000 years. Note the rapid southward displacement of the front 11,000–10,000 BP. [From Ruddiman and McIntyre, 1981a] *Reprinted with permission from Quaternary Research.*

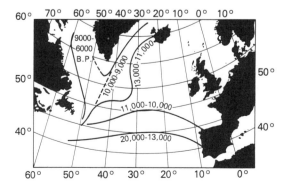

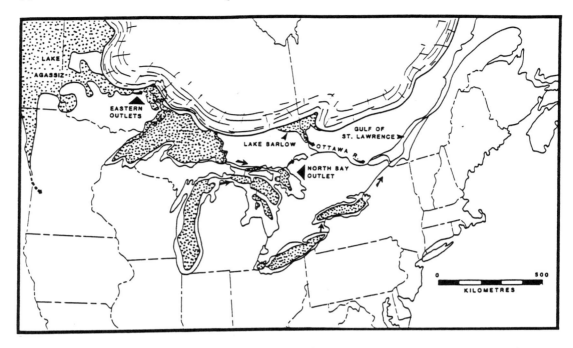

Fig. 3.17. Example of proglacial lake extent during deglaciation—the Great Lakes about 9500 BP. Ice of the Marquette advance had just retreated from the Superior basin, reopening the eastern outlets of glacial Lake Agassiz and allowing the 2 million km² western drainage basin to be integrated with the Great Lakes system. Lakes Erie and Ontario overflowed northeastward through the St. Lawrence River. [From Teller, 1987] *Reprinted with permission of The Geological Society of America. From J. T. Teller, In North America and Adjacent Oceans During the Last Deglaciation, W. F. Ruddiman and H. E. Wright, Jr., eds., The Geology of North America, v. K-3, 1987, pp. 39–69.*

ing was especially strong in the circum-subpolar North Atlantic basin (Broecker et al., 1988a). There is some evidence for a Younger Dryas oscillation in other parts of the world. Abrupt changes occur in an upwelling record from a marginal Caribbean basin (Overpeck et al., 1989). A rapid change in Ethiopian lake levels occurs around this time (Gillespie et al., 1983; cf. Fig. 5.2). Antarctic ice core $\delta^{18}O$ records show a two-step deglacial pattern (Lorius et al., 1979; Jouzel et al., 1987a), but the ice core chronology has not yet been definitively tied to the Northern Hemisphere chronology. ^{14}C dated records on Tierra del Fuego (Heusser and Rabassa, 1987) indicate a climate oscillation in this region that correlates with the Younger Dryas. However, some other regions (Tasmania, New Zealand) do not show evidence of a Younger Dryas cooling (e.g., McGlone, 1988; Heusser, 1989). The areal extent of the Younger Dryas event needs to be better documented.

The Younger Dryas may be due to meltwater-induced changes in the atmosphere–ocean cir-

culation. Much of the melting ice initially discharged into lakes (Fig. 3.17) adjacent to the ice sheets ("proglacial" lakes). Discharge from ice-age lakes was sometimes catastrophic. For example, glacial Lake Missoula (Montana, western United States) was dammed by ice, which catastrophically failed and unleashed a wall of water 600 m high onto the Columbia Plateau of eastern Washington—by far the largest known flood in the geologic record. The resulting devastated landscape has been graphically named the "channeled scablands" (e.g., Buetz, 1923). These colossal glacier outbursts (termed "jökulhlaups") may have been repeated more than 40 times during the deglaciation (Waitt, 1985). Catastrophic drainage of glacial Lake Bonneville in Utah/Idaho (western United States) resulted in the second largest flood known to have occurred (Jarrett and Malde, 1987).

$\delta^{18}O$ records from the Gulf of Mexico (Fig. 3.18) indicate that during the early stage of deglaciation meltwater from the Laurentide Ice Sheet emptied primarily into the Mississippi

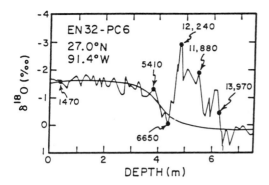

Fig. 3.18. AMS ^{14}C ages of deglacial events in the Gulf of Mexico. The very negative δ^{18}O excursion reflects a meltwater plume from the Mississippi River. [From Broecker et al., 1988a] *Courtesy of American Geophysical Union.*

River (Kennett and Shackleton, 1975; Emiliani et al., 1978; Leventer et al., 1982; Broecker et al., 1988a). By about 11,000 BP, the ice margin had retreated sufficiently to open up drainage through the St. Lawrence River system (Broecker et al., 1989). The subsequent outflow of meltwater from large lakes dammed by the ice sheets could have caused the low-salinity lens in the subpolar North Atlantic (cf. Fig. 3.16).

There was a corresponding depression of SST almost back to glacial maximum conditions (Ruddiman and McIntyre, 1981b). Note that the above explanation may apply even though the time period 11,000–10,000 BP was not the time of maximum meltwater flux (Fig. 3.14); location of the meltwater plume may be as important as volume of outflow (cf. Fig. 4.14 and Fairbanks, 1989).

The outflow of low-salinity water into the subpolar North Atlantic may also have affected the rate of deep-sea mixing. Worthington (1968) first predicted some of these effects when he suggested that the low-salinity water would not be dense enough to sink through the pycnocline (density gradient at ~100 m depth). The lack of overturn might result in greatly diminished production rates of NADW. Geologic evidence now supports some of this scenario. There is geochemical evidence (Fig. 3.19) for decreased NADW production during the polar front readvance (Berger and Vincent, 1986; Boyle and Keigwin, 1987). Biogenic fluxes to the seafloor indicate very low values during this time interval (Ruddiman et al., 1980; Ruddiman and McIntyre, 1981b). The combination of low-oxygen conditions on the seafloor and low food

Fig. 3.19. Evidence that Younger Dryas cooling affected production rates of North Atlantic Deep Water (note Cd/Ca changes at about 10,500 BP coincide with increase of cold-water foraminifera *N. pachyderma*). [From Boyle and Keigwin, 1987] *Reprinted by permission from Nature 330, 35–40, copyright © 1987, Macmillan Magazines Ltd.*

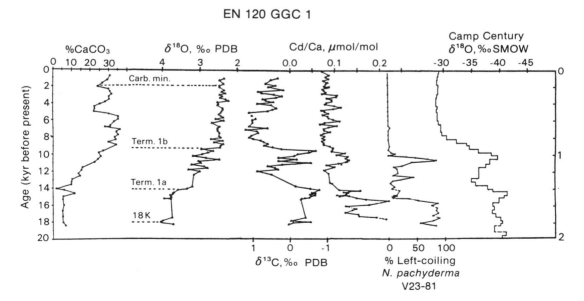

supply from the surface resulted in ecological conditions on the seafloor so impoverished as to be almost unsupportive of life.

The oceanic feedback loop apparently triggered by meltwater outflow may even have affected circulation in the equatorial Atlantic. For example, it is known that production of NADW results in an export of cold deep water from the North Atlantic basin and an import of warm South Atlantic waters across the equator in the North Atlantic (e.g., Hastenrath, 1980; cf. Fig. 2.6). A reduction in NADW production rates at the termination might be expected to result in decreased flow of South Equatorial Current water across the equator, cooling waters north of the equator and warming them to the south. Studies of equatorial Atlantic cores indicate that one mode of SST variability which occurs during the deglacial may reflect such a response (Mix et al., 1986). The combination of relatively warm waters in the South Atlantic and cold waters in the north should have resulted in enhanced atmospheric convection in the south and decreased convection in the north equatorial regions (see Folland et al., 1986), with the net effect being drying in North Africa (Street-Perrott and Perrott, 1990; cf. Fig. 5.2).

3.2.4   Second Stage of Warming

The second major stage of deglaciation began about 10,000 BP. There is evidence for a remarkably rapid shift in aerosol concentrations (Fig. 3.20) in the Dye 3 Greenland core (Herron and Langway, 1985). The decrease indicates a rapid change in the basic state of the atmosphere (at least in the North Atlantic sector). Detailed sampling of this transition indicates that it may have occurred in the astonishingly short interval of 20 years (Dansgaard et al., 1989; cf. Herron and Langway, 1985).

3.2.5   CO₂-Chemistry Changes

As we will see in several places in the book (Chaps. 4 and 7), the causes of the glacial-interglacial climate changes is a subject of great interest and ranks as a major unsolved scientific problem. Since CO_2 was lower during the last ice age, changes in CO_2 may have been very important for triggering or accelerating the deglacia-

tion. It is therefore critically important to determine when CO_2 started to increase. The presumed timing of CO_2 changes with respect to the deglaciation has changed somewhat in the last few years. Initially a "CO_2 proxy" index was proposed for deep-sea sediments that was based on measuring changes in the vertical gradient of $\delta^{13}C$ in the ocean (cf. Fig. 6.15)—a change related to ocean fertility and the "biological pump" (Broecker, 1974; Shackleton et al., 1983). Analysis of this "$\Delta\delta^{13}C$" record suggested that "CO_2" led ice volume changes (Shackleton and Pisias, 1985) and thus helped force climate change.

However, reanalysis of a "cleaned" $\Delta\delta^{13}C$ record suggests that it could not be used as a proxy CO_2 index (Curry and Crowley, 1987; cf. Barnola et al., 1987). This conclusion was supported by direct measurements of CO_2 and $\delta^{18}O$ in the Byrd (Antarctica) ice core, showing that CO_2 lagged $\delta^{18}O$ at the glacial-interglacial transition (Fig. 3.21). Until the Byrd ice core $\delta^{18}O$ chronology can be conclusively tied into an absolute time scale, we cannot be sure that CO_2 lagged some of the changes discussed above.

A further complication involves CO_2 studies from the Dye 3 (Greenland) ice core (Stauffer et al., 1985a). In this record CO_2 apparently in-

Fig. 3.20. Aerosol variations (Cl⁻, SO₄²⁻, and NO₃⁻ concentrations) versus depth in the Dye 3 (Greenland) ice core, indicating a very rapid change (<100 years) at the glacial-interglacial transition at ~10,000 BP (1786 m). [From Herron and Langway, 1985] *Courtesy of American Geophysical Union.*

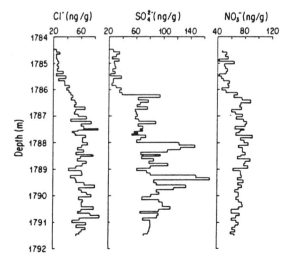

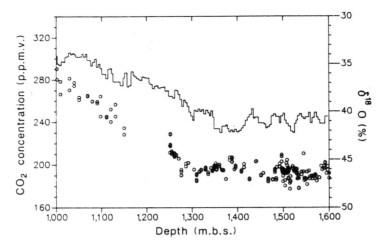

Fig. 3.21. $\delta^{18}O$ and CO_2 record 1000–1600 m below surface for the Byrd (Antarctic) ice core, illustrating that CO_2 lags $\delta^{18}O$ changes during the deglaciation. The solid line corresponds to the age-adjusted $\delta^{18}O$ curve. [From Neftel et al., 1988] *Reprinted by permission from Nature 331, 609–611, copyright © 1988, Macmillan Magazines Ltd.*

creases around 13,000 BP and then decreases again during Younger Dryas time. Unfortunately, this record may have been affected by summer melting (Stauffer et al., 1985a), which seriously compromises the quality of the CO_2 record. Until we have a higher quality CO_2 record with unambiguous time control the detailed relationship between CO_2 and climate during the last deglaciation remains a matter of some uncertainty.

The glacial-interglacial transition affected other aspects of chemical cycling in the ocean. Cores in both the eastern equatorial Atlantic and eastern equatorial Pacific have manganese-rich layers at the glacial-Holocene transition (Berger et al., 1983; Boyle, 1983). These have been interpreted in terms of decreased oceanic fertility possibly due to decreased rates of oceanic overturn (variable delivery rates of organic matter to the seafloor affect oxidation levels and mobization of manganese in the sediment; Berger et al., 1983).

There is also widespread evidence in deep-sea sediments for enhanced calcium carbonate preservation at the boundary (Berger, 1977), which may also reflect decreased productivity (cf. Broecker, 1971). Additionally, there should also have been an increased accumulation of carbonate on continental shelves during the sea level rise (Berger and Keir, 1984). Changing carbonate preservation and productivity patterns should in turn have affected atmospheric pCO_2 levels. Thus there are some intricate and fascinating (but not completely understood) connections between climate, ocean circulation, oceanic productivity, and atmospheric pCO_2 during the last deglaciation.

3.3 Holocene

3.3.1 Changes in Temperature

Although remnants of the Laurentide Ice Sheet did not disappear until about 7000 BP, the early Holocene (approximately 4500–10,000 BP) has often been considered as warmer than the last 4500 years. As will be discussed in more detail in the modeling section (Chap. 4), some of the early Holocene warmth may primarily represent seasonal (summer) warmth rather than year-round warmth.

Conclusions about early Holocene warmth are based on several lines of evidence—latitudinal displacements of vegetation zones in eastern North America and western Europe (Bernabo and Webb, 1977; Wijmstra, 1978; Davis et al., 1980; Ritchie et al., 1983; Overpeck, 1985) and vertical displacements of vegetation and/or mountain glaciers in western North America, Alpine Europe, and New Guinea (Hope et al., 1976; Kearney and Luckman, 1983; Porter and Orombelli, 1985). A particularly dramatic ex-

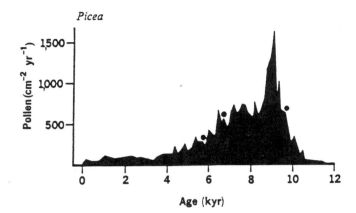

Fig. 3.22. Pollen influx values for *Picea* (spruce) plotted against radiocarbon age, illustrating early Holocene peak warmth for the Northwest Territories, Canada. [From Ritchie et al., 1983] *Reprinted by permission from Nature 305, 126–128, copyright © 1983, Macmillan Magazines Ltd.*

ample involves a pulse of Holocene warmth in northern Canada around 9000 BP (Fig. 3.22; Ritchie et al., 1983). A number of sites in the mid- and high latitudes of the Southern Hemisphere also record warmer conditions during the early Holocene (e.g., Jouzel et al., 1987b; Clapperton et al., 1989; Heusser, 1989).

Quantitative estimates of Holocene temperature changes are available from some regions (e.g., Webb et al., 1987; COHMAP, 1988). In the midwestern United States, mean annual early Holocene temperatures were about 2°C warmer than the present (Webb, 1985), with most of the changes due to increased summer warming (Bartlein et al., 1984). Summer temperatures also increased about 2°C in Europe (Huntley and Prentice, 1988). Vertical migration in regions suggests warming of as much 2°C in New Guinea (Hope et al., 1976), 4°C in the Alps (Porter and Orombelli, 1985; Huntley and Prentice, 1988), and about 1°C in the western United States (Scuderi, 1987). Temperatures were 0.5–1.0°C warmer than present on South Georgia Island (54°S) and Antarctica (Jouzel et al., 1987b; Clapperton et al., 1989; cf. Fig. 5.8).

Over the ocean, there is less information about differences between the early and late Holocene. This is for two reasons. Sedimentation rates of many deep-sea cores are sufficiently low as to sometimes prevent reliable time discrimination. Furthermore, coring techniques used to retrieve deep-sea sediments very often "blow off" the top 20–30 cm of the core—an interval that happens to encompass the late Holocene. (This problem is now being rectified by different

coring techniques, but as of this date, many of the published records potentially have this problem.)

There is sufficient information from deep-sea cores to draw the following conclusions about differences between the early and late Holocene. Faunal displacements in south polar regions indicate that warm waters penetrated into higher latitudes in the early Holocene (e.g., Hays et al., 1976b). The evidence for greater warmth in the early Holocene is much more ambiguous in the northern Atlantic (Kellogg, 1980; Ruddiman and McIntyre, 1984), a conclusion which seems to be born out by negligible differences between early and late Holocene $\delta^{18}O$ values in the Greenland ice cores (Dansgaard et al., 1984). However, waters were warmer along parts of the eastern boundary current of the North Atlantic (Kipp, 1976; Crowley, 1981).

The transition at about 3500–4500 BP marks the return to cooler and drier climates of the late Holocene. The late Holocene cooling is sometimes known as the Neoglaciation. Since that time, the North Atlantic circulation has been approximately at its present position (Ruddiman, 1968).

3.3.2 Early Holocene Precipitation Changes

The first half of the Holocene was also marked by significant differences in precipitation patterns (Fig. 3.23). For example, there was an eastern extension of the "Prairie Peninsula" in the Great Plains of North America (Webb and Bryson, 1972) with an estimated 20% decrease in precipitation (Webb, 1985). Western North America was also drier (an interval known as the Altithermal in this region).

The most pronounced changes in precipitation patterns occurred in the monsoon belt of Africa and Asia (Fig. 3.23). Lake levels rose to a maximum across much of Africa, from the Namib Desert at 26°S to the northern tropics. Regions now well within the hyperarid core of the Sahara (e.g., Timbuktu, Tibesti Massif) indicate moist conditions and a northward shift of the summer rain (monsoon) belt by at least 600 km (Ritchie and Haynes, 1987; Fabre and Petit-Maire, 1988). Crocodile, giraffe, elephant, gazelle, and hippopotamus remains have been found along presently dry streambeds (wadis). Paleolithic (see Fig. 6.3 caption) artifacts are common (e.g., Bradley, 1985). Papyrus, which presently is restricted to the southern half of the Sudan, extended 1500 km northward into Egypt. The increased moisture contributed to the very extensive groundwater deposits under the Sahara (Sonntag et al., 1980).

Increased early Holocene moisture, indicative

of a stronger southwest (summer) monsoon, extended eastward across Saudi Arabia, Mesopotamia, and into the presently dry Rajastan region of northwestern India (Bryson and Swain, 1981). There is also evidence of a stronger monsoon in offshore sediments. In the western Arabian Sea, off the coast of Saudi Arabia, early Holocene fauna indicate greater upwelling (Prell, 1984b), caused by stronger southwest flow (and therefore increased Ekman transport of cool subsurface waters to the surface). In the northern Bay of Bengal, a foraminifera species indicative of low-salinity water increased in abundance (Cullen, 1981). This increase can be interpreted in terms of more outflow from the Ganges and Brahmaputra river systems. Low-salinity surface layers in the eastern Mediterranean and Red seas, probably resulting from increased runoff, prevented overturn, leading to organic-rich sediments in the Mediterranean and accumulation of metal-rich brines in the Red Sea (e.g., Thunell and Lohmann, 1979; Adamson et al., 1980; Rossignol-Strick et al., 1982; Thunell et al., 1984; Rossignol-Strick, 1987).

Fig. 3.23. Reconstruction of lake levels during the early part of the present Holocene interglacial (6000 BP), illustrating that many areas in the tropics were wetter than at present. [From Street-Perrott and Harrison, 1985] *Reproduced with permission of F. A. Street-Perrott and S. P. Harrison, Lake levels and climate reconstruction in "Paleoclimate Analysis and Modeling," A. D. Hecht (Ed.). Copyright 1985, John Wiley and Sons.*

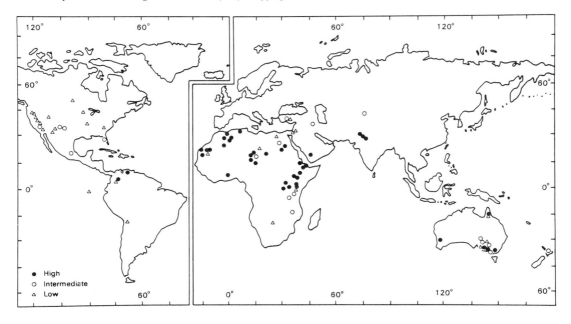

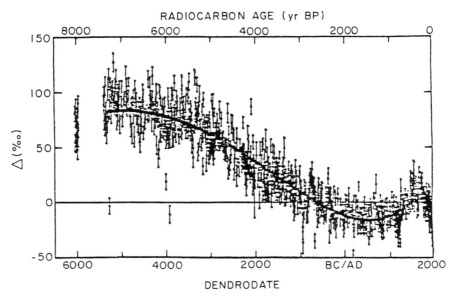

Fig. 3.24. Atmospheric ¹⁴C variations over the last 10,000 years. [From Neftel et al., 1981] *Reproduced with permission of Elsevier Science Publishers.*

3.3.3   ¹⁴C Changes

Another feature of the Holocene involves changes in atmospheric ¹⁴C levels (Fig. 3.24). It has long been known that there is a secular drift in atmospheric ¹⁴C records (e.g., Damon et al., 1978; Suess, 1980). However, until recently it was not possible to satisfactorily distinguish among several hypotheses—geomagnetic drift, changes in solar output, or changes in climate (e.g., oceanic overturn) which might affect the atmospheric ¹⁴C reservoir. For example, varying rates of deep-water production could vary the rates at which "old" ¹⁴C-depleted water is upwelled and exposed to the atmosphere. Andree et al. (1986) utilized a geochemical box model to estimate that the large atmospheric ¹⁴C changes would require a decrease in the rate of oceanic overturn by about one-half. Such a pattern would cause an increase in the age of deep-Pacific waters. However, AMS ¹⁴C dating of surface and benthic forams in the deep western Pacific shows little evidence for a greater age difference between surface and deep waters in the early Holocene (Andree et al., 1986).

The above measurements suggest that "reservoir effects" are not the principal source of the secular variations in the atmospheric ¹⁴C record.

These results are consistent with ¹⁰Be variations (Beer et al., 1988). ¹⁰Be is another cosmogenic nuclide, and its concentration can be measured in ice cores. ¹⁰Be variations compare favorably to the ¹⁴C record (Beer et al., 1988). Variations in the earth's geomagnetic field may be primarily responsible for fluctuations illustrated in Fig. 3.24 (cf. Damon et al., 1978). This interpretation is supported by direct paleomagnetic measurements (McElhinny and Senanayake, 1982).

3.4   Summary

A wealth of information has been uncovered for the last 20,000 years. There were radical changes in climate in many regions, with expanded ice cover, tropical precipitation changes, changes in atmospheric composition, and abrupt transitions being some of the most dramatic examples. Of all the changes discussed, perhaps two stand out as objects for further scrutiny: (1) the 18,000 BP tropical climate—high SST conundrum (Sections 3.1.4 and 3.1.7), and (2) the magnitude, extent, and rate of deglacial change at ~13,000 BP (Section 3.2.2). No doubt future studies will uncover more fascinating insights and problems regarding this most thoroughly studied climate interval in earth history.

4. MODELING THE CLIMATE OF THE LAST 20,000 YEARS

In the last few years there has been a great deal of progress in modeling the climates of the late Quaternary. This topic is of interest for several reasons. First, it gives us an opportunity to test models with radically different boundary conditions. Second, it can increase our insight into some of the processes responsible for large changes in climate. Third, it allows us to test our understanding of climate and thereby determine the level of confidence in extrapolating these ideas to older time intervals or the future.

Although the initial focus of modeling studies was on the last glacial maximum, interests have broadened to include fluctuations throughout the last 20,000 years and earlier time intervals of interest. This chapter will discuss results from climate models that determine equilibrium solutions to the climate at key intervals during the last 20,000 years (see Chap. 3). The time-dependent evolution of Pleistocene climate will be discussed in Chaps. 6 and 7.

4.1 Last Glacial Maximum Experiments with CLIMAP Boundary Conditions

Several groups have been involved in general circulation modeling (GCM) of the climate at 18,000 BP (e.g., Williams et al., 1974; Gates, 1976a,b; Manabe and Hahn, 1977; Kutzbach and Otto-Bliesner, 1982; Hansen et al., 1984; Kutzbach and Guetter, 1984, 1986; Manabe and Broccoli, 1985a,b; Rind and Peteet, 1985; Lautenschlager and Herterich, 1990a,b). Many of these studies utilized CLIMAP (1976, 1981) boundary conditions, which stipulated changes in sea ice, ice sheets (area and elevation), albedo, SST, and, in some cases CO_2. Orbital insolation values for January and July were within 1% of present values for all latitudes and seasons at the glacial maximum, so this effect is relatively minor for 18,000 BP climate. This section will discuss results of such experiments.

4.1.1 North America–European Sector

Ice sheets and sea ice had a significant effect on atmospheric circulation in northern midlatitudes (Kutzbach and Wright, 1985; Kutzbach, 1987; COHMAP, 1988). For example, the Laurentide Ice Sheet occupies 77 5° × 5° grid boxes in the NCAR GCM (Kutzbach and Guetter, 1986) and considerably surpassed in extent and average elevation the effect of the Rocky Mountains. GCM experiments indicate that the great dome of the North American Ice Sheet split the jet stream (Fig. 4.1), with one branch flowing around the northern edge and another, stronger branch around the southern edge of the ice front. Splitting of the jet was caused by both ice sheet orography and midtropospheric cooling, although the orographic effect seems to dominate (Rind, 1987; Cook and Held, 1988). The effect of ice sheet orography on circulation depends in part on the elevation of an ice sheet, which in this case was estimated to be in equilibrium at the glacial maximum (Hughes et al., 1981). However, some workers argue that this assumption is not valid (e.g., Andrews, 1982; Wu and Peltier, 1983). The volume of the Laurentide Ice Sheet may have been less than mapped by CLIMAP. It would be desirable to reexamine the jet stream splitting using an ice sheet of lower elevation.

The disturbed upper air flow pattern exerted (via the jet stream) a significant influence on surface flow (Fig. 4.2). Model-generated southerly air flow in the Alaskan sector resulted in generally mild conditions. The cyclonic flow of the wintertime Aleutian low was so persistent that the south-southwesterly surface flow patterns were reflected in dune fields and loess deposits in Alaska (Hopkins, 1982). Along the northern edge of the ice sheet prevailing westerlies replaced the prevailing easterlies of the present (Kutzbach and Guetter, 1986). This re-

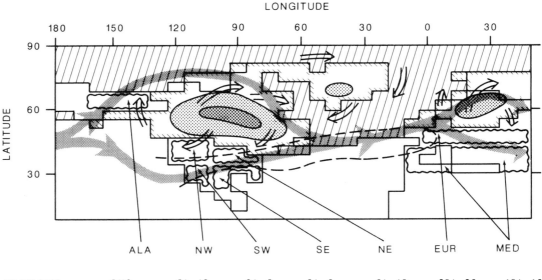

	ALA	NW	SW	SE	NE	EUR	MED
ΔT (K) JAN	5/10	−5/−10	−0/−5	0/−5	−5/−10	−20/−25	−10/−15
ΔP (%) JAN	−25/−30	~0	30/35	~0	~0	−10/−15	−25/−30
Δ(P-E) (%) ANN	−45/−50	−15/−20	80/85	−25/−30	−10/−15	−70/−75	0/−5
ΔT (K) JUL	0/−5	−5/−10	−0/−5	0/−5	−5/−10	−0/−5	0/−5
ΔP (%) JUL	−40/−45	−25/−30	−25/−30	−25/−30	−10/−15	−5/−10	5/10

Fig. 4.1. NCAR climate model (CCM) simulations in northern midlatitudes at 18,000 BP and comparison with geologic data: winds aloft (stippled arrows), surface winds (double-shafted arrows). Location and highest elevations of the North American, Greenland, and European ice sheets are shown with hatched lines. The dashed lines along the southern jet stream enclose the band of increased precipitation. Area average values of departures (18,000 BP-minus present) of temperature (°K), precipitation (%), and precipitation minus evaporation (%) are shown for seven areas: Alaska (ALA), northwest USA (NW), southwest USA (SW), southest USA (SE), northeast USA (NE), Europe (EUR), and the Mediterranean (MED). The values are expressed by the nearest 5°K or 5% increments that bracket the area average. For example, in area NW in January 18,000 BP, the area average temperature was 6°K below present; the table entry is −5/−10. Note that this is not necessarily the range of temperatures over the area. Underlining indicates that the departure from the model control was significant above the 90% level compared to the model's natural variability. [From Kutzbach and Wright, 1985] *Reprinted with permission from Quaternary Science Reviews 4, 147–187, J. E. Kutzbach, and H. E. Wright, "Simulation of the climate of 18,000 yr BP: Results for the North American/North Atlantic/European sector and comparison with the geologic record," Copyright 1985, Pergamon Press plc.*

sulted in a reversal of the present anticyclonic flow pattern in the Arctic Basin to a cyclonic flow pattern.

With bifurcation of the jet stream by the ice sheet (Fig. 4.1), the southern branch of the Pacific jet was located over the southwestern United States. Pacific low-pressure systems steered by the jet traversed regions of relatively high SST (Fig. 3.11), so their moisture level was relatively high. The southern branch of the jet also accelerated as it crossed North America because the winter locus of the jet core was over the North Atlantic. Acceleration in this "jet entrance" region typically is associated with low-level convergence, rising air, and precipitation (Kutzbach and Wright, 1985).

Altered circulation in the southwestern United States had a significant effect on precipitation. The combination of jet location, relatively high moisture levels, and a favorable site for low-pressure formation resulted in one of the few areas where model-generated precipitation increased at 18,000 BP. This is in agreement with relatively high lake levels at 18,000 BP (Fig. 3.6). Combined with lower air temperatures, there was an increase of about 80% in net precipitation minus evaporation (P-E) in the southwestern United States (Kutzbach and Wright, 1985; Fig. 4.1). This value has the correct sign of the observed changes, but it may be slightly low in magnitude; Benson and Thompson (1987) estimate perhaps a net increase of 140%

for some Pleistocene lakes in the region. Since model-generated temperature reductions (2–3°C) for the Southwest are less than estimated by data (~6°C; cf. Spaulding and Graumlich, 1986), reconciliation of model and data P-E changes may ultimately depend on understanding the reason for differences in temperatures.

There were significant circulation changes in the North Atlantic sector. Immediately downwind of the Laurentide Ice Sheet, the increase in net P-E along the eastern seaboard of North America was 100% (Fig. 4.3). This response reflects a jet stream "anchored" by the ice sheet, with low-pressure systems intensifying in this area of strong temperature gradient. Farther north, strong winds advected extremely cold air down the ice divide separating the Laurentide and Greenland ice sheets (Fig. 4.1). The strong winds and frigid air caused a pronounced cooling of northern Atlantic surface waters and con-

Fig. 4.2. Surface winds for July 18,000 BP (top), 9000 BP (middle), and the 0 BP control (bottom). Arrows denote direction and magnitude of winds (key at top figure). [From Kutzbach and Guetter, 1986] *Reprinted from Journal of the Atmospheric Sciences with permission of the American Meteorological Society.*

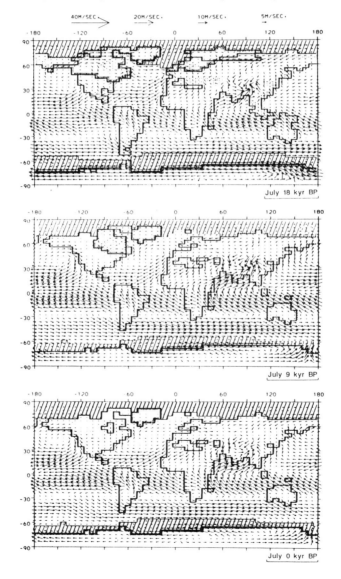

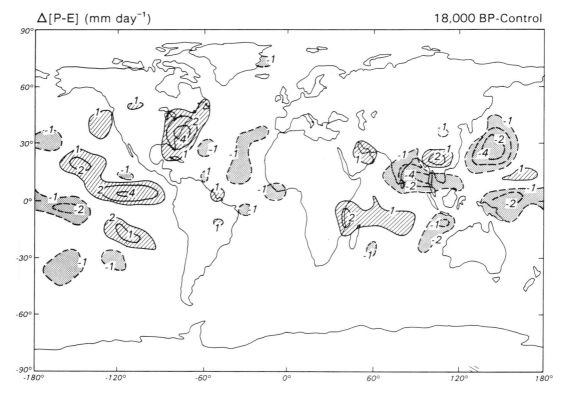

Fig. 4.3.   Mean annual net precipitation minus evaporation (mm/day) at 18,000 BP, as calculated by the NASA/GISS GCM. [Redrawn from Rind and Peteet, 1985] *Reprinted with permission from Quaternary Research.*

tributed to the significant equatorward exten-sion of the polar front margin (Manabe and Broccoli, 1985b; see below, Fig. 4.8).

Expansion of North Atlantic sea ice in turn affected European climate. The extensive area of winter sea ice upwind of Europe (Fig. 3.2) re-sulted in a marked change of conditions from the present, for air can cool greatly over sea ice. Advection of the cold air and local cooling from the Fennoscandian Ice Sheet resulted in winter temperature anomalies (20–25°C) in Europe larger than for any other land area (Kutzbach and Wright, 1985); cf. Lautenschlager and Her-terich, 1990b). Farther south, high lake levels in northwestern Africa and parts of the Mideast (cf. Fig. 3.6) may have been caused by winter storms tracking along the southern margin of the sea ice (cf. Fig. 3.2).

With the summer retreat of the oceanic polar front to high latitudes there was an advection of relatively warm air into northwestern Europe. The summer jet axis shifted from the ocean to the southern margin of the North American Ice

Sheet (Kutzbach and Wright, 1985). Decelera-tion in the "jet exit" region caused a component of air to converge on the high-pressure (south) side of the jet, with sinking aloft. The drier con-ditions in the southeastern United States during this time may have been a consequence of this process.

Several GCM simulations indicate that con-siderable summer melting took place along the southern margin of the ice sheets (Gates, 1976b; Manabe and Broccoli, 1985b; Kutzbach, 1987). The surface melt resulted from intense temper-ature inversions near the surface during the warm season (Kutzbach, 1987). Release of la-tent heat through condensation of water vapor was also important. This suggests that the air near the Laurentide Ice Sheet was relatively warm and moist (Manabe and Broccoli, 1985b)—its southern terminus was near 42°N and summer surface flow was southerly. The simulations are supported by the narrow strip of tundra in eastern North America and close proximity of vegetation to the ice sheet.

Modeling studies also estimated $\delta^{18}O$ variations at 18,000 BP (Covey and Haagenson, 1984; Joussaume and Jouzel, 1987). This suggests a $\delta^{18}O$ for the Laurentide Ice Sheet of about $-30‰$, approximately agreeing with geologically based estimates (e.g., Dansgaard and Tauber, 1969; Mix, 1987; Hillaire-Marcel and Causse, 1989). The studies are consistent with a Laurentide Ice Sheet oceanic $\Delta\delta^{18}O$ contribution of about 0.9‰ (difference between the present and 18,000 BP) to marine records.

In summary, models give a great deal of insight into the origin of regional climate change in the North American–European sector. Much of the perturbed circulation across the area can be traced to the system response to the enormous Laurentide Ice Sheet. Overall, agreement between models and data is very good. Some discrepancies remain, however. For example, "observed" temperatures in the southeastern United States suggest a winter temperature decrease of almost 20°C and a summer temperature decrease of only about 5°C (see Chap. 3). However, GCM calculations indicate comparable changes for both seasons ($\leq 5°C$), with the winter departure significantly less than the data suggest (Kutzbach and Wright, 1985). This discrepancy may not be as serious as it seems. For example, we know that there should be significant katabatic (density) outflow from the large ice sheets; this flow might then flood the area south of the ice sheet with a dense pool of frigid air. We also know that such features are not well simulated in GCMs, presumably in part because of coarse vertical resolution. Thus the model–data disagreement, although significant, may not be too damaging to the overall assessment of GCM behavior in the northern mid- and high latitudes.

4.1.2 Changes in the Tropics

In general, model-generated atmospheric circulation changes were much less in the tropics. Because tropical SST decreases at 18,000 BP were usually less than 2°C (CLIMAP, 1981), and atmospheric temperatures are effectively "anchored" by the temperature and high heat capacity of the lower boundary, simulated temperature decreases are also relatively small. For example, NASA/GISS simulations (Rind and Peteet, 1985) suggest relatively high tropical temperatures for all seasons and elevations (Fig. 4.4). Because there are more tropical temperature data for high altitudes (Fig. 3.3), the most significant discrepancies occur for this level. One way to reconcile the GCM–data differences is to postulate a change in ice-age lapse rate, which would in turn require a change in the relative humidity profile of the tropical atmosphere. However, this option is not supported by observations (Webster and Streeten, 1978; Rind and Peteet, 1985) or by common practice in GCM simulations (Manabe and Wetherald, 1967). Reassurances aside, it would be instructive if modelers reexamined this assumption to consider how robust it is under altered boundary conditions. For example, based on evidence from South America, van der Hammen (1985) suggested lowland temperature reductions of about 2–3°C but upland changes of 5–6°C. Similarly, estimated temperature changes (Porter and Orombelli, 1985; Huntley and Prentice, 1988) in the early Holocene are larger for the Alps ($+4°C$) than for lowland Europe ($+2°C$).

Because CLIMAP SSTs in the tropics changed little, trade wind velocities were not much different from present (Gates, 1976b; Manabe and Broccoli, 1985b; Kutzbach and Guetter, 1986; Lautenschlager and Herterich, 1990b). In some cases model winds decreased— results that run counter to various geologic inferences about ice-age trade winds (Chap. 3). These results also contravene conventional geologic wisdom regarding relations between wind and temperature. The explanation for this response requires understanding the role of moisture content in the tropical marine atmosphere. Ascending motion in the tropics is strongly influenced by the moisture content of the air. Even slightly lower ice-age SSTs significantly reduced atmospheric moisture content. Because vertical tropical motions are "fueled" by the latent heat released from condensing water vapor, an ice-age decrease in water vapor caused a decrease in the amount of ascending motion in the tropics (Kraus, 1973; Gates, 1976b; see Fig. 4.5). Lower SSTs may also have substantially weakened the intensity of ice-age hurricanes and tropical storms (Hobgood and Cerveny, 1988).

Adjustments of the circulation also affected precipitation in the tropics. There are coherent

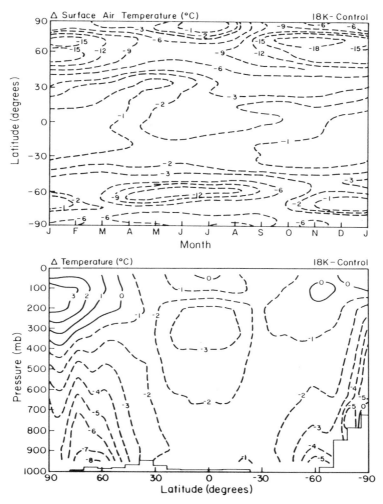

Fig. 4.4. NASA/GISS GCM temperature differences (ice age minus modern control) for month, latitude, and altitude. [From Rind and Peteet, 1985] *Reprinted with permission from Quaternary Research.*

areas of both increased and decreased net P-E. Net P-E increased about 50% in the eastern Pacific and South Indian oceans (Fig. 4.3). In one GCM (Lautenschlager and Herterich, 1990a,b) increased precipitation in these regions offset reductions elsewhere, so that globally-averaged precipitation at 18,000 BP was not significantly differently from the present. The model response presumably reflects minor (and sometimes positive) SST variations at 18,000 BP (see Fig. 3.11). There are also two large areas that show net drying. In the equatorial Pacific, the drying may reflect enhanced cooling along the equatorial divergence at 18,000 BP (cf. Chap. 3). The other area is in southeast Asia and east of

Japan, and may reflect in part lower CLIMAP SSTs in the western equatorial Pacific (Fig. 3.11). Since a revised version of the CLIMAP SST field (Prell, 1985) does not produce such strong cooling, this result may be sensitive to initial boundary conditions. Modeled precipitation decreases in the western Pacific may also reflect decreased rainfall caused by a heavier snowcover in Asia. This is consistent with suggestions (Flohn, 1968), observations (Hahn and Shukla, 1976), and calculations (Manabe and Hahn, 1977; Barnett et al., 1988; cf. Fig. 5.12) that an important factor affecting the strength of the Asian summer monsoon is snowcover in central Asia (especially the Tibetan Plateau).

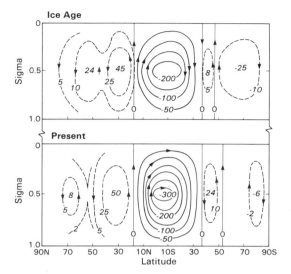

Fig. 4.5. The meridional mass transport streamfunction (10^{12} g sec) simulated for the ice-age July (above) and present July (below). This index measures vertical overturn in the atmospheric circulation. Note the weakened Hadley cell at 18,000 BP. This calculation used CLIMAP's 1976 SST field, which was colder than the 1981 boundary conditions. One might therefore expect lesser differences for an updated figure, but the principle relating vertical overturn with SST remains the same. Note: Sigma equals pressure divided by surface pressure. [From Gates, 1976b] *Reprinted from Journal of the Atmospheric Sciences with permission of the American Meteorological Society.*

There are some indications that calculated rainfall reductions are too small to account for inferred vegetation changes in tropical rainforests (Hansen et al., 1984; Guetter and Kutzbach, 1990). Relatively high simulated rainfall occurs in some tropical rainforest areas (e.g., the Amazon Basin) where geologic data suggest drier conditions (Section 3.1.4). This, along with estimates of tropical temperature change, suggests a difficulty in reconciling tropical land data with CLIMAP tropical SST estimates (Rind and Peteet, 1985).

Some of the tropical precipitation results may be model-dependent and may not include some important feedbacks. For example, the high-resolution version of the NASA/GISS GCM simulates a 20% reduction in Amazon Basin rainfall at 18,000 BP (Rind, 1988), whereas the low-resolution version and the NCAR GCM simulate very little change (Hansen et al., 1984; Guetter and Kutzbach, 1990). Furthermore, as discussed in Section 3.1.4, an important biofeedback is

not included in these models. Observations indicate that as much as 50% of Amazon rainfall is recycled by plants transpiring water vapor to the atmosphere (e.g., Salati, 1985). This feedback is not included in most GCMs and is still in the experimental stage of development (e.g., Dickinson, 1984; Rind, 1984; Sellers et al., 1986). Initial calculations of the vegetation feedback effect suggest that total precipitation changes vary from 0% to −20% as the vegetation cover is reduced (Dickinson and Henderson-Sellers, 1988; Lean and Warrilow, 1989). Further work on this important problem is warranted.

4.1.3 High-Latitude Southern Hemisphere

Less attention has been given to 18,000 BP model–data comparisons in the high latitudes of the Southern Hemisphere. The most important observations involve expanded sea ice (CLIMAP, 1981), stronger winds (DeAngelis et al., 1987), and central Antarctic temperatures 8–10°C less than present (Jouzel et al., 1987a). GCM simulations yield stronger winds in the Southern Ocean, but we have not been able to determine if the simulated winds increased as much as observations imply (i.e., about a 7 m/sec or ∼70% increase). GCMs simulated central Antarctic temperatures about 5–6°C less than present. (Hansen et al., 1984; Broccoli and Manabe, 1987), a decrease partially attributable to an estimated 400- to 500-m increase in the elevation of the East Antarctic Ice Sheet (CLIMAP, 1981). Given a presumed nearly dry adiabatic lapse rate of about 8°C/km for this region, the elevation effect translates into about 3–4°C—greater than one-half of the actual simulated difference. The remaining GCM changes were presumably due to expanded sea-ice cover in the Southern Hemisphere.

GCM estimates of Antarctic Ice Sheet temperature changes may require reassessment. The greater elevation of the East Antarctic Ice Sheet was based on a glaciological assumption that the sheet had come to equilibrium with its new boundary conditions (Hughes et al., 1981). This assumption no longer appears valid (Section 3.1.1). Thus, if the elevation effect were removed, the GCM-simulated 5–6°C decrease would presumably be reduced by about one-half and there would be a significant model–data discrepancy in the southern high latitudes.

4.1.4   Further Comments on Model–Data Discrepancies for the Tropics at 18,000 BP

To summarize results from the experiments with 18,000 BP boundary conditions, analysis of GCM results has led to an impressive increase in our understanding the 18,000 BP world, especially in the high-latitude Northern Hemisphere. However, there are some significant model–data differences, particularly in the tropics. In this section we will discuss the discrepancies more thoroughly.

As discussed in Section 4.1.2, the relatively small model response at 18,000 BP in the tropics can be traced to the relatively warm SSTs. If this critical boundary condition did not change

much, the model response will not change much either. One way to get a larger model response, reconciling simulations with observations, is to decrease SSTs by 2°C (Rind and Peteet, 1985). However, most attempts to validate the CLIMAP SST estimates indicate that they are approximately correct (Prell, 1985; Brassell et al., 1986; Broecker, 1986). The CLIMAP SSTs have also been "theoretically validated." A detailed comparison of EBM and GCM (Kutzbach and Guetter, 1986) simulations of temperature fluctuations over the last 18,000 years (Hyde et al., 1989) indicates very good agreement between the two models (Fig. 4.6a). Since the NCAR model utilized prescribed SSTs from CLIMAP, whereas the EBM calculates the temperatures, results imply that CLIMAP SSTs are in approximate energy balance with the ocean–atmosphere system. The energy balance changed at 18,000 BP as a result of increased albedo from

Global Annual Average Temp Departures (Ocean)

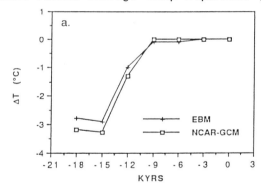

Fig. 4.6. (a) Comparison of EBM and NCAR GCM global mean annual temperature departures over the ocean from 18,000 BP to the present. (b) Same comparison with 15,000 and 18,000 BP EBM temperatures lowered by 0.5°C to include ice-age CO_2 and ice sheet elevation–infrared effects. Since the NCAR model used prescribed temperatures from CLIMAP, whereas the EBM temperatures are calculated. The good agreement suggests that CLIMAP SSTs are consistent with the altered energy balance of the 18,000 BP world. [From Hyde et al., 1989] *Reprinted from Journal of Climate with permission of the American Meteorological Society.*

Global Annual Average Temp Departures (Ocean)

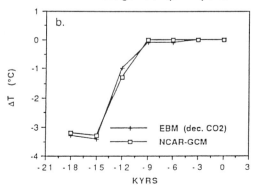

ice sheets and sea ice, with the infrared sensitivity term [B of Eq. (1.1)] magnifying the altered energy balance in the EBM. Since the CLIMAP SSTs presumably include a change due to lower atmospheric CO_2 levels, a correction of the EBM estimates (Fig. 4.6b) still leaves good agreement.

The above discussion therefore suggests that CLIMAP SSTs are valid. Yet the tropical SST/ snowline elevation argument also seems correct. What, then, is missing? The models do not account for increased ice age aerosols (Section 3.1.4). Higher aerosol concentrations may decrease temperatures perhaps 2–3°C (Harvey, 1988), with the effect presumably being greater over land than over sea because of higher dust concentrations over land.[1] Increased aerosols may also significantly lower tropical precipitation because the stability of the atmospheric column would be increased (cf. Coakley and Cess, 1985). Lower CO_2 levels could also reduce plant productivity, thereby enhancing the "appearance" of drier conditions (Wigley and Jones, 1985). At present we do not have an answer for the discrepancy; it represents an important unsolved problem in paleoclimatology.

4.2 18,000 BP Sensitivity Experiments

Since there is at least some agreement among altered boundary conditions, model response, and independent "observational verification" by data, it is of interest to inquire what processes may have been most responsible for model response to altered boundary conditions. This section will discuss sensitivity experiments that address some of the observed changes.

4.2.1 Global Average Temperature

Several different climate models indicate that the global average temperature at 18,000 BP was 3.5–4.0°C less than present (Hansen et al., 1984; Kutzbach and Guetter, 1986; Hyde et al., 1989; Lautenschlager and Herterich, 1990a). This value is of particular interest when measured against estimates of global temperature increases

1. [Note added in proof] Recent work suggests that the temperature effect of dust variations may be overestimated by almost an order of magnitude (Anderson and Charlson, 1990).

due to a doubling of CO_2 (2.8–5.2°C, with a mean of 3.9°C; Schlesinger, 1989). There may also have been changes in dust cover and clouds over the last 20,000 years (Sections 3.1.3 and 3.1.9). These processes could enhance the 4°C global average temperature difference between the ice age and present, perhaps making it 6°C (see Harvey, 1988). Given the magnitude of the ice-age changes associated with a global temperature change of at least 4°C, it is evident that the greenhouse warming may well produce a future climate change of very significant levels (see Chap. 14).

A series of control experiments (Hansen et al., 1984) has determined the relative contribution of different altered boundary conditions to the GCM's predicted global average temperature change of ~4°C (Fig. 4.7). In the NASA/GISS simulation, lower SSTs were responsible for about 45% of the change (partly through decreased water vapor greenhouse effect, partly through altered cloud levels); ice-albedo effects contributed about 33%; albedo-induced changes in vegetation about 6%; and CO_2 about 16%. These authors note that the model is still somewhat out of radiative balance (as it is for the present; Hansen et al., 1983). The deficit could be accommodated by either lower ice-age SSTs or modification of the cloud feedback effect. Experiments with a higher resolution version of the GISS model show a significant reduction of this radiative imbalance (Rind, 1988).

4.2.2 Modeling the "Southern Hemisphere Connection"

A question of long-standing interest in climate theory involves reasons for the approximate synchroneity of climate change in the northern and southern hemispheres. Modeling studies have shed light on the nature of this connection. For example, experiments with the GFDL GCM indicate that the Northern Hemisphere ice sheets had little direct temperature effect on climate in the tropics and in the Southern Hemisphere (Manabe and Broccoli, 1985b). In this study, SSTs were calculated in order to determine the hemispheric and global scale response to the cooling effect of ice sheets. The results were very enlightening (Fig. 4.8). Calculated ocean temperature changes were large in the vi-

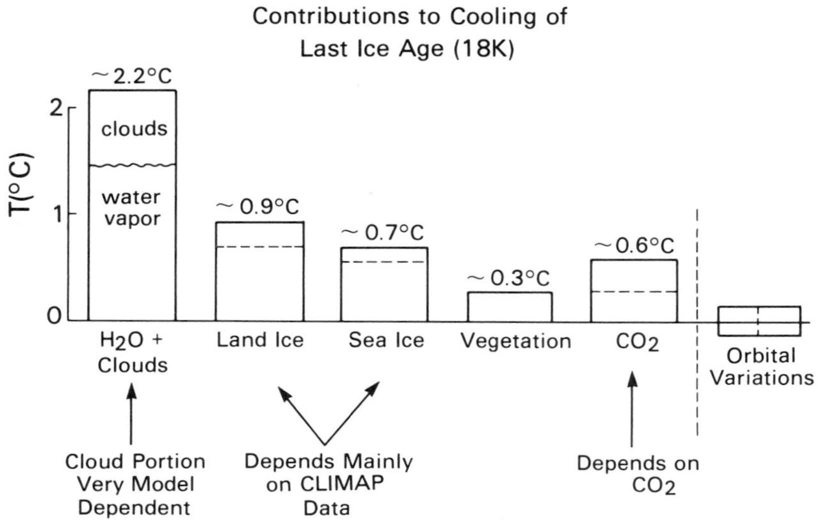

Fig. 4.7. Contributions to the global mean temperature differences between 18,000 BP and today's climate as evaluated with the NASA/GISS GCM and assumed boundary conditions. The cloud and water vapor portions were not separated but, based on other GCM experiments, the cloud part is estimated at 30–40% of the sum. For land ice the dashed line refers to the "minimal extent" model of CLIMAP (1981), and for sea ice the dashed line refers to the reduced sea ice cover discussed in Hansen et al. (1984). The solid line for CO_2 refers to a change of about 100 ppm and the dashed line to a change of about 50 ppm. [From Hansen et al., 1984] *Courtesy of American Geophysical Union.*

cinity of the ice sheets and resemble those mapped by CLIMAP. However, temperature changes in the tropics or the Southern Hemisphere were minor. This result suggests that, if no other mechanisms were operating, the tropics or the Southern Hemisphere would not "know" an ice age was taking place in the Northern Hemisphere. However, there may have been significant planetary wave interactions between the midlatitudes and the tropics (cf. Hartmann, 1984) that are not manifested as large temperature changes. For example, Rind (1987) notes an increase in midlatitude Southern Hemisphere eddy activity in a sensitivity run with only Northern Hemisphere ice sheets.

An analysis of hemispheric heat budgets in the above experiment indicates that relatively low surface temperatures induce only a relatively small increase in heat transport from the warmer atmosphere in the Southern Hemisphere because the modest increase in exchange of dry sensible heat (due to increased temperate gradients) is almost offset by the decreased transfer of latent heat (due to lower SST). Temperature adjustments in northern latitudes are more a consequence of local radiative effects. The de-

crease in incoming shortwave radiation is almost entirely compensated by a reduction in outgoing longwave radiation (due to lower surface temperatures). The radiative compensation in the Northern Hemisphere is much more effective than the thermal adjustment through the interhemispheric heat exchange (cf. North, 1984). Temperatures therefore changed little in the tropics and Southern Hemisphere. A similar EBM calculation by Hyde et al. (1989) yielded the same conclusion.

Further studies incorporating lower ice-age CO_2 levels (Manabe and Broccoli, 1985a; Broccoli and Manabe, 1987) resulted in significantly lower SST in the tropics and in the Southern Hemisphere (Fig. 4.9). These studies therefore suggest that CO_2 is an important factor linking climate fluctuations globally. Before this explanation can be accepted, however, it is necessary to consider other possible mechanisms. For example, the direct effect of Milankovitch insolation changes on Southern Hemisphere sea ice was relatively minor, either at 18,000 BP or at times of greater variation (Crowley and Parkinson, 1988a; cf. Lemke, 1987). The small response can be explained by a number of factors: ice and cloud albedo effects reduce the insolation perturbation at the surface, some of the shortwave radiation entering the ocean in sea ice leads (open areas) contributes to bottom ablation rather than lateral melting of the ice (cf. Maykut and Perovich, 1987), and the radiation perturbation at the upper surface of the ice must partly go to warming the surface to the melting point before melting ensues. The amount remaining for ice growth or melting is therefore relatively small.

Another process that could affect Southern Hemisphere climate involves variable North At-

lantic Deep Water (NADW) production rates. This mechanism has long been popular (Weyl, 1968; Broecker, 1984) because upwelling of warm subsurface waters (partly of NADW origin) at present appears to play a key role in seasonal meltback of Antarctic sea ice (Gordon, 1981). Since NADW input into the Southern Ocean may have been near zero at 18,000 BP (Oppo and Fairbanks, 1987), the mechanism could be quite important.

Initial attempts at evaluating the "NADW/ Antarctic Connection" (Fig. 4.10) indicate that

Fig. 4.8. Sensitivity test for the effect of ice sheets on February monthly mean SST (°K; stippling indicates positive difference). *Top:* Difference between ice sheet and standard experiments. *Bottom:* Difference between 18,000 BP and present (as reconstructed by CLIMAP, 1981). Areas covered by sea ice in the ice sheet experiment and at 18,000 BP are indicated by black shading. [From Manabe and Broccoli, 1985b] *Courtesy of American Geophysical Union.*

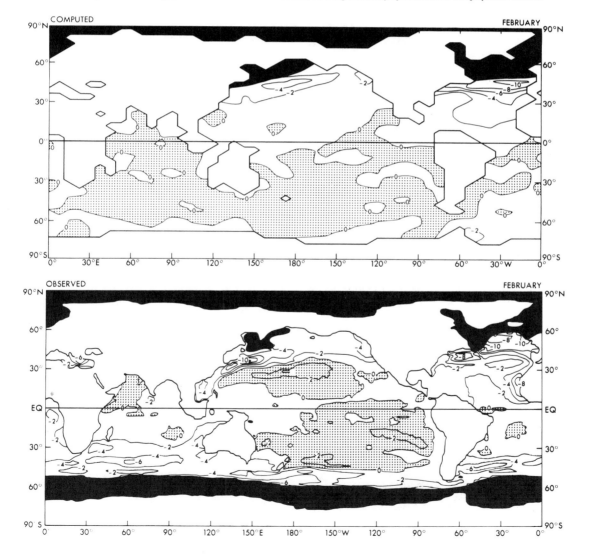

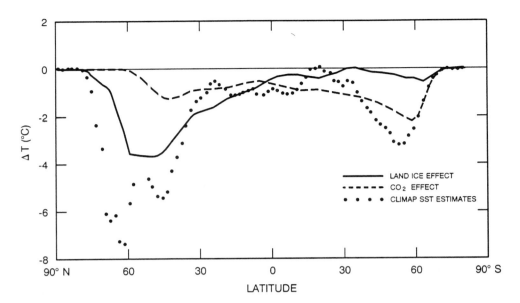

Fig. 4.9. Latitudinal distribution of the difference in zonally averaged annual mean sea surface temperature between the late glacial maximum and the present, as estimated by CLIMAP (dots) and as simulated by a GCM (Broccoli and Manabe, 1987). The solid line represents the zonally averaged SST difference (ice age minus control) due to changes in land ice as prescribed by CLIMAP (1981). The dashed line represents the SST difference due to lower ice-age CO_2 levels. [From Crowley and Parkinson, 1988b; redrawn and relabeled from Broccoli and Manabe, 1987] *Reproduced with permission of Springer-Verlag.*

Fig. 4.10. Schematic of Southern Ocean summer ice conditions in the South Atlantic at the last glacial maximum, suggesting that past changes in sea ice cannot be accounted for by changing production rates of NADW. Illustrated are locations of calculated ice edge for a "no-NADW" simulation and for a larger bottom-water temperature change (0 W/m²). For reference, the present winter maximum ice edge is shown also [based on satellite observations of Zwally et al. (1983), the present summer ice edge is at the coast of Antarctica]. All values represent estimated limits of 85% ice coverage. Note that different surface oceanographic conditions deliver different sediment types to the seafloor [IRD = ice-rafted detritus; FORAM/RAD = biogenic sediments composed of foraminifera and radiolaria; DIATOM/SILT = sediment type found south of a sediment boundary presently related to sea ice (see discussion in Crowley and Parkinson, 1988a)]. Polar front and sea ice positions at the glacial maximum are from Hays et al. (1976a), Burckle et al. (1982), and Cooke and Hays (1982); iceberg locations are from Labeyrie et al. (1986). [From Crowley and Parkinson, 1988b] *Reproduced with permission of Springer-Verlag.*

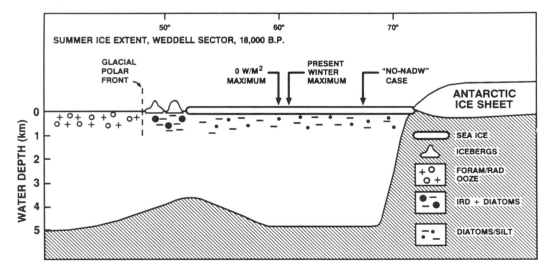

the effect can only explain about 20–30% of the glacial-interglacial changes in Antarctic sea ice (Crowley and Parkinson, 1988b). This study did not explicitly model NADW flow but calculated the effect of changes in the transport of warm deep waters to the Southern Hemisphere by altering the parameterization of ocean heat flux to the base of the sea ice layer in a thermodynamic-dynamic model of sea ice (Parkinson and Washington, 1979). The conclusion has been validated by analysis of independent sets of data (Crowley and Parkinson, 1988b). The conclusion is also consistent with the calculations of Manabe and Broccoli (1985a,b), who suggested that the dominant influence on Southern Hemisphere sea ice is CO_2.

Despite the convergence of calculations suggesting CO_2 as an important if not dominant factor explaining temperature changes in the high latitudes of the Southern Hemisphere, there are still some notable gaps in our understanding. As discussed in Section 4.1.4, the reasonably good model-data agreement over Antarctica is due to an erroneously high elevation to the East Antarctic Ice Sheet. Lowering the elevation creates a model-data discrepancy. Furthermore, there are some significant CO_2/temperature phase offsets during other parts of the last glacial cycle. During the deglaciation, the Southern Hemisphere warmed about 14,000 BP even though CO_2 values had not reached interglacial levels (cf. Figs. 3.15 and 3.21). At the end of the last interglacial (120,000 BP) temperatures decreased on Antarctica significantly before CO_2 levels started to decrease (Fig. 6.14). Both of these observations demonstrate that our understanding of climate change in the high latitudes of the Southern Hemisphere is still incomplete.

4.3 Ocean Circulation at 18,000 BP

Work is just starting on 18,000 BP paleo-ocean modeling. To date we have just pieces of the puzzle rather than whole-ocean simulations.

4.3.1 North Atlantic Surface Circulation

The huge Laurentide Ice Sheet may have influenced the surface circulation of the North Atlantic (Keffer et al., 1988). Modeling studies (e.g., Held, 1983) demonstrated that topography has

a significant effect on the upper atmosphere's stationary waves, and that this effect extends to the ice-age case (Cook and Held, 1988). It is well known that the prevailing upper atmosphere wind direction exerts a significant influence on surface flow. Modification of the prevailing wind field by the Laurentide Ice Sheet at 18,000 BP should therefore have "steered" the Gulf Stream/North Atlantic Current system (cf. Fig. 2.5) from its present path directed toward the northeastern Atlantic to one directly eastward. Comparison of a calculated altered path for the Gulf Stream/North Atlantic Current with 18,000 BP observations indicates good agreement (Keffer et al., 1988).

4.3.2 Poleward Ocean Heat Transport

Initial calculations suggest that poleward ocean heat transport decreased in the North Atlantic (Fig. 4.11) and increased in the North Pacific at 18,000 BP (Miller and Russell, 1989). This work was based on an indirect method, as it did not explicitly simulate the ocean circulation but rather determined transport changes as a residual (cf. J. Miller et al., 1983). If total outgoing radiation at the top of the atmosphere is known, then transport in the ocean can be calculated by subtracting the atmospheric component (determined from a GCM) from the total. This method represents only one step involved in assessing the role of the ocean, but it is a useful exercise.

The work of Miller and Russell (1989) agrees somewhat with a previous calculation by Manabe and Bryan (1985) utilizing a sector version (i.e., idealized geography) of the NOAA/GFDL coupled ocean–atmosphere model, i.e., a model that explicitly calculates currents and the thermohaline circulation, but utilizes simplified geography. This study focused on the relation between changing CO_2 levels and ocean feedbacks. Results (see Fig. 8.14) indicate that overall levels of thermohaline circulation remain relatively constant for higher CO_2 levels (this will be discussed further in Chap. 8) but decreased CO_2 levels cause a significant reduction in the thermohaline circulation and the magnitude of poleward ocean heat transport. The latter result is consistent with the "indirect" method discussed above and with considerable geologic

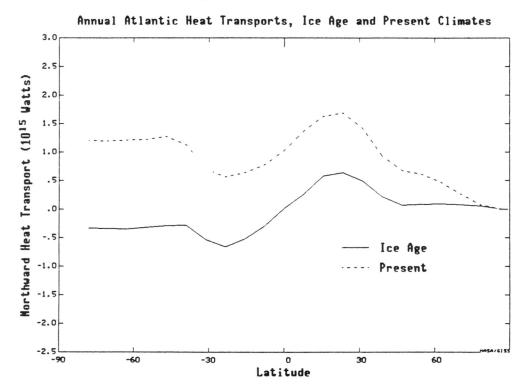

Fig. 4.11. Poleward ocean heat transport in the North Atlantic for the present and ice-age case, as calculated from NASA/GISS atmospheric GCM results (see text). [From Miller and Russell, 1989] *Courtesy of American Geophysical Union.*

data for decreased NADW production and older Pacific deep waters at 18,000 BP (Section 3.1.8).

The enhanced sensitivity of the coupled model to reduced CO_2 levels results from the effect of lower CO_2 on sea ice, which expanded greatly for reduced levels (Manabe and Bryan, 1985). The opposite effect occurs to a diminishing extent with higher CO_2 because sea ice area converges on zero with ever higher values of CO_2. The net effect is that the increased sea ice cover suppresses the energy exchange between ocean and atmosphere in high latitudes, and weakens the thermohaline circulation. This in turn decreases poleward heat transport, providing a positive feedback and further expanding sea ice.

These initial experiments with ocean heat transport may provide some insight into and perhaps even partial resolution of the main disagreement between the atmospheric models and data, i.e., the discrepancies in the tropics at 18,000 BP (see Section 4.1.4). The postulated reduction in North Atlantic poleward ocean heat

transport may result in compensating increases in the atmospheric circulation, adjustments which may not be manifested in atmospheric GCM runs with prescribed SSTs. One consequence for coupled runs might involve stronger trade wind strength than in uncoupled runs. This feature is apparently required by geologic data (see Chap. 3) and is manifested in the coupled runs by Manabe and Bryan (1985; cf. Fig. 4.12). Trade wind-induced changes may also affect net precipitation minus evaporation—another apparent discrepancy in the area of tropical rainforests (see Chap. 3). Finally, decreases in ocean heat transport should maintain higher SSTs in tropical regions than might otherwise occur with expanded ice cover. This could help explain the relatively stable glacial-interglacial tropical SSTs mapped by CLIMAP (1981). All of these possibilities will remain just that until coupled ocean–atmosphere calculations are actually made with realistic conditions, but possible payoffs may motivate modelers to accelerate their efforts toward that goal.

4.4 Deglaciation

4.4.1 Ice Melt and Global Warming

Three different GCMs indicate that at 18,000 BP the mass balance of the ice sheets was negative (see Section 4.1.1). These results suggest that complete melting would occur over a period of a few thousand years (Manabe and Broccoli, 1985b; Kutzbach, 1987; Rind, 1987). With increased summer insolation during the termination (see text following), this mass imbalance would have increased. Ice sheet decay may also have been affected by a number of processes discussed in Chap. 7 (Sections 7.3.1 and 7.3.2). Calculations indicate that CO_2-induced air temperature changes were large enough to cause disintegration of an extensive marine-based ice sheet on Eurasia (Lindstrom and MacAyeal, 1989; see Fig. 3.2). Ice sheet disintegration resulted from extensive crustal depression from ice loads that led to "ungrounding" of the marine-based ice sheet (see Section 7.3.1). This in turn resulted in rapid flow toward ablation areas. Surface warming also enhanced ablation. Simulated ice sheet disintegration was accomplished in about 4000 years.

Is melting of the ice sufficient to explain the observed rate of climate change during the deglaciation? Broecker and Denton (1989) argue that much of the planet had warmed considerably by about 12,000 BP (Section 3.2.2), when the Laurentide Ice Sheet was still of significant size (cf. Fig. 3.15). Thus, some type of abrupt transition may have been responsible for the warming (cf. Flohn, 1979; Broecker et al., 1985; Flohn, 1986). Broecker and Denton suggest that changes in the coupled ocean–atmosphere circulation in the North Atlantic were responsible for the changes (Fig. 4.13). However, changes in ocean–atmosphere heat transport generally do not increase planetary temperatures; they only alter the distribution of heat on the globe. For example, Manabe and Stouffer (1988) demonstrate that when the North Atlantic is in an in-

Fig. 4.12. The intensity of the low-latitude Hadley circulation in a coupled ocean–atmosphere model as plotted for six experiments with different CO_2 levels (cf. Fig. 8.14). Units are in 10^{12} g/sec. Note the increased overturn (and presumably stronger winds) for the lower CO_2 cases. The coupled model appears to generate a stronger low-latitude circulation than for an uncoupled model (cf. Fig. 4.5). [From Manabe and Bryan, 1985] *Courtesy of American Geophysical Union.*

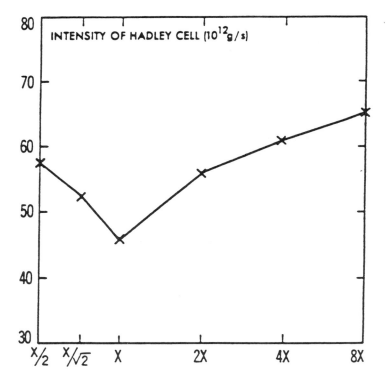

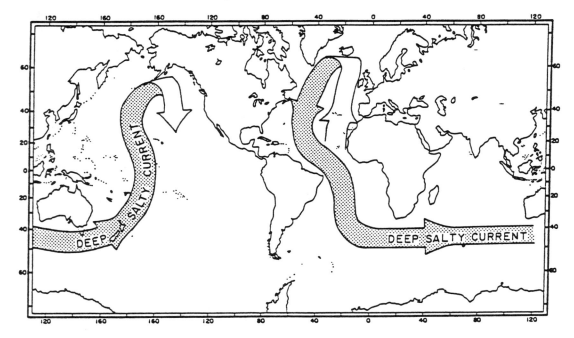

Fig. 4.13. A mechanism for effecting rapid climate shifts. A large-scale salt transport system operating in today's ocean compensates for the transport of water (as vapor) through the atmosphere from the Atlantic to the Pacific. Salt-laden deep water formed in the northern Atlantic flows down the length of the Atlantic around Africa through the southern Indian Ocean and finally northward in the deep Pacific. This coupled atmospheric water vapor transport and ocean salt transport system may be self-stabilizing. Records from ice and sediment suggest that it was disrupted during glacial time and replaced by an alternate mode of operation. [From Broecker et al., 1988a] *Courtesy of American Geophysical Union.*

terglacial mode (warm North Atlantic and active NADW formation), Southern Hemisphere temperatures are lower than when the circulation is in a "glacial" mode. This response reflects the fact that an interglacial mode circulation in the North Atlantic is exporting cold NADW across the equator and importing heat via warm Southern Hemisphere surface waters (Fig. 2.6). During the glacial mode, export of warm waters from the Southern Hemisphere decreases; temperatures south of the equator rise accordingly.

There are at least two conditions which allow global-averaged temperature changes with altered ocean–atmosphere circulation. One involves changes in ice cover due to enhanced heat transport into high latitudes. The change in albedo will change the planetary radiation balance. For example, onset of an interglacial mode North Atlantic circulation should cause a retreat of North Atlantic sea ice and a significant local increase in temperatures (cf. Rind et al., 1986). However, the effect of North Atlantic changes

on global averaged temperatures is small [0.3°C for the energy balance model discussed in Hyde et al. (1989) and illustrated in Fig. 4.6]. Changes in the ocean–atmosphere system could also have affected global temperatures through a CO_2 feedback. For example, Boyle (1988b,c) discusses how changes in the thermohaline circulation could affect the vertical distribution of ΣCO_2 in the ocean (see Section 6.4.5). However, this model requires a several thousand year CO_2 lag in response to the deep-circulation changes.

CO_2 changes could also have caused global warmth. At present the timing of CO_2 change does not support this scenario, but there is just enough uncertainty in the CO_2 measurements that we cannot eliminate this possibility with confidence. What, then, caused the system to warm? At present we do not have a good answer. There may also have been changes in dust cover and clouds over the last 20,000 years (Sections 3.1.3 and 3.1.9). Volcanism has been proposed (Flohn, 1979; Bryson, 1989), but we doubt that the effect would be significant enough. It is not clear whether any of the above mechanisms can actually explain the magnitude of warming at 13,000 BP. The explanation for this warmth is an important unsolved question in paleoclimatology.

4.4.2 *Effect of Meltwater Plumes*

Modeling studies are just starting on events associated with the termination of the last ice age. As for the previous section, we have just bits and pieces to report on. As discussed in Section 3.2.3, ice sheet melting produced meltwater plumes first in the Gulf of Mexico and later in the subpolar North Atlantic. These plumes may have affected the atmosphere–ocean circulation in several ways. Local cooling in the Gulf of Mexico would increase high pressure and strengthen Atlantic trades. The calculated response is consistent with observed changes in the Caribbean (Overpeck et al., 1989) and possibly explains drying in North Africa (cf. Fig. 5.2 and Section 3.2.3). The cold water also caused a decreased intensity of low-pressure systems downstream along the eastern seaboard of North America and in the North Atlantic, with significant decreases in precipitation in these regions (Oglesby et al., 1989). [At present the Gulf of Mexico is the major source of water vapor for much of the precipitation over eastern North America and the North Atlantic (Peixoto and Oort, 1983).] Perhaps the relatively low P-E conditions in the subpolar North Atlantic resulted in surface waters salty enough to get the ocean–atmosphere system to flip from its glacial to its interglacial state (see previous section).

A number of significant climate effects were associated with conjectured meltwater accumulation in the subpolar North Atlantic at about 11,000 BP. The low-salinity lens allowed for the expansion of sea ice and significant cooling. A buildup of high pressure in the subpolar North Atlantic blocked advection of warm air from the North Atlantic into northwestern Europe (Rind et al., 1986), leading to a radical drop in temperatures in this region (the Younger Dryas). However, the cooling effect from expanded sea ice was restricted to northern midlatitudes. This result is consistent with sensitivity experiments for 18,000 BP (Fig. 4.8).

It has long been conjectured that a glacial meltwater lid in the subpolar North Atlantic would suppress NADW production rates (Worthington, 1968). Experiments with several types of ocean models provide some support for this conjecture (F. Bryan, 1986; Crowley and Häkkinen, 1988; Maier-Reimer and Mikolajewicz, 1989; cf. Birchfield et al., 1990). In a set of experiments with an ocean GCM, Maier-Reimer and Mikolajewicz (1989) tested the time-dependent effects of variable outflow of low-salinity water from the Mississippi and St. Lawrence Rivers. Results indicate that once sufficient levels of meltwater are produced, NADW production and North Atlantic heat transport will decrease (Fig. 4.14). However, more Mississippi outflow is required to achieve the same level of reduction because some of the Mississippi water is mixed with more saline North Atlantic surface waters as it transits toward the subpolar North Atlantic. Thus, a Younger Dryas oscillation may not necessarily reflect the time of maximum meltwater flux; rather it could indicate a more localized meltwater plume due to breaching of the ice in the St. Lawrence River Valley (cf. Broecker et al., 1989).

4.5 Early Holocene Climate

4.5.1 *Effects of Orbital Forcing on Temperature and Precipitation*

GCMs have been very helpful in understanding the origin of some dramatic changes in early Holocene climate. Many of the changes are due to significant orbitally induced insolation changes in the seasonal cycle (Kutzbach and Otto-Bliesner, 1982; Kutzbach and Guetter, 1984, 1986; Kutzbach and Street-Perrott, 1985). At its maximum 11,000 years ago, July insolation changes were more than 40 W/m^2 at some latitudes (Fig. 4.15). These values translate into a difference of about 8% in seasonal insolation receipt. The changes are large in comparison to forcing typically used in GCM experiments (Kutzbach and Guetter, 1986). For example, a 1% change in the solar constant corresponds to a global average solar radiation change of about 3.4 W/m^2 and a doubling of atmospheric CO_2 level corresponds to a 4.3 W/m^2 increase in the radiative heating at the earth's surface (Ramanathan, 1988). However, forcing in the latter case

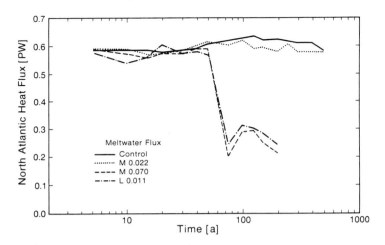

Fig. 4.14. Effect of low-salinity meltwater plumes on North Atlantic heat transport (heat transport plotted vs. time in model years from beginning of integration). Heat transport varies as a function of NADW production—as NADW production decreases, so too does heat transport. Figure illustrates results from several sensitivity experiments with an ocean GCM. "L" refers to outflow from the St. Lawrence Rivers (in Sverdrups; 1 Sv = 1 × 10⁶ m³/sec), "M" refers to outflow from the Mississippi. Note that low values of outflow are insufficient to reduce NADW production and that larger values of Mississippi outflow are required to achieve the same level of reduction as occurs with the St. Lawrence outflow. [After Maier-Reimer and Mikolajewicz, 1989] *Reproduced with the permission of the authors.*

is mean annual, while it is only seasonal in the former, and climate sensitivity significantly increases as the frequency of forcing decreases (North et al., 1984; cf. Fig. 1.11).

Experiments with GCMs of varying levels of complexity indicate that increased July insolation caused large changes in summer temperatures over Northern Hemisphere land masses and small changes over the ocean. The largest change occurred on the largest land mass and was about 2–4°C (Kutzbach and Guetter, 1984; Schneider et al., 1987). The temperature changes appear to represent a linear response to changes in forcing (see Fig. 1.10).

Increased heating caused enhanced low-level uplift and decreased surface pressure over the center of Eurasia. A schematic diagram (Fig. 4.16) illustrates the effect of these changes on the African–Asian monsoon system. Increased uplift caused increased upper level easterlies. The enhanced overturn of the thermally direct cell resulted in increased low-level southwesterlies, which increased moisture flux from the ocean to the land. Convergence of moisture in the Intertropical Convergence Zone resulted in a 10–20% increase in the amount of monsoon precipitation (Kutzbach and Guetter, 1986). Stronger

Fig. 4.15. Solar radiation departures (past minus present, in W/m²) for July and January as a function of latitude and time (18,000 BP to present). The numbers in parentheses are the departures from present expressed in percent. [From Kutzbach and Guetter, 1986] *Reprinted from Journal of the Atmospheric Sciences with permission of the American Meterorological Society.*

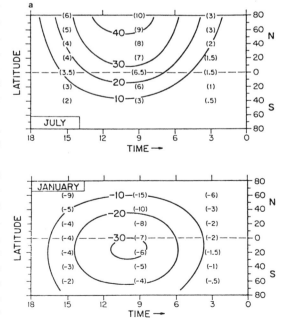

southwesterly winds also increased upwelling in an ocean model (Luther et al., 1990), a result confirming geologic data from this region (e.g., Prell, 1984b).

Although summer insolation increases are more or less offset by winter insolation decreases (Fig. 4.15), there was an annually averaged increase in tropical precipitation because precipitation is already very low in January and small reductions in the small winter value cannot offset large increases in the summer. A comparison of net P-E values indicates an early Holocene increase. Incorporation of these changes into hydrologic models of lake level fluctuations shows good agreement (Fig. 4.17) between calculated and observed lake level fluctuations (Kutzbach and Street-Perrott, 1985; COHMAP, 1988). There is also fairly good agreement between modeled P-E and lakes in eastern North America (Harrison, 1989), although in this case precipitation was both monsoonal and midlatitude in nature.

Even though increased Northern Hemisphere summer insolation is offset by decreased winter insolation, model experiments with an interactive mixed-layer ocean and sea ice (cf. Chap. 2) yield some changes in mean annual quantities

(Kutzbach and Gallimore, 1988; Mitchell et al., 1988). Two different GCM experiments both show net annual decreases in sea ice because of greatly enhanced summer heating. In high latitudes large changes in summer are not offset by small absolute changes in winter insolation. Sea ice decreases annually and SSTs increase in high latitudes, a result in agreement with geologic data (Chap. 3).

The Kutzbach–Gallimore experiment also suggests that tropical SSTs decreased by 0.1–0.3°C at 9000 BP because there was a net mean annual decrease in low-latitude insolation due to obliquity changes. A similar response was found in EBM calculations (Hyde et al., 1989; cf. Fig. 4.6). Overall changes in global-averaged temperatures were very minor (~0.1°C).

The above result and an analogous reasoning applied to the 125,000 BP interglacial (Chap. 6) has some bearing on greenhouse gas predictions (Chap. 14) for the future. An increase in global mean annual temperatures has sometimes been compared to the early Holocene or last interglacial—(Hansen et al., 1988). In fact, although seasonal warming was locally impressive, there may have been at most a few tenths of a degree warming in mean annual global temperatures

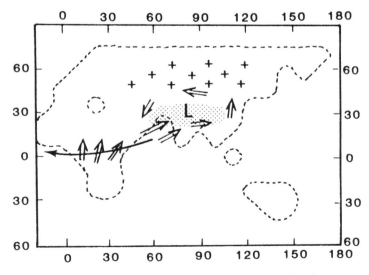

Fig. 4.16. Schematic diagram of temperature and circulation changes that resulted from greater seasonal forcing of the African–Asian monsoon during the early part of the present interglacial. Plus signs mean warmer surface temperatures; L, intensified monsoon low; broad arrows, stronger low-level winds; narrow arrows, stronger upper level wind; shaded area, increased precipitation. [From Kutzbach and Otto-Bliesner, 1982] *Reprinted from Journal of the Atmospheric Sciences with permission of the American Meteorological Society.*

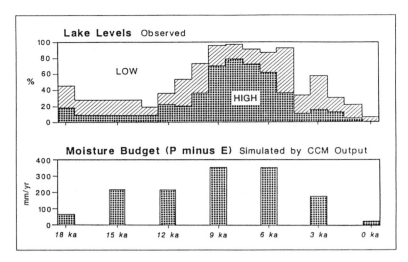

Fig. 4.17. Observed status of lake levels (upper row) and model-simulated moisture budget, precipitation minus evaporation (lower row) for each 1000 years (observations) or each 3000-year interval (model) from 18,000 BP to present. The observations and the simulated hydrologic budgets from the NCAR CCM are for the latitude belt 8.9–26.6°N. The observations and simulations agree in placing the highest lake levels and largest excess of precipitation over evaporation in the interval 9000–6000 BP. [From COHMAP, 1988] *Reproduced from Science 241:1043–52, 1988, copyright 1988 by the AAAS.*

during these times. Even minimum greenhouse scenarios project warming larger than this (Hansen et al., 1988). The analogy between future and past warmings in the Quaternary therefore does not appear to be appropriate.

There are some model–data disagreements for the early Holocene. For example, eastern North America was drier at 6000 BP than at 9000 BP (Harrison, 1989; the model predicts the opposite). Observations in Europe suggest Alpine summer temperature increases of about 4°C (Porter and Orombelli, 1985; Huntley and Prentice, 1988), whereas GCM simulations for Europe are about 2°C (Kutzbach and Guetter, 1986; note that GCMs do not adequately resolve climate change over "small" areas such as the Alps). The above result has some striking similarities to model–data discrepancies in the tropics at 18,000 BP, where high-elevation areas seem to be more sensitive than low-elevation areas. At present we do not have an adequate explanation for this phenomenon.

Another apparent model–data difference occurs in the Southern Hemisphere, in which there is some evidence for warmer conditions than the present between at least 45°S and Antarctica (Section 3.3.1). In highest latitudes the estimated changes (0.5–1.0°C) can be explained by

changes in orbital forcing at the obliquity period, which is in phase with Northern Hemisphere changes (Crowley, 1990). However, temperature changes at lower latitudes cannot be explained by this mechanism—the effect of obliquity is primarily manifested in high latitudes (Fig. 7.7) and precessional effects are out-of-phase between the northern and southern hemispheres. Before we conclude that the model–data differences are significant, we need to have more quantitative estimates of early Holocene temperature change in the lower latitudes of the Southern Hemisphere.

4.5.2 Tropical Phase Lags

The generally good agreement between forcing and response at 9000 BP cannot explain some detailed phase lag discrepancies between the two. Geologic records indicate that monsoon precipitation lagged forcing by a few thousand years (Street and Grove, 1979; Prell, 1984b). Two-dimensional EBM calculations (cf. Chap. 1) provide part of the explanation for the latter. The temperature response in equatorial regions might lag June solstice by as much as 3000 years (Short and Mengel, 1986; Short et al., 1990). This pattern results because heating in the equatorial strip is sensitive to both the annual and semiannual cycles in solar forcing (the latter reflecting the twice yearly passage of the sun across the equator). Calculations indicate that an orbital configuration with perihelion (Chap. 7) about 6 weeks after summer solstice produces the most favorable conditions for warmest summers.

4.6 Summary

A summary of the GCM studies for the last 20,000 years (Fig. 4.18) indicates that in general we have a reasonably satisfactory explanation for the steady-state climate in the northern high latitudes at 18,000 BP and for the tropics at 9000 BP. Progress is being made in the high latitudes of the Southern Hemisphere, but there are still some significant question marks. There is a more significant model–data discrepancy in the tropics at 18,000 BP that has provoked a number of possible explanations, none of which are completely satisfactory at this time. Experiments with ocean circulation models and for the deglaciation are just beginning.

Perhaps more for this chapter than any other chapter in the book, we see the benefits of drawing together the two fields of observational paleoclimatology and climate modeling. The nu-

" Paleo-Modeling Scorecard"	
Region and Time	Model-Data Agreement
N. High-Lat. (18k)	Very Good
Low-Lat. (18k)	Poor
Low-Lat. (9k)	Very Good
S. High-Lat. (18k)	Fair
Ocean Circ. (18k)	Insufficient Work

Fig. 4.18. Status of model–data agreements for various time and space slices over the last 20,000 years. Evaluation represents the authors' judgments, but there is considerable support for each rating. [After Crowley, 1989] *Reprinted by permission Kluwer Academic Publishers.*

merous modeling studies on climate of the last 20,000 years greatly enhance our understanding of processes over this time interval and also enable us to assess the behavior of climate models. In some areas there is good agreement between models and data. In other areas there are significant discrepancies which provide a focal point for further research. Although there are still a number of outstanding unsolved problems, it is clear that the studies discussed demonstrate significant progress toward the goal of better understanding past climate fluctuations.

5. HISTORICAL CLIMATE FLUCTUATIONS

Although the climate of the last few thousand years technically falls within the Holocene, we have chosen to place the discussion in a different chapter because the evidence is of a somewhat different nature, the mechanisms appear to be different, and there is an inevitable overlap of what happened most recently (the Little Ice Age) with what is happening at present and indeed what may happen in the future as a result of a greenhouse warming. This blending of implications means that the material in this chapter is not just strictly relevant to climates of the past (see Chap. 14).

5.1 Decadal to Millennial Scale Climate Change

5.1.1 General Patterns

In Chap. 3 we saw that climate of the last 20,000 years can be divided into ~10,000 year units (interglacial and glacial). However, a growing body of literature indicates that some significant higher frequency trends (10^3 and 10^2 years) are superimposed on the 10^4 year fluctuations of the last 20,000 years. This evidence is based on a diversity of proxy indices, e.g., lake records, ice cores, and glacier fluctuations over the last few thousand years. Over the last thousand years these records have been augmented by tree rings and historical documents. Since the records of the last 200 years overlap some long instrumental records, there is an opportunity for direct calibration of the different indices. The reader is referred to Bradley (1985, 1990) and Grove (1988) for additional comprehensive discussions of methodologies and results of historical climate studies.

As an example of the available evidence, Fig. 5.1 illustrates the millennial scale fluctuations of glaciers for the last 20,000 years (Mayewski et al., 1981). Although the illustrated glacier fluctuations could result from some type of "inter-nal" glacier dynamics, similar scale fluctuations (e.g., Fig. 5.2) have also been found in some deep-sea records, $\delta^{18}O$ variations of Greenland ice cores, pollen records from Europe, river downcutting and alluviation in North America, African lakes, charcoal from soil in the Amazon basin, and lake level fluctuations in western Australia (Pisias et al., 1973; Churchill et al., 1978; Woillard and Mook, 1982; Gillespie et al., 1983; Knox, 1983, 1985; Dansgaard et al., 1984; Sanford et al., 1985; Keigwin and Jones, 1989). These data clearly indicate 10^2–10^3 year variance in the climate system. The results are very interesting from a modeling viewpoint, as their origin is not well understood.

There seems to be some correlation in timing of millennial scale fluctuations in different regions. For example, Röthlisberger (1986) compiled records of glacier fluctuations in both hemispheres over the last 10,000 years (Fig. 5.3). Although there are some differences in fluctuations, there are also a number of similarities. For example, the widespread "Neoglacial" cooling between 3500 and 4500 BP signals the return to the generally cooler conditions of the late Holocene (cf. Section 3.3.1). The "Little Ice Age" cooling of the last few hundred years is also quite extensive (see next section). Glacial advances in the Southern Hemisphere were more extensive for the 500–1000 AD cooling, whereas they are more extensive for the Little Ice Age cooling in the Northern Hemisphere.

There are some indications that the glacier advances coincide with some of the other millennial scale climate oscillations previously noted. For example, cooling at about 8000 BP (cf. Begét, 1983) correlates with a desiccation event in some African lakes (Fig. 5.2) and a decrease in atmospheric CO_2 by about 20 ppm (Fig. 3.10). In this case the pattern of the linkage is the same as at the glacial maximum (i.e., cold-dry-low CO_2). However, it is unlikely that this simple picture applies to all millennial scale

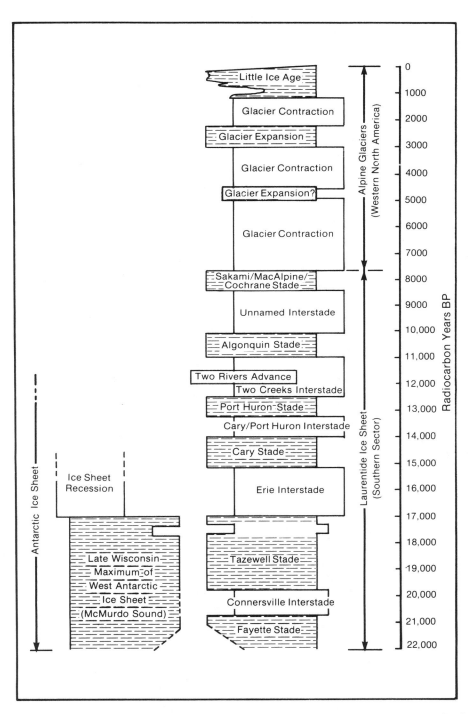

Fig. 5.1. Evidence for millennial scale climatic fluctuations during the last 20,000 years. A schematic diagram of pulses of glacier activity in North America is superimposed on major late Wisconsin and Holocene climatic changes. Antarctic information is from Stuiver et al. (1981). The time scale is in radiocarbon years before present. [From Mayewski et al., 1981] *Reproduced with permission from P. A. Mayewski, G. H. Denton and T. J. Hughes, in: "The Last Great Ice Sheets," G. H. Denton and T. J. Hughes (Eds.), copyright 1981, John Wiley and Sons.*

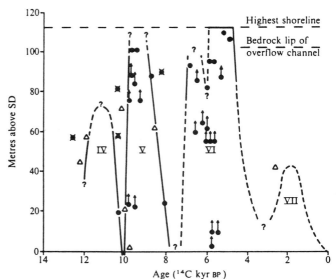

Fig. 5.2. Water level curve for paleolake Ziway-Shala in Ethiopia, 14000–0 BP, based on available stratigraphic, paleoecological, and ^{14}C data, illustrating millennial scale fluctuations in tropical aridity. Arrows indicate minimum elevation for corresponding lake level; dashed line denotes uncertainty. [From Gillespie et al., 1983] *Reprinted by permission from Nature 306:680–683, copyright © 1983, Macmillan Magazines Ltd.*

fluctuations. For example, the cooling at about 5000 BP (Fig. 5.1) coincides with a dry phase in the Amazon basin (Sanford et al., 1985) but a wetter phase in Africa (Fig. 5.2). Unfortunately, we have relatively little information on how many of these Holocene oscillations correlate, so that more general statements will have to await further analysis.

In Europe the cool period that marked the onset of the Neoglaciation about 4500 BP ended around 2500 BP (Fig. 5.3). As this interval overlaps with the development of civilization, it is sometimes called the Iron Age cold epoch (Gribbin and Lamb, 1978). A climate amelioration began near the dawn of the Roman Empire. Perhaps symbolically, the decline and fall of the empire, and the advent of the Dark Ages, coincided with a return to colder climates (500–1000 AD). Thence followed the Medieval optimum (~1100–1300 AD), in which European temperatures reached some of their warmest levels of the last 4000 years (it is not clear whether the warming of the 1980s exceeded the Medieval warming). Wine grapes were harvested in England (Le Roy Ladurie, 1971) and the Arctic pack ice retreated. Iceland and Greenland were settled around this time, and Alpine passes were open between Germany and Italy.

The Medieval warm period may not be of global extent, at least in terms of its relative warmth vis-a-vis the 20th century. Historical records in China and an ice core record on Antarctica indicate that maximum warmth of the last 1500 years occurred earlier in these regions, about 600–1000 AD (Zhu, 1973; Morgan, 1985). Also, Greenland was relatively cool when Europe was warm (Johnsen et al., 1970). Some of these regional differences may reflect changes in the atmospheric circulation (see next section). The geographic extent of the Medieval warmth is an important problem that requires further study.

5.1.2 Little Ice Age

Beginning about 1450 there was a marked return to colder conditions. This interval is often called the "Little Ice Age." The term was first introduced by Matthes (1939) and was used to describe an "epoch of renewed but moderate glaciation which followed the warmest part of the Holocene." Although its use has been criticized because it was not considered an event of global significance (Landsberg, 1985), we feel that the scale of the cooling is in fact large enough to justify its usage.

The Little Ice Age is considered to end around 1890, although different authors might choose slightly different dates. As we will see, there is considerable evidence that the Little Ice Age consisted of two main cold stages of about a century's length. These occurred in the seventeenth

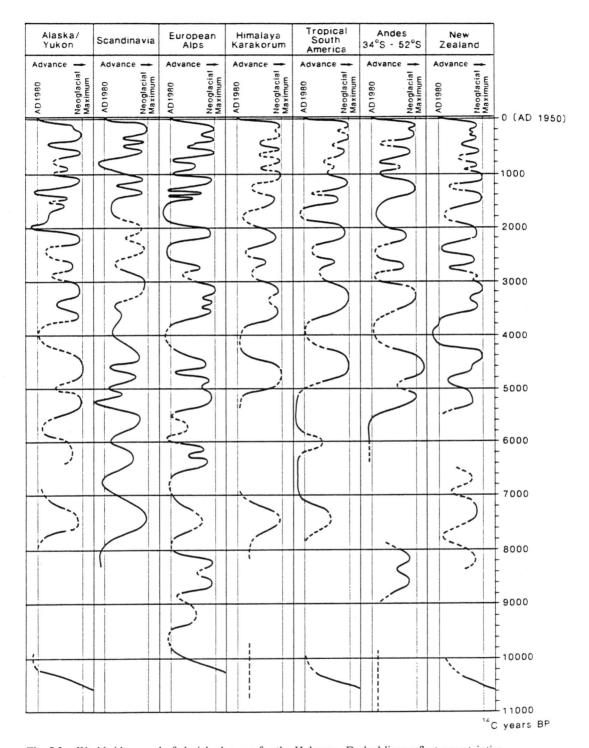

Fig. 5.3. Worldwide record of glacial advances for the Holocene. Dashed lines reflect uncertainties. [From Röthlisberger, 1986] *Reproduced by permission Aarau-Verlag.*

and nineteenth centuries. Most records indicate that the coldest decades occurred in the mid-late 1600s, the early 1800s, and the late 1800s.

Because the Little Ice Age was so recent, there is a substantially greater amount of information available about this cool period than about some of the other oscillations in the Holocene. Studies can also make use of tree rings and historical records, which are generally much less available for the older periods. In addition, the Little Ice Age overlaps the beginning of the use of meteorological measurements (early 1700s), so changes can sometimes be directly calibrated with temperatures. For these reasons we will take a more in-depth look at the Little Ice Age in order to draw some generalizations that may be applicable to other cool oscillations. This information will also prove useful in developing models attempting to explain the oscillations.

As extensively summarized by Grove (1988; cf. Hastenrath, 1984), there is a very large amount of evidence indicating that during the Little Ice Age alpine glaciers advanced in virtually all mountainous regions of the world—the Alps, the Sierra Nevada and Rocky Mountains, eastern Africa, the Himalayas, New Guinea, the southern Andes, and New Zealand. In the European Alps especially, the advances of glaciers into alpine valleys has been repeatedly documented in literature, with maps, prints, drawings, photos, and oil paintings.

In addition to the early instrumental record, which clearly records colder temperatures (e.g., Manley, 1974; van den Dool et al., 1978), evi-dence for cooling involves, for example, observations of expanded snowcover in the Scottish highlands, freezing of lakes and rivers, and changes in agricultural trends. Much of this information falls in the category of "phenological data" (refers to phenomena such as flowering times of trees, migration dates of warm-loving biota, harvest times, etc.).

When sifted properly historical data can yield more quantitative information about past climates. For example, Le Roy Ladurie and Baulant (1980) demonstrated that the grape harvest dates in France between 1370 and 1879 were inversely related to April–September temperatures in Paris ($r = 0.86$ for an 83 year period of overlap with the instrumental record). A reconstruction of Paris summer temperatures back to 1370 yields the surprising result that, except for the nineteenth century, Little Ice Age cooling is not manifested in summer records (Fig. 5.4). This conclusion agrees with the more comprehensive analysis of Pfister (1985) that in Europe the Little Ice Age was predominantly a winter-time phenomenon. It remains to be determined whether this same conclusion applies elsewhere.

In addition to the glaciological record, there are substantial proxy data from other parts of the world that provide additional support for Little Ice Age climate changes. Sea ice expanded greatly around Iceland (Bergthorsson, 1969) and essentially cut off the colony established in Greenland during the preceding warm period (Dansgaard et al., 1975). In addition to the advance of mountain glaciers in North America,

Fig. 5.4. Reconstructed summer temperature for Paris, 1370–1879. [From Bradley, 1990; based on data from Le Roy Ladurie and Baulant, 1980] *Reproduced by permission Elsevier Scientific Publishing Co.*

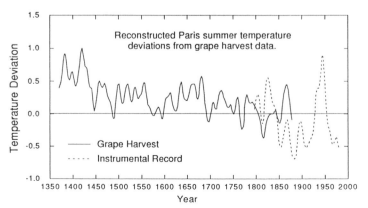

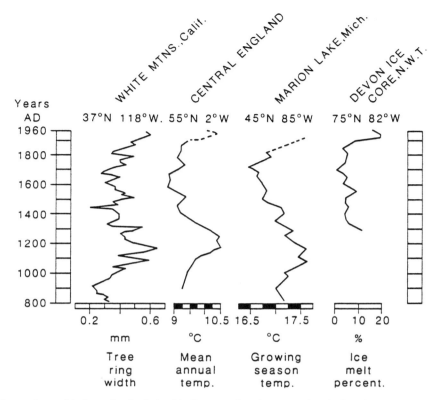

Fig. 5.5. Comparison of independently derived indicators of environmental and climatic change covering the last 1000 years. Tree ring record from LaMarche (1974); central England temperatures record after Manley (1974); Michigan lake record from Bernabo (1981); ice core record from Koerner (1977). [From Bernabo, 1981] *Reprinted with permission from Quaternary Research.*

altered temperatures are detected in tree ring records from the Sierras (e.g., Fig. 5.5) and northern Canada (LaMarche, 1974; Jacoby and Cook, 1981). Records from fur trappers in Canada indicate that the freezing dates at Moose Factory at the southern shores of James Bay (southern Hudson Bay) were generally a week earlier and break-up dates 10 days to 2 weeks later (Catchpole et al., 1976).

There is also substantial evidence for Little Ice Age climate change in the United States. Winter records show evidence for cooling (e.g., Fig. 5.5; Bernabo, 1981; Baron et al., 1984). There is a large amount of tree ring data from lower elevations which have a considerable amount of inter-regional variability (e.g., Stockton et al., 1985). Many of these records seem to be monitoring summer drought conditions and have distinctive decadal-scale oscillations (e.g., Cook and Jacoby, 1979; Stahle et al., 1985, 1988; Stahle and Cleveland, 1988).

Additional information about the Little Ice Age is available from east Asia. The climate inferences are based mainly on historical records, with the Chinese historical record being especially large (e.g., Wang and Zhang, 1988; J. Zhang, 1988; Zhang and Crowley, 1989). Results indicate that winter temperatures in China and Japan were markedly colder than present (e.g., Yamamoto, 1971; Maejima and Tagami, 1984; D. Zhang, 1988). The inferred temperature trend from historical records is consistent with a tree ring record from western China (Fig. 5.6). Historical records also indicate that there was a more frequent occurrence of droughts and dust storms in China (Zhang, 1984) during the Little Ice Age (Fig. 5.6). Chinese and ice core records clearly show the decadal scale variability of Little Ice Age climates (Fig. 5.6).

In addition to the expansion of glaciers in low latitudes and the Southern Hemisphere, additional evidence for Little Ice Age climate change

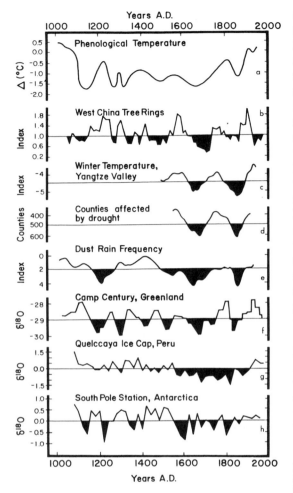

Fig. 5.6.   Examples of decadal and centennial-scale climate variability. (a) The phenological temperature in China (after Zhu, 1973). This index is based on timing of recurrent weather-dependent phenomena, such as dates of flowering of shrubs or arrivals of migrant birds, or distribution of climatically sensitive organisms. (b) The growth ring index of a juniper tree from western China (after Wang et al., 1983). (c) The winter temperature index in the lower reaches of the Yangtze River (after Zhang, 1980). (d) The number of Chinese counties affected by drought (after State Meteorological Administration, 1981). (e) The frequency curve of dust rains in China (after Zhang, 1984). (f) The $\delta^{18}O$ record from Camp Century, Greenland (after Johnsen et al., 1970). (g) The $\delta^{18}O$ record from Quelcaya Ice Cap, Peru (Mosley-Thompson et al., 1990). (h) The $\delta^{18}O$ record from the South Pole (Mosley-Thompson et al., 1990). Shading equals cool intervals. [From E. Mosley-Thompson; modified from Zhang and Crowley, 1989 and Ren, 1987] *Courtesy E. Mosley-Thompson.*

can be found in these regions. $\delta^{18}O$ variations in a Peru ice core from 14°S suggest changes synchronous with the Northern Hemisphere (Fig. 5.7; Thompson et al., 1986). The changes prob-

ably reflect circulation and air mass changes rather than temperature (Grootes et al., 1989). Similarly, cave deposits in New Zealand, geomorphic features on South Georgia Island in the sub-Antarctic (54°S), and ice cores on East Antarctica (Fig. 5.8; cf. Fig. 5.6) record changes broadly synchronous with the Northern Hemisphere (Wilson et al., 1979; Mosley-Thompson and Thompson, 1982; Morgan, 1985; Clapperton et al., 1989). An index of upwelling in the Ross Sea suggests stronger winds during the Little Ice Age (Leventer and Dunbar, 1988). These authors conjecture that the stronger winds may have resulted in greater upwelling and more open water around Antarctica. Such a conjecture is consistent with (limited) observations (diaries of Captain Cook) indicating no consistent evidence for expansion of sea ice around Antarctica (Parkinson, 1990). Sea ice coverage was generally greater in the Weddell Sea but less in the Ross Sea.

Other Southern Hemisphere changes are related to precipitation. Lake levels in eastern Australia were higher in the nineteenth century than at present (Churchill et al., 1978), whereas conditions in southern Chile were drier than present. Lamb (1969; cf. Sanchez and Kutzbach, 1974) suggests that in the Atlantic sector

Fig. 5.7.   Comparison of Northern and Southern Hemisphere records over the last 400 years. Decadal temperature departures (from the 1881–1975 mean) in the Northern Hemisphere from 1580 to 1975 (from Groveman and Landsberg, 1979) compared with decadal average $\delta^{18}O$ values for Quelccaya (Peru) ice cores. The dashed line is the 1880–1980 mean. [From Thompson et al., 1986] *Reprinted with permission from Science 234:361–364, copyright 1986 by the AAAS.*

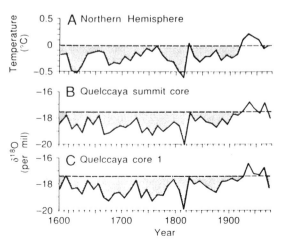

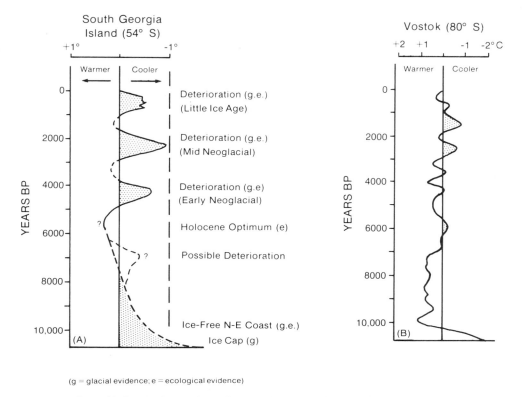

(g = glacial evidence; e = ecological evidence)

Fig. 5.8. Comparison of inferred climate history from (a) South Georgia Island in the sub-Antarctic (54°S) with (b) smoothed Vostok ice core (Antarctica) temperature record (from Jouzel et al., 1987b). [From Clapperton et al., 1989] *Reprinted with permission from Quaternary Research.*

Little Ice Age changes involved southward shifts of the atmospheric circulation in both hemispheres. Thus Chile would be exposed more frequently to the dry subtropical high of the South Pacific.

An exception to the overall cooling pattern in the Southern Hemisphere involves South Africa and the Antarctic peninsula (cf. Fig. 10.15; Tyson, 1986; Mosley-Thompson et al., 1990). Although cooling occurred in South Africa during the early stages of the Little Ice Age, nineteenth century conditions were as mild as those of the twentieth century. Tyson (1986) suggests that this pattern may reflect warm SSTs in the eastern South Atlantic that have been mapped by Lamb (1969). Warming on the Antarctic peninsula between 1600–1830 occurred at the same time as cooling at the South Pole (Mosley-Thompson et al., 1990). The asymmetric pattern may reflect an atmospheric circulation response to stronger zonal westerlies (Mosley-Thompson et al., 1990).

Although climate was generally cooler in the centuries preceding the present, there are some indications that fluctuations over the last 1000 years may not be exactly synchronous. As discussed earlier, the warming in China during the eleventh century precedes by 200 years the peak warming in Europe during the Middle Ages. Peak cold events in the Little Ice Age occurred at the end of the seventeenth century in Europe and in the mid-seventeenth century in China (Lamb, 1979; Zhang and Crowley, 1989). There is a cool event in the early 1800s in North America and Greenland (Johnsen et al., 1970; Jacoby et al., 1988) that lags changes in China (Zhang and Crowley, 1989). Although it is desirable to document more thoroughly the spatial "irregularities" in the temperature trends, they have tentatively been interpreted as indicating a more meridional structure to the atmospheric circulation during the Little Ice Age (see text following). The more heterogeneous pattern of climate change during the Little Ice Age appears to be different from the warming pattern of the last century, which appears to have been more uniform globally (Bradley et al., 1987; Hansen and Lebedeff, 1987).

There have been a number of attempts to estimate temperature decreases during the Little Ice Age (especially the peak cold periods). Utilizing diverse methods, most estimates of Little Ice Age temperature decreases are generally on the order of 1.0–1.5°C (e.g., Allison and Kruss, 1977; Lamb, 1977; Smith and Budd, 1981; Druffel, 1982; Begét, 1983; Zhang and Crowley, 1989). Lamb (e.g., 1979) suggests that some areas of the North Atlantic may have been warmer during part of the Little Ice Age (Fig. 5.9). This response could reflect wind-induced changes in the surface circulations that modify

upwelling or changed advection rates of warm water from low latitudes.

There is less information available about precipitation variations during the Little Ice Age. Although there was an enhanced frequency of dust storms in China, there is also some evidence for enhanced flooding (Zheng and Feng, 1986). Similar evidence for enhanced variability can be found in frost records from eastern New England (Baron et al., 1984). Conditions in sub-Saharan Africa were wetter from the sixteenth to the eighteenth century (Nicholson, 1978). This result is somewhat similar to precipitation trends elsewhere, suggesting that there may have been two phases of precipitation variations dur-

Fig. 5.9. Reconstructed sea surface temperature patterns for 1879 in the subpolar North Atlantic, as based on departures from modern values. [From Lamb, 1979] *Reprinted with permission from Quaternary Research.*

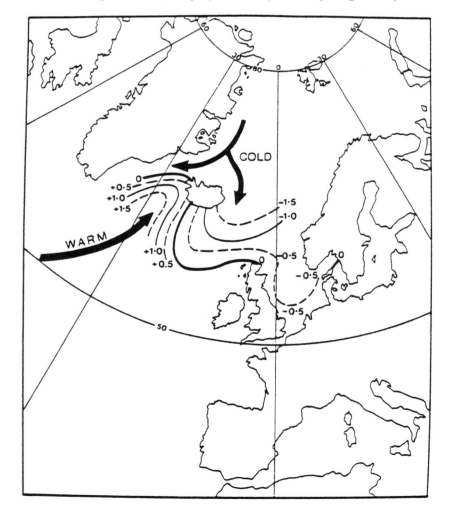

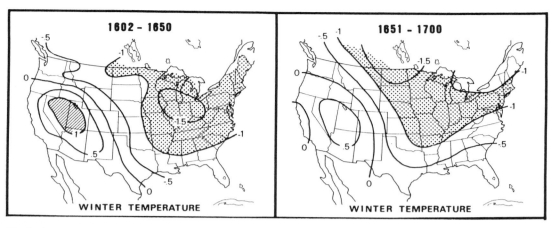

Fig. 5.10. Mean reconstructed anomalies (from tree rings) in winter temperature for the United States during Little Ice Age periods 1602–1650 and 1651–1700, expressed as departures in °C and percentages of the 1901–1970 mean of the observational record. [From Fritts et al., 1979] *Reprinted with permission from Quaternary Research.*

ing the Little Ice Age. During the postulated first phase (1450–1700), conditions were cool and moist in the Andes and western North America (LaMarche, 1974; Thompson et al., 1985). Dust levels did not significantly increase in China until the end of this period. Conditions were cool and dry during the following period (~1700–1880). It would be desirable to test this tentative conclusion with data from other regions before accepting it as a generalization.

We know much less about what the Little Ice Age was like in oceanic regions. In general we have information from only two types of records: very high sedimentation rate cores found in coastal regions and corals whose annual bands can be interpreted much like tree rings. The limited information available indicates that SSTs were lower and winds stronger in the eastern Pacific (Pisias, 1979; Juillet-Leclerc and Schrader, 1987). The stronger winds are consistent with historical data compilations for the North Atlantic, in which some periods (e.g., 1690s) were very stormy (Lamb, 1979). A $\delta^{18}O$ record of coral from the western Atlantic suggests a sea surface temperature decrease of ~1°C during the Little Ice Age (Druffel, 1982).

Because the Little Ice Age was relatively recent, we have better geographic control than for earlier periods. We can therefore examine features that make possible a greater understanding of the climate of this cool epoch. For example, studies in China, western North America, and Greenland suggest a more meridional compo-

nent to the atmospheric circulation (Fig. 5.10; Zhu, 1973; Fritts et al., 1979; Kelly et al., 1987). The pattern over North America and Europe is similar to one that occurs today about 20% of the time (van Loon and Rogers, 1978). It is associated (Fig. 5.11) with an elongated upper level ridge over western North America, a trough in eastern North America, another ridge over the North Atlantic (bringing relatively warm air to parts of Greenland), and colder conditions again over Europe (cf. Crowley, 1984). As stated earlier, some of the nonsynchronous changes between different regions may be due to altered flow patterns such as that illustrated in Fig. 5.11.

To date there have been no modeling studies of Little Ice Age climate fluctuations. However, studies on one aspect of present Asian moisture fluctuations may provide some insight into regional dynamics during this time. GCM sensitivity experiments indicate that increased Eurasian snowcover decreases the strength of the subsequent summer monsoon (Barnett et al., 1988). The area of model response (Fig. 5.12) is the same area as maximum ice-age aridity in China in the seventeenth century. The GCM response reflects albedo, temperature, and soil moisture feedbacks associated with the expanded snow cover (Barnett et al., 1988). A similar explanation may apply to the Little Ice Age (Zhang and Crowley, 1989). However, the ultimate reason for the winter cooling has yet to be explained (see text following).

Fig. 5.11. Upper atmosphere winter circulation pattern consistent with a number of regional anomalies during part of the Little Ice Age. Amplified ridge over North America and Greenland consistent with warm and dry conditions in the western United States (Fig. 5.10) and out-of-phase temperature trends between Greenland and Europe. Expanded trough in eastern Asia is also consistent with colder temperatures in this region. Dotted area refers to regions of warm SST anomalies (due to southerly flow; cf. Fig. 5.9). Horizontal dashed lines refer to postulated area of glacial inception at the beginning of the last ice age—the Little Ice Age pattern may also be applicable to glacial inception. [From Crowley, 1984] *Reprinted with permission from Quaternary Research.*

5.2 Mechanisms

The causes of centennial and millennial scale climate fluctuations are not well understood. In this section we will examine the three most frequently cited explanations: volcanoes, solar variability, and internal interactions in the coupled ocean–atmosphere system.

5.2.1 *Volcanism*

For some time volcanoes have been cited as an explanation for decadal scale cooling. However, only recently has there been hard evidence to support some type of volcano–climate connection. Detailed studies of volcanic events of the last 100 years indicate that significant cooling occurs after a large eruption for a period of about six months (Sear et al., 1987; Bradley, 1988; cf. Self, et al., 1981). Such cooling has been detected as frost-freeze events in tree rings (Fig. 5.13; LaMarche and Hirschboeck, 1984). Of particular interest is the great eruption of Tambora in Indonesia in April 1815 (e.g., Stothers, 1984), which has often been linked with exceptionally cold temperatures over the next 2 years (e.g., Stommel and Stommel, 1981; Stothers, 1984)—1816 being known as the "year without a summer" in New England and Canada. Oenologists should note that late grape

Fig. 5.12. Effect of increased Eurasian snow cover on monsoon precipitation the subsequent summer. Figure illustrates differences in total precipitation (in cm) for a doubled Eurasian snowfall general circulation model experiment minus the control. [Redrawn from Barnett et al., 1988] *Reprinted with permission from Science 239:504–507, copyright 1988 by the AAAS.*

June Precipitation Departures (cm) for Enhanced Eurasion Snowfall

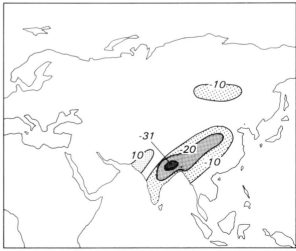

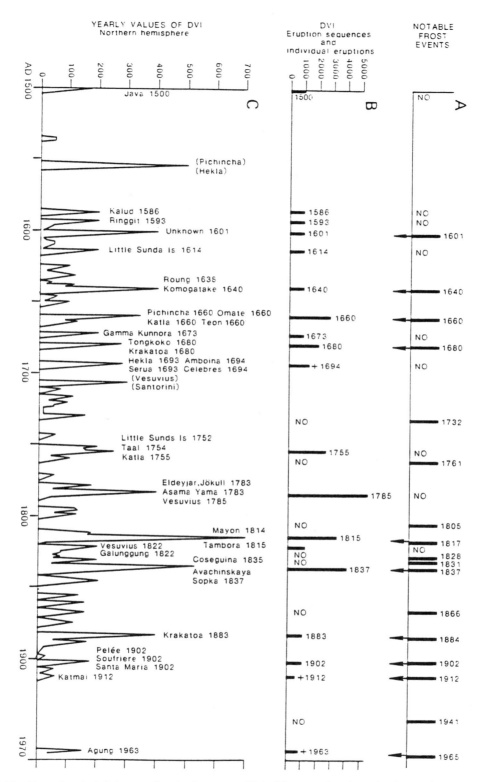

Fig. 5.13. Frost rings in bristlecone pines in the western United States and major volcanic eruptions: (A) Dates of notable frost ring events; arrows indicate associated eruptions. "No" indicates absence of major frost events at time of major eruption. (B) Dust veil index (Lamb, 1970) and dates of eruption or eruption sequences of 1000 or greater. "No" indicates apparent absence of major eruption corresponding to frost ring events shown in (A). (C) Integrated yearly dust veil index, with names of volcanoes and dates of major eruption. [After LaMarche and Hirschboeck, 1984] *Reprinted with permission from Nature 307:121–126, copyright © 1984 Macmillan Magazines Limited.*

harvests tend to follow large volcanic eruptions (Stommel and Swallow, 1983)!

In an important study, Porter (1986) demonstrated a significant correlation between glacial advances of the last 1400 years and frequency of volcano eruptions as estimated by changes in acid content in one of the Greenland ice cores (Fig. 5.14; acidity changes were assumed due to atmospheric sulphate changes from volcanic eruptions). A similar agreement occurs between the Greenland acidity record and central England temperatures, with a correlation coeffi-

cient of −0.52, significant at the 95% level (Hammer et al., 1980). Because this correlation only explains ∼25% of the total variance in the records, we must also consider other causes.

A caveat is required concerning the above correlations. Although sulfate can be derived from volcanic activities, it is now known that changes in ocean productivity can also release dimethyl sulfide to the atmosphere, which after conversion to sulphate can be deposited in ice cores (Charlson et al., 1987; Legrand et al., 1988a; cf. Section 3.1.9 and Fig. 6.17). Thus, the

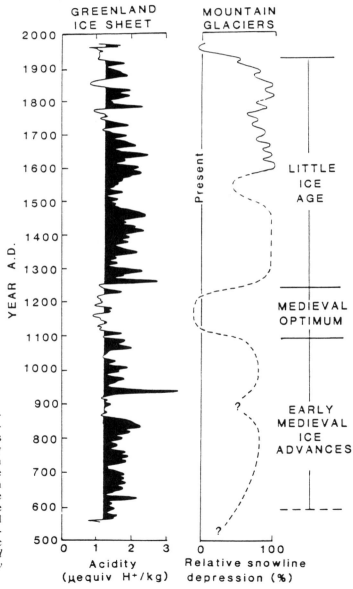

Fig. 5.14. Inferred correlation of Greenland Ice Sheet acidity record with evidence of late Holocene glacier variations in the Northern Hemisphere. For many midlatitude mountain ranges, maximum snowline depression during the Little Ice Age amounted to 100–200 m. Although fluctuations of glaciers appear to have been generally synchronous, the relative extent of individual advances differed among glaciers or groups of glaciers because of local geographic and climatic factors. [From Porter, 1986] *Reprinted with permission from Quaternary Research.*

background acidity level in the ice core could reflect a response to climate change rather than a cause of climate change. However, individual spikes in acidity records probably represent volcanic eruptions.

In order to clarify the role of volcanism it is necessary to "clean" the sulphate record in ice cores of the ocean productivity signal. As marine plankton also produce methanesulfonic acid, which can be measured in ice cores (Saigne and Legrand, 1987), this goal seems achievable. Another problem with the volcano–climate connection involves the observation that volcanic events, even very large ones, often do not have a global-scale effect. Observations from ice cores indicate that only eruptions in equatorial latitudes are transmitted by the upper atmosphere winds to the high latitudes of both hemispheres (Legrand and Delmas, 1987). Eruptions in the high latitudes of each hemisphere tend to stay in that hemisphere. Unless we call for equatorial eruptions to play the main role in causing the changes in climate, the volcano explanation may have a severe constraint. On the other hand, one could argue that some of the interhemispheric differences in climate records could reflect the more localized forcing in one hemisphere.

It is also necessary to examine from a modeling viewpoint how episodic volcanic impulses are transmitted into the lower frequency part of the climate spectrum; observations from both the instrumental record and ice cores indicate that the volcanic signal and the detectable temperature effect are manifested in only the first 1–2 years after an eruption (Legrand and Delmas, 1987; Bradley, 1990). One possible explanation involves some type of ice-albedo feedback (e.g., Robock, 1978), in which short-term cooling events affect sea ice cover. Due to the longer time constant of sea ice, the effect may then be felt over decades. It is also possible that closely spaced multiple eruptions may have some cumulative effect that has not been fully examined.

5.2.2 Solar Variability

Another popular mechanism for longer term climate variations involves solar variability (cf. Crowley, 1983b; Sofia, 1984). It has long been known that the solar magnetic cycle involves periods of 11 and 22 years. There is an ∼88 year cycle in sunspot, aurora, and ^{14}C variations (e.g., Gleissberg, 1966; Feynman and Fougere, 1984; Stuiver and Braziunas, 1988), plus evidence for a fundamental solar period of about 420 years, with significant harmonics at 140 and 220 years (Stuiver and Braziunas, 1989).

In addition to sunspot variability, there have also been discussions about possible changes in the diameter of the sun (and of solar output), although these conclusions have been frequently disputed (e.g., Sofia et al., 1983; Ribes et al., 1987; Morrison et al., 1988). An 80-year periodicity is a commonly cited number (Gilliland, 1981; Parkinson, 1983) and one that in fact corresponds to $\delta^{18}O$ fluctuations in one of the Greenland ice cores (Dansgaard et al., 1971).

The most direct way for solar variability to affect climate involves changes in solar irradiance (which in turn may be due to changes in sunspot activity; cf. Sofia, 1984). For example, satellite measurements indicate that the solar "constant" is not constant and has varied in output by about 0.09% between 1980 and 1985 (Kyle et al., 1985). This amount alone appears insufficient to affect climate significantly, for it translates into only a 0.3 W/m² change in forcing at the top of the atmosphere. With a 30% planetary albedo, this causes a 0.2 W/m² change in forcing at the surface. The resultant temperature change is ∼0.1°C—one-tenth or less of the Little Ice Age temperature departures (Section 5.1.2). We know enough about the sensitivity of climate models to postulate a change in the solar constant on the order of 0.5–1.0% would be required to directly cause climate change of the observed magnitude during the Little Ice Age; in other words, a change in forcing approximately ten times greater than the observed range over the last 15 years. One model of solar variability (Schatten, 1988; cf. Reid, 1987) suggests that decreased solar output during the seventeenth century Maunder Minimum may have been significantly larger than recent changes, but it is still not clear as to whether it was large enough to cause the observed changes.

The solar–terrestrial connection can be tested by extending the sunspot observational record with proxy data. For example, changes in solar variability can be estimated by measuring tem-

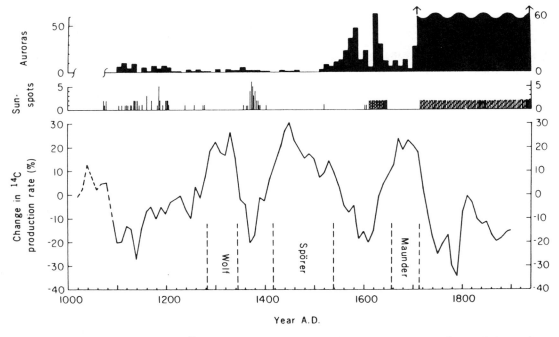

Fig. 5.15. Evidence for atmospheric ^{14}C changes attributable to a variable sun. Frequency of auroral observations, sunspots, and changes in ^{14}C production rate are calculated from a carbon reservoir model. The number of naked eye sunspot observations per decade before 1610 is given by the vertical bars. The cross-hatched area denotes the existence of sunspots as observed by telescope. [From Stuiver and Quay, 1980] *Reprinted with permission from Science 207:11–19, copyright 1980 by the AAAS.*

poral fluctuations in the cosmogenic nuclides ^{14}C and ^{10}Be. Changes in solar output are associated with changes in the heliosphere (outermost solar atmosphere), which in turn modulates the cosmic ray flux at the top of the earth's atmosphere, causing significant variations in production rates of ^{14}C and ^{10}Be. Both of these isotopes can be measured in geologic records. A comparison of ^{14}C measurements in tree rings with the historical record of auroras and sunspots indicates good agreement (Fig. 5.15). Because the ^{14}C record is also susceptible to perturbations by climate fluctuations (i.e., is involved in the global carbon cycle), it is desirable to check it against ^{10}Be, which can be measured in ice cores, and which is less susceptible to terrestrial modification. A comparison of ^{14}C and ^{10}Be estimates of ^{14}C fluctuations over the last 5000 years (Fig. 5.16) yields a high degree of correlation (Beer et al., 1988), suggesting that, at least over this time interval and at this time scale, the two records are in fact recording variations in solar activity.

A key question relating to solar variability in-

volves whether the solar fluctuations correlate with climate over the last 1000 years. On this account, evidence is mixed as to whether solar variability is related to climate change. For many years there have been numerous heroic efforts to establish a sunspot–climate relationship (see Pittock, 1983). Most attempts have been either unsuccessful or successful because of the faulty use of statistics (some possible exceptions are discussed below). However, there was renewed interest in the sun–climate connection when Eddy (1976) noted that a well-known minimum of sunspot activity at the end of the seventeenth century coincided with one of the coldest periods of the Little Ice Age.

There is at present some "flickering" support for a possible solar–climate connection. For perhaps the first time, meteorological records (upper air circulation) have been convincingly correlated with solar variability (Labitzke and van Loon, 1988). Initially, studies of tree ring records from the central United States indicated a 22-year drought rhythm, which correlates with the 22-year Hale double-sunspot cycle (Mitchell et al., 1979). Many of the major glacial advances of the last 10,000 years appear to have occurred near times of solar variability (Wigley, 1988). There is also some evidence for periods of ~80, 114, 140, and 210 years in both ^{14}C and some climate records—glacial advances (80 and 140 years; cf. Fig. 5.3), tree rings (114 and 210 years), and an ice core (~80 years; Johnsen et al., 1970; Sonett and Suess, 1984; Stuiver and Braziunas, 1989).

One problem with most of the above analyses is that the solar peaks do not consistently occur in all records. There are additional problems with some of these sun–climate correlations. Comparisons of other proxy records with solar variability indicates a less satisfactory level of agreement (Stuiver, 1980). The central United States tree ring record has been reanalyzed as having an 18.6-year lunar tidal cycle rather than a 22-year cycle (Currie, 1984). Lunar tidal effects have also been reported for Asian monsoon fluctuations (Campbell et al., 1983; Hameed et al., 1983). A Precambrian climate record once thought to indicate solar forcing (Williams, 1981; cf. Crowley, 1983b) has also been reinter-preted in terms of lunar tidal forcing (Williams, 1988 for further discussion of tidal effects see Kvale et al., 1989; Smith et al., 1990). The statistical significance of some analyses in terms of either solar or lunar forcing has also been questioned (see Clegg and Wigley, 1984). Finally, even if the sun–climate calculations are statistically significant, overall coherencies (cf. Appendix C) are relatively low (0.4–0.5; cf. Fig. 5.17)—much lower than coherencies obtained for many Pleistocene correlations (~0.8; cf. Chap. 7). Furthermore, the amount of variance in the climate records that can be explained by solar forcing is usually less than 10% (e.g., Mitchell et al., 1979; Cook and Jacoby, 1979). The matter clearly requires more investigation before we can draw any definitive conclusions about the sun–weather connection.

5.2.3 "Internal" Variations in the Ocean-Atmosphere System

A third mechanism for causing decadal to millennial scale climate fluctuations involves the ocean circulation. Because the surface mixed

Fig. 5.16. Comparison of measured Δ^{14}C and calculated Δ^{14}C (based on ^{10}Be variations), indicating that at periods of 10^2–10^3 years, the former is primarily driven by solar variations (see text). (a) Illustration for the period 3000 BC to 2000 AD after removal of long-term trends. The upper curve represents the measured Δ^{14}C variations; the lower shows calculated Δ^{14}C based on ^{10}Be data. The lower curve is shifted by 10‰ or easier comparison. Both curves are slightly smoothed. (b), Cross-correlation coefficient between the measured and calculated Δ^{14}C curves as a function of lag time. [From Beer et al., 1988] *Reprinted by permission from Nature 331:675–680, Copyright © 1988, Macmillan Magazines Ltd.*

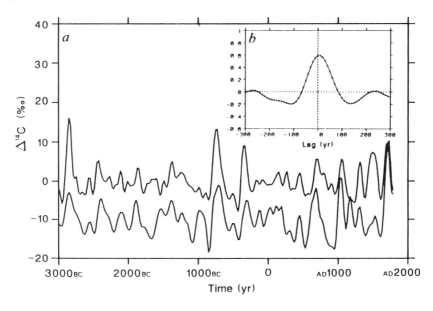

ATMOSPHERIC ^{14}C VS COMPOSITE MORAINE RECORD
(−2355 TO +1735)

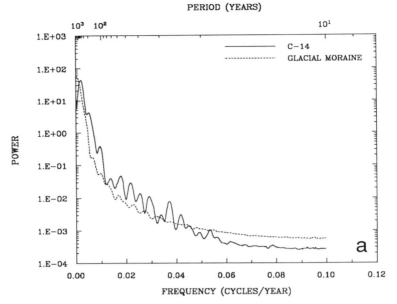

ATMOSPHERIC ^{14}C VS COMPOSITE MORAINE RECORD
(−2355 TO +1735)

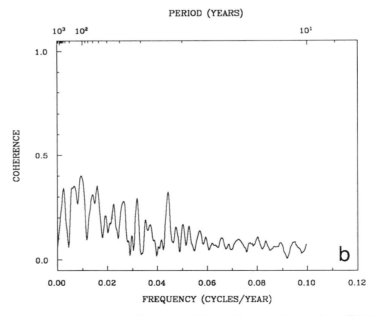

Fig. 5.17. Comparison of records of solar and climate variability. (a) Spectra of atmospheric ^{14}C (cf. Fig. 5.16) from Stuiver and Braziunas (1988) and of composite moraine record from seven locations (cf. Fig. 5.3). (b) Coherence (cf. Appendix C) between the two spectra. Note that maximum coherences are in the range of 0.4—significantly less than coherences for ice-age spectra (e.g., Fig. 7.8).

layer and deep ocean have different response times, essentially "random" atmospheric forcing of the lower medium may generate responses at low frequencies (Mitchell, 1966; Hasselmann, 1976; Frankignoul and Hasselmann, 1977; Watts and Hayder, 1983; Gaffin et al., 1986). Modeling studies suggest that such low-frequency variations can generate patterns similar to observations (cf. Fig. 14.6).

To date geologic data are inadequate for definitively testing this mechanism. There is a suggestion of millennial scale fluctuations of North Atlantic deep water (NADW) at about 4000 and 2000 BP (Boyle and Keigwin, 1987; Fig. 3.19), with possible decreased NADW production associated with colder events. Similarly, instrumental records indicate that the deep North Atlantic has warmed over the same time interval (1957–1981) in which the surface has cooled (cf. Roemmich and Wunsch, 1984). Whether these fluctuations are real and represent internal oscillations remains to be determined.

5.3 Summary

Millennial and centennial scale fluctuations occur during the Holocene and in some cases are of the same "sign" as fluctuations at the last glacial maximum (e.g., cold–dry). But there seems to be more variability to the regional response for these smaller climate oscillations as well as indications that they are not all in phase.

At least three processes may be contributing to climate fluctuations on this time scale: volcanoes, solar variability, and internally driven oscillations of the ocean–atmosphere system. But the relative importance of these mechanisms needs more study. In Chap. 14 we will further discuss modeling studies that assess the relative importance of these mechanisms for explaining temperature trends in the twentieth century.

From the above discussion the reader may now understand why a better understanding of the historical record is so important. Because the results involve new types of information (tree rings, historical climate records), overlap with the instrumental record, and even have a bearing on our predictions of the future course of climate, we have chosen to discuss the climate fluctuations in historical times in a separate chapter. It is also evident that, although some progress has been made on the study of climate fluctuations on this time scale, there is a pressing need for more quantitative testing of the relative importance of different mechanisms. It is also imperative that records from different areas be carefully correlated in order to determine the level of regional variance superimposed on a postulated global scale signal. This testing is not only important for understanding the climate fluctuations of the past 1000 years, but also for increasing our understanding of how climate may change in the future (cf. Chap. 14).

6. TEMPORAL TRENDS IN PLEISTOCENE CLIMATES

Records from time intervals beyond the last 20,000 years provide additional valuable information about the nature of climate fluctuations during the Pleistocene. Some of the more interesting results will be presented in this and the following chapter. Emphasis will be placed on general trends in the time domain and responses that may not be detectable through time series analysis. In the following chapter, spectral analysis techniques will be utilized to discuss variability in the frequency domain.

6.1 Pleistocene Chronology

The chronology and history of Pleistocene climates has seen several periods of stasis and change. Before the widespread use of deep-sea cores it was known that there had been a number of fluctuations of Pleistocene glaciers, with the number of major ice ages usually given as four (e.g., Flint, 1971). In some cases the geographic extent of the most recent Wisconsin/Würm ice sheets was fairly well known (e.g., Fig. 3.2). However, the chronology and detailed history was essentially unknown beyond the range of ^{14}C dating (about 30,000–40,000 years), and the number of ice ages was uncertain, because each succeeding glacial advance tended to erode much of the preexisting evidence.

In 1955 Cesare Emiliani made a historic contribution to Pleistocene climate research when he showed from $\delta^{18}O$ analyses of deep-sea cores that there were many more ice ages than had previously been suspected. This method was first suggested by Urey (1947), who proposed that carbonates precipitated from different water temperatures should have different ratios of ^{18}O to ^{16}O. Epstein et al. (1953) verified that theory with laboratory mollusks grown at different temperatures. A 1.0‰ change in $\delta^{18}O$ results from a 4°C change in water temperature or a ~2.0‰ change in salinity (cf. Broecker, 1989). The reader is referred to Duplessy (1978)

for a fuller discussion of principles and techniques.

From an analysis of windblown loess deposits (see Chap. 3) George Kukla (1970) first demonstrated a similar pattern of climate change on land. An updated comparison of Chinese loess variations with the deep-sea $\delta^{18}O$ record (Fig. 6.1) indicates very good agreement (Liu et al., 1985; Kukla, 1987; Kukla et al., 1988; cf. Hovan et al., 1989). Detailed comparisons of the Czech loess sequence with Alpine moraines used to define the classical four glacial advances indicated that the moraine sequence was incomplete and should not be used as an index of continuous climate change (Kukla, 1977). Despite this analysis, a number of scientists continue to employ the old terminology based on discontinuous land sequences in describing Pleistocene climate change. Unless the reader is intimately involved in working on land sequences, he or she is advised against using the classical sequence terms to describe climate change in other parts of the world.

The question still remained as to the chronology of the Pleistocene glaciations. For many years the matter could not be addressed by direct radiometric dating methods; radiometric techniques such as uranium series dating yield unsatisfactory results for deep-sea sediments because the methods are beset with questionable assumptions about constancy of sedimentation rates of key constituents.

Development of an absolute chronology for deep-sea cores received impetus from two different directions, one focusing on the late Quaternary and one that addressed the timing of older events. In a study of uplifted coral reef terraces on Barbados, Mesolella et al. (1969) utilized uranium series methods on isotopes trapped in the carbonate crystal lattice of corals to determine that the ages of some key interglacial reef horizons centered on about 82,000, 105,000, and 125,000 BP. The pattern of sea level change

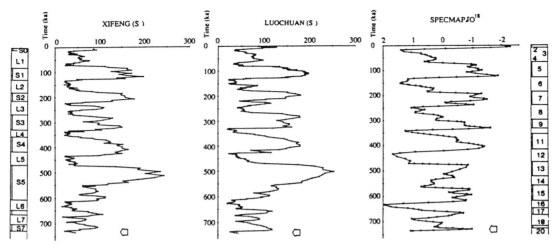

Fig. 6.1. Comparison of marine and land records for the last 700,000 years. SPECMAP oxygen isotope record, tuned to astronomic chronology (Imbrie et al., 1984), compared to magnetic susceptibility in Xifeng and Luochuan (China) loess sequences, which are plotted on susceptibility time scale. Brunhes/Matuyama boundary (730 Ka) marked with arrow. 1–20 are oxygen isotope stages; S0–S7 are soil units; L1–L6 are loess units. [From Kukla et al., 1988] *Reprinted with permission of Geological Society of America. From G. Kukla, F. Heller, L.-X. Ming, X.-T. Chun, L.-T. Sheng, and A.-Z. Sheng, 1988, Geology 16, 811–814.*

was similar to (but not identical to) $\delta^{18}O$ changes in deep-sea sediments (e.g., Fig. 6.2), and thus the $\delta^{18}O$ records were correlated to the three Barbados peaks, providing key time lines for the last 125,000 years. Numerous dates on coral reefs from other regions provided a basic support for the chronology of events of the last 125,000 years (e.g., Bloom et al., 1974; Ku et al., 1974;

Fig. 6.2. Sea level history derived from analyses of marine $\delta^{18}O$ record compared with the record derived from uplifted coral reefs on New Guinea. [From Shackleton, 1988b] *Reprinted with permission from Quaternary Science Reviews 6, 183–190, N. J. Shackleton, "Oxygen isotopes, ice volume, and sea level," copyright 1988, Pergamon Press plc.*

Oxygen Isotopes and Sea Level

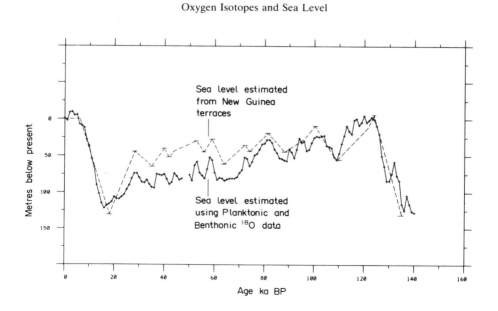

Aharon and Chappell, 1986; Edwards et al., 1987), a time interval known as the last glacial cycle, and one comprising the last interglacial (118,000–125,000 BP) and the subsequent buildup of glaciation that culminated in the glacial maximum at 18,000 BP. A typical $\delta^{18}O$ record for the last glacial cycle is illustrated in Fig. 6.3, along with key stratigraphic terms and Pleistocene hominid evolutionary events for reference.

Beyond 125,000 BP the chronology of glacial events remained in dispute until the first $\delta^{18}O$ record penetrated the Brunhes–Matuyama magnetic reversal boundary at 730,000 BP (Shackleton and Opdyke, 1973). Correlations were later extended to the entire Pleistocene (Shackleton and Opdyke, 1976; Prell, 1982; cf. Fig. 6.4). As a first assumption, the ages of samples between 125,000 and 730,000 BP were determined by linear interpolation. Later, with the

publication of another historic paper (see Chap. 7) linking glacial advances to orbital forcing (Hays et al., 1976b), the ages of intervening samples were determined by tuning the $\delta^{18}O$ record to the orbital time scale (Imbrie et al., 1984; Martinson et al., 1987), which can be accurately calculated from celestial mechanics considerations (Milankovitch, 1941).

With some exceptions the problem of Pleistocene chronology did not undergo another round of change until only recently (1985–1990). The publication of a long $\delta^{18}O$ record from Vostok, Antarctica (Lorius et al., 1985; Jouzel et al., 1987b) resulted in a chronology for the last interglacial (140,000–120,000 BP) different from that indicated by the deep-sea record. The warm period began about 10,000 years earlier in the ice core chronology. The latter was based on a flow model of ice, so it is theoretically subject to uncertainty. For example, the inter-

Fig. 6.3. Some key late Cenozoic terms, with hominid evolutionary events included for reference. Magnetic epoch and event boundaries are from Mankinen and Dalrymple (1979). The curve on the right is a high-resolution $\delta^{18}O$ record for the last 150,000 years from a core in the eastern tropical North Atlantic (from Shackleton, 1977). $\delta^{18}O$ stages are from Emiliani (1955); substages are from Shackleton (1969). Boundaries at 11,000 and 128,000 years BP are called terminations I and II (Broecker and van Donk, 1970). The interval between these two terminations is known as the last glacial cycle. Evolutionary and migratory events for man (some of which are overlapping) are from Patrusky (1980), Rensberger (1980), Bischoff and Rosenbauer (1981), Guidon and Delebrias (1986), and Valladas et al. (1988). For reference, some stages in man's cultural evolution are Paleolithic [Acheulian (1.5–~0.15 Ma), Mousterian (~40–120 Ka), Upper Paleolithic (~10–40 Ka)] and Neolithic [~5–10 Ka]. About 5000 BP the Age of Metals began in Mesopotamia (based on information from McAlester (1977), Bye et al. (1987), and Bischoff et al. (1988). [After Crowley, 1983b] *Courtesy of American Geophysical Union.*

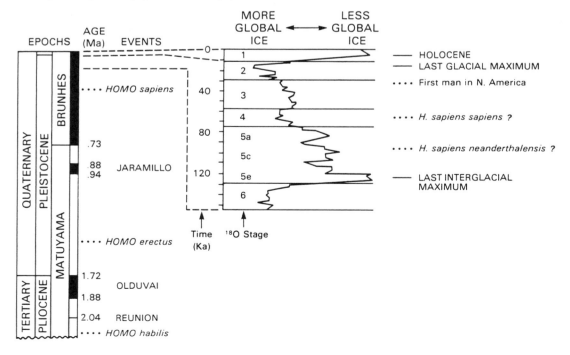

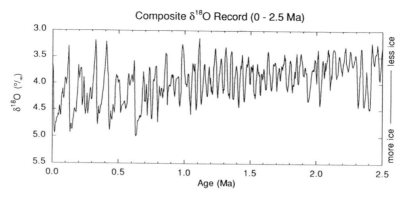

Fig. 6.4. Composite record of $\delta^{18}O$ fluctuations for the last 2.5 million years, based on averaging measurements from three hydraulic piston cores taken by the Ocean Drilling Program. [Based on data in Raymo et al., 1990b]

glacial interval could be an expanded section due to high accumulation rates of snow. However, ^{10}Be measurements in the same core (Raisbeck et al., 1987), which can be used as an index of snow accumulation, do not indicate any such change.

The chronology picture was further complicated by uranium series dates on a $\delta^{18}O$ carbonate record from the western United States (Winograd et al., 1988), which yielded a $\delta^{18}O$ record very similar to the deep-sea record but with a chronology for the beginning of the last interglacial more in agreement with the Vostok record. Since that time the marine chronology has been supported by independently dated loess and cave deposits (Kukla et al., 1988; van den Bogard et al., 1989; Li et al., 1989). In addition, direct correlation of the ice core and marine records (via inferred atmospheric dust layers in both sections) suggests that, despite the ^{10}Be measurements, the ice core chronology should be revised to be in approximate agreement with the deep-sea chronology (Petit et al., 1990). Despite this progress in reconciling differences in marine and terrestrial chronologies, the western United States ^{18}O record has sufficiently good radiometric control to lead us to conclude that there may still be some unresolved questions concerning dating of late Pleistocene events (cf. Winograd, 1990).

6.2 General Trends in Ice Volume Records

Inspection of $\delta^{18}O$ records for the last 2.5 million years (Fig. 6.4) indicates that, in addition to numerous oscillations of glaciers, there are some

important long-term trends. Distinct cyles of 100,000 years duration are only present for the last 700,000 years. Interglacials as warm as the present occurred for only about 10% of the time in the late Quaternary (Emiliani, 1972). Distribution of glacial moraines on North America and Europe indicates that ice extent during some of these advances was greater than the last glacial maximum (Fig. 6.5; Flint, 1971).

Before the late Pleistocene, glacial fluctuations had a periodicity of about 40,000 years (Shackleton and Opdyke, 1976; cf. Chap. 7). The transition between the modes may have required several hundred thousand years (Ruddiman et al., 1989b). Median values and variances for the two intervals are also different. These data show that glacial events in the early Quaternary were less extreme than late Quaternary stages. However, early Quaternary interglacials may have had more ice than later Quaternary interglacials (Fig. 6.4).

6.3 Bistable Climate States in the Pleistocene?

Analysis of the $\delta^{18}O$ record in ice cores suggests two basic climate states (Fig. 6.6), one glacial and one interglacial (Newell, 1974; Dansgaard et al., 1984; cf. Broecker et al., 1985). Evidence for an abrupt transition can also be found in a continuous pollen sequence from western Europe (Woillard and Mook, 1982). Broecker et al. (1985) suggested that such modes could possibly be maintained by changes in factors which control the production rate of deep water (surface salinity in the subpolar North Atlantic). Ex-

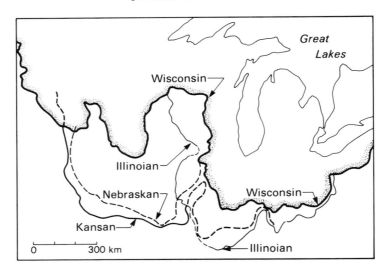

Fig. 6.5. Extent of continental ice in central North America during the Nebraskan, Kansan, Illinoian, and Wisconsin glacial advances. [Modified from Flint, 1971] *Reprinted with permission from "Glacial and Quaternary Geology," Flint, R. F., copyright 1971, John Wiley and Sons.*

Fig. 6.6. $\delta^{18}O$ profiles along the Dye 3 and Camp Century (Greenland) ice cores, plotted on a common linear time scale, illustrating a tendency for the system to have bistable modes, especially in the Pleistocene. [From Dansgaard et al., 1984] *Courtesy of American Geophysical Union.*

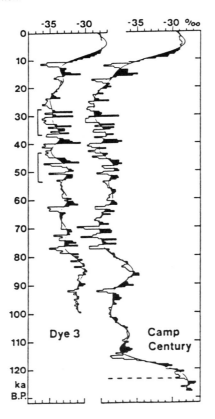

periments with a coupled ocean–atmosphere model provide some support for this hypothesis (F. Bryan, 1986; Manabe and Stouffer, 1988).

There are also some indications of rapid CO_2 fluctuations (Fig. 6.7) during the interstadial period of 30,000–40,000 BP in the Dye 3 (Greenland) ice core (Oeschger et al., 1985). If these patterns are real they imply that the ocean–atmosphere system is capable of switching between modes in a very short time interval. However, the large oscillations have not yet been detected in the Byrd (Antarctica) ice core (Neftel et al., 1988; cf. Fig. 3.10). This disparity could reflect one of two explanations. The "closure time" for gas bubbles could be too long in Antarctica to capture the rapid CO_2 fluctuations (the closure time is a function of accumulation rate, which is low in Antarctica). Alternatively, the Greenland record may be susceptible to summer melting, which would greatly increase CO_2 concentrations in the ice. The latter is a distinct possibility, as the core is from southern Greenland, which experiences some summer melting at present. In order to definitively resolve this important climate problem, we need to have ice core measurements from an area with relatively high accumulation rates and no summer melting. A joint U.S.–European ice core project is being planned for the summers 1989–1991 for the Greenland summit, so we should have the

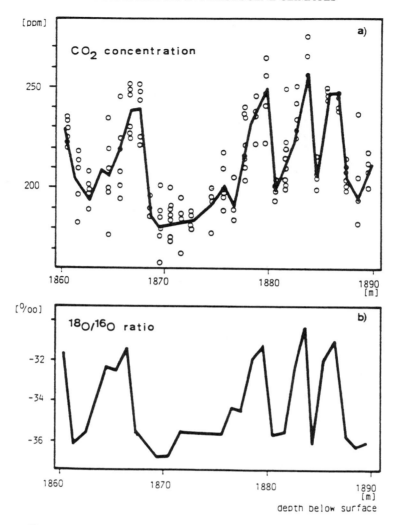

Fig. 6.7. CO_2 and $\delta^{18}O$ values measured on the Dye 3 (Greenland) ice core. The 30-m increment corresponds to about a 10,000-year interval between 30,000 and 40,000 BP. Note the very rapid shifts from glacial to interglacial values. The CO_2 fluctuations have not yet been substantiated by other ice core records (see text). [From Oeschger et al., 1985] *Courtesy of American Geophysical Union.*

answer to this problem within a relatively short period of time.

6.4 Evolution of Climate Over the Last Glacial Cycle (Last 150,000 Years)

The detailed description of patterns over the last 150,000 years provides insight into the evolution of climate over an entire glacial cycle and the penultimate glacial maximum. Emphasis is placed on relations between different climate indices in order to determine the coupling between different components of the climate system. The narrative discussion represents a useful

introduction to Chap. 7, which will examine climate fluctuations in the frequency domain during the last few hundred thousand years.

6.4.1 A Superglacial at 150,000 BP?

There are some small but significant differences between different glacial and interglacial periods. For example, the Warthe moraines of northern Europe may have extended farther south than their Wisconsin (Weichsel) counterpart (Kukla, 1977). Compared to the last glacial maximum, subpolar waters penetrated about 5° of latitude closer to the equator in the North At-

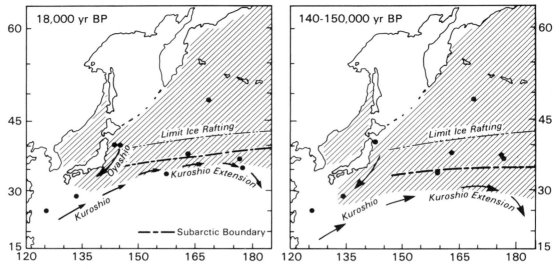

Fig. 6.8. Core locations and distribution of cold-dwelling planktonic foram *N. pachyderma* in northwestern Pacific for the last and previous glaciations, illustrating that maximum cooling in this region occurred during the penultimate glacial maximum. Left: 18,000 BP position of subarctic boundary based on percentage of left-coiling *N. pachyderma* and ice-rafted detritus [From Kent et al., 1971]. Right: 140,000–150,000 BP position of subarctic boundary. [After Thompson and Shackleton, 1980] *Reprinted by permission from Nature Vol. 287, pp. 829–833, Copyright © 1980 Macmillan Magazines Ltd.*

lantic, North Pacific (Fig. 6.8), and South Pacific (Luz, 1973; Thompson and Shackleton, 1980; Crowley, 1981; Thompson, 1981). The latitudinal displacements in the North Atlantic may have contributed to the very wet conditions in North Africa at ~150,000 BP (Causse et al., 1989) by displacing the storm tracks of midlatitude low pressure systems (cf. Section 4.1.1 and Figs. 3.6 and 4.1). $\delta^{13}C$ levels in bottom waters (Fig. 6.9) also reached more extreme values than at 18,000 BP (Curry and Lohmann, 1982; Duplessy and Shackleton, 1985; Shackleton and Pisias, 1985; Boyle and Keigwin, 1985/1986). These data, together with the glacial moraine record from North America (Fig. 6.5), suggest that the most extreme climate conditions of the Pleistocene have yet to be systematically modeled. Initial calculations indicate that very large ice sheets cause an enhanced meridional flow pattern in some sectors of the Northern Hemisphere (cf. Fig. 4.1), resulting in warmer temperatures than at 18,000 BP for some regions (Shinn and Barron, 1989).

6.4.2 Last Interglacial

The rapid deglaciation (Termination II) at about 130,000 BP marked the end of a long glaciation.

Examination of various $\delta^{18}O$ records suggests that this deglaciation may have even been more complete (in terms of removal of ice) and more rapid than the one which ended the last glacial maximum (e.g., the interval in cores is shorter and there is no indication of a pause in glaciation). As in Termination I, there is an interval of very low accumulation of sediments in the high-latitude North Atlantic and virtually lifeless conditions at the sediment–water interface (Duplessy and Shackleton, 1985). The authors postulated that NADW production may have temporarily ceased during that time (see Section 3.2.3).

Studies from coral reef terraces indicate that during the last interglacial maximum (125,000 BP), global sea level was about 6 m higher than present (Mesolella et al., 1969; Bloom et al., 1974; Ku et al., 1974), a fact attributed by some to the melting of the West Antarctic Ice Sheet (e.g., Mercer, 1968; Stuiver et al., 1981). Alternatively, the Greenland Ice Sheet may have been extensively reduced in size (Koerner, 1989). It has been suggested that one consequence of a future CO_2-induced warming would be a melting of the West Antarctic Ice Sheet (e.g., Mercer, 1978; Stuiver et al., 1981). It would be desirable to acquire more evidence to determine whether the Greenland or the West Antarctic Ice Sheet was responsible for the sea level rise during the previous interglacial.

Temperature estimates for the last interglacial are available from a few regions, but it is not clear whether we have enough information to infer global changes in warmth. Land records, mainly from western Europe, suggest temperatures at least 1–3°C warmer during the last interglacial (van der Hammen et al., 1971; Woillard, 1978; Keen et al., 1981; Mangerud et al., 1981; G. Miller et al., 1983; de Vernal et al., 1986). Nearshore marine sediments in The Netherlands are more representative of waters similar to those found at present off the coast of Portugal (Kukla, 1977). A startling last interglacial find in England has yielded fossil remains of hippopotamus, lion, rhinoceros, and elephant

(Gascoyne et al., 1981). Temperatures were 2–3°C warmer than present in East Antarctica (Jouzel et al., 1987b; see below, Fig. 6.14). It has yet to be determined whether these warm events were globally synchronous.

Summer precipitation in the African and east Asian sectors of the monsoon was apparently greater than present (Rossignol-Strick, 1983; Liu et al., 1985; Pokras and Mix, 1985; Petit-Maire, 1986). The record is more ambiguous in the Indian sector of the monsoon, where indices used to infer monsoon fluctuations suggest a somewhat weaker monsoon (Prell, 1984b).

As contrasted to the (limited) land record, open-ocean SST reconstructions for 120,000 BP (CLIMAP, 1984) are generally similar to the present interglacial ocean (Fig. 6.10). However, there is some evidence for slightly cooler condi-

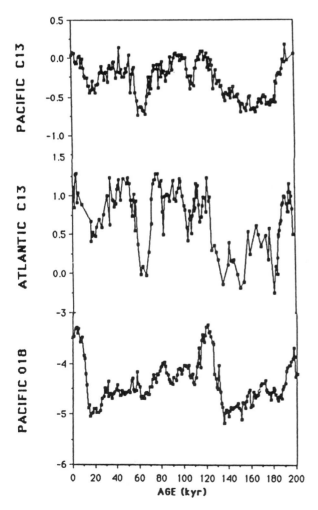

Fig. 6.9. Time variations of benthic $\delta^{18}O$ and $\delta^{13}C$ values for two deep-water sites. The North Atlantic site monitors NADW, the eastern equatorial Pacific site monitors Pacific Deep Water (PDW). ^{13}C changes can reflect both deep-water circulation changes and changes in the amount of the continental carbon biomass transfered to the deep-ocean (terrestrial and marine carbon sinks extract different levels of ^{13}C from their sources). Simultaneous changes in both basins are probably dominated by reservoir effects. Note that ^{13}C changes during glacial stages at ~70,000 and ~150,000 BP are greater than changes during the last glacial maximum (~18,000 BP).

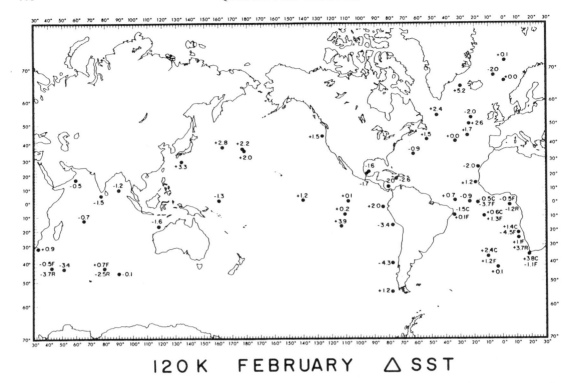

120K   FEBRUARY   △ SST

Fig. 6.10. Difference between modern sea surface temperature and estimated February sea surface temperature (in °C) at the last interglaciation 120,000 years ago. Negative values mean that the last interglacial ocean was colder than today. Note that most SST values are similar to present [present minus 120,000 BP = −0.05 ± 2.06°C (n = 52)]. Samples with more than one estimate reflect use of more than one transfer function (F = foram, R = radiolaria, C = coccolith). [From CLIMAP, 1984] *Reprinted with permission from Quaternary Research.*

tions in the central Norwegian Sea (Kellogg, 1980) and along the eastern boundary currents of the North Atlantic and North Pacific (Crowley, 1981; Muhs and Kyser, 1987). Geochemical studies of bottom-dwelling organisms provide intriguing indications of some differences between the present and last interglacial (Duplessy et al., 1984; Boyle and Keigwin, 1985/1986). δ^{13}C data can be interpreted in terms of either greater Antarctic Bottom Water Production (AABW) or less NADW production (Duplessy et al., 1984). The decreased NADW interpretation is consistent with Cd/Ca data from the North Atlantic (Boyle and Keigwin, 1985/1986). Overall changes in the carbon inventory of deep-sea sediments show a shift of about 0.15‰, which can be interpreted as resulting from about a 7% reduction in continental biomass (Duplessy et al., 1984). The same pattern could also result from decreases in deep-ocean overturn (cf. Broecker and Peng, 1987).

Since early Holocene (6000–10,000 BP) increased monsoon rainfall was apparently due to enhanced summer isolation forcing (see Section 4.5.1), it is of interest to inquire as to monsoon intensity at 125,000 BP, as seasonal forcing was larger at 125,000 BP (cf. Fig. 7.3). GCM experiments simulate a large part of Eurasia as 4°C warmer in the summer at 125,000 BP (Prell and Kutzbach, 1987). These results are consistent with greater monsoon intensity (Fig. 6.11). Interior Eurasia summer warming was about 1.0–1.5°C greater than at 9000 BP (Prell and Kutzbach, 1987).

Additional results from modeling studies indicate that increased uplift over land (due to greater heating) was accompanied by increased atmospheric subsidence over the ocean. Simulated wind fields over the central North Atlantic increased significantly (Prell and Kutzbach, 1987). This configuration should have spun up the North Atlantic gyre, a result consistent with geologic data (Crowley, 1981) indicating that the radius of the central waters was less at 120,000 BP than during the Holocene (due to conservation of angular momentum, a gyre spin-up should result in radial shrinkage). The gyre spin-up should have increased Gulf Stream transport into the subpolar North Atlantic but also increased equatorward transport of cool waters along the eastern boundary current. The former should increase SSTs along the coast of western Europe; the latter should decrease SSTs along the eastern boundary current. Both features are observed in the geologic record (see preceding text). It would be desirable to make a more quantitative comparison of the 120,000 BP ocean circulation with geologic data.

6.4.3 Glacial Onset

Following the interglacial, the onset of glaciation is marked by two stages of ice volume growth at 115,000 and 75,000 BP. There were two warm periods and two cold periods during the first phase (Fig. 6.3), with ice volume perhaps one-half glacial maximum values. The location of the ice sheet is uncertain, although

Boulton et al. (1985) have suggested that early ice growth was associated primarily with polar ice sheets (Fig. 6.12), whereas later more significant ice growth involved the Laurentide and Fennoscandian Ice Sheets (cf. Andrews and Barry, 1978). Onset of glaciation may have occurred in as short a time as 3000–4000 years (Frenzel and Bludau, 1987).

There is some information on regional climate patterns over the interval 80,000–110,000 BP. There is some cooling in long European pollen sequences (Fig. 6.13). Wolverine remains are found in England during the second cold period (Sutcliffe et al., 1985). However, sea level and pollen records indicate the subsequent interstadials were relatively warm (Fig. 6.13; e.g., Woillard and Mook, 1982; Stringer et al., 1986; de Vernal et al., 1986; Guiot et al., 1989).

Although the Norwegian Sea cooled rapidly after the last interglacial (Kellogg, 1980; Belanger, 1982), SST decreases lagged ice volume growth in the subpolar northern Atlantic (Ruddiman and McIntyre, 1979, 1981c). However, there were rapid and large temperature changes in the Southern Hemisphere after the last interglacial with temperature decreases on Antarctica

Fig. 6.11. Departures (experiment minus control) of surface temperature (*T*), winds (arrows), and precipitation (*P*) for the Indian Ocean sector for July 126,000 BP, illustrating the effect of increased summer insolation on *T* and *P*. [After Prell and Kutzbach, 1987] *Courtesy of American Geophysical Union.*

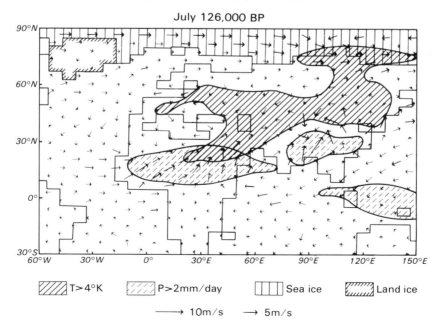

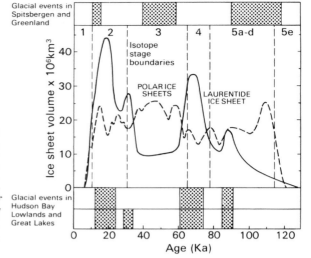

Fig. 6.12. Postulated changes in the volume of the Laurentide Ice Sheet through the last glacial cycle. Dashed line indicates inferred changes in high-latitude ice volume needed to reconcile land and marine records. The required high-latitude changes appear to correlate with Arctic glacial events. [After Boulton et al., 1985] *Reproduced by permission of the Geological Society from "Glacial geology and glaciology of the last mid-latitude ice sheets," G. S. Boulton, G. D. Smith, A. S. Jones, and J. Newsome, 1985, Jour. Geol. Soc. (London) 142, 447–474.*

Fig. 6.13. Estimated annual average precipitation and temperature changes for two long pollen records from France. [From Guiot et al., 1989] *Reprinted by permission from Nature Vol. 338, pp. 309–313, Copyright © 1989 Macmillan Magazines Ltd.*

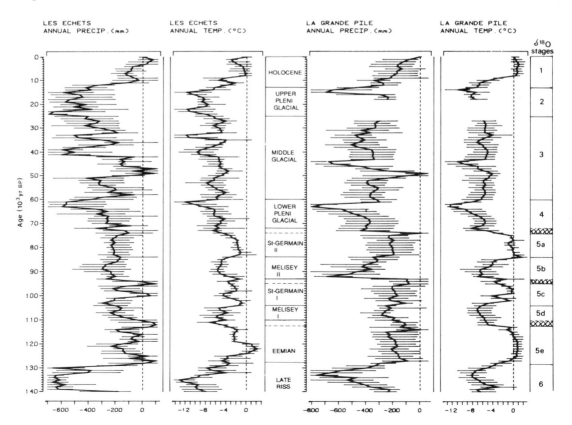

clearly leading CO_2 changes in the same record (Fig. 6.14). Temperatures on Antarctica and in the Southern Ocean reached near-glacial maximum levels by 110,000 BP.

Additional features of the initial ice growth interval involve tropical aridity and possibly deep-water temperature changes. Records of African aridity suggest that early ice buildup was a time of extreme aridity in northwest Africa but one in which moist conditions prevailed south of the equator (Pokras and Mix, 1985). In fact, these authors note that only during the last 30,000 years have aridity variations been synchronous across Africa; most of the time they are out of phase. Atlantic trade winds were relatively weak during these early growth periods (Hooghiemstra and Agwu, 1988).

It has been suggested that deep-water temperatures decreased by 1.0–2.0°C at the time of glacial inception (Chappell and Shackleton, 1986; Labeyrie et al., 1987). It is difficult to reconcile such decreases with decreased rates of oceanic mixing because geothermal heat flux from the ocean floor should have warmed deep waters (Mix and Pisias, 1988). This matter may require more study.

6.4.4 Modeling Glacial Onset

Since the development of the astronomic theory of glaciation (Milankovitch, 1930, 1941; cf. Chap. 7), climatologists have long conjectured that a seasonal cycle configuration with "cool summer" orbits would be very important, perhaps critical, for glacial inception (Köppen and Wegener, 1924). The rationale centered on the assumption that it is not possible to initiate ice cap formation if it gets too hot in the summer, and that cool summer orbits are needed to lower temperatures in critical areas of ice growth, such as northeastern Canada. A number of GCM sensitivity experiments indicate that the cool summer orbit at ~115,000 BP would lower temperatures 2–3°C (J. Royer et al., 1983, 1984; Rind et al., 1989).

It is not clear whether the calculated temperature changes are enough to initiate glacial inception. The only part of North America where these changes could depress values to freezing would be along the coast of Arctic Canada (Rind et al., 1989). This region is particularly sensitive

S. HEMISPHERE CLIMATE RECORDS

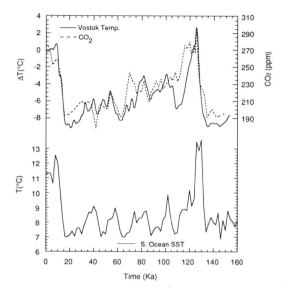

Fig. 6.14. Comparison of three high-latitude records from the Southern Hemisphere. Note that the Vostok (Antarctica) temperature and SST records reach full glacial values by 110,000 BP and that both records change before CO_2 fluctuations. [Vostok records plotted according to the revised chronology of Petit et al. (1990).] The CO_2 record therefore may not be the only factor causing sea ice-related SST changes in the Southern Ocean. [Vostok temperature records from Jouzel et al., 1987b; CO_2 record from Barnola et al., 1987; SST record after Martinson et al., 1987]

to climate change, as studies on Baffin Island suggest significantly expanded permanent snow fields during the Little Ice Age (Andrews et al., 1976; Locke and Locke, 1977). However, most of North America remains snow-free in other sensitivity experiments, even with expanded snow cover and more extreme changes in boundary conditions (Rind et al., 1989). The main problem is that the continent warms up too much in the summer to allow for the preservation of snow.

The GCM results contravene experiments with ice sheet models (with simpler geography and atmospheric models). These models suggest that ice sheets can grow under altered 115,000 BP boundary conditions (e.g., Andrews and Mahaffey, 1976; Hyde and Peltier, 1985). The above GCM results suggest either that (1) GCMs are not sensitive enough to altered boundary conditions; (2) there was much less ice than geologists concluded (a conclusion not supported by $\delta^{18}O$ records, cf. Ruddiman and McIntyre, 1982); or (3) the ice sheet grew vertically and could move south only after reaching elevations above the equilibrium snow line (cf. Fig. 7.18). The first possibility has considerable implications for validation of climate models (Rind et al., 1989), for they suggest that models may be too insensitive to change in forcing—a disturbing result when one considers that the same models are used for greenhouse doubling scenarios.

6.4.5 Carbon Dioxide Changes

CO_2 also decreased after the last interglacial (Fig. 6.14). Since the availability of a CO_2 record for the entire glacial cycle enables us to more thoroughly examine proposed mechanisms for the changes, this section will evaluate some of the different explanations.

Hypotheses as to the causes of CO_2 changes, and the phase relationships of CO_2 and ice volume, have passed through many stages in the last few years. Initially, Broecker (1982a,b) suggested that transfer of nutrients from eroding continental shelfs would raise the productivity level of the ocean, with the increased biomass in the surface waters acting as a "biological pump" to draw down atmospheric CO_2. Broecker (1974) had previously proposed that past productivity levels in the ocean could be monitored by measuring the differences in $\delta^{13}C$ between the surface and deep waters. At present this difference (Fig. 6.15) results from extraction of isotopically light carbon to build the soft tissue of plankton. The soft tissue would then be remineralized after the remains had fallen through the water column. Both of these processes would cause the observed surface-deep fractionation in $\delta^{13}C$.

Broecker (1974) reasoned that if more nutrients were available for photosynthesis, there would be a greater level of fractionation between surface and deep waters, and that by measuring past changes in the vertical $\delta^{13}C$ profile (termed $\Delta\delta^{13}C$), we could infer past changes in oceanic productivity. In 1982 Broecker coupled this ingenious idea with the observations of ice-age CO_2 decreases to propose the nutrient shelf hypothesis as the source of the CO_2 changes. However, past changes in whole-ocean nutrient levels appear to be too small to account for changes in the biological pump (Boyle and Keigwin, 1985/1986; Boyle, 1988a). Nevertheless, the $\Delta\delta^{13}C$ profile can still theoretically provide information about past changes in oceanic fertility.

Initial attempts to measure past $\Delta\delta^{13}C$ changes supported the concept of increased glacial level productivity (Shackleton et al., 1983) and furthermore suggested that ice-age productivity (and CO_2) changed before global ice volume and therefore CO_2 was contributing to cli-

Fig. 6.15. Profiles of ^{13}C content of dissolved inorganic carbon ($\delta^{13}C$), dissolved oxygen (O_2), and total dissolved inorganic carbon (ΣCO_2) from a typical midlatitude Pacific ocean station (17°S, 172°W). This figure, redrawn from Kroopnick et al. (1977), depicts a typical surface-to-bottom carbon isotope gradient in the marine environment. [From Hsü and McKenzie, 1985] *Courtesy of American Geophysical Union.*

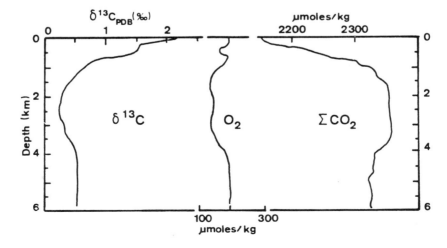

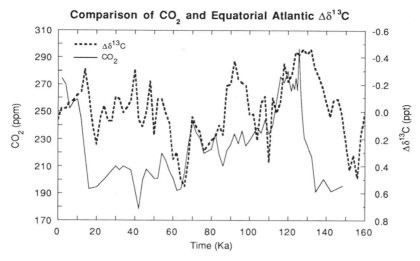

Fig. 6.16. Comparison of ice core CO_2 (Barnola et al., 1987) and oceanic equatorial Atlantic $\Delta\delta^{13}C$ record of marine productivity (Curry and Crowley, 1987) for the last 160,000 years. The former was plotted with the revised ice core chronology of Petit et al. (1990), the latter with the deep-sea chronology of Imbrie et al. (1984). Although one theory suggests that productivity changes should drive CO_2 variations, this figure demonstrates that the relationship is more complicated.

mate change (Shackleton and Pisias, 1985). However, a "cleaned" $\Delta\delta^{13}C$ does not agree well with observed CO_2 variations (Fig. 6.16; Curry and Crowley, 1987), so this conclusion may not be justified.

What, then, is responsible for the ice-age CO_2 decreases? The answer is unclear. For example, the cleaned $\Delta\delta^{13}C$ index illustrated in Fig. 6.16 does not correlate as well with the ice-core CO_2 record as measurements using samples that should theoretically not be as good as the cleaned samples (cf. Shackleton, 1988a). For some reason the latter may be better. There may also be sources of variance not adequately measured by the $\Delta\delta^{13}C$ index. For example, a different index of marine productivity (non-sea salt sulfate in ice cores) does not vary in the same manner as the illustrated $\Delta\delta^{13}C$ curve. This material is derived from dimethyl sulfide in marine waters, which varies with productivity level in the ocean (Bates et al., 1987). The compound is degassed to the atmosphere, converted to sulfate, and deposited in ice cores (Legrand et al., 1988a). If oceanic productivity is measured with this index (Fig. 6.17), then it increases over the same time interval (15,000–65,000 BP) that the $\Delta\delta^{13}C$ index illustrated in Fig. 6.16 decreases. To our knowledge, the origin of the different responses for the two indices is not understood.

A number of other factors could influence atmospheric pCO_2 levels. Some examples are:

· Changes in the organic carbon/carbonate rain ratio of biogenic debris falling out of the upper water column (Broecker and Peng, 1982; Dymond and Lyle, 1985). This ratio might change due to, for example, ecological changes associated with enhanced glacial productivity in the equatorial oceans (Sarnthein et al., 1988; Mix, 1989). The productivity would increase as a result of increased upwelling due to inferred stronger glacial trade winds (cf. Sect. 3.1.5). In the ocean, high-productivity regions are characterized by siliceous plankton whereas carbonate-secreting plankton are more common in low productivity regions. Thus, enhanced upwelling might cause a shift in the dominant phytoplankton group, which in turn would cause a change in the organic carbon/carbonate rain ratio. Less carbonate would be falling out of the upper water column. Increased carbonate ion concentration would increase the alkalinity of the surface waters (see text following). Since pCO_2 is very sensitive to alkalinity (cf. Bender, 1984; Sarmiento et al., 1988b), atmospheric CO_2 values would decrease.

· Enhanced utilization of nutrients in high-latitude surface waters (Ennever and McElroy,

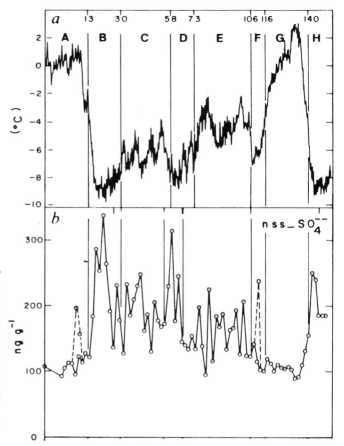

Fig. 6.17. A different index (cf. Fig. 6.16) of marine productivity plotted versus time (as calculated from the Vostok deuterium/temperature record). Variations with time of (a) Vostok isotope temperature record (in °C; from Jouzel et al., 1987b); (b), non-sea-salt sulfate (in ng/g), an index of marine productivity. Dating is indicated (in Ka) in the upper part of the figure. Note that in this record productivity is distinctly higher during the glacial interval 15,000–65,000 BP (cf Fig. 6.16). [After Legrand et al., 1988a] *Reprinted by permission from Nature, Vol. 334, pp. 418–420, Copyright © 1988 Macmillan Magazines Ltd.*

1985; Wenk and Siegenthaler, 1985; Toggweiler and Sarmiento, 1985). Not all nutrients are utilized in high latitudes because exchange rates with deep waters are fast enough to prevent equilibration between phytoplankton fertility and nutrient levels. Any change in production rates of deep water could therefore enable increased efficiency of nutrient utilization without affecting the whole-ocean nutrient reservoir. Atmospheric CO_2 levels would decrease accordingly.

· Changes in vertical fractionation in the deep-ocean circulation (Boyle, 1988b,c; Keir, 1988; Sarmiento et al., 1988b). There is at present (Fig. 6.18) a difference in the vertical profiles of ΣCO_2 and alkalinity in the ocean [alkalinity is a measure of the excess positive charge in the ocean that must be balanced by carbonate (CO_3^-) and bicarbonate (HCO_3^-) ions; pCO_2 of the atmosphere is in turn a function of ΣCO_2 and the alkalinity; as the alkalinity changes, so

too must the relative balance between HCO_3^- and $CO_3^=$, an adjustment which in turn affects $[pCO_2]_{aq}$; see Broecker and Peng, 1982, pp. 149–158]. Changes in deep- and intermediate-water production rates during the last glacial maximum (Section 3.1.8) would cause a redistribution of this profile (Boyle, 1988b,c). Older deep waters would increase ΣCO_2, which would result in enhanced carbonate dissolution on the seafloor. This response would increase alkalinity in the deep ocean. When the deep waters are returned to the surface in high latitudes, the altered ΣCO_2 and alkalinity ratios will result in a drop in atmospheric pCO_2. This mechanism requires several thousand years to cause a pCO_2 response (Boyle, 1988c) because it is constrained by the 3000- to 6000-year lag in response of carbonate sediments to altered deep-ocean ΣCO_2.

From the above discussion it is evident that a diversity of ideas exists as to the origin of ice-age CO_2 fluctuations. It should also be evident that the exercise involves some fascinating interactions between geochemistry, physical oceanography, and paleoclimates. Working on the carbon cycle on this time scale is like trying to piece together a giant puzzle for which some of the pieces are missing and some of the rules not thoroughly understood. But the progress that has been made over the last few years is impressive, and we are optimistic that a revisitation of this subject in a few years will indicate considerable advances over what has been presented above.

6.4.6   Main Glacial Stage (14,000–70,000 BP)

The ice growth event at about 75,000 BP ushered in the main phase of glaciation during the last cycle. Subsequent to that time the climate was mainly in a glacial mode. The pattern of warm SST in the subpolar North Atlantic (Section 6.4.3) also occurs for the major ice growth event at 75,000 BP (Ruddiman and McIntyre, 1981c). This time was associated with a massive transfer of carbon from the terrestrial reservoir to the ocean (Fig. 6.9). Approximately 40–50% of all the carbon stored in the terrestrial biosphere was involved in this transfer. A global dissolution event in deep-sea carbonates was approximately coincident with the carbon transfer (Crowley, 1983a).

In high latitudes of the North Atlantic, the main phase of glacial advance corresponded to significant equatorward expansion of the polar front (Ruddiman and McIntyre, 1976). Large SST changes occurred repeatedly in this area

(Fig. 6.19) over the last several hundred thousand years (Ruddiman and McIntyre, 1984). A major change in the rates of ice-rafted deposition (Ruddiman, 1977) also occurred after 70,000 BP in the subpolar North Atlantic.

In lower latitudes, temperatures stayed relatively stable (Fig. 6.19), especially in gyre centers (Crowley, 1981). The effect of equatorward migrations of cold polar waters against a central water mass that varied little is illustrated in Fig. 6.20. Repeated migrations of polar water resulted in glacial intervals of very large SST gradients, a feature that might have caused increased transport of warm Gulf Stream waters (and increasing SST; Fig. 6.21) in the central North Atlantic (Crowley, 1981).

Study of impurities in ice cores provide additional information about the main glacial phase. Sodium concentrations in the Vostok ice core (see Section 3.1.5) indicate that glacial age winds were significantly higher (DeAngelis et al., 1987), especially for the glacial maximum (Fig. 6.22). Acidity changes, which can be used as a proxy index of volcanic activity, do not correlate with the ice core temperature record (Legrand et al., 1988b)—a result different from correlations on time scales of 10^2 and 10^3 years (cf. Section 5.2.1).

The low-latitude monsoon also evolved through the glacial cycle. As discussed in Section 4.5.1, changes in the monsoon are strongly related to changes in the seasonal cycle of solar heating. Modeling studies (Prell and Kutzbach, 1987) suggest that the simulated time evolution of the monsoon over the glacial cycle is com-

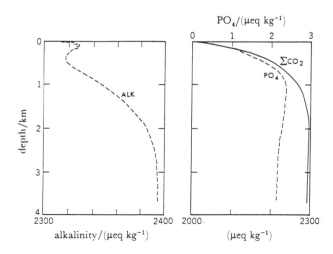

Fig. 6.18. Global average profiles of alkalinity, ΣCO_2, and dissolved phosphate in the ocean. Changes in the vertical profiles can affect atmospheric CO_2 levels. [From Sarmiento et al., 1988b] *Reprinted with permission of The Royal Society.*

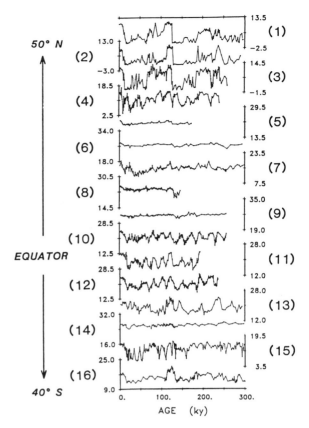

Fig. 6.19. Variations of foraminiferal estimates of sea surface temperature (cold season) versus age ranging from the high-latitude North Atlantic (core 1, top) to the South Atlantic (core 16, bottom). This transect illustrates the spatial variability of the SST response to climate change. [From Imbrie et al., 1989] *Reproduced by permission of Kluwer Academic Publishers, Copyright © 1989 D. Reidel Publishing Co.*

parable to the observed pattern, although the sensitivity of the system is somewhat less during glacial periods. Comparison of "simulated time series" of precipitation versus various proxy indices of monsoon fluctuations shows an overall good agreement and indicates that during interglacials summer insolation forcing is most important for strong monsoons, while during glacials, existence of the ice sheets both decreases and regionally alters the sensitivity.

Although the transition at 70,000 BP was marked by a general shift to more glacial conditions, there were some notable departures from this trend during ^{18}O Stage 3 (see Fig. 6.3; approximately 27,000–60,000 BP). This interstadial period is marked by both glacial and interglacial features. Sea level records (Chappell, 1983) suggest that a substantial amount of the ice formed at 70,000 BP had disappeared, for levels were only 30–40 m below present (Fig. 6.2). There were apparently four sea level (and Antarctic temperature) fluctuations during Stage 3 (Aharon and Chappell, 1986; Jouzel et al., 1989), with an average interval of 10,000

years—unusual for Pleistocene climate records (see Chap. 7). There is also some evidence for abrupt climate shifts during this interval (Section 6.3).

While the Laurentide Ice Sheet had diminished in size during Stage 3, it was still large enough to occupy the lowlands of the St. Lawrence River Valley (McDonald and Shilts, 1971; Andrews et al., 1983). The much reduced size of the Laurentide Ice Sheet also implies that there was inadequate time for the ice sheet at the glacial maximum to reach an equilibrium profile (Andrews, 1982) and that CLIMAP ice sheet reconstructions may be too high (cf. Section 3.1.1). Much warmer, interglacial level climates may have prevailed at the same time in western North America, eastern Europe, western Siberia, and Beringia (Dreimanis and Raukas, 1975; Hopkins, 1982). Thus, there may have been a very marked longitudinal asymmetry of warm and cold temperatures in high latitudes.

During early Stage 3 surface temperatures in the subpolar North Atlantic increased and NADW production rates were more interglacial in character (Boyle and Keigwin, 1982; Crowley, 1983a). Toward the end of the interstadial (28,000–30,000 BP) conditions cooler than the glacial maximum occur in a number of lower latitude cores in the North Atlantic (cf. Fig. 6.21; Crowley, 1981). This result may have some implications for the model–data discrepancies at 18,000 BP in the tropics (Sections 3.1.4 and 4.1.4). Rind and Peteet (1985) proposed that CLIMAP SSTs may be too warm by 2°C in order to explain tropical snowline depressions. Since some of these snowlines are not well dated and only assumed to be glacial maximum, it is possible that part of the discrepancy may be explained by chronology differences.

6.5 Additional Long-Term Trends in Pleistocene Records

6.5.1 Other Climate Records

Prior to the mid-1970s, most of the information on longer Quaternary climate records (i.e., older than about 150,000 years) was restricted to ocean sediments. Since that time there has been a small but growing number of long records from continental areas. This information comes from sources as diverse as pollen deposits in lakes (Japan, Spain, Greece, and Colombia), loess sequences in China and central Europe, lake sediments in western North America, and continentally derived material in deep-sea cores (e.g., Florschutz et al., 1971; van der Hammen et al., 1971; Kukla, 1977; Woillard, 1978; Rossignol-Strick, 1983; Hooghiemstra, 1984; Kanari et al., 1984; Smith, 1984; Janecek and Rea, 1985; Liu et al., 1985; Lorius et al., 1985; Pokras and Mix, 1985). Significant climate fluctuations are indicated in all these records. However, the level of climate inference is often impeded by a lack of satisfactory time control. A notable exception involves the 3.5 million year record from the Andes (Hooghiemstra, 1984).

Some of the best land records are the windblown loess deposits of China and central Europe. There are many similarities in the marine

Fig. 6.20. Diagram of time fluctuations of North Atlantic plankton over the last 150,000 years, illustrating that repeated advances against a central water mass that did not vary much in temperature (cf. Fig. 6.19) resulted in enhanced zonal temperature gradients in midlatitudes, which should have been translated into increased transport for the Gulf Stream system. [From Crowley, 1981] *Reprinted with permission from Elsevier Science Publishers.*

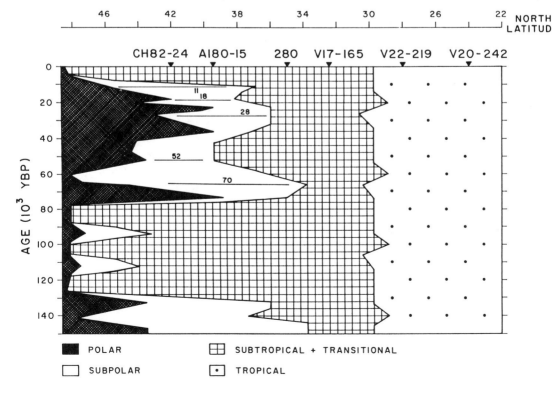

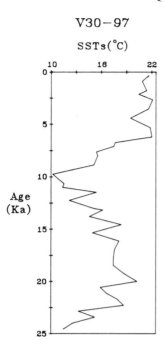

Fig. 6.21. Estimated summer sea surface temperature in core V30–97 from the northern subtropics near the southern edge of the glacial polar front. Note that SSTs increased from 25,000 to 18,000 BP. Similar changes have been found in some other North Atlantic records (Crowley, 1981). [From Ruddiman, 1987] *Reprinted with permission of Geological Society of America. From W. F. Ruddiman, 1987, in "North America and Adjacent Oceans During the Last Deglaciation," W. F. Ruddiman and H. E. Wright, Jr. (Eds.), The Geology of North America, vol. K-3, pp. 137–154.*

and land records. For example, the loess sequence of China records the same number of glacial-interglacial fluctuations as $\delta^{18}O$ in deep-sea cores (Fig. 6.1). In fact, it is quite easy to correlate between them. The same is true for loess records from central Europe (Kukla, 1977).

Interglacial soils in the loess sequence provide some clues as to the magnitude and nature of the warm events (Fig. 6.1). Variations in the level of weathering of interglacial soils in the Chinese sequence have been interpreted in terms of variations in the intensity of the summer monsoon (most of the precipitation in China occurs during the summer). For example, the last interglacial soil horizon suggests warmer temperatures and a more intense summer monsoon (Liu et al., 1985). Changes through time may reflect large variations in the intensity of monsoon forcing through time. Climate model

calculations suggest that maximum summer temperatures in Eurasia varied by 11°C during the last few hundred thousand years (Short et al., 1990).

Another indication of long-term variations in monsoons involves organic carbon-rich muds (sapropels) in deep-sea cores from the eastern Mediterranean. Accumulation of organic carbon is thought to be due to oxygen stagnation in bottom waters (e.g., Thunell et al., 1984). Since the stagnation may result from episodic intrusions of low-salinity water into the eastern Mediterranean (the low salinity prevents overturning of the water column), sapropel layers may be due to changes in Nile River outflow, which in turn is affected by variations in monsoon rainfall (e.g., Adamson et al., 1980; Rossignol-Strick, 1983). Numerous occurrences of sapropel layers may therefore indicate periodic increases in the intensity of the African/Asian monsoon (Fig. 6.23).

6.5.2 Regimes of Climate on Time Scales of Several Hundred Thousand Years?

The purpose of this section is to examine longer records in order to discuss evidence for possible significant decoupling between different components of the climate system. Particular attention will be given to differences between ice and other climate records. Some of these differences are so large as to suggest that the earth experienced several regimes of climate during the Pleistocene.

Evidence for divergences between climate and ice volume varies by region. In areas influenced by ice sheets (e.g., the subpolar North Atlantic), SST trends closely parallel the ice sheets. Most of the evidence for divergence is in regions somewhat distant from the cooling effect of the ice sheets. For example, Briskin and Berggren (1975) identified long-term changes in SST in the equatorial Atlantic that have a time scale of about 400,000 years (Fig. 6.24). Significantly lower SSTs between 450,000 and 600,000 BP found in this core also occur in the equatorial Atlantic, Caribbean, and possibly the South Atlantic (Ruddiman, 1971; Imbrie and Kipp, 1971; Lohmann, 1978; Embley and Morley, 1980).

The lower SSTs in the early Brunhes (cf. Fig.

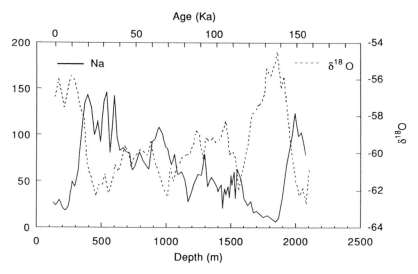

Fig. 6.22. Comparison of the marine salt (Na_m) and ^{18}O records in the Vostok ice core. The marine salt record has been interpreted in terms of changes in wind over the ocean. [Redrawn from DeAngelis et al., 1987] *Reprinted by permission from Nature Vol. 325, pp. 318–321, Copyright © 1987 Macmillan Magazines Ltd.*

6.24) may be responsible for generally cooler climate in the northern Andes at this time (Hooghiemstra, 1984). A long record of lake level fluctuations in the Great Basin of western North America shows time scale fluctuations similar to those of the Atlantic record (Smith, 1984). There are also long-term changes in global carbonate preservation in deep-sea sediments (Hays et al., 1969; Crowley, 1985; Peterson and Prell, 1985).

A number of other paleoclimatic transitions also occurred in the mid-Brunhes (350,000–450,000 BP; Jansen et al., 1986). Ice volume lev-els of interglacial events may have been greater than earlier interglacial events (Fig. 6.4), a conclusion supported by direct measurements on European shoreline fluctuations (Hearty et al., 1986). The $\delta^{18}O$ differences translate into approximately 20 m of sea level equivalent ice volume change. Ice cover in the Artic Ocean was also more persistent after ~400,000 BP (Scott et al., 1989).

Long-term climate trends are not uniform in sign. Although equatorial Atlantic SSTs were generally lower prior to 450,000 BP, northern Pacific SST were higher (Sancetta and Silvestri,

Fig. 6.23. Comparison of freshwater events (black bars) in the eastern Mediterranean with the generalized open-ocean oxygen isotopic record for the last 350,000 years (the latter after Emiliani, 1966). The standardized late Quaternary organic-rich mud (sapropel) stratigraphy (after Cita et al., 1977). Note that only sapropel S_6 is associated with a glacial stage. [From Thunell et al., 1983] *Reproduced by permission of The Cushman Foundation.*

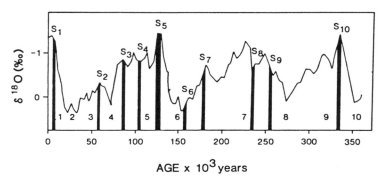

VI6 - 205

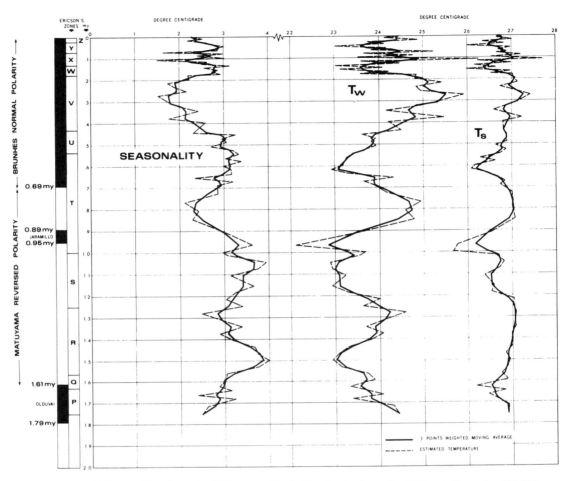

Fig. 6.24. An SST record for the entire Pleistocene, illustrating the tendency for SST variations at ~400,000-year periods. Graph shows the faunal indices for winter temperature (T_w), temperature (T_s), and seasonality ($T_s - T_w$), plotted versus time. The dashed line represents the estimated temperature and the solid line is a three-points-weighted moving average of the estimated temperatures. Paleomagnetic boundaries and foraminiferal faunal zones are inserted near the time scale in millions of years. Four roughly symmetric climatic cycles are detected in T_w. [From Briskin and Berggren, 1975] *Reprinted by permission of American Museum of Natural History.*

1986). The latter data agree with loess records from both China and central Europe, which also indicate different conditions prior to 450,000 BP (e.g., Kukla, 1977; Liu et al., 1985). In China, the older regime is dominated by spores and pollen, while herbs and shrubs are more common in the younger layers (Liu et al., 1985). This trend can be interpreted in terms of generally drier conditions in the younger interval. A similar boundary in central Europe (Kukla, 1977) may separate older interglacial soils of Mediterranean types from younger soils typical of inter-

glacial conditions presently occurring in Europe.

There are also some notable faunal transitions that occur near the mid-Brunhes regime shift. Faunal transitions occur in European vertebrates (Kukla, 1977) and for several types of marine plankton—coccoliths (Thierstein et al., 1977), radiolaria (Hays and Opdyke, 1967; Sachs, 1973), and diatoms (Sancetta and Silvestri, 1986). There may also have been changes in deep-water circulation. In the eastern South Atlantic, a common species of benthonic foraminifera is found only in samples from the younger regime (Embley and Morley, 1980).

In summary, there is fairly widespread evidence of approximately synchronous changes in climate that occur on a time scale of several hundred thousand years. The origin of the changes is not clear. Kukla (1977) notes that the transition in the European loess sequence occurs near a time of increased Alpine uplift, and suggests that tectonics may play a role. However, the existence of 400,000-year power in some of the records (Briskin and Harrell, 1980; Moore et al., 1982) implicates eccentricity variations as partially responsible (cf. Chap. 7); Berger (1976) noted that the most important term in the series expansion for eccentricity occurs at this period.

Limited information is available about the amount of variance in Pleistocene climates that is attributable to the postulated regimes. Moore et al. (1982) calculate that 400,000-year variations in carbonate dissolution account for about 50% of the variance in Pleistocene dissolution. Regimes of Pleistocene climate may therefore represent a major feature of past climate.

There is evidence for regime-type climate changes at other intervals of the late Quaternary. At ~250,000–300,000 BP, there was a shift to warmer conditions in the equatorial Atlantic (Ruddiman, 1971), decreased wind variations over the North Pacific (Janecek and Rea, 1985), and changes in upwelling patterns and a 5° northward shift of the Intertropical Conver-

gence Zone in the eastern equatorial Pacific (Schramm, 1985; Rea et al., 1990). The North Pacific cooling initiated at about 450,000 BP became more severe at this time (Sancetta and Silvestri, 1986), and there was a significant reduction in accumulation rates of planktonic foraminifera in Baffin Bay (Mudie and Aksu, 1984).

A regime shift at 900,000 BP included both SST and ice (e.g., Ruddiman et al., 1986b). Large faunal changes occurred around 900,000 BP during the transition to large ice volumes and 100,000-year cycles. This event in Europe is noted as the time of transition from the Villafranchian to Biharian fauna (e.g., Kukla, 1977). Since the faunal changes are some of the most significant within the Pleistocene, regime shifts appear to play a significant role in evolution on this time scale.

6.6 Summary

A considerable amount of work has been done on climate fluctuations prior to ~20,000 BP. Some of the more interesting results involve indications for bistable climates and regimes of climate. Modeling studies are limited, but results suggest that it is difficult to understand how glacial inception began at 115,000 BP. A substantial amount of work remains before we can understand Pleistocene CO_2 changes.

7. TIMES SERIES ANALYSIS OF PALEOCLIMATE RECORDS

As illustrated in Chap. 4, significant progress has been made in understanding the equilibrium climate response to altered boundary conditions. These experiments do not address the time evolution of climate, especially processes responsible for formation and destruction of ice sheets. This entirely separate problem is one of fundamental interest in climate dynamics and can be examined in two separate ways, one by examining fluctuations of geologic records as a function of time, the other by examining the same fluctuations in the frequency domain. This latter method employs basic statistical techniques of time series analysis and can provide very powerful insight into the dynamics of climate change. In this chapter we review the principal findings from spectral analysis for both the Pleistocene and pre-Pleistocene. The reader is referred to Appendix C for a brief description of some terms used in time series analysis and in this chapter.

Substantial progress has been made in understanding the evolution of climate in the frequency domain. Much of this progress is either directly attributable to or inspired by a key paper linking ice volume fluctuations to orbitally induced changes in solar insolation—the astronomical theory of glaciation. Before discussing this paper, however, it is necessary to outline the general features of the astronomical theory of glaciation.

7.1 Nature of Astronomical Theory

7.1.1 Historical Background

As discussed by Imbrie and Imbrie (1979) in a historical review of the ice-age problem, the astronomical theory of glaciation was first proposed in 1842 by the French mathematician Joseph Adhémar and was more fully developed by the Scottish geologist James Croll (1864, 1867a,b). Following a period of heightened interest, the subject fell into disfavor. It was resus-citated by the Serbian astronomer Milutin Milankovitch. During World War I he was captured by the Austro-Hungarian army and eventually released, but was confined to Budapest for the duration of the war. During his 4-year stay, Milankovitch utilized principles from celestial mechanics to calculate the effect of planetary perturbations on the geographic receipt of solar insolation (1930, 1941). He consulted two other famous scientists—the geographer Köppen and the geologist Alfred Wegener (of continental drift fame)—to develop his postulate that decreased summer insolation in northern high latitudes was the most critical variable for glacial inception. The latter incorporated Milankovitch's earlier work into their investigations of past climate fluctuations (Fig. 7.1). The Köppen–Wegener work sparked more interest in the problem, but again it fell into disfavor because it did not seem to be supported by geologic data. Even as late as 1961, A. Nairn, editor of the first modern book on paleoclimatology, wrote that "the effect of the varying distance between the earth and the sun from perihelion to aphelion, the basis of Croll's theory of the origin of ice ages, is not now thought to be a significant factor in climate."

A third wave of interest in the Milankovitch problem started in the late 1960s, when Wallace Broecker and John Imbrie introduced new data providing more support for the astronomical theory of glaciation (Broecker, 1966; Mesolella et al., 1969; Imbrie and Kipp, 1971). However, it was not until 1976 that the orbital theory became much more widely accepted. This was a result of a landmark paper (Hays et al., 1976b) demonstrating statistically significant ice volume fluctuations at all the orbital frequencies.

7.1.2 Primary Orbital Periods

Gravitational effects of planetary bodies cause orbital perturbations that periodically vary the geographic distribution of incoming solar radi-

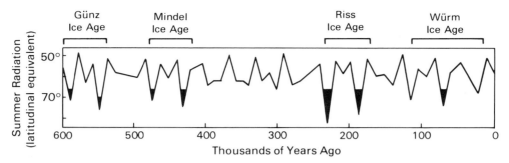

Fig. 7.1. Comparison of the Milankovitch radiation curve for 65°N with Alpine glacial advances. Köppen and We-gener (1924) suggested that the main Alpine glacial advances corresponded with low points in the orbital insolation curve. Although it is now known that there are many more than four glacial advances, and that the chronology of the four glacial events may be in error, the four time periods identified in this figure actually correspond to the largest amplitude fluctuations of the ^{18}O record. [Redrawn from Köppen and Wegener, 1924]

ation (Fig. 7.2). There are three main perturba-tions, with five primary periods (Fig. 7.3): (1) The eccentricity of the orbit varies between near-circularity and slight ellipticity ($e \approx 0.06$ max) at periods of about 100 KY and 400 KY (period in thousand years). The most important

term in the series expansion for eccentricity oc-curs at 413 KY (e.g., Berger, 1976); (2) The tilt (obliquity) of the earth's axis varies between about 22 and 25° at a period of about 41 KY. Obliquity perturbations tend to amplify the sea-sonal cycle in the high latitudes of both hemi-

Fig. 7.2. Schematic diagram illustrating effect of planetary forces on the earth's axis and orbit. These forces cause changes in the eccentricity or ellipticity of the orbit **(a)**, the tilt of the rotational pole (ϵ), and the gyroscopic spin of the planet (precession). The precession effect is illustrated more fully in Fig. 7.4. [Modified from Vernekar, 1968]

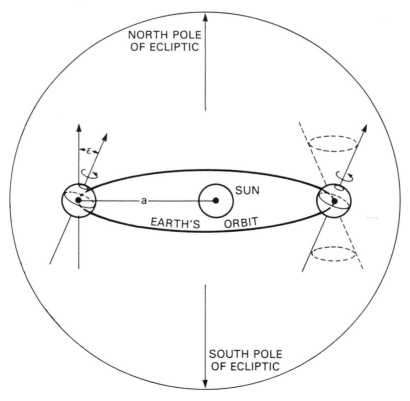

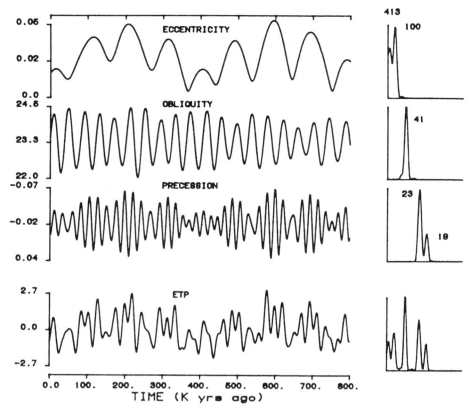

Fig. 7.3. Variations in eccentricity, obliquity (tilt), and precession index ($\Delta e \sin \omega$) over the past 800,000 years. Left: the three upper time series are from the work of Berger (1978). These have been normalized and added to form the curve labeled ETP. The scale for obliquity is in degrees; for ETP, in standard deviation units. Right: variance spectra calculated from these time series, with the dominant periods (KY) of conspicuous peaks indicated. [From Imbrie et al., 1984] *Reproduced by permission of Kluwer Academic Publishers, copyright (©) 1984 D. Reidel Publishing Co.*

spheres simultaneously. The strength of the effect is small in the tropics and maximum at the poles.

The third effect is precession, which changes the distance between the earth and the sun at any given season. Precession (Fig. 7.4) has two components: (1) axial precession, in which the torque of the sun and planets on the earth's equatorial bulge causes the axis of rotation to "wobble" like that of a spinning top. The net effect is that the North Pole describes a circle in space (Fig. 7.4a) with a period of 26,000 years; (2) elliptical precession (Fig. 7.4b), in which the elliptical figure itself is rotating about one focus. The two effects together result in what is known as the "precession of the equinoxes," in which the equinox (March 20 and September 22) and solstice (June 21 and December 21) shift slowly around the earth's orbit, with a period of 22,000 years. This term is modulated by eccentricity

which splits the precession frequency. The periods of the modulated effect are 19,000 and 23,000 years. These are the expected periods that would actually be recorded in the record.

The precession effect can cause warm winters and cool summers in one hemisphere while doing the opposite across the equator. The strength of the effect is largest at the equator and tapers off smoothly toward the poles. Presently, perihelion (the point of closest approach of the earth to the sun) occurs in northern winter (January; cf. Fig. 7.4c). Thus portions of the winter hemisphere receive as much as 10% more insolation than they will 11,000 years later, when perihelion occurs in the summer.

7.1.3 Effect on Planetary Energy Balance

The orbital perturbations have a significant effect on planetary radiation receipt. The precession effect causes a latitudinal and seasonal redistribution of solar radiation at the top of the atmosphere. This effect is most important at 0–60°N (Fig. 7.5) and results in large changes in the seasonal insolation forcing which cancel out in the global annual mean (however, the presence or absence of high-albedo ice could make some difference in energy absorbed). For example, over the last 100,000 years, midlatitude Northern Hemisphere insolation values at summer solstice have varied by about 8% around the mean ($\sim$40 W/m^2; cf. Fig. 4.15). For comparison, the radiative forcing from a doubling of

CO_2 is about 4 W/m^2 (however, CO_2 forcing is year-round and sensitivity to mean annual forcing is several fold greater than sensitivity to seasonal forcing; cf. Fig. 1.11). Over longer time intervals, values at the summer solstice exceed 13% of the mean. Any increase in summer insolation is offset by an equivalent decrease in winter insolation, so there is no change in the annual mean insolation receipt due to precession.

It is often assumed that there is also no annual change in the insolation receipt at the tilt frequency, which dominates in polar latitudes.

Fig. 7.4. Components of precession: (a) axial precession akin to that of a spinning top; (b) precession effect due to changes in elliptical orbit; (c) combined effect of the two results in a slow shift of the equinox through the earth's elliptical orbit. [From Pisias and Imbrie, 1986/87] *Courtesy of Oceanus Magazine (©) 1986 by Woods Hole Oceanographic Institution.*

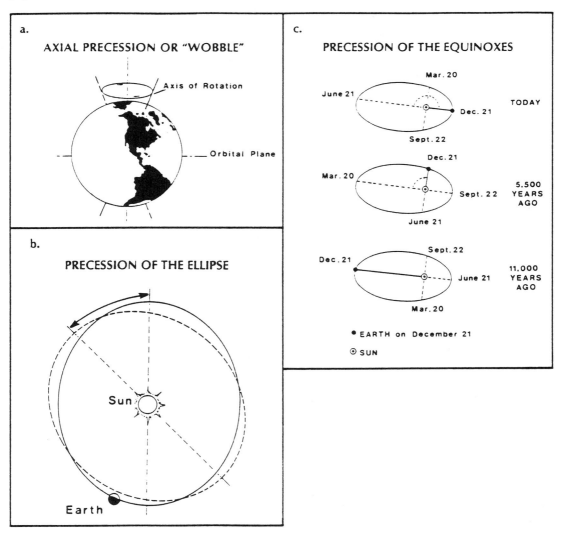

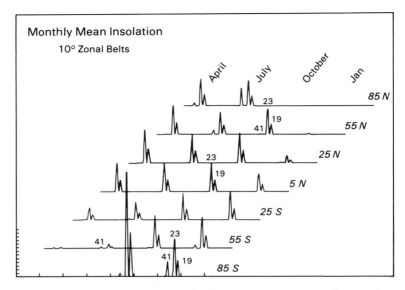

Fig. 7.5. Spectral analysis of mid-month insolation for four seasons and seven latitudes. The vertical scale is in relative units and is proportional to the amount of variance explained by each frequency band (periods given in KY). [From Berger and Pestiaux, 1984] *Reproduced by permission of Kluwer Academic Publishers, copyright © 1984 D. Reidel Publishing Co.*

This is not quite the case. In polar regions, an increase in summer insolation cannot be offset by a decrease in winter insolation because insolation receipt is already very low in the winter (polar night). The net annual change in insolation receipt increases toward the poles and can reach maximum values of 17 W/m². This value is large enough to have a significant climate effect.

Eccentricity variations produce a maximum change of about 0.2% in solar insolation receipt. This number translates to about 0.7 W/m² at the top of the atmosphere. With a 30% albedo (Chap. 1), the net effect on the planetary radiation budget is about 0.5 W/m². This number is about one-third of the radiative forcing from ice-age CO_2 changes (which with feedbacks cause about a 1–2°C change in global average temperature; cf. Chaps. 3 and 4). The maximum eccentricity effect (with feedbacks) will therefore be on the order of 0.5°C.

The surface temperature response to orbital insolation variations is a function of land–sea distribution (Short et al., 1990). Over the open ocean, precession effects are generally small because the large heat capacity of water suppresses any seasonal temperature change. The lower heat capacity of land could cause larger changes

in the seasonal cycle of temperature. Since the greatest area of land occurs in the northern mid- and high latitudes, the thermal response to Milankovitch 23-KY forcing is greatest at these latitudes (Fig. 7.6).

There is also a hemispheric asymmetry to the polar response to obliquity forcing. In higher latitudes, net temperature changes can occur over water as a result of net annual changes of tilt-induced radiation (see text preceding). Even though the tilt effect is synchronous in both hemispheres, the response is greater in the Southern Hemisphere (Fig. 7.7) because there is more land in the Southern Hemisphere at highest latitudes (Short et al., 1990).

7.2 Observational Studies

7.2.1 Documentation of Milankovitch Forcing in the Geologic Record

In a landmark paper, Hays et al. (1976b) studied geologic time series from the southern Indian Ocean and for the first time demonstrated that (1) significant spectral peaks in the ice volume record occurred at values of 19 KY, 23 KY, 41 KY, and 105 KY; (2) the 105-KY contribution to the total variance far exceeds that expected from a simple linear relationship between insolation and ice volume; and (3) there is a fairly consistent phase relationship between insolation, SST, and ice volume—each preceded the next by 2–4 KY.

Although the phase lead between SST and ice volume has been questioned (Howard and Prell, 1984), subsequent work demonstrated that it is real (Imbrie et al., 1989; Morley, 1989). The significance of the 100-KY cycle in $\delta^{18}O$ has also been amply repeated in a number of other deep-sea cores (Fig. 7.8). Its phase locking with tilt and precession (Fig. 7.9), even to the point of matching variable amplitudes of peaks, indicates a primary orbital influence on ice volume fluctuations (Imbrie et al., 1984). Additional evidence has been added for significant 400-KY power in some SST and carbonate records (e.g., Briskin and Harrell, 1980; Moore et al., 1982). Berger (1976) showed that the most important term in the series expansion for eccentricity oc-curs at this period. Thus, all the significant periods of orbital variation have been found in geologic time series, and there is a large amount of power at the 100-KY period. Other work in the last decade has unearthed a multitude of evidence for a strong astronomical influence on climate fluctuations (e.g., Berger et al., 1984).

7.2.2 Strategy for Analyzing the Geologic Record

Since the establishment of an orbital connection, research efforts have shifted to two primary topics: analysis of geologic time series to determine how the orbital signal is translated into a climate response of glacial magnitude, and modeling the time evolution of Pleistocene ice sheets. Most of the remaining part of this chapter will focus on specifics of these two thrusts.

Given the evidence for Milankovitch cycles in the geologic record, it is possible to develop a

Fig. 7.6. Effect of precession variations on the earth's surface temperature (northern summer minus northern winter) based on energy balance model calculations. This figure illustrates the differences in maximum summer temperatures for the two extreme orbital configuration for the Pleistocene that are due to eccentricity modulation of the precessional forcing. Because of lower heat capacity, land areas are much more sensitive than water to seasonal changes in precession forcing. [From Short et al., 1990] *Reprinted with permission of Quaternary Research.*

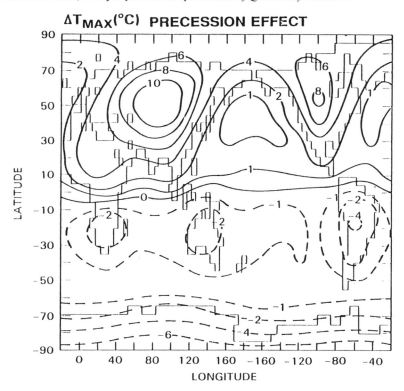

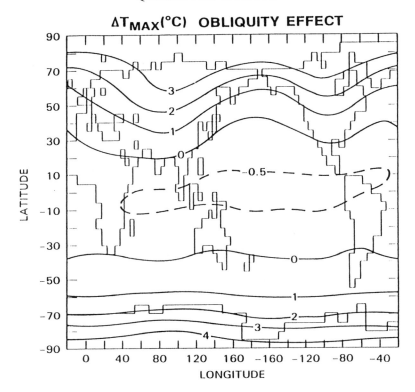

Fig. 7.7. Effect of maximum and minimum tilt (obliquity) variations on maximum summer temperature as calculated from energy balance model. Note that tilt is most important at high latitudes and affects both land and water (cf. Fig. 7.6). [From Short et al., 1990] *Reprinted with permission of Quaternary Research.*

strategy to determine how the complex climate system incorporates orbital insolation variations (e.g., Imbrie et al., 1989). One strategy is to partition the total climatic response into frequency components whose phase (and amplitude) can be compared at different sites with the geographic pattern of radiative forcing. Examination of the climate spectra enables us to assess both linear and nonlinear responses to radiation variations. Analysis of phase relationships can provide information about the response time of different variables and their functional relationship (e.g., whether one might be forcing the other). Analysis of the geographic distribution of the response provides information about which regions play the most critical roles in climate change. Taken together, this information can provide great insight into processes responsible for climate change on glacial-interglacial time scales. This research effort has been the goal of the SPECMAP group (Spectral Mapping Project), and much of what will be discussed in

the following sections is either directly attributable to this group or inspired by their effort.

7.2.3 High-Latitude Response to Orbital Forcing

The first area we will examine involves climate fluctuations in the North Atlantic sector—the main area of ice sheet growth. Analysis of geologic time series (cf. Fig. 6.19) from the subpolar North Atlantic (Fig. 7.10) indicates that SST fluctuations, a monitor of the oceanic polar front, varied with a 41,000-year period, a response consistent with a dominant influence by the high-latitude tilt signal (Ruddiman and McIntyre, 1984). Farther south, along the latitude of the southern terminus of the Laurentide Ice Sheet (about 45°N), SST varied on a 23-KY period (Fig. 7.10). SST in this region was presumably affected by 23-KY precession fluctuations, which are more important in low latitudes. The precise physical mechanisms

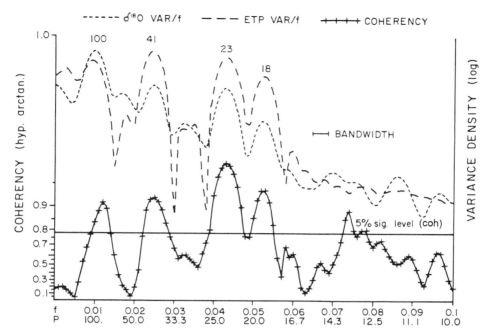

Fig. 7.8. Coherency and variance spectra calculated from records of climatic and orbital variation spanning the past 780,000 years. Two signals have been processed: (1) ETP, a signal formed by normalizing and adding variations in eccentricity, obliquity, and precession (cf. Fig. 7.3); and (2) a composite $\delta^{18}O$ record. Top: variance spectra for the two signals are plotted on arbitrary log scales. Bottom: coherency spectrum (cf. Appendix C) plotted on a hyperbolic arctangent scale and provided with a 5% significance level. Frequencies are in cycles per thousand years. [From Imbrie et al., 1984] *Reproduced by permission of Kluwer Academic Publishers, copyright © 1984 D. Reidel Publishing Co.*

responsible for SST fluctuations are not completely understood, but may partially involve meltwater outflow during ice decay phases of peak summer insolation in precession. These "cold spikes" at very regular intervals result in very strong power at the 23-KY period. Ruddiman and McIntyre (1981c) further suggest that

variations in ocean heat transport may also have been important at the 23-KY period.

Analysis of the 23-KY record at 45°N indicates that SST lagged ice volume by about 8000 years (Ruddiman and McIntyre, 1981c). This lag is partly due to the occurrence of coldest SST (meltwater) during deglaciation and partly due

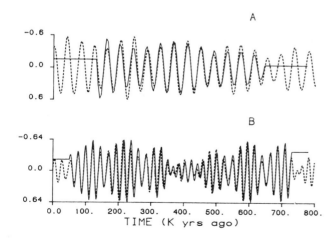

Fig. 7.9. Variations in obliquity, precession, and the corresponding frequency components of $\delta^{18}O$ over the past 800,000 years. Dashed lines are phase-shifted versions of obliquity (A) and precession (B) curves. Solid lines are filtered versions of the stacked $\delta^{18}O$ record plotted on the SPECMAP time scale. Note that the filtered $\delta^{18}O$ record tracks closely in amplitude the variations of the forcing. The very high coherence (cf. Appendix C) is strong support for the orbital theory of the ice ages. [From Imbrie et al., 1984] *Reproduced by permission of Kluwer Academic Publishers, copyright © 1984 D. Reidel Publishing Co.*

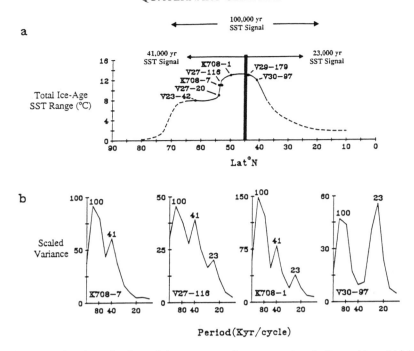

Fig. 7.10. Variance density plots of estimated August sea surface temperature in four mid- and high-latitude North Atlantic cores. The 100,000-year rhythm is strong in all regions; the 41,000-year power is strongest in the three northern cores but yields to 23,000-year power in midlatitude core V30-97. [From Ruddiman and McIntyre, 1984] *Reprinted with permission of Geological Society of America from "Ice-age thermal response and climatic role of the surface Atlantic Ocean, 40°N to 63°N," Ruddiman, W. F. and A. McIntyre, 1984, Geol. Soc. Amer. Bull. 95, 381–396.*

to warm conditions in the subpolar North Atlantic during glacial growth. Such a pattern could cause an oceanic feedback for ice volume fluctuations (Ruddiman and McIntyre, 1981c). Open water during glacial growth might enhance moisture flux (snow) to the ice sheets. Low SST during glacial decay would suppress moisture supply and therefore provide a positive feedback.

Milankovitch cycles can also be found in mid- and high-latitude pollen records from western Europe and Japan, eolian sediments from Alaska and eastern Asia, fluctuations of the North Pacific subarctic front and the North Atlantic Deep Water (NADW), and Antarctica (Kanari et al., 1984; Molfino et al., 1984; Pisias and Leinen, 1984; Janecek and Rea, 1984; Boyle and Keigwin, 1985/1986; Jouzel et al., 1987b; Begét and Hawkins, 1989). Over the North Pacific, a long time series of eolian transport from interior eastern Asia records significant fluctuations at the 100-KY and 41-KY periods. Janecek and Rea (1984) suggest that the 41-KY fluctuations result from tilt variations

that cause changes in equator-to-pole temperature (and hence wind) gradients. Young and Bradley (1984) note that times of ice buildup coincide with times of strong insolation gradients between low and high latitudes, and vice versa for ice decay. These gradients may affect atmospheric transport.

In addition to a 41-KY signal found in the above high-latitude record, a pollen series from Grande Pile, France (Woillard, 1978) yields spectra dominated by 23-KY precession power (Molfino et al., 1984). There are also indications of significant higher frequencies, with the most common being around 8.8–9.2 KY (Fig. 7.11). These peaks may be harmonics of precession forcing.

A particularly interesting series of analyses indicate that NADW also fluctuated at 41-KY periods (Boyle, 1984; Boyle and Keigwin, 1985/1986), with NADW lagging obliquity by about 8000 years. This conclusion is supported by analysis of benthic $\delta^{13}C$ records, which also provide information about deep water (Mix and Fairbanks, 1985), and which further indicate

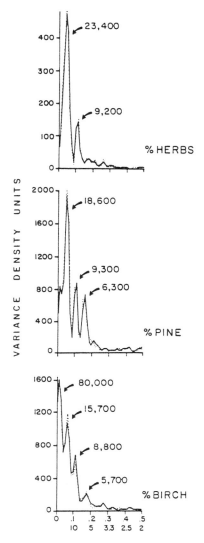

Fig. 7.11. High-resolution spectra of pollen relative abundance data from Grande Pile, France. Note the presence of peaks (harmonics?) shorter than the orbital periods. [After Molfino et al., 1984] *Reproduced by permission of Kluwer Academic Publishers, copyright © 1984 D. Reidel Publishing Co.*

that NADW fluctuations are in phase with the passage of the North Atlantic polar front. These linkages are apparently due to open water in the subpolar North Atlantic during ice growth (see text preceding), which causes the 8000-year phase shift. Subsequent advance of the polar front may reduce NADW formation due to imposition of a low-salinity lid on the surface (this tends to suppress convection).

Variations in NADW may further have af-

fected climate by changing the export of heat across the equator into the Southern Ocean. Crowley and Parkinson (1988b) calculated that such changes could account for about one-fourth of glacial-interglacial fluctuations in Antarctic sea ice. Cross-correlation of NADW and Southern Ocean SST records supports this calculation, since only about 20% of the variance of the two records are coherent and in phase (Crowley and Parkinson, 1988b). This comparison furthermore suggests that the predominant NADW–Antarctic connection is at the 41-KY period.

7.2.4 Low Latitudes

In lower latitudes, climate fluctuations have more 23-KY power on both land and sea. In the Arabian Sea an upwelling index related to the Asian monsoon has 23-KY fluctuations that lag precession by about 5000 years (Prell, 1984b). As the monsoon should respond virtually instantaneously to precessional forcing (Sect. 4.5.1), a full explanation for the origin of this lag is missing, although part of the lag in equatorial regions might be explained by the twice-yearly crossing of the equator by the sun (Short and Mengel, 1986; cf. Section 4.5.2).

An index of African aridity has been developed based on abundances of African freshwater diatoms in equatorial Atlantic sediments (diatom abundances increase during aridity as a result of lake evaporation and wind deflation). There are diatom fluctuations off northwestern Africa with clear 23-KY power, and also significant harmonics (Pokras and Mix, 1987). As discussed in Section 4.5.1, monsoon fluctuations should have a strong 23-KY signature. However, part of the 10 KY precession peak may result from the twice-yearly passage of the sun across the equator and may not be a true harmonic in the strict sense (i.e., a nonlinear effect; Short et al., 1990). The other harmonics may reflect a nonlinear response of erosion to monsoon fluctuations as opposed to nonlinearity in the monsoon itself. Pokras and Mix (1987) suggest that primary input of diatoms to sediments occurs during times of lake lowering, when the diatom deposits are initially exposed to erosion. Such a spiked input should produce harmonics.

In the eastern equatorial Atlantic, oceano-

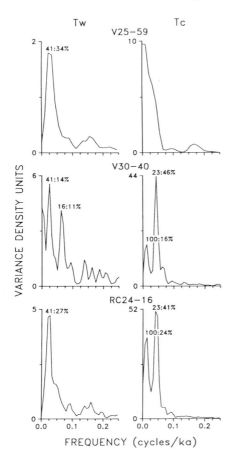

Fig. 7.12. Variance spectra of SST from a west–east transect in the equatorial Atlantic. Dominant periods in KY of significant peaks (and variance explained) are indicated. T_c = cold season estimates; T_w = warm season estimates. Note the occurrence of strong 23-KY and 41-KY power in the equatorial records. [From McIntyre et al., 1989] *Courtesy of American Geophysical Union.*

graphic changes also fluctuated with a 23-KY rhythm. McIntyre et al. (1989) found very strong 23-KY power in a foraminifera record from this region (Fig. 7.12). The signal may record 23-KY fluctuations of SST due to variations in advection of cool water from the higher latitude South Atlantic (McIntyre et al., 1989) and also local changes in the depth of the thermocline as a result of variations in trade wind-induced Ekman pumping at precession periods. There is also 41-KY power in this series (Fig. 7.12), a response that may reflect a tilt peak at the equator (Short et al., 1990).

The eastern equatorial Pacific records are different from equatorial Atlantic records.

Whereas a strong 23-KY signal is found in Atlantic records, the spectrum in the Pacific is more complicated—equatorial divergence and wind records have primary power at 31 KY (Pisias and Rea, 1988). This unusual period may be a cross-product of precession and tilt (Pisias and Rea, 1988). The 31-KY linkage also provides direct evidence for a significant correlation between grain size variation (a function of wind speed) and productivity in the equatorial Pacific.

7.2.5 Atlantic Synthesis

Spectral analysis of a number of Atlantic time series (Fig. 6.19) illustrates the spatial dependence of the frequency response of SST records (Fig. 7.13). When expressed in the frequency domain, most of the climate records show distinct concentrations in the primary Milankovitch bands (Imbrie et al., 1989). The largest SST response (longest vectors in Fig. 7.13) occurs near 50°N in the North Atlantic and reflects passages of the Gulf Stream/North Atlantic Current and North Atlantic polar front. In the obliquity band, there is relatively little response in low latitudes, except in a few records near the equator. In the precession band there is a suggestion of a phase shift across the equator; South Atlantic temperature changes lead oxygen isotopes (ice volume) and North Atlantic responses are either in phase or lag ice volume. The Southern Hemisphere phase lead also occurs in the 100-KY band (Fig. 7.13). With the exception of one site, all SST phases at 100 KY are within 45° (12,000 years) of ice volume and eccentricity maxima. The above results are consistent with the findings of Hays et al. (1976b) for two South Atlantic cores, but the new SPECMAP findings provide much more information about the regional coherence of SST response.

7.2.6 Carbon Cycle

There is a growing body of information available as to fluctuations of various components of the carbon cycle. For example, the ice core CO_2 record has significant power in Milankovitch bands, especially at ~21 KY. As Broecker (1982a,b) discussed (see also Section 6.4.5), the ice core CO_2 record may require explanations in

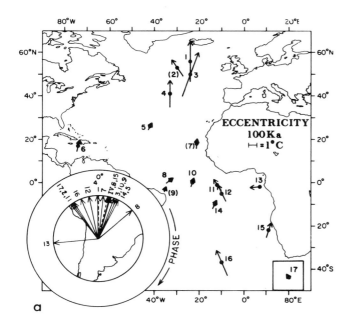

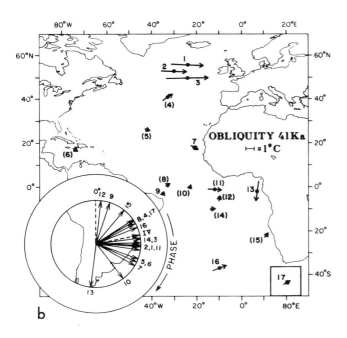

Fig. 7.13. SPECMAP analysis of Atlantic SST records, illustrating Milankovitch cycles of cold season sea-surface temperature at 16 sites in the Atlantic (cf. Fig. 6.19) for (a) eccentricity; (b) obliquity; and (c) precession. The figures also illustrate the "SPECMAP phase wheel," which should be read as follows. 12 o'clock on each of the diagrams illustrates when Northern Hemisphere is on the "hot end" with respect to orbital forcing. Arrows on the diagrams illustrate the phase relationship of the SST records with respect to this forcing. For example, a 41-KY record at 3 o'clock lags forcing by 90°, which is equivalent to about 10,000 years (90° is ¼ of 360° and ¼ of 41,000 equals approximately 10,000 years). The coherent amplitude of each cycle in °C is indicated by the length of the arrow, according to the scale shown. The phase is indicated by the direction of the arrow. Vectors plotted in the phase wheel represent the phase of sites where the coherency is significant at the 80% level. Dashed vector on the wheel shows the phase of minimum ice volume (IV). [From Imbrie et al., 1989] *Reproduced by permission of Kluwer Academic Publishers, copyright © 1989 D. Reidel Publishing Co.*

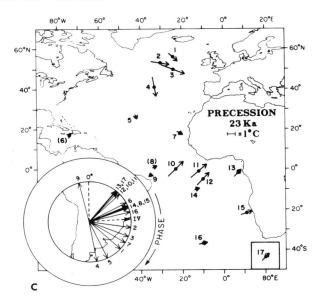

Fig. 7.13.   *(continued)*   **C**

terms of marine productivity changes. There have been a number of observed changes in components of the carbon cycle in the deep sea, but none which unambiguously explain the ice core patterns. For example, Shackleton and Pisias (1985) demonstrated that benthic $\delta^{13}C$, a measure of whole-ocean carbon storage, increased in the glacials, with the fluctuations coherent and in phase with ice volume changes. They interpreted the carbon changes as originating from transfer of terrestrial carbon to marine reservoirs during times of sea level lowering and increased continental aridity. Subsequent work suggests that part of this benthic $\delta^{13}C$ signal is due to changes in vertical fractionation of carbon in the ocean (Boyle, 1988b; Duplessy et al., 1988).

There are additional orbital insolation signals in components of the ocean carbon cycle, with different components recording different types of response. For example, organic carbon accumulation rates in the equatorial oceans all have the dominant Milankovitch periods (Lyle, 1988), whereas calcium carbonate dissolution changes and equatorial Atlantic productivity indices have 100-KY and 41-KY signals, but no 23-KY power (Peterson and Prell, 1985; Curry and Crowley, 1987). Other measures of ocean

productivity have significant 23-KY power (Shackleton and Pisias, 1985; Legrand et al., 1988a). However, analysis of "whole-Pleistocene" $\delta^{13}C$ records of NADW variations indicate significant fluctuations at non-Milankovitch periods between 250 and 350 KY (Raymo et al., 1990b). It seems as if a considerable amount of additional research must be done before we can satisfactorily link all of the observed changes in the ocean carbon cycle with the ice core CO_2 record. The reader is referred to Section 6.4.5 for additional discussions on carbon cycle changes and CO_2.

7.2.7   Nonstationary Ice Volume Changes

In addition to late Pleistocene changes discussed previously, other Pleistocene changes of note involve a transition in the subpolar North Atlantic from dominant 100-KY and 41-KY fluctuations in the late Pleistocene to 41-KY fluctuations (Figs. 7.14 & 7.15) prior to about 900,000 BP (Ruddiman et al., 1986a,b), i.e., prior to the initiation of 100,000-year ice volume fluctuations (cf. Shackleton and Opdyke, 1976; Prell, 1982). The close linkage between SST and $\delta^{18}O$ is probably due to the very strong influence of surrounding ice sheets on local surface conditions in the North Atlantic (cf. Sections 4.1.1 & 4.2.2).

The strong linkage between $\delta^{18}O$ and SST at high northern latitudes may not hold in regions removed from the direct influence of the ice sheets. For example, there is significant 400-KY power in an equatorial Atlantic SST record that is not reflected in ice volume fluctuations (Briskin and Harrell, 1980; cf. Fig. 6.24). Relatively low SST in this record between about 400,000 and 600,000 BP can also be found in higher resolution Caribbean and South Atlantic cores (Imbrie and Kipp, 1971; Morley and Hays, 1981). The pattern may reflect a decoupling of low- and high-latitude climates at low frequencies. This result is not unexpected, since Manabe and Broccoli (1985b) showed that there is relatively little direct radiative influence of ice sheets in regions more than 10–20° of latitude removed from the ice sheets (cf. North, 1984 and Section 4.2.2). Thus, these other regions may be "free" to fluctuate differently.

7.2.8   Pre-Pleistocene Milankovitch Cycles

There is increasing evidence for Milankovitch cycles throughout the Phanerozoic for both glacial and presumed nonglacial states. Calculations (Berger, 1989) suggest that over the Phanerozoic, changes in the Earth–Moon distance, the Earth's rotation rate, and its moment of inertia could cause changes in precessional periods of ~10% and in obliquity periods of ~20% (earlier time intervals have shorter periods).

Pre-Pleistocene fluctuations may reflect in part some type of insolation forcing of the low-latitude monsoonal circulation. There are 100-KY and 400-KY fluctuations in calcium carbonate from the Atlantic and Pacific (Moore et al., 1982; cf. Crowley, 1985). A comparison of late Miocene (5.8–8.5 Ma) and Pleistocene vari-

Fig. 7.14.   Area/variance composite of spectral analysis of winter SST record from four Pleistocene record segments in the subpolar North Atlantic. Note the time evolution of the dominant peak. [From Ruddiman et al., 1986b] *Reproduced by permission of the Geological Society from "North Atlantic sea-surface temperatures for the last 1.1 million years," Ruddiman, W. F., N. J. Shackleton and A. McIntyre, 1986, Geol. Soc. (London) Spec. Pub. 21:155–173.*

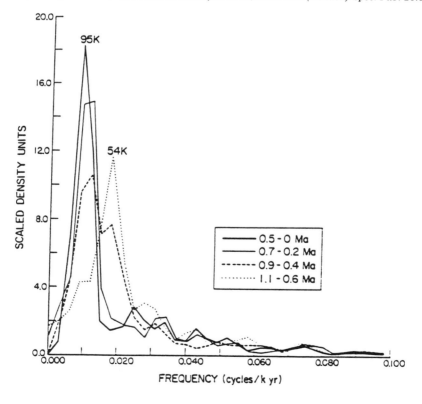

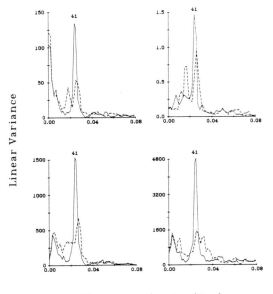

Linear Variance

Frequency(cycles/Kyr)

Fig. 7.15. Spectral analysis of four early Pleistocene time series, illustrating that climate fluctuations during this interval were primarily at ~41-KY (tilt) periods. Solid lines equal spectra for modified time scale. [From Ruddiman et al., 1986a] *Reprinted with permission of Elsevier Science Publishers.*

ations in the Pacific (Fig. 7.16; Moore et al., 1982) notes the strong presence of 400-KY power even before the assumed time of Northern Hemisphere ice initiation. Moore et al. (1982) suggest that the discrepancy between carbonate and $\delta^{18}O$ spectral records may lie in the fact that the different variables are monitoring different parts of the climate system (oceans and ice sheets).

Many pre-Pleistocene records are from regions where the absolute chronology is not as precise as in the Pleistocene (cf. Hallam, 1986a). However, the evidence is still compelling enough to suggest that orbital forcing may be involved. The evidence usually involves very rhythmic variations in sediment types. Time control is sufficiently good to estimate with a moderately high degree of confidence that the characteristic time scale of fluctuations have Milankovitch periods (Arthur et al., 1984; de-Boer and Winders, 1984; Hardie et al., 1986; Herbert and Fischer, 1986; Olsen, 1986). In some cases this assumption can be checked by analysis of varved sediments, which presumably record annual variations in sediment deposition. For example, counting of varve layers in the Permian Basin evaporite deposits of western Texas also reveals precession time scale periods (Fig. 7.17; Anderson, 1984).

In addition to recording precessional periods in low-latitude deposits, there may be tilt and eccentricity signals in the pre-Pleistocene record. For example, low-latitude records shift from predominantly tilt-controlled signals in the early Cretaceous (Herbert et al., 1988) to precession-controlled signals in the mid-Cretaceous (Herbert and Fischer, 1986). Additionally, sea level variations in Carboniferous cyclothem deposits of the North American interior (280–296 Ma) have a characteristic time scale of several hundred thousand years (cf. Fig. 11.13; Heckel, 1986). Fluctuations may reflect sea level changes induced by glaciation on Gondwanaland (cf. Section 11.1.5). There are also 100,000-year pe-

Fig. 7.16. Comparison of Pleistocene and pre-Pleistocene spectral records of carbonate fluctuations. This figure illustrates the time dependency of the 100-KY peak and the relative stability of the 400-KY peak. [From Moore et al., 1982] *Reprinted with permission of Elsevier Science Publishers.*

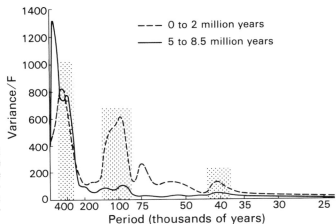

--- 0 to 2 million years
— 5 to 8.5 million years

Variance/F

Period (thousands of years)

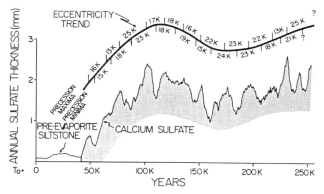

Fig. 7.17. Smoothed plot of absolute thickness of varved calcium sulfate deposits in the Permian Castile Formation of Texas. Note the dominant oscillations in the period range of the precession and eccentricity cycles. The intervals between the successive maxima and minima of the sulfate thickness are given in thousands of years along the heavy line ("ecc. trend") and marked as "precession maxima" and "precession minima." [From Anderson, 1984] *Reproduced by permission of Kluwer Academic Publishers, copyright © 1984 D. Reidel Publishing Co.*

riods in Triassic and Cretaceous sediments (e.g., Herbert and Fischer, 1986; Olsen, 1986). These variations occur during a time of presumed ice-free climates. Yet the 100,000-year periodicity in Pleistocene records is usually ascribed to non-linearities in ice volume growth and decay (see text following). There is at present no satisfactory explanation for its presence during times of presumed ice-free conditions.

The above results are also of interest because they suggest that the orbital cycles are surprisingly regular over very long intervals of time. This result is not expected by theory, which predicts either secular drift of the periods with time or even chaotic behavior of orbits (e.g., Berger, 1984; Bretagnon, 1984; Buys and Ghil, 1984; Laskar, 1989). Perhaps astronomers can use the "Milankovitch metronome" in the geologic record to improve their understanding of planetary motions!

7.3 Modeling Ice Volume Fluctuations

Since the publication of the historic Hays et al. (1976b) paper, a number of modeling efforts have attempted to explain the relation between astronomical forcing and climate change. For example, simulations of monsoon variability (e.g., Kutzbach and Street-Perrott, 1985; Prell and Kutzbach, 1987) have provided insight into possible controls of 23-KY peaks in low latitudes. However, most modeling studies have focused on the origin of the 100,000-year cycle in ice volume. These studies have been motivated by the recognition that the amount of insolation perturbation at 100 KY is not enough to cause a climate change of ice-age magnitude. The non-

linear origin of the 100-KY peak has prompted a series of explanations, none of which have unequivocal preeminence at the present time. Specific explanations for the large 100-KY variance generally fall into two classes. Each class usually incorporates both orbital input and terrestrial nonlinearities, but the explanations differ in emphasis. In the first case, orbital variations are essential to ice volume fluctuations, but nonlinear interactions in the air–sea–ice system modify the signal. In the second case, ice volume fluctuations result from inherently nonlinear interactions in the air–sea–ice system, with orbital variations serving only to phase-lock the variations at the preferred time scales.

7.3.1 Ice Volume Fluctuations Primarily Driven by Orbital Forcing

Success in modeling ice volume fluctuations has varied by time scale. For example, Imbrie et al. (1984) showed that at the 23-KY and 41-KY periods, ice volume ($\delta^{18}O$) responds linearly to orbital forcing (cf. Fig. 7.9). At 100 KY the effect is nonlinear. This linear–nonlinear couplet has some parallels with progress in modeling ice volume fluctuations. For example, Suarez and Held (1976, 1979) incorporated orbital parameters into a seasonal energy balance model, keyed to ice-albedo feedback (cf. Section 1.3.2), and produced an output qualitatively similar to the geologic record of the last 150,000 years. Pollard (1978) and Pollard et al. (1980) added Weertmann's (1976) ice sheet model to an energy balance model and also generated some encouraging results, especially with respect to the 23- and 41-KY cycles. However, the magnitude of ice volume response, and the dominant 100-KY cycle, were not well simulated by any of these models.

Although models differ in the specifics of how 100-KY power is generated, many "successes" are related to the observation that 100-KY power can be generated by transmission of 19-KY and 23-KY frequencies through a nonlinear system (Wigley, 1976), producing substantial power in both harmonics and 100-KY subharmonics. Specific details of this nonlinear interaction vary considerably, but many mechanisms focus on the abrupt terminations of ice ages, which occur over an interval of about 10,000 years. There are several different proposals as to how the ice sheets may rapidly decay. Ruddiman and McIntyre (1981c) note that terminations coincide with summer maxima in both precession and obliquity. Maximum melting amplified by ocean feedback loops may drive the system toward full deglaciation. Catastrophic meltback of ice streams may further enhance the deglaciation process (cf. Hughes et al., 1977; Hughes, 1987b). (Ice streams represent lines of convergence in glacier flow. As the name implies, flow in these streams is significantly greater than in other parts of the glacier. There are various nonlinear mechanisms for accelerating ice drainage. If these mechanisms are triggered, they could cause catastrophic glacial collapse.) Experiments with a sea ice model provide some support for the ocean feedback mechanisms (Ledley, 1984).

Other approaches to the 100-KY cycle focus on nonlinear interactions between accumulation/ablation, ice sheet flow, elastic lithosphere, and viscoelastic mantle (Oerlemans, 1980, 1981, 1982a; Birchfield et al., 1981, 1982; Pollard, 1982; Hyde and Peltier, 1985). Satisfactory models for the abrupt deglaciations produce variance at 100-KY periods. There are good theoretical and empirical reasons to expect asymmetric growth of ice sheets (e.g., Weertmann, 1964; Broecker and van Donk, 1970). Imbrie and Imbrie (1980) estimated that the time constant for growth and decay differ by about a factor of 4. The next two paragraphs briefly describe how some of the above interactions occur.

The relaxation time of the mantle strongly influences the amount of time that an ice sheet remains below what is known as the equilibrium snow line (a line demarcating ablation versus net accumulation). In turn, the slope and mean elevation of the snow line on an ice sheet has a strong influence on the mass balance of the system (Fig. 7.18). The elevation of the snow line can be affected by high ground in areas of glacier formation. During early ice growth, most of the ice sheet is above that line, and so it is only modestly affected by insolation variations.

As the ice sheet grows, the cumulative weight depresses the mantle by as much as 1 km. At this stage, the ice sheet is very vulnerable to insolation forcing, for much of it is "trapped" beneath the equilibrium snow line even if significant melting occurs because of the slow relaxation time of the mantle. Thus, wastage can be very rapid. Ice decay can be further enhanced if bedrock depressions are subsequently filled with glacial meltwater in proglacial lakes, thereby enhancing calving (Andrews, 1973; Pollard, 1982). If the ice sheets are near sea level, ice streams can undergo catastrophic wastage (cf. Hughes et al., 1977). Any of these feedbacks would increase power in the 100,000 band.

Models incorporating many of the above concepts (e.g., Fig. 7.19) had some success reproducing many of the features of the ice volume curve (e.g., Budd, 1981; Pollard, 1984; Hyde and Peltier, 1985, 1987). However, almost all of these models encounter some difficulties. For example, the Hyde–Peltier model works very well for the most recent 500,000 years, but less so for the preceding 300,000 years. Imbrie (1985) suggested that there has been a change in sensitivity to forcing over the same time period. Since the Hyde–Peltier model is sensitive to small variations in adjustable parameters (compare upper and lower panels of Fig. 7.19), and the transition occurs at the same time indicated by empirical studies (Imbrie, 1985), there may be some justification for Imbrie's conjecture.

7.3.2 Ice Volume Fluctuations Modulated by Orbital Forcing

An entirely different approach to modeling the low-frequency climate variability involves the hypothesis that glacial-interglacial fluctuations are a consequence of nonlinear "internal" interactions in a highly complex system (e.g., Sergin, 1980; Saltzman et al., 1981; Nicolis, 1984; Saltz-

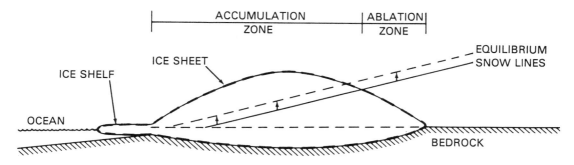

Fig. 7.18. Schematic diagram of ice sheet models used in modeling Pleistocene ice volume fluctuations. This figure illustrates several features which have been invoked to explain how ice sheet fluctuations can enhance the 100,000-year cycle. Changes in accumulation and ablation can affect the equilibrium snow line—the line separating net accumulation from net ablation (note that in this figure ablation increases to the right because the axis is assumed to represent latitude, with north on the left and south on the right). The equilibrium snow line can change due to a number of factors. For example, if orbital forcing caused warmer summers, the ablation zone would increase. However, the very long time constants involved in bedrock rebound would effectively trap a larger area of the ice sheet in the ablation zone, thus providing a positive feedback. If the ice sheet is drained by "ice streams" (see text) which feed into ice shelves, then the unstable behavior of the ice shelves/ice streams can cause a further positive feedback. Both of these feedbacks would tend to amplify variance in the 100-KY record.

man, 1985). As a simple example of this concept, it can be demonstrated that low-frequency (red noise) variance (see Appendix C) can be produced when white noise forcing, from a time series with a short response time, is applied to another system with a relatively long response time. For example, variable atmospheric winds on the sea surface can create SST anomalies on a longer time scale (Frankignoul and Hasselmann, 1977; Herterich and Hasselmann, 1987).

Fig. 7.19. Top: a scaled version of the last 800,000 years of an ice volume model (solid line) compared with the SPECMAP $\delta^{18}O$ curve (dashed line). Bottom: same model response with slightly altered boundary conditions. [After Hyde and Peltier, 1987] *Reprinted from Journal of Atmospheric Sciences with permission of the American Meteorological Society.*

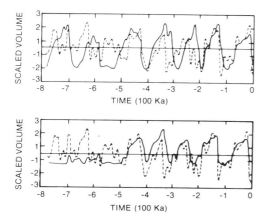

In addition, low-frequency variations in soil moisture can be generated by essentially random (white noise) variations in precipitation (Delworth and Manabe, 1988). From the above perspective, a low-frequency peak (e.g., 100 KY) in the $\delta^{18}O$ record would be generated via nonlinear interactions in the air–sea–ice system.

Theoretical studies suggest that the slope of a variance spectrum for such "stochastically driven" ice volume fluctuations might lie between -1 and -2, depending on the interactions involved (Hasselmann, 1976; Lemke, 1977). In fact, log-log plots of the variance spectrum of the $\delta^{18}O$ record (Fig. 7.20) yield a red noise spectrum with a slope of -2 (Kominz et al., 1979; cf. Kominz and Pisias, 1979) between 100 KY and 12 KY. This red noise pattern is common for many types of geophysical phenomena (Båth, 1974). The trend is consistent with that predicted by white noise forcing from time scales shorter than 12 KY. The "slope break" in Fig. 7.20 at about 100 KY can, under certain circumstances, be interpreted in terms of the system's response time (e.g., North et al., 1981, section 8; cf. Leith, 1975, 1978; Bell, 1980). Further analysis of the $\delta^{18}O$ record suggests that four to six independent variables are required to model its variation (Maasch, 1989).

Internally driven glacial-interglacial fluctuations can also be modeled more explicitly. For example, Ghil and colleagues (Ghil, 1981; Ghil

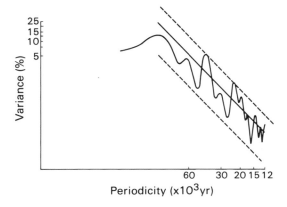

Fig. 7.20. Log-log plot of the variance spectrum of the $\delta^{18}O$ record. The solid straight line has a slope of -2, which conforms to some models of a strong stochastic element to climate change (see text). The dashed lines represent the 80% confidence interval about that line. [From Kominz et al., 1979] *Reprinted with permission of Elsevier Science Publishers.*

and LeTreut, 1981; LeTreut et al., 1988) modeled internally generated glacial-interglacial fluctuations which, when orbitally forced, produce characteristic Milankovitch periods, including harmonics (e.g., 10 KY) and subharmonics (e.g., 100 KY). In another example, Saltzman (1985, 1987; Saltzman and Sutera, 1987) incorporated a small number of variables (CO_2, surface and deep-water temperatures, and ice volume) in a model and was able to reproduce with some success both the 100-KY cycle and the transition from dominant 41-KY ice volume fluctuations prior to 900,000 BP to dominant 100-KY ice volume fluctuations after 900,000 BP (Fig. 7.21).

The latter transition may also reflect the existence of "multiple equilibrium" states, whereby slowly changing boundary conditions can cause an abrupt transition in the climate state (e.g., North et al., 1983; Watts and Hayder, 1984b; North and Crowley, 1985; cf. Section 1.3.4).

Saltzman's (1985) approach is fundamentally different from some other schemes. It is not based on the standard deductive modeling method, in which a model is derived from the fundamental statements of conservation of mass, energy, and momentum, and energy by successive averaging and parameterization of flux processes. Rather it is based on an inductive method because Saltzman maintains that it is difficult to impossible to determine the various phenomenological coefficients needed for a normal deductive modeling approach (e.g., the average annual global radiation imbalance during ice growth and decay is on the order of 0.1 W/m² — far below our level of adequate detection or modeling). The inductive approach requires formulation of a closed set of equations to yield known output. In other words, the results are utilized to formulate the equations. Saltzman argues that this approach is not just an exercise in curve matching. He points out that the fundamental laws of both physics and meteorology were formulated based on such observations.

An entirely different approach to modeling "internal" glacial-interglacial fluctuations involves the postulate that ice sheet nonlinearities are not responsible for the main glacial-intergla-

Fig. 7.21. Solution of a three-component system for global ice volume illustrating that internal variations in the climate system can generate low-frequency variance with steplike transitions. [From Saltzman and Sutera, 1987] *Reprinted from Journal of the Atmospheric Sciences with permission of the American Meterological Society.*

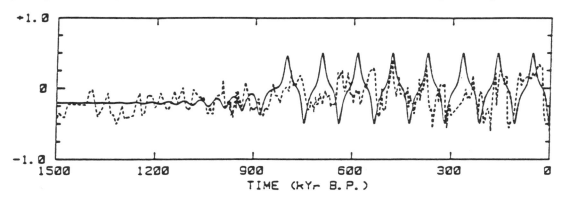

cial oscillations and that instabilities in the ocean–atmosphere system are the answer (Broecker et al., 1985, 1988a; see discussions in Sections 3.2.1, 3.2.2, 4.4.1, and 4.4.2). Work is still continuing on this alternate approach. To our knowledge it has not yet been cast into a frequency domain model.

7.4 Summary

There is a wealth of data supporting the astronomical theory of climate change. The principal focus in recent years has now shifted toward determining how the astronomical signal is transmitted through the climate system, and how the astronomical signal interacts with "internally" generated climate change.

Observations indicate that the global ice volume signal records the dominant Milankovitch periods. Fluctuations at 23 KY and 41 KY appear to be linearly related to orbital forcing; the 100-KY signal is clearly nonlinear. Prior to 0.8–0.9 Ma, Pleistocene ice volume fluctuations were primarily at the 41-KY period. Another global signal (CO_2) also appears to fluctuate at Milankovitch periods of 23 and 100 KY. Other proxy indices record more local signals and support the connection between tilt-dominated forcing in high latitudes and precession-dominated forcing in low latitudes. Pre-Pleistocene records also record Milankovitch periods, with some showing unexpected 100-KY power during presumed ice-free periods. There are some indications of harmonics in the record as well as periods that may represent cross-products of different Milankovitch periods (e.g., 31-KY periods in the equatorial Pacific that may be combinations of precessional and tilt forcing).

Modeling studies provide some insight into lower latitude monsoonal forcing that may generate power in some precipitation records at 23 KY. The primary emphasis in modeling studies, however, involves efforts at reproducing global ice volume, with the main emphasis on understanding the nonlinear 100-KY cycle. At present there are many different viewpoints on this subject.

Given the plethora of plausible approaches to modeling the 100-KY cycle, it is necessary to inquire as to which explanation (if any) is correct. The models are so fundamentally different that they cannot all be right. Unfortunately, the field is not yet at the stage where we have made much progress in sifting through the different models in order to make some attempt to rank them. The outlines of the ranking approach are becoming clear, however. For example, one criterion to grade models involves their robustness of solution (Saltzman, 1988). If a limited number of parameters with a relatively broad range of values can explain a certain phenomenon, then that model might be preferable to one that can only explain a phenomenon over a limited range of parameter space. The opposite claim can be made that a model operative under a limited set of parameter space is more testable. At the present stage, we probably need both approaches. But it is clear that the field has gone through the first stage of its evolution, where it has explored different possibilities, to where we hope that it is now entering a second stage (rigorous testing of the different models).

PART III
PRE-QUATERNARY CLIMATES

8. MID-CRETACEOUS CLIMATE

One of the most important paleoclimate conclusions based on study of the geologic record is that certain periods of earth history were substantially warmer than the present. The mid-Cretaceous (Aptian–Albian–Cenomanian, about 120–90 Ma (million years ago); see Appendix A for a geologic time scale) represents one of the last times for which there is frequently cited evidence for a warmer and perhaps ice-free state. This time period is of special interest in climate theory because the relatively widespread distribution of fauna, flora, and rocks (on land, in nearshore marine sediments, and in ocean cores) provides good spatial detail, which is necessary for developing a reasonably clear picture of the environmental setting. This information is valuable for testing various theories of climate change. This chapter will focus on the mid-Cretaceous as an example of an ice-free period and will examine both relevant evidence and modeling results.

8.1 Nature of the Record

8.1.1 Paleogeography

Any discussion of Cretaceous paleoenvironments first requires a description of land–sea distribution, which was significantly different from the present (Fig. 8.1). North America was still connected to Europe, as was South America to Africa and Australia to Antarctica. India was an island continent in the southern subtropics. Sea level was 100–200 m higher than present and flooded about 20% of the continental area (Barron et al., 1980).

Two of the most dramatic responses to the sea level changes involved the equatorial Tethys Seaway and the midcontinent of North America. The Tethys Sea flooded large parts of western Europe and North Africa (Fig. 8.1). A near-equatorial current system may have circumscribed the planet, partially obstructed only by a narrow peninsula in southeastern Asia. The bulge of Africa displaced the seaway into the Northern Hemisphere, so parts of Europe were subjected to very warm ocean currents (e.g., Gordon, 1973; Berggren and Hollister, 1974). In North America, a shallow sea extended north from the Gulf of Mexico and sometimes merged with another seaway extending southward from the Arctic Ocean. At times this Cretaceous interior seaway stretched for more than 5000 km through the heartlands of the continent. The rearranged land–sea patterns resulted in a hemispheric distribution more symmetric than that of the present (Fig. 8.2).

Paleontological data also suggest a land connection between Asia and North America during parts of the Cretaceous (Colbert, 1973; Hickey, 1981). The conclusion is based on biogeographic analysis of distributions of plants and the Ceratopsian Order of dinosaurs (*Triceratops* was the most notable genus of this order). The dinosaur group only occurs in deposits from eastern Asia and western North America (the plants had a slightly larger distribution). A land connection must therefore have been utilized for population dispersal of fauna and flora.

8.1.2 Evidence for Warmth

There is a considerable amount of evidence for warmer temperatures in high latitudes during the mid-Cretaceous (Fig. 8.3). Coral reefs extended 5–15° poleward of their present warm-water habitat (e.g., Habicht, 1979). There were also significant latitudinal displacements of a number of other invertebrates—large foraminifera, gastropods, rudist bivalves, ammonoids, and belemnoids (e.g., Sohl, 1969; Gordon, 1973; Kauffman, 1973; Lloyd, 1982). Dinosaurs of presumed warm-weather affinity ranged north of the Arctic Circle (Colbert, 1973). Floral provinces expanded as much as 15° poleward of their present location (Barnard, 1973; Vakhrameev,

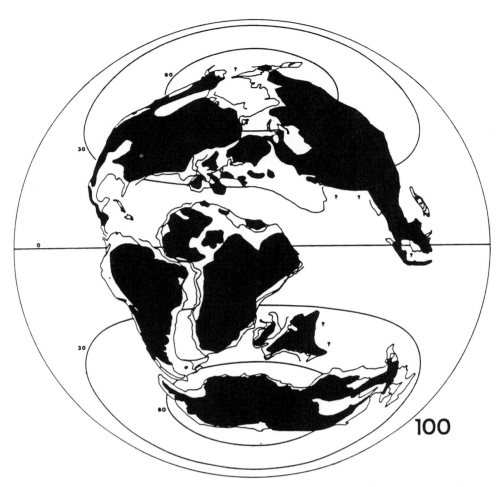

Fig. 8.1. Paleogeographic reconstruction at 100 Ma. Light areas on continents indicate regions flooded by shallow seas (maximum depth 100–200 m). [After Barron et al., 1980] *Reprinted with permission of Elsevier Science Publishers.*

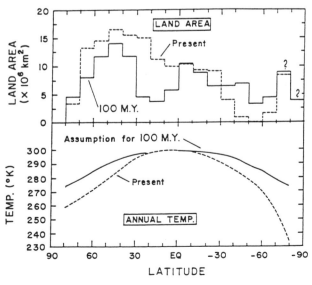

Fig. 8.2. Area of land in each 10° latitude belt for the present and 100 Ma (Barron et al., 1980). Present day mean annual surface air temperature with respect to latitude and the temperature assumption for the Cretaceous. [From Thompson and Barron, 1981] *Reproduced by permission of the University of Chicago Press. From "Comparison of Cretaceous and present earth albedos: Implications for the causes of paleoclimates," S. L. Thompson and E. J. Barron, Jour. Geol. 89:143–167, copyright 1981, University of Chicago Press.*

156

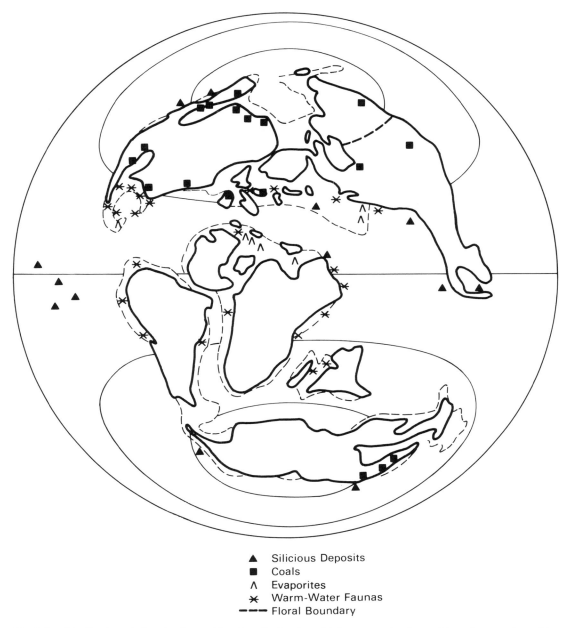

Fig. 8.3. Distribution of climatically sensitive deposits for the mid-Cretaceous (Aptian–Albian–Cenomanian, 120–90 Ma). Data from many sources (see text).

1975) and diverse assemblages at high latitudes indicate mild/warm temperatures (Smiley, 1967; Taylor, 1972; Krassilov, 1973, 1981). Breadfruit trees have been reported (Nathorst, 1911) from the Late Cretaceous of Greenland (55°N paleolatitude). This paleontological information, along with sedimentologic data from coals and laterites/bauxites (the latter develop-

ing in seasonally warm, moist climates), suggests that climates in high latitudes were generally warmer than at present.

Oxygen isotope records also provide information about Cretaceous climates. $\delta^{18}O$ measurements of bottom-dwelling organisms provide information about the history of deep waters through the last 100 million years (e.g.,

Emiliani, 1961; Savin, 1977) and indicate that Cretaceous intermediate-deep waters were between 15 and 20°C (cf. Fig. 10.3), i.e., about 15° warmer than present. This result seems incompatible with presence of extensive high-latitude ice in the Cretaceous.

The Cretaceous deep-water circulation may also have been different from that of the present. Brass et al. (1982a; cf. Chamberlin, 1906) suggested that warm, saline bottom waters originated in the subtropics as a result of high evaporation in the broad, shallow extensions of the Tethys Sea. A bottom water similar to the present day Mediterranean Intermediate Water would have formed (this water mass starts out with a density greater than Antarctic Bottom Water, but mixing significantly reduces its density). It is important to emphasize that geologic records have yet to provide solid support for this idea.

$\delta^{18}O$ measurements also suggest that mid- and high-latitude surface temperatures were warmer than those of the present (see Barron, 1983). However, there are some problems with these measurements. Many published records have not been sufficiently examined to test for alteration effects (diagenesis), which significantly

modify $\delta^{18}O$ values (e.g., Killingley, 1983). Furthermore, the depth habitat of some of the measured organisms (e.g., belemnites) is not well known, and $\delta^{18}O$ values can vary by depth. $\delta^{18}O$ values are also significantly modified in coastal regions, and many of the measurements (Fig. 8.4) are from shallow-water marine realms.

8.1.3 Some Significant Uncertainties Concerning Cretaceous Warmth

Although there is good evidence that climates were warmer in high latitudes, there are two significant uncertainties about Cretaceous paleoclimate data. First, conclusions are more ambiguous about the presence or absence of permanent ice, i.e., whether the warmth was year-round. Although unequivocal tillites (consolidated glacial drift) of Cretaceous age have yet to be found (Hambrey and Harland, 1981b), there is some scattered evidence (Fig. 8.5) for early Cretaceous (130–110 Ma) ice-rafted deposits in both the Northern and Southern Hemispheres (Frakes and Francis, 1988). These data suggest at least seasonally cold conditions. Terrestrial biota from the Cretaceous (105–130 Ma) of south Australia (paleolatitude ~75°S) may also indi-

Fig. 8.4.   Isotopic paleotemperature estimates of surface water temperature, 100 Ma. [From Barron and Washington, 1982b] *Reprinted with permission of Elsevier Science Publishers.*

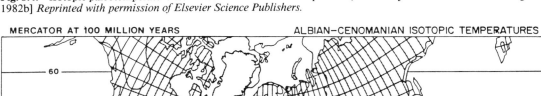

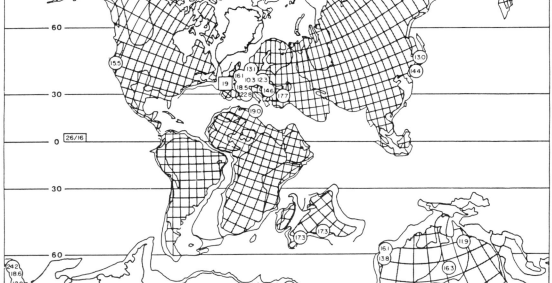

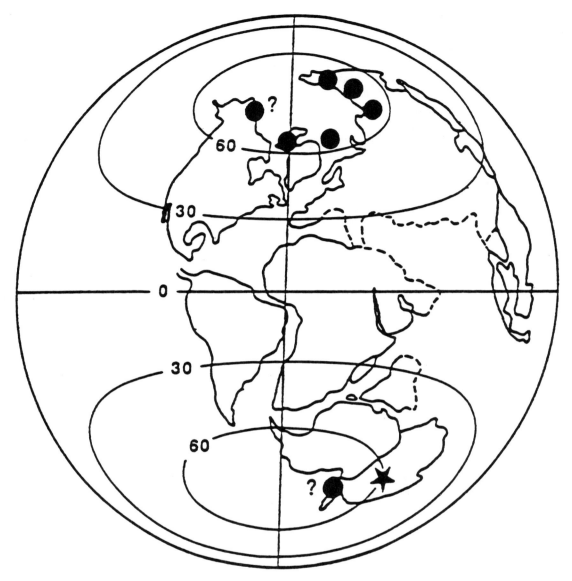

Fig. 8.5. Postulated distribution of Early Cretaceous ice-rafted deposits. [From Frakes and Francis, 1988] *Reprinted by permission from Nature, Vol. 333, pp. 547–549, Copyright (©) 1988 Macmillan Magazines Ltd.*

cate strong seasonality, with mean annual temperatures perhaps as low as −6°C (Rich et al., 1988). Likewise, analysis of late Cretaceous (∼70 Ma) fossil plants on the North Slope of Alaska (paleolatitude 70–80°N) suggests that they may have been deciduous, which is also evidence for seasonality (Spicer and Parrish, 1986).

It also may not be justified to infer year-round warmth from reptile deposits in high latitudes.

Dinosaurs may have been warmblooded (Bakker, 1975, 1980) and able to tolerate colder winters than present mammals. Some dinosaurs may also have seasonally migrated to warmer climes (e.g., Axelrod, 1984; Parrish et al., 1987). A rate of 1 km/hr for 12 hr would have allowed North Slope dinosaurs to migrate to the Arctic Circle—a distance of about 2000 km (Parrish et al., 1987). [Elephants presently range as far as 500–650 km in search of food and water (Sikes,

1971)]. However, some of the dinosaurs, such as the hadrosaurs ("duck-billed dinosaur"), may not have been sufficiently mobile to migrate.

Another uncertainty with respect to the Cretaceous involves the question of low-latitude sea surface temperatures (SSTs). Although warm or warmer equatorial temperatures in the mid-Cretaceous have long been assumed, the evidence is not nearly as compelling as high-latitude evidence. Siliceous deposits in the equatorial Pacific (Fig. 8.3) indicate that equatorial upwelling was present (Drewry et al., 1974), as it is today. Furthermore, $\delta^{18}O$ data from low latitudes are not extensive. Some measurements (Fig. 8.4) are from nearshore locations and might be affected by salinity changes. One set of measurements from a site located in the equatorial Pacific suggests temperatures of 25–27°C—values comparable to the present (Douglas and Savin, 1975). It would be very desirable to acquire more measurements from open-ocean sites, for climate models are greatly dependent on tropical SST as a boundary condition.

The absolute value of Cretaceous low-latitude SSTs is quite significant when viewed from the perspective of proposed mechanisms for Cretaceous warmth (see modeling section at the end of this chapter). Increased levels of carbon dioxide is the most popular explanation for Cretaceous warmth. Yet climate model experiments suggest that CO_2 levels required for high-latitude warmth also produce low-latitude SST increases of 4–5°C (Manabe and Bryan, 1985). This increase should produce about 1.0‰ decrease in $\delta^{18}O$ records (i.e., more negative values). When this change is coupled with about a 1.0‰ decrease due to elimination of permanent ice sheets, Cretaceous low-latitude oxygen isotopes should be about 2.0‰ lighter than present (i.e., about −3.0‰ to −4.0‰). Such changes should be easily detectable in climate records. It is not clear whether any such light values have been reported.

8.1.4 Carbon Cycle Variations

There were very significant changes in the carbon cycle during the Cretaceous. Deep-Sea Drilling Project results from the Atlantic, Pacific, and Indian Oceans reveal widespread occurrences of Cretaceous black shales (e.g., Fig.

8.6). The total organic carbon content of the shales may exceed that of all known coal and hydrocarbon reservoirs (Ryan and Cita, 1977). Some of the Pacific events were relatively brief, lasting less than one million years (Sliter, 1989).

The large amount of organic matter probably reflects low-oxygen conditions in intermediate or bottom waters. This condition could result from low rates of overturn and relatively low dissolved oxygen levels of deep water (the partial pressure of O_2 in warm water is less than in cold water). Low sedimentation rates in deep-ocean cores support this interpretation (Bralower and Thierstein, 1984). Alternatively, low-oxygen levels can result from high productivity in the source regions of bottom water formation (Sarmiento et al., 1988a).

In addition to higher accumulation rates of organic carbon in deep-sea sediments, approximately 60% of all known oil reserves are from the Cretaceous (Irving et al., 1974) with the most important locations being the Persian Gulf and Middle America. This massive sequestering of organic carbon and hydrocarbons caused an overall shift of $^{13}C/^{12}C$ ratios in deep-sea carbonates—the removal of isotopically light organic carbon enriched the ratios in seawater from which calcareous shells were secreted (Scholle and Arthur, 1980).

Changes in organic carbon burial should also affect atmospheric oxygen levels (O_2 increases if carbon is sequestered in the crust; cf. Berner, 1989). Initial attempts to verify this theory seemed successful; "fossil air" trapped in amber yielded higher O_2 levels (Berner and Landis, 1988). However, further attempts to verify these measurements have not been successful (Cerling, 1989).

8.1.5 Higher Resolution Cretaceous Records

There are also higher frequency fluctuations in mid-Cretaceous climate (e.g., Arthur et al., 1984; deBoer and Winders, 1984; Fischer and Schwarzacher, 1984; see also Section 7.2.8). Most of these records are from tropical marine sites. These fluctuations are very rhythmic and occur in a number of different types of sediment. Although there is some uncertainty in the absolute time scale, time control is sufficiently good to estimate with a relatively high degree of

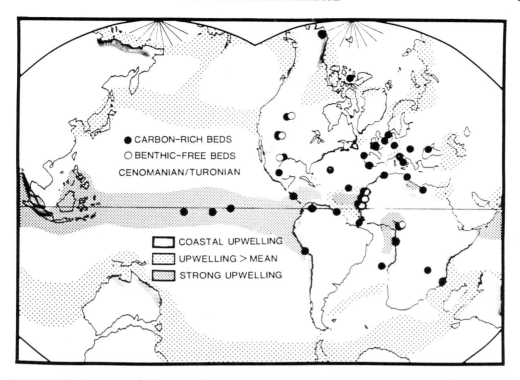

Fig. 8.6. Relation between organic carbon-rich deep-sea sediments and predicted areas of upwelling for the mid-Cretaceous (cf. Fig. 8.11), plotted on a Cenomanian–Turonian (~92 Ma) paleocontinental reconstruction. Different shadings represent types or intensity of upwelling. Black dots are known organic carbon-rich Cenomanian–Turonian localities with the inclusion of additional Deep Sea Drilling Project sites. [From Arthur et al., 1987] *Reprinted with permission of the Geological Society Publishing House. From "Marine Petroleum Source Rocks," M. A. Arthur, S. O. Schlanger and H. C. Jenkyns, 1987, J. Brooks and A. J. Fleet (Eds.), Geol. Soc. Spec. Publ. (London) 26, 401–420.*

confidence that fluctuations have periods of about 20,000 and 100,000 years, i.e., the precession and eccentricity periods (deBoer and Winders, 1984; Herbert and Fischer, 1986). A predominantly tilt-driven period (41,000 years) occurs in the early Cretaceous (Herbert et al., 1988). Experience with modeling early Holocene climates (see Section 4.5.1) suggests that orbitally driven variations in monsoon intensities might account for some of the 20,000-year alternations in sediment types (cf. Barron et al., 1985; Glancy et al., 1986; Oglesby and Park, 1989).

8.2 Modeling Studies

8.2.1 Overview

The good evidence for increased high-latitude warmth in the mid-Cretaceous has motivated climate modelers to understand the origin of

such a large climate change (e.g., Barron et al., 1981a; Berner et al., 1983; Barron and Washington, 1984; Schneider et al., 1985). In particular, Barron and colleagues conducted an extensive set of sensitivity experiments to determine which factors may have been important for increased high-latitude warmth.

The discussion on origin of high-latitude warmth is sufficiently involved that it is perhaps useful to orient the reader ahead of time by revealing the main conclusions. Early ideas suggesting that increased high-latitude temperatures are due to changes in land–sea distribution are only partially supported by climate-modeling studies. Preliminary results suggest that changes in oceanic heat transport also may be insufficient to account for inferred above-freezing temperatures in winter at high latitudes. However, this conclusion requires more scrutiny. Satisfactory comparisons of model output with geologic data are hampered because most data are from coastal and nearshore marine environments, and geologic estimates of winter temperatures in high latitudes may therefore be biased toward warm values.

If the standard paleoclimate scenario of year-round warmth at high latitudes is valid, higher atmospheric CO_2 levels may have been responsible. Climate models suggest that changes in atmospheric CO_2 concentrations by a factor of 4–8 may have to be invoked in order to account for higher temperatures. Geochemical models provide some justification for higher atmospheric CO_2 levels. However, there are insufficient data from the mid-Cretaceous to rigorously test for higher CO_2 levels. Some preliminary results from the late Cretaceous (80–65 Ma) and Cenozoic (65–15 Ma) suggest that CO_2 levels may have changed at most by a factor of 2 during this time, but these results are in turn open to alternative interpretations. Since the late Cretaceous was cooler than the mid-Cretaceous, extrapolation of the "low-CO_2" conclusion to the warmer, antecedent period may not be valid.

The following topics are most important for clarifying the question of high-latitude warmth: (1) improved estimates of tropical SST (in order to constrain climate models and determine whether ocean heat transport may have changed); (2) further testing of the effect of possible changes in ocean heat transport on high-latitude climates; (3) further evaluation of geologic data from continental interiors to discriminate seasonal and year-round warmth; and (4) further analysis of geologic data to test the hypothesis of higher CO_2 levels in the mid-Cretaceous.

8.2.2 Geographic Effects

It has long been thought that changes in land–sea distribution played an important role in the evolution of past climates. Changes in sea level or drifting of continents into high latitudes (e.g., Frakes and Kemp, 1972; Donn and Shaw, 1977) supposedly modified heating patterns sufficiently (via changes in ice-albedo feedback) to prevent formation of permanent ice caps. Modeling studies only partially support this scenario (Barron and Washington, 1984; Barron et al., 1984). A version of the NCAR GCM with an-

nually averaged insolation and a "swamp" ocean (cf. Section 2.3.2.1) indicates that changes in land–sea distribution caused a global average temperature change of 4.8°C (Barron et al., 1984; Barron and Washington, 1984; Barron, 1985a; cf. Fig. 8.7). About one-third of this change is due to removal of the polar ice caps, resulting in a net increase in radiation receipt. The remaining two-thirds is due to changes in the land–sea distribution. In the model increased land area in high latitudes produces a net cooling because it allows for the preservation of snow. As we discuss in Section 10.2.1, computations with mean annual forcing do not allow for seasonal warming on the larger high-latitude land masses typical of the late Cenozoic (Crowley et al., 1986). This warming might eliminate winter snows and erase some of the cooling effects.

Simulated climate changes for seasonal experiments do not result in temperature patterns compatible with geologic data (Fig. 8.8); temperatures are still too low in high latitudes. Further sensitivity experiments with Cretaceous geography (Barron and Washington, 1984) indicate only minor differences in average temperatures as a result of changes in sea level or to-

Fig. 8.7. Comparison of zonally averaged surface temperatures (°K) with respect to latitude for a present day control simulation and a Cretaceous geography sensitivity experiment. [From Barron and Washington, 1985] *Courtesy of American Geophysical Union.*

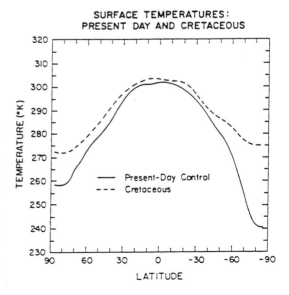

SURFACE TEMPERATURES:
PRESENT DAY AND CRETACEOUS

CRETACEOUS GEOGRAPHY — PREDICTED SURFACE TEMPERATURES

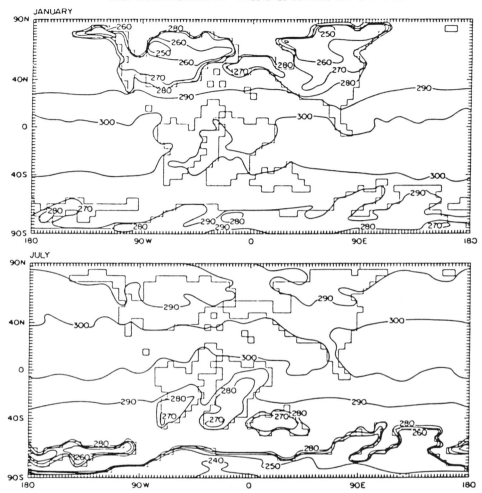

Fig. 8.8. The time mean (30-day average) temperature at the ground (mixed-layer temperature at ocean grid points) averaged over the diurnal cycle for January and July solar insolation and mid-Cretaceous geography with specified warm-polar seas. The temperatures are given in °K. Note the subfreezing temperatures in winter. [From Barron and Washington, 1982b] *Reprinted with permission of Elsevier Science Publishers.*

pography (Fig. 8.9). This latter response is at first puzzling, for high sea levels in general correlate with warm climates (e.g., Damon, 1968), suggesting a causal link. However, as we discuss in Chap. 1, water mainly controls the amplitude of the annual cycle rather than mean annual temperatures. Since there is only a small difference in the albedo of land and water in low latitudes (both about 0.10–0.15, except in desert areas), different explanations for the linkage are required.

Even though changing land–sea distribution modified global temperatures, calculated changes were not large enough (especially in winter) to explain the latitudinal displacements of fauna and flora. Utilizing an assumed sea-surface temperature distribution, Barron and Washington (1982b) compared modeled winter temperature estimates (Fig. 8.8) with the distribution of presumed tropical flora in eastern Asia (Mongolia). This site is one of the few locations that might not have been influenced by maritime warmth in coastal regions. Results indicate that winter temperature estimates are still too low to maintain tropical flora. It is possible that the ecological tolerance of mid-Cretaceous flow-

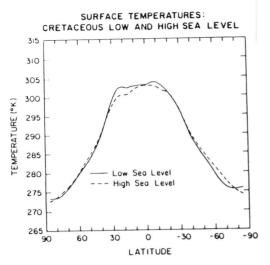

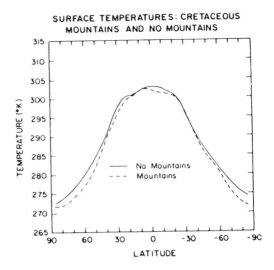

Fig. 8.9. Cretaceous climate model sensitivity simulations to changes in solid-earth boundary conditions. (Left) Comparison of zonally averaged surface temperatures for Cretaceous simulations for high and low sea level [in low sea level experiment, shallow seas (cf. 8.1) do not flood the continents]. (Right) Comparison of zonally averaged surface temperatures for simulations with and without mountains. [From Barron and Washington, 1984] *Courtesy of American Geophysical Union.*

ering plants was different from their recent counterparts because they were just evolving during this time and may not have completely defined the range of their ecological niches. Nevertheless, Cretaceous crocodiles also occur in Mongolian deposits (Lefield, 1971) and there are numerous examples of Cretaceous dinosaurs from the Gobi Desert which support the conclusion of relatively warm temperatures. Thus, the effect of altered land–sea distributions alone on the atmosphere does not seem adequate to explain the distribution of fauna and flora.

8.2.3 Ocean Changes

It is of interest to inquire whether higher polar temperatures in the mid-Cretaceous may be due to changes in poleward ocean heat transport. At present, the ocean is responsible for about one-third of the poleward heat transport (von der Haar and Oort, 1973; Oort and von der Haar, 1976; cf. Section 2.1.2) and past changes could perhaps have modified temperatures in high latitudes. Because we are still in the early stages of paleoclimate application of ocean circulation models, we will discuss a number of approaches that have attempted to bypass some of the modeling limitations.

The first attempt at Cretaceous ocean circulation used a rotating tank model of the Northern Hemisphere, complete with reconstructed continental configurations (Luyendyk et al., 1972). The experiment was an extension of a technique introduced by von Arx (1952, 1957) when he reproduced the major features of the present ocean circulation. The primary result of interest to changes in high-latitude temperatures is that the Gulf Stream penetrated to the latitude of Nova Scotia before turning eastward toward Europe (this feature is verified by presence of tropical microfossils in mid-Cretaceous sediments from this region; Berggren and Hollister, 1974).

A different approach to evaluating the role of the ocean was taken by Barron and Washington (1982b) and Schneider et al. (1985). These authors imposed higher SST as a boundary condition and changed the SST gradient from pole to equator; thus they could simulate the effects of varying oceanic heat transport. In the first case, Barron and Washington (1982b) utilized a "perpetual" version (cf. Sec. 2.3.2.2) of the GCM and stipulated that Cretaceous SST should not fall below 10°C. Temperatures were still below freezing in continental interiors (Fig. 8.8).

Schneider et al. (1985) examined the case of uniform ocean temperatures in order to test the extreme assumption of infinitely efficient poleward heat transport. SST everywhere was stipulated at 20°C. Rather than using the ensemble average value of modeled winter temperatures, individual "weather" events (i.e., components of the average) were analyzed to determine whether below-freezing temperatures occurred at any time during the model run. This approach was motivated by the knowledge that isolated freezes are often sufficient to prevent establishment of permanent tropical flora. Individual realizations of the model are of course less believable than the ensemble average. Nevertheless, the exercise is instructive. Results indicate that certain weather events would still at times produce below-freezing temperatures in high latitudes. This result does not disprove the importance of the ocean's role in maintaining an ice-free state; it merely states that it is more difficult to demonstrate its importance.

An analysis of model output clarified why increased ocean heat transport did not mitigate winter freezing. Schneider et al. (1985) concluded that a very weak latitudinal temperature gradient weakens the baroclinic component of atmospheric circulation, particularly transport of heat from the ocean to continental interiors. In the extreme case, radiative cooling in continental interiors cannot be compensated by weak inflow of warm air from the oceans; winter temperatures therefore drop below freezing. An additional consequence of enhanced ocean heat transport is that temperatures decrease in the tropics (Covey and Thompson, 1989)—a result that cannot be verified at the present time (cf. Section 8.1.3).

An ocean GCM simulation for the Cretaceous utilized results from an atmospheric simulation to force the ocean circulation (Barron and Peterson, 1989). One of the surprising results of this experiment (Fig. 8.10) was that surface circulation in the equatorial Tethys Seaway was from west to east, opposite that inferred by geologists (e.g., Gordon, 1973). The differences stem from the fact that earlier reconstructions assumed the equatorial current followed geography. In the ocean GCM the northern rim of Tethys was in the westerly wind belt (which did not migrate northward; see Section 8.2.6). The combination of westerly winds at midlatitudes and easterly winds in the tropics caused a gyral circulation in the Tethys. Barron and Peterson (1989) note that this simulation has considerable implications for proposed pathways of dispersal of marine organisms.

Although ocean circulation changes may not provide the explanation for Cretaceous high-latitude warmth, understanding the Cretaceous ocean may be very important for examining other processes. For example, Barron (1985b) simulated likely zones of coastal upwelling based on model-generated wind fields (Fig.

Fig. 8.10. Model-simulated surface currents for the mid-Cretaceous. [From Barron and Peterson, 1989] *Reprinted with permission from Science 244, 684-686, copyright 1989 by the AAAS.*

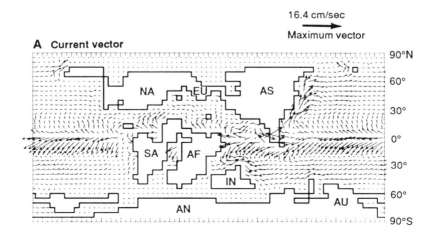

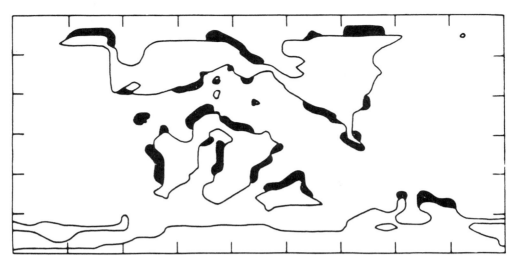

Fig. 8.11. Predicted areas of coastal upwelling in the mid-Cretaceous. Prediction based on simulations of ocean surface divergence fields as forced by simulated atmospheric winds. [From Barron, 1985b] *Reprinted by permission from American Association of Petroleum Geologists.*

8.11). Results bear some resemblance to areas of organic carbon deposition (cf. Fig. 8.6). Additionally, Parrish et al. (1984) summarize work on oceanographic features of the Cretaceous western interior seaway of North America (Fig. 8.1). Tidal fluctuations in the shallow seaway may occasionally have been very significant (cf. Klein and Ryer, 1978; Slater, 1981). The prevailing westerly wind field was favorable for localized upwelling along the western edge of the seaway, especially during storms.

8.2.4 Role of CO_2

To summarize results of the above modeling studies, all reasonable (and some unreasonable) changes in model boundary conditions cannot explain greater warmth in high latitudes, especially in the winter. This result leads to one of two possible conclusions: either the geologic data are open to major reinterpretation or some other factor has to be considered. It is difficult to conclude that the many different types of data discussed in the first part of this section could conspire so consistently to mislead us about the nature of mid-Cretaceous climates. There is, however, one very significant deficiency in the data. Almost all of them are from maritime regions—either nearshore marine or coastal areas (Barron and Washington, 1982b). Winter

temperatures in such regions are notably milder than in continental interiors, and extrapolation of these conditions to continental interiors may not be justified. As noted earlier, there are only a handful of records from continental interiors in the mid-Cretaceous (Barron and Washington, 1982b). It would clearly be of interest to examine more records from such regions.

In order to explain Cretaceous warmth it may be necessary to invoke higher atmospheric CO_2 concentrations (e.g., Rubey, 1951; Budyko and Ronov, 1979). Calculations with the NCAR GCM (Barron and Washington, 1985) suggest that a quadrupling of CO_2 levels could bring latitudinal and seasonal temperature profiles more in line with minimum estimates from geologic data (Fig. 8.12). Values six to eight times present concentrations would be necessary to explain maximum Cretaceous temperatures (Barron and Washington, 1985). It is interesting to note that flowering plants, which first appeared in the mid-Cretaceous, photosynthesize at optimum rates of CO_2 concentration that are up to five times higher than present levels (Leopold, 1964).

Higher Cretaceous atmospheric CO_2 levels would be due to increased tectonic activity. Berner et al. (1983; see also Lasaga et al., 1985) calculated that CO_2 levels in the mid-Cretaceous may have been as much as 10 times greater than the present (cf. Fig. 10.26). Levels are higher for two reasons: volcanism was greater (volcanism is the primary source for atmospheric CO_2) and the area of continents was smaller (atmospheric CO_2 is removed by the weathering of silicates on continents; since higher sea levels covered 20% of the continents, this rate of removal may have decreased).

In addition to possibly "closing the circle" between model and data, higher atmospheric CO_2 levels may account for the strong correlation between high sea level and warm climate (Damon, 1968). Rather than directly causing warm climates, high sea level may be due to increased magma upwelling and higher plate velocities, with concomitant thermal expansion of the crust and upper mantle, displacing oceanic waters onto the continents (Hays and Pitman, 1973; Pitman, 1978).

If the CO_2 hypothesis is valid, Barron and Washington (1985) calculated that there may have been a significant strengthening of the hydrologic cycle in the Cretaceous. This is because higher atmospheric temperatures would result in higher concentrations of water vapor in the atmosphere. The increased availability of latent heat could have led to more severe storms during the Cretaceous (Barron, 1989).

Weathering of material on the continents may also have varied with CO_2 levels. Although GCM experiments involving a future CO_2 warming decreased precipitation on continents with the present configuration (Washington and Meehl, 1984), Cretaceous simulations with the same model increased runoff (Barron and Washington, 1985), because there is substantially less Cretaceous land area in subtropical regions dominated by the downward branches of the Hadley circulation (Fig. 8.13). An ocean GCM experiment indicates that higher evaporation rates in the subtropics caused increased formation rates of warm, saline bottom water (Barron and Peterson, 1990). Removal of atmospheric CO_2 by increased weathering may also have partially offset the postulated lower cycling rates due to less land area (cf. Berner et al., 1983).

Higher CO_2 would cause two feedbacks that would amplify or maintain high polar temperatures. Because warm air holds more water vapor than cold air, there would be an increased latent heat transport to high latitudes (Barron and Washington, 1985). However, this did not quite compensate for a decreased sensible heat transport due to a smaller latitudinal temperature gradient. A similar pattern develops in climate models that simulate the effects of a future CO_2-induced warming (Manabe and Wetherald, 1980).

In another experiment with a coupled ocean–atmosphere model with simplified geography, Manabe and Bryan (1985) found that with higher CO_2 ocean heat transport was comparable to that of the present (Fig. 8.14). Even though latitudinal temperature gradients decreased, transport remained high because increased CO_2 caused higher tropical SSTs. Since the coefficient of thermal expansion of seawater

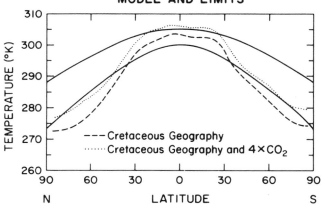

Fig. 8.12. Cretaceous zonally averaged surface temperature limits (°K) in comparison with Cretaceous model-derived surface temperatures for the geography and geography-plus-CO_2 quadrupling experiments. Solid lines equal estimated range of Cretaceous temperatures based on data. [From Barron and Washington, 1985] *Courtesy of American Geophysical Union.*

CRETACEOUS PRECIPITATION SENSITIVITY

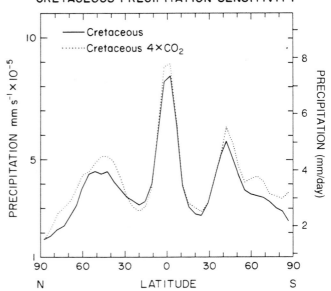

Fig. 8.13. Latitudinal sensitivity of GCM simulated precipitation rate for a quadrupling of CO_2 with Cretaceous geography. [From Barron and Washington, 1985] *Courtesy of American Geophysical Union.*

Fig. 8.14. Streamlines illustrating the meridional circulation in a coupled ocean–atmosphere model for six differnt levels of CO_2 forcing. Units are in Sverdrups (10^{12} g/sec). Note that the overall mass transport tends to remain relatively stable for different levels of increased CO_2 but decreases significantly for lower CO_2. [From Manabe and Bryan, 1985] *Courtesy of American Geophysical Union.*

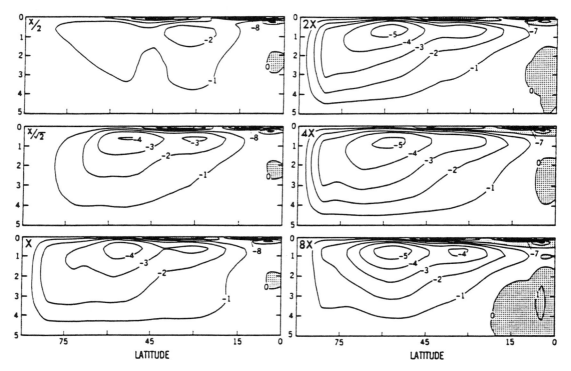

is very sensitive to changes involving warm SSTs, higher values reduced tropical densities, resulting in approximately the same horizontal buoyancy flux as at present.

8.2.5   Evidence for Higher CO_2 Levels?

Although the CO_2 model is appealing, it is important to verify the model with evidence for increased volcanism and proxy geochemical data for CO_2 fluctuations. The former problem is potentially difficult to evaluate, for it is well known that volcanism is episodic and the spatial distribution of volcanos very heterogeneous. Sampling problems involving space and time scales must therefore be addressed.

Despite the potential sampling obstacles, there is a rather consistent set of diverse geologic data supporting the conclusion of higher degrees of mantle activity in the mid-Cretaceous. For example, large regions of the western Pacific (the "Darwin Rise") are populated with volcanically formed seamounts that originated during this time (Arthur et al., 1985b; e.g., Ribe and Watts, 1982). Analysis of tectonic terrains in western North America has identified mid-Cretaceous fragments that originated in the South Pacific and were transported to their destination at seafloor spreading rates several times greater than the present maximum (e.g., Alvarez et al., 1980b; Tarduno et al., 1986). Average rates of the velocities of major plates were higher in the Late Cretaceous and ocean ridge volumes were greater (cf. Fig. 10.27; Davis and Solomon, 1981; Kominz, 1984). Subduction and volcanism would therefore be expected to increase. Finally, there was a Phanerozoic peak in the emplacement of rocks of deep-crustal origin (e.g., kimberlites, carbonatites), which are an index of convective activity (MacIntyre, 1971; 1973). Thus, despite sampling obstacles, geologists and geophysicists have established with a fair degree of confidence that rates probably did change.

A major problem with the CO_2 scenario is that insufficient proxy evidence has been gathered to rigorously support the CO_2 model. Some progress has been made, however. For example, increased frequency of storm-bed thicknesses in the Cretaceous (Brandt and Elias, 1989) are consistent with higher CO_2 levels (cf. Section 8.2.4).

Arthur et al. (1985a; cf. Dean et al., 1986) note that $^{13}C/^{12}C$ ratios of Cretaceous organic carbon are 5–7‰ lighter (i.e., more negative) than Holocene organic matter. They suggest that increased isotopic fractionation could be due to relatively low overall oceanic productivity coupled with higher partial pressures of CO_2. Initial calculations based on $\delta^{13}C_{org}$ fractionation suggest mid-Cretaceous CO_2 levels at least three times greater than the Holocene (Rau et al., 1989). Preservation of different species of carbonate minerals also support higher mid-Cretaceous CO_2 levels (cf. Fig. 11.5; Sandberg, 1983). One calculation estimates maximum CO_2 levels about seven times greater than the present (Berger and Spitzy, 1988).

Although the above calculations are encouraging, the reality of higher CO_2 levels is so important that it needs to be tested with as many different proxy indices as possible. There is a need for intercomparison of different CO_2 proxy estimates and for continued use of the refined CO_2 estimates in additional climate model simulations. Full evaluation of the CO_2 paradigm will only be possible after these steps are taken.

8.2.6   Implications of Modeling Studies for Some Venerable Geologic Assumptions

Some results of GCM calculations contest conventional geologic wisdom about the nature of circulation in warm climates. For example, it is often assumed that winds are weaker during warm periods (due to decreased thermal gradients). However, Barron and Washington (1982a) calculated that zonally averaged winds increased in their Cretaceous simulation. In retrospect, this result is not surprising. Atmospheric motion is greatly dependent on the amount of latent heat released by tropical convection. Since saturation vapor pressure is nonlinearly related to surface temperatures, warm or warmer Cretaceous SST will cause higher levels of latent heat release in the atmosphere (cf. Kraus, 1973). In terms of the "hierarchy of believability" of climate model results (Chap. 2), the conclusion about wind strength is fairly reliable because tropical SST exerts a fundamental constraint on the atmospheric circulation. One strong, direct link between SST and the atmosphere involves the temperature/saturation vapor pressure relationship.

Stronger winds also caused an *equatorward* displacement of subtropical highs in the NCAR GCM (Barron and Washington, 1982a). This result is consistent with evidence from evaporite deposits (Gordon, 1975; cf. Fig. 13.5) indicating an equatorward displacement of this rock type in the Cretaceous (this sediment most commonly forms in regions of high net evaporation). Both sedimentological and modeling results therefore oppose conventional hypotheses of a uniform poleward displacement of atmospheric circulation systems during warm periods.

A further sensitivity experiment eliminated another possible explanation for high-latitude warmth—that the obliquity of the earth's axis (now 23.5°) was less in the past (e.g., Wolfe, 1978). This mechanism receives little support from either celestial mechanics considerations (Ward, 1982) or climate models (Barron, 1984). In the climate sensitivity experiment, Barron demonstrated that a reduced obliquity actually reduces annual insolation receipt in high latitudes, with the net result being a substantial cooling.

8.3 Summary

Geologic evidence indicates greater warmth in mid- and high latitudes during the Cretaceous. However, seasonal cooling, perhaps even freezing, may have occurred. Although warmer low latitudes are often assumed, there is insufficient information to support this conclusion. Modeling studies indicate that changes in land–sea distribution are only partially responsible for Cretaceous warmth. Changes in ocean heat transport also cannot explain the warmth. Higher CO_2 is a likely candidate, with atmospheric levels increasing due to enhanced mid-Cretaceous volcanism. There is a considerable need for additional multiple-proxy evidence to test the hypothesis of higher Cretaceous CO_2 levels.

9. ENVIRONMENTAL CONSEQUENCES OF AN ASTEROID IMPACT

The end of the Cretaceous (66 Ma) marks one of the most spectacular events in earth history. The catastrophic extinction which was used to define the end of a geologic era coincided with the elimination of an estimated 75% of all living species (Russell, 1979). Some of the extinctions were very abrupt. For example, some biotic turnover occurred over a sedimentary interval of only a few centimeters (Smit and Hertogen, 1980; Preisinger et al., 1986). A number of groups were adversely affected (e.g., Russell, 1982; Alvarez et al., 1984b; Surlyk and Johansen, 1984), with some reduced in diversity (e.g., brachiopods, bryozoans, planktonic foraminifera) and others completely eliminated (e.g., dinosaurs, pterosaurs, plesiosaurs, some groups of Mesozoic mammals and other reptiles, ammonoids, rudist bivalves, some other mollusk groups, and some plants). However, a number of scientists continue to point out that some extinctions occurred before the end of this period (e.g., Maurasse, 1988; Ward and MacLeod, 1988; Keller, 1989) and that some groups, such as the benthonic foraminifera, were not affected at all (Thomas, 1988).

In this chapter we will review the asteroid impact hypothesis for the Cretaceous–Tertiary (K-T) extinction and recount some environmental and climate phenomena postulated to have occurred. We will also briefly discuss evidence for the event. The latter is somewhat of a departure from the theme of the book, but the conjectured environmental consequences are so severe (and so interesting) that it is useful to inquire into the strength of the evidence. Although the asteroid hypothesis has been challenged on a number of grounds (e.g., Officer and Drake, 1983, 1985; Hallam, 1987; Courtillot et al., 1988; Duncan and Pyle, 1988), we concur with Alvarez et al. (1984a) and Alvarez (1987) that the weight of evidence is in favor of an impact. This argument does not prove the point, however. We therefore adopt the attitude that we

will write the chapter *as if* an asteroid impact had occurred. We prefer to remain more noncommittal on the somewhat related subject of periodic impacts (e.g., Raup and Sepkowski, 1984, 1986; cf. Stigler and Wagner, 1987; Rampino and Stothers, 1988). The paleontological record clearly illustrates that extinctions occur every 20–30 million years (Fig. 9.1), but it is not clear whether the fluctuations are truly periodic (i.e., whether the variance is explained by a narrow spectral band; cf. Section 9.3.1).

9.1 Evidence

In 1980, Alvarez et al. (1980a) published their famous paper proposing that the K-T extinctions were at least in part caused by the impact of a 10-km asteroid (cf. Hsü, 1980; Smit and Hertogen, 1980). Evidence was based originally on high abundances of iridium (Ir) in K-T boundary layer clays (e.g., see below, Fig. 9.4), abundances that could only be attributable to material of extraterrestrial origin. [Although Ir has also been found in a Kilauea (Hawaii) volcanic eruption (Zoller et al., 1983), overall assessment of the evidence suggests a nonvolcanic source for the K-T peak. Furthermore, there is no historical evidence of Kilauea-type eruptive plumes penetrating the stratosphere—a condition necessary for worldwide distribution.]

Insufficient evidence for elevated amounts of Plutonium-244 eliminates the possibility of a supernova origin for the Ir peak (Alvarez et al., 1980a), and the authors suggested that the most likely candidates involved a class of chondritic meteorites with earth-crossing orbits. Since the publication of the 1980 paper, our appreciation of the composition of earth-crossing objects has increased considerably. The number of earth-crossing comets is more than four times greater than the number of earth-crossing asteroids. Production of large (>150 km) craters, the size produced by a 10-km impact, is probably dom-

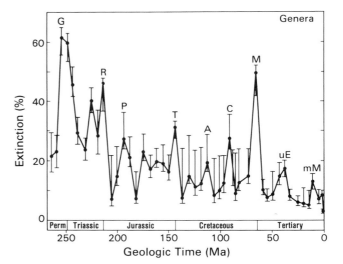

Fig. 9.1. Percentage of extinction of marine animal genera in stratigraphic intervals between the mid-Permian and the Recent. Letters represent different geologic ages; the "M" (for Maastrichtian) marks the K-T boundary. [After Raup and Sepkowski, 1986] *Reprinted with permission from Science 231, 833–836, copyright 1986 by the AAAS.*

inated by comets (Shoemaker et al., 1988). The mean rate of collision of 10-km earth-crossing objects is on the order of once every 100–200 million years (Shoemaker et al., 1988). The collision rate for smaller objects is correspondingly higher. In 1983, the comet IRAS-Araki-Alcock, with a major axis of 9.3 km, came within 4.7 × 10⁶ km of the earth (Shoemaker et al., 1988). Its collision might have produced some of the scenarios we will discuss below.

Since the initial paper by Alvarez et al., (1980a), the iridium anomaly has been mapped as a worldwide event. It is a very rare event in earth history (Fig. 9.2; Kyte and Wasson, 1986). Some Ir anomalies occur in the Late Eocene (~40 Ma; Hut et al., 1987) and Late Miocene (~10 Ma; Asaro et al., 1988), but they are not nearly as prominent as the K-T event. In fact, in terms of magnitude and sharpness, the K-T Ir peak may be unique in the entire Phanerozoic (Orth and Attrep, 1988).

Further support for the extraterrestrial origin of the Ir anomaly comes from a variety of ingenious efforts. Additional siderophile (iron-rich) elements typical of chondritic meteorites (scandium, titanium, chromium, rhodium) have been found in anomalous abundance at the boundary (Gilmore et al., 1984). ¹⁸⁷Osmium/¹⁸⁶Osmium ratios and rhodium abundance in boundary layer sediments are also more typical of extraterrestrial origin (Luck and Turekian, 1983; Bekov et al., 1988). Two types of amino acids have been found that are rare on earth but

more abundant in carbonaceous chondrites (Zhao and Bada, 1989).

Other evidence supports an impact origin for some altered terrestrial material. Quartz grains with shock-metamorphic features (Bohor et al., 1984) bear mute witness to the event. A very dense form of quartz (stishovite), which forms only under very high pressures typical of an impact, has been found in K-T boundary sediments from the southwestern United States (McHone et al., 1989). However, one line of supporting evidence for an impact has been reinterpreted. Sanidine spherules (potassium feldspar) were initially thought to be altered impact droplets of basaltic composition (Smit and Klaver, 1981; Montanari et al., 1983). These spherules now appear to have an entirely terrestrial explanation (Izett, 1987).

Where did it hit? Quartz and feldspar abundances suggest that the impact took place at least in part on a continent (Bohor et al., 1984, 1987; Owen and Anders, 1988). There is some intriguing evidence pointing to the North American sector as an impact site. Floral extinctions reached their greatest percentage in northern midlatitudes, particularly northwestern North America and eastern Asia (Hickey, 1981). Analysis of boundary layer sediments of the Raton Basin of southeastern Colorado identified impact ejecta (Fig. 9.3) which, based on analogies with sizes associated with other craters, suggests impact within 2000 km of the site (Bohor et al., 1987). Seismic studies identified a 35-km crater

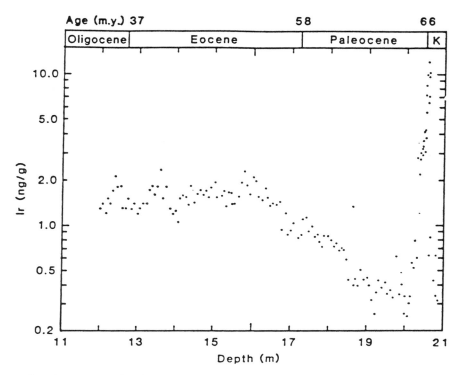

Fig. 9.2. Profile of Ir concentrations in a 9-m section of piston core from the central Pacific gyre. The only anomalously high Ir peak is observed at the K-T boundary. [From Kyte and Wasson, 1986] *Reprinted with permission from Science 232, 1225–1230, copyright 1986 by the AAAS.*

Fig. 9.3. The maximum size (mm) of shock-metamorphosed quartz grains in K-T boundary sediments plotted on a paleocontinental map (Early Paleocene time). The size of locality dots indicates that shock-metamorphosed minerals are many orders of magnitude more abundant at North American sites relative to elsewhere in the world. [From Izett, 1987] *Reprinted with permission of Geological Society of America. From G. A. Izett, "Authigenic 'spherules' in K-T boundary sediments at Caravaca, Spain, and Raton Basin, Colorado and New Mexico, may not be impact derived," 1987, Geol. Soc. Amer. Bull. 99, 78–86.*

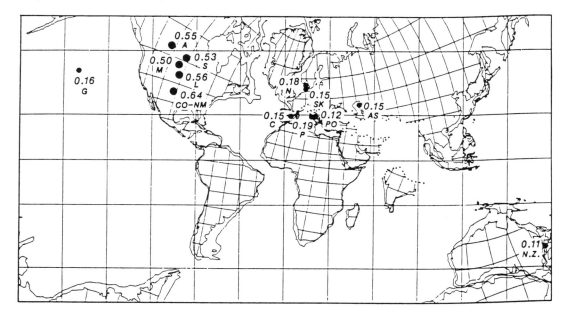

buried beneath Pleistocene drift in Manson, Iowa. Radiometric dates yield an age indistinguishable from the boundary (Kunk et al., 1989). However, this crater does not seem large enough to explain the global scale perturbations. There may have been more than one impact due to the breakup of the bolide. Evidence for a large tsunami (and therefore oceanic impact) in some sections of the U.S. Gulf Coast supports this hypothesis (Bourgeois et al., 1988; Hildebrand and Boynton, 1988).

9.2 Environmental Perturbations

9.2.1 Atmospheric Effects

Environmental consequences of an impact read like a litany from the Apocalypse. Although this subject is relatively open to conjecture, some K-T scenarios are on fairly solid ground. Passage of a bolide through the atmosphere should generate about 10^{30} ergs of energy (O'Keefe and Ahrens, 1982). The shock wave generated by a 20 km/sec impact probably had a devastating effect on forests and large animals over a very large area (Russell, 1982). In parts of western North America and Japan, pollen records (Fig. 9.4) indicate a drastic reduction in angiosperm pollen, with fern spores reaching 99% of the sample (Pillmore et al., 1984; Tschudy et al., 1984; Saito et al., 1986). This distribution implies an open vegetation with a lack of large plants and a large scale to the perturbed region.

The impact fireball far exceeded in explosive power the 10^{26} ergs assumed for nuclear war (e.g., Turco et al., 1983). The channel swept out

Fig. 9.4. Coincidence of iridium enhancement and drop in the ratio of fossil pollen to fern spores. These data, from the Raton Basin of New Mexico, indicate that plants as well as animals felt the effect of a bolide impact. The drop in the ratio of angiosperm pollen to fern spores indicates a shift in the flora from Cretaceous species to Tertiary species. The extrema of the two peaks occur in the same centimeter of stratigraphic height. [From Alvarez, 1987] *Reproduced with permission by the American Institute of Physics.*

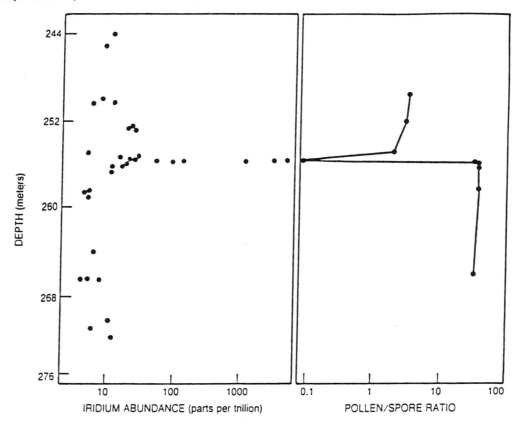

through the atmosphere by the incoming meteorite must have been filled with extremely hot, low-density plasma, which afforded an outlet for venting the impact fireball to space (Lewis et al., 1982). Turbulent inflow into both the low-density column and the fireball would have produced winds with a characteristic speed of Mach 1 (Lewis et al., 1982). Much of the dust generated by the impact may have escaped through the hole in the atmosphere created by the bolide with global dispersal occurring in the hours-to-days timeframe along suborbital trajectories.

There may have been significant chemical effects related to an impact. Heat generated in the atmosphere would cause reactions between atmospheric O_2, N_2, and H_2O to produce nitric acid (Lewis et al., 1982; Prinn and Fegley, 1987). The amount generated would strongly depend on the composition of the bolide and size of the bolide (Prinn and Fegley, 1987). If it was a very large ice-rich comet, the pH could have been reduced to ~0–1.5 globally. If it was an iron-rich asteroid, the pH would be ~0–1 near the impact and 4–5 globally. Concentrated acid rain would pollute freshwater reservoirs, significantly affecting the calcareous plankton (these organisms were among the most devastated of fossil groups). The effect of acid fallout would be to some extent selective. Silica-secreting organisms, or those living in alkaline lakes, in burrows, or the deep sea would be less affected by the acid rain (Prinn and Fegley, 1987). Concentrated acid rain, and intense continental weathering, may also explain a $^{87}Sr/^{86}Sr$ pulse at the K-T boundary (MacDougall, 1988). The concentrated acid rain scenario may be too extreme. It seems unlikely that tropical plants, which do not have the same protective hard seed coating as high-latitude plants, could have survived at a 50% rate (cf. Hickey, 1981) if the global pH was 0.0–1.5. Additional calculations using a more realistic bolide diameter and impact velocity indicate much smaller acidity changes (Thompson and Crutzen, 1990).

High atmospheric NO_x would also defoliate plant leaves and asphyxiate lung-respiring animals. It would scavenge ozone from the stratosphere, permitting harmful ultraviolet radiation to penetrate to the surface (Lewis et al., 1982; Thompson and Crutzen, 1990). If the impact was in the Northern Hemisphere, ozone destruction may have spread only slowly to the Southern Hemisphere. In addition to acidity changes, high trace element levels (e.g., nickel) in the meteorite may have led to dissolved values in lakes, rivers, and the sea many times present levels and possibly reaching toxic concentrations (Erickson and Dickson, 1987).

There could have been a "nuclear winter" component to the K-T event. Concentrations of elemental carbon in the boundary layer are equivalent to 5–10% of the present biomass of the earth (Wolbach et al., 1985, 1988). This implies that much of the earth's vegetation burned and/or that substantial amounts of fossil fuels were ignited (again, this estimate seems to contravene tropical plant survival rates, although distinction between biomass reduction and species loss may alleviate the problem). This interpretation is supported by organic geochemical evidence for a type of hydrocarbon that is characteristic of combustion (Venkatesan and Dahl, 1989). The pyrotoxins formed during a "global fire" would be harmful to land life. Carbon monoxide levels would have been distinctly toxic. Co-occurrence of soot and Ir suggests that the fires were generated by the impact rather than following sometime later as a result of massive burning of dead vegetation (Wolbach et al., 1988).

9.2.2 Climate Effects

The impact was probably associated with some climate anomalies, which may have varied by time scale (O'Keefe and Ahrens, 1982). Initially, the fraction of bolide energy transferred to the atmosphere could have increased temperatures perhaps 30°C (O'Keefe and Ahrens, 1982), with the characteristic time for a radiative perturbation of this type being about 30 days. In addition, very substantial amounts of thermal radiation would be produced by the ballistic re-entry of ejecta condensed from the vapor plume of the impact (Melosh et al., 1990). Calculations suggest that the global radiation flux emitted from the descending ejecta could reach values 50–150 times the solar output for periods ranging from one to several hours. These power levels are comparable to that obtained in a domestic oven set at 'broil' (Melosh et al., 1990).

The combined heating of the atmosphere

from kinetic energy transfer through the shock wave and the radiative flux from heated ejecta could have been responsible for the ignition of global wildfires (see previous section). The temperature increase alone may have been intolerable to many biota, especially large vertebrates (Emiliani et al., 1981). No terrestrial vertebrate larger than about 25 kg is known to have survived the extinction event (Russell, 1982).

The huge dust cloud generated by the impact would block out significant amounts of solar radiation over a large but undetermined area for a period of several weeks to several months (the maximum gravitational settling time for such particles). Photosynthesis could have been severely inhibited (Alvarez et al., 1980a). Early calculations suggested that the K-T cloud could have lowered average surface temperature by several tens of degrees (Toon et al., 1982; Pollack et al., 1983). In fact, the calculations resemble initial estimates of the climate effect from nuclear winter smoke clouds (the initial nuclear winter calculations were done by this same group using a modification of the same model; cf. Turco et al., 1983). However, these results are from a very idealized model, and the estimated magnitude of the warming is susceptible to considerable modification (see text following).

The nuclear winter analogy can be extended to derive a first-order picture of regional variations in the temperature response. If the impact did in fact occur in northern midlatitudes (the same assumptions as for a nuclear winter smoke cloud), current GCM simulations would predict greater temperature decreases over land than over sea, but a significantly smaller temperature response than in the simpler models (e.g., Malone et al., 1986; Schneider and Thompson, 1988; Turco et al., 1990). The regional pattern is primarily due to the different heat capacities of land and sea (see Chap. 1) and is therefore a moderately reliable estimate. Experience with nuclear winter models suggests that temperature decreases might also be larger in summer than in winter because insolation is sufficiently low in winter to mute the effects of increased optical depth.

Temperature decreases in GCM runs for an asteroid impact (Fig. 9.5) are consistent with the nuclear winter results—the regional pattern and magnitude of the cooling is much less than in the simpler models. Thus, the GCM results do not support the concept of an "asteroid winter" (if dust dispersal was along suborbital trajectories, this conclusion could be modified). However, the altered vertical temperature profile and enhanced atmospheric stability (due to massive

Fig. 9.5. Simulated near-maximum surface temperature response to increased atmospheric dust loading from an asteroid impact. Note that large changes are almost entirely restricted to landmasses, that the changes are patchy, and that the magnitude of such changes is considerably smaller than was earlier calculated from simpler models. [From Thompson and Covey, 1990] *Reprinted with permission of the Geological Society of America. From S. Thompson and C. Covey, in V. L. Sharpton and W. Ward (Eds.), "Global Catastrophes in Earth History." Geol. Soc. Amer. Spec. Paper, 1990, in press.*

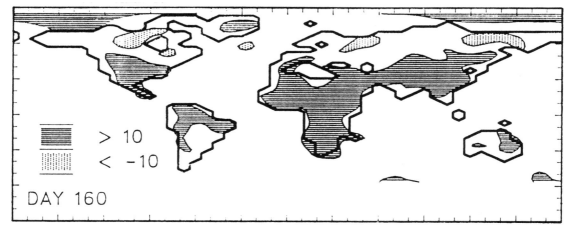

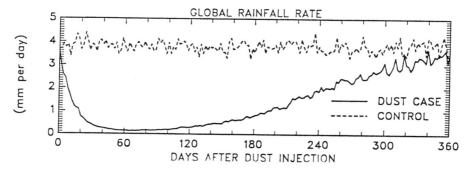

Fig. 9.6. Time variations of global rainfall after an impact. Note the collapse of the global hydrologic cycle for approximately 1 year. [From Thompson and Covey, 1990] *Reprinted with permission of the Geological Society of America. From S. Thompson and C. Covey, in V. L. Sharpton and W. Wards (Eds.), "Global Catastrophes in Earth History." Geol. Soc. Amer. Spec. Paper, 1990, in press.*

dust loading) caused almost a complete collapse of the hydrologic cycle in a GCM (Fig. 9.6; Thompson and Covey, 1990). This reduction may have contributed to extinctions on land.

Regional patterns in North American plant fossils suggest that a brief, low-temperature excursion may have occurred (Wolfe and Upchurch, 1986; cf. Fig. 9.4). In the northern high plains, plant extinction rates are 50–60% (Hickey, 1981), whereas rates are much lower in Colorado (Wolfe and Upchurch, 1986). However, the occurrence in the northern high plains of presumably cold-blooded, "survivor" reptiles (e.g., turtles, crocodiles) in the earliest Paleocene also suggests that the cold excursion was not so severe or prolonged as to wipe out refugia (see Wolfe and Upchurch, 1986). These data are essentially consistent with the GCM simulations (Fig. 9.5).

9.2.3 Aftermath

Despite the severity of the K-T event, analyses of plant distributions immediately above the horizon indicate that some features of the earth's environment returned rather quickly to the background state. For example, in south-central Saskatchewan (Canada), relatively high abundances of species of palm and screw pine, similar to those presently living in southeastern Asia, indicate that no profound and lasting paleoclimatological change accompanied the K-T event (Nichols et al., 1986). Other data suggest a more gradual return to equilibrium and that there may have been a greenhouse aftermath (Hsü

and McKenzie, 1985). Analysis of $\delta^{13}C$ in deep-sea sediments shows a significant excursion at the boundary (e.g., Fig. 9.7), with the magnitude of the effect varying according to the species an-

Fig. 9.7. Carbon isotope gradient in the earliest Tertiary. The anomalous carbon isotope gradient of dissolved carbonate in the earliest Tertiary may reflect a drastically reduced level of surface water biomass (cf. Sec. 6.4.5). The altered surface ocean carbon budget may in turn have increased atmospheric CO_2 levels. [From Hsü and McKenzie, 1985] *Courtesy of American Geophysical Union.*

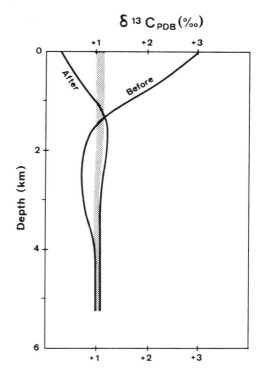

alyzed (Stott and Kennett, 1989). The changes can be interpreted in terms of decreased biomass returning isotopically light $\delta^{12}C$ to the ocean, thereby altering $^{13}C/^{12}C$ ratios (see Section 6.4.5). Similar patterns are known at some other extinction boundaries (Magaritz, 1989).

Reductions of carbon in the organic carbon sink should have been associated with increases in atmospheric CO_2. A greenhouse effect might have ensued, with atmospheric CO_2 levels increased by a factor of approximately 2–3 (Hsü and McKenzie, 1985). Other processes may also have increased global average temperatures. For example, if an impact was in a shallow-marine carbonate environment, vaporization (plus feedbacks) could increase CO_2 levels several fold (O'Keefe and Ahrens, 1989). Additional heating could result from biota-induced changes in cloud cover (Rampino and Volk, 1988; cf. Section 3.1.9). The combined effect of these processes could exceed 10–15°C (global average temperature).

The above scenarios present a serious problem when tested against the geologic record. Although plant remains in North America support a ~10°C mean annual temperature increase for perhaps 0.5–1.0 million years after the K-T boundary (Wolfe, 1990), the deep-sea $\delta^{18}O$ record is much more ambiguous in its support of the greenhouse model. Some $\delta^{18}O$ records indicate a slight warming immediately following the K-T event (Boersma, 1984). However, others do not or, if they do, they have been interpreted as a diagenetic overprint resulting from subsequent interactions with pore waters (Zachos and Arthur, 1986; Margolis et al., 1987). Save for one $\delta^{18}O$ record, there is no reliable evidence for either warming or cooling across the K-T boundary (Zachos and Arthur, 1986; Zachos et al., 1989). This is a severe discrepancy between theory and observation, and one that has not been adequately addressed.

By analogy with calculations of the future greenhouse effect, a single CO_2 excursion would last at least a few thousand years before being buffered by ocean carbonates (e.g., Broecker and Takahashi, 1977; Sundquist, 1985). However, the virtual elimination of carbonate–secreting organisms in the deep sea implies that the relaxation time would have been significantly longer during the Paleocene. For example, carbonate production and marine productivity remained low for 500,000 to a million years after the K-T (Zachos and Arthur, 1986; Zachos et al., 1989). Hsü and McKenzie (1985) postulate that lowermost Tertiary nannoplankton blooms (cf. Perch-Nielsen et al., 1982), commonly associated with population explosions of opportunistic species in an unstable environment, may have caused repeated rapid fluctuations in CO_2 exchange between the ocean and the atmosphere. Detailed sampling above the K-T boundary may help to further clarify the nature of any longer term effects of the impact. Sampling strategies must also take into consideration the fact that considerable aliasing of data can result from irregular sample intervals (Pisias and Prell, 1985); switching from low-resolution to high-resolution sampling can artificially introduce variability that may not exist in the original record.

Some additional work on planktonic foraminifera indicates highly fluctuating variations after the K-T (Gerstel et al., 1987). However, it may not be justified to interpret biotic fluctuations in terms of changing environments when the populations are not in equilibrium. Population size can theoretically vary greatly during a period of low competition, i.e., during nonequilibrium situations. As predator–prey relations approach equilibrium, fluctuations may vary much less as a consequence of biotic interactions. The time series for such a population might resemble fluctuations of a damped harmonic oscillator (Fig. 9.8).

Although there may have been a very severe environmental disruption at the K-T, and perhaps a CO_2 aftermath lasting tens of thousands of years, there does not seem to be any strong evidence that the impact affected the long-term evolution of the climate of the earth. The "mean background climate state" of the latest Maastrichtian (~67 Ma, i.e., just before the impact) does not seem greatly different from the "postaftermath" early Paleocene (~64 Ma).

9.3 Complications with Impact Theories

9.3.1 Further Comparisons with Extinction Events

As pointed out at the beginning of this chapter, there is substantial evidence supporting an asteroid impact at the K-T. There are also some

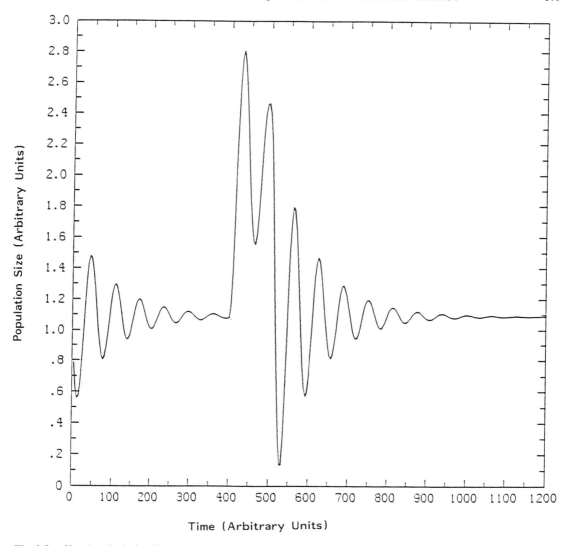

Fig. 9.8. Simple calculation illustrating that population size (y axis) can vary as a function of prey abundance, with the latter varying with time (x axis) as an ecosystem recovers from a catastrophic perturbation (beginning at $t = 400$). The calculation assumes that reduction in predators (due for example to an impact) leads to a rapid expansion of prey, which overshoot their equilibrium population level. Rapid expansion of predators in turn leads to a reduction in prey population below their equilibrium level. The system gradually converges on the original equilibrium population values. The relationship is described in terms of two first-order differential equations and resembles the fluctuations of a damped harmonic oscillator. *Courtesy of William T. Hyde.*

vexing problems that are difficult to explain with the asteroid theory. Clear evidence of extinctions and cooling (Fig. 9.9) prior to the boundary (and any Ir anomaly) indicates that more than one process is operating in the system (Johnson and Hickey, 1988; Zachos et al., 1989; Stott and Kennett, 1990), as does a long-term sea level fall which does not appear to be related to changes at the boundary (Hallam, 1987). A very significant outpouring of lava on the Dec-

can Plateau (India) may also have contributed to climate change over the K-T interval (e.g., McLean, 1985; Courtillot et al., 1988; Duncan and Pyle, 1988). Thus, the total amount of Late Cretaceous biotic variance explained by a K-T asteroid impact has not been fully clarified, and there is possibly a need to integrate the effect of an asteroid impact with changes due to a terrestrially driven biotic turnover.

It is also not clear how much asteroid impacts

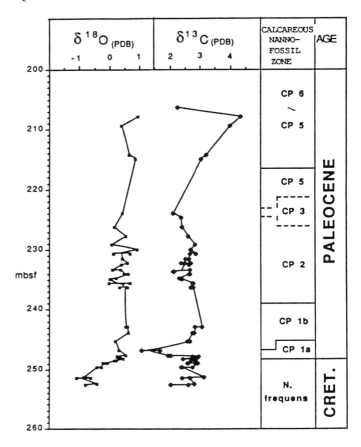

Fig. 9.9. Evidence for $\delta^{18}O$ changes prior to the K-T. Results are from Ocean Drilling Program samples from the Southern Ocean (Maud Rise). mbsf = meters below sea floor. [From Stott and Kennett, 1988; based on data in Stott and Kennett, 1990.]

may have contributed to other extinction events or to the long-term evolution of climate. As noted earlier, there is some statistical evidence for periodic extinctions. This result suggests extraterrestrial forcing. However, we believe that additional work is required in order to determine whether the 20- to 30-million year signal is truly periodic or represents a characteristic time scale in the terrestrial system. For example, if one examines maps of continental distributions every 20 million years (e.g., Barron et al., 1981b), there is the impression that about 20–30 million years is required for the system at time $t + 1$ to look significantly different from the system at time t. This "self-similarity" time scale is set by the rates of plate motion (about 2–5 cm/year) and could result in a similar time scale of climate variance related to changing solid-earth boundary conditions. (In Section 11.2.3 we discuss another example of how plate tectonic-driven processes can impart cli-

mate variance into the system on this time scale.)

Thus, in order for the system to be truly periodic, it would have to have a narrow-band spectrum. The standard deviation of "errors" (differences between predicted extinction peaks and observed extinction peaks) in the Raup and Sepkowski (1984) periodicity study are 3.85 and 5.65 million years (two different time scales were used). Is this a narrow or a broad band? To our knowledge the distinction has not yet been tested.

There are other difficulties associated with extrapolating the asteroid hypothesis to a more general cause of extinctions. Although there are some indications of smaller Ir anomalies near some other extinction events, sampling of the geologic record has so far not yielded convincing evidence that other extinction events are associated with iridium anomalies nearly as dramatic as the K-T (Orth and Attrep, 1988; Zhou and Kyte, 1988; Holser et al., 1989). The environmental effect of smaller impacts must be examined more closely, particularly with respect to whether global scale perturbations can result. Not surprisingly, initial calculations (Covey et al., 1990) suggest that smaller impacts do have a smaller effect on temperatures than is illustrated in Fig. 9.5. A very promising time interval to examine in this respect is ~73 Ma, for there is good evidence for two large (~70 km) craters associated with an impact in northern U.S.S.R. (Koeberl et al., 1990). Most of the environmental scenarios developed for the K-T should also be applicable to this time interval. In some respects it may eventually prove more interesting than the K-T, because here we are starting with the "smoking gun" of a known impact, whereas all the craters for the postulated K-T event have yet to be found (see Section 9.1).

It is conceivable that smaller impacts discussed above could have triggered a climate change poised to occur, i.e., when the system was unstable and highly sensitive to small perturbations (cf. Section 1.3.4 and Crowley and North, 1988). For example, there is evidence for a small (0.5-km) ocean impact in the Late Pliocene (2.3 Ma). Kyte et al. (1988) suggested that this impact may have triggered the Late Pliocene transition in Northern Hemisphere ice volume (Section 10.1.10). However, the most recent date for initial cooling now seems to be about 2.6 Ma (Cieselski and Grinstead, 1986; Jansen et al., 1988). Thus, we are not convinced that smaller but more frequent asteroid impacts significantly affected the long-term evolution of climate.

Another difficulty with generalizing the asteroid–extinction connection involves the observation that some extinctions represent localized population declines. For example, the Pliensbachian (195 Ma) and Tithonian (150 Ma) extinctions are apparently due to local sea level lowering rather than an asteroid impact (Hallam, 1986b). The Cenomanian–Turonian (90 Ma) extinction appears closely related to an ocean anoxic event (Arthur et al., 1988). However, in many cases marine invertebrates show largely congruent rises and falls in extinction intensity throughout the Phanerozoic (Raup and Boyajian, 1988). The most likely explanation for this congruence is that extinction is physically rather

than biologically driven and that it is dominated by the effects of geographically widespread environmental perturbations influencing most habitats (Raup and Boyajian, 1988).

9.3.2 Other Endogenous Mechanisms for Abrupt Change

Considering the mixed evidence for a dominant control of asteroid impacts on climate and biota, it is useful to consider "earth-bound" explanations, such as volcanoes (McCartney et al., 1990) or abrupt change due to terrestrially induced climate instabilities (Crowley and North, 1988). In this book we have discussed models and data that support this concept (Chaps. 1 and 10; cf. Crowley and North, 1988). Elsewhere we have summarized evidence that abrupt climate changes and extinctions often coincide (Crowley and North, 1988). Climate instability near "critical points" in the climate system (cf. Fig. 1.14) may also disperse unstable behavior over a longer but still geologically brief interval of time and possibly account for the "stepwise" nature of some extinctions, i.e., that rather than being a sharp transition, extinctions are characterized by a number of steps occurring within a geologically brief interval.

Abrupt climate shifts could potentially be very rapid. For example, the rate of change of $\delta^{18}O$ at the end of the last interglacial was comparable, in terms of geologic resolution (5–10 cm) to the Ir anomaly at the K-T (cf. CLIMAP, 1984, Fig. 2). The actual elapsed time would be very different—a few thousand years versus an instant. In another example, the mean background state of the atmosphere (at least in the North Atlantic sector) may have changed from glacial to interglacial in 20 years (Dansgaard et al., 1989; cf. Herron and Langway, 1985; and Fig. 3.20).

It has sometimes been argued that climate change did not cause the extinctions because very few extinctions occurred in the late Pleistocene. This argument fails to consider the time aspect involved in extinctions (cf. Cronin, 1985). For example, the three major climate transitions in the Cenozoic (Late Eocene/Early Oligocene, Mid-Miocene, Late Pliocene) all coincide with extinction events. Stanley (1988) argued that many of the forms sensitive to extinctions were eliminated at these times. In other words, the ecological niches for a number of groups that were very sensitive to extinctions had already been removed millions of years before the Pleistocene. Furthermore, organisms continue to be affected by late Cenozoic climate change. For example, van der Hammen et al. (1971) illustrate that the great reductions in late Cenozoic plant diversity in Europe are related to cooling conditions of the late Cenozoic. However, the reductions did not all occur at once; rather, it can be demonstrated that each cooling event was associated with the removal of some forms. The cumulative effect of many cooling events was to remove many forms.

There is little evidence for an ice volume increase or thermohaline instability at the K-T. In fact, deep-water environments did not change much (Thomas, 1988)—a trend very different from Pleistocene ice volume changes (cf. Chap. 6). However, the climate transition prior to the K-T (Fig. 9.9) could reflect a climate instability. At present we are still at an exploratory stage on the subject. We believe that the connection between climate instabilities and extinctions should be kept in mind as we further explore the interactions between climate, life, volcanos (?), and asteroids.

9.4 Summary

There is a growing body of evidence for an asteroid/comet impact at the K-T boundary. The environmental consequences of such an impact are quite severe. Present calculations suggest that chemical changes may have been more extreme than the climate effect. Calculations also suggest that there was a CO_2 aftermath to the extinction. This prediction is presently not supported by $\delta^{18}O$ records.

The extent to which extraterrestrial impacts affect biota is a matter of healthy debate. A number of scientists continue to point out that some extinctions occur before the K-T boundary. Extinctions could be due to some combination of terrestrial and extraterrestrial causes. It has also been suggested that extinctions were periodic—a suggestion that would strongly implicate extraterrestrial forcing. This conclusion is also a matter of debate. There continues to be a stream of alternate explanations for rapid climate change and biotic turnover. Future work will no doubt shed more light on this fascinating topic.

10. THE LAST 100 MILLION YEARS

So far we have discussed climates of the warm mid-Cretaceous (Chap. 8) and of the Quaternary (Chaps. 3–7). This chapter will examine the evolution of climate from the mid-Cretaceous to the significant expansion of Northern Hemisphere ice cover at about 2.4 Ma. As we will see there is a considerable amount of information available about evolution of climate over this time interval. Some of the points we will discuss involve (1) general trends in climate; (2) fluctuations in climate, one of which resulted in the warmest interval of the Cenozoic (reexamines the issue of ice-free climates); (3) evidence for cooling; (4) trends in aridity and deepwater circulation; (5) some severe swings in climate in the late Miocene/Pliocene; (6) onset of midlatitude Northern Hemisphere glaciation; (7) progress in modeling climate change over the last 100 million years, including separate discussions on atmosphere, ocean, and geochemical models; and (8) summary, including a discussion of major problems and opportunities.

10.1 Geologic Evidence

10.1.1 Paleogeographic Changes

Before embarking on the nature of the evidence, it is useful to briefly review plate tectonic changes that occurred over this time interval (Fig. 10.1). The widening of the North and South Atlantic continued after initial rifting in the Jurassic. Westward motion of North and South America caused increased convergence along their western margins, with resultant formation of the western Cordillera–Andes mountain chain. Likewise, the convergence of Africa against Europe and the Arabian and Indian subcontinents against Asia resulted in formation and uplift of the Alpine–Himalayan mountain chain. Further constriction of the circumglobal equatorial Tethys Seaway (see Sec. 8.1.1) resulted from northward movement of Australia/

New Guinea and emergence of the Central American isthmus.

In polar regions, important seaways developed between North America and Europe and in the Southern Ocean. Initial rifting in the Norwegian Sea began about 60 Ma, but final separation did not result until about 40 Ma (Talwani and Eldholm, 1977). In the Southern Hemisphere, final separation of Australia from Antarctica was completed by about 40 Ma. The development of a Circumpolar Current was completed with the opening of the Drake Passage about 10 million years later (Barker and Burrell, 1977). The opening was initially shallow water and did not widen to allow deep transport until sometime later (the exact timing of Drake Passage opening requires more work.)

10.1.2 Late Cretaceous/Early Cenozoic Warmth

Although late Cretaceous temperatures (e.g., Fig. 10.2) were cooler than the mid-Cretaceous thermal maximum (~90–120 Ma), values remained relatively high into the early Cenozoic (Lowenstam and Epstein, 1954; Gordon, 1973; Douglas and Woodruff, 1981; Wolfe and Upchurch, 1987; Parrish and Spicer, 1988). In the Northern Hemisphere, the Arctic Ocean was probably at least seasonally ice-free during the Late Cretaceous (e.g., Clark, 1988), for siliceous organisms suggest seasonal upwelling (Kitchell and Clark, 1982). Campanian (75 Ma) black mud with terrestrial plant material reflects greater levels of plant fertility in the circum-Arctic (Clark et al., 1986). This interpretation is supported by plant fossils from the North Slope of Alaska (about 75° paleolatitude; Parrish and Spicer, 1988). In northern Alaska, Parrish and Spicer (1988) estimated that mean annual temperatures were warmer than present but about 5°C lower than the mid-Cretaceous thermal peak (Chap. 8). Abundant dinosaur remains can

20 MILLION YEARS

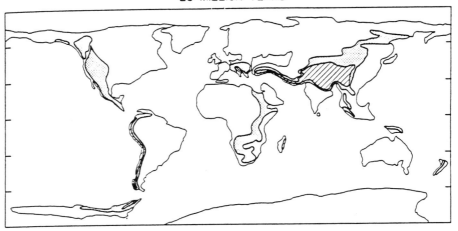

40 MILLION YEARS

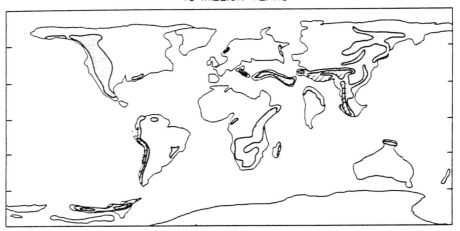

60 MILLION YEARS

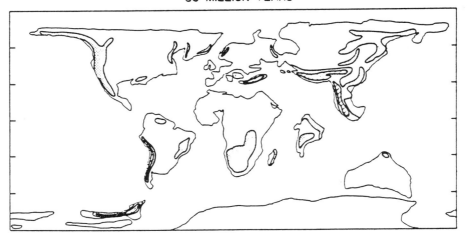

Fig. 10.1. A series of paleogeographic maps for Paleocene (60 million years), Eocene (40 million years), and Miocene (20 million years). Each map indicates shoreline position, paleolatitude, and topography (low = no shading; middle = stipple; and high = lines). [From Barron, 1985a] *Reprinted with permission of Elsevier Science Publishers.*

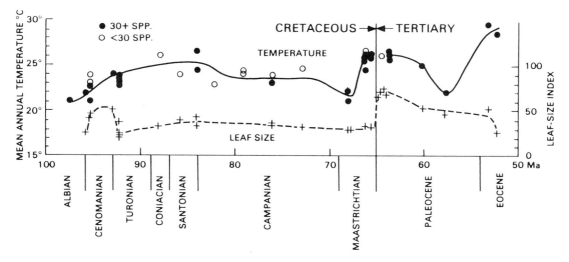

Fig. 10.2. Evidence for temperature changes during the late Cretaceous and early Cenozoic. Plots are for annual mean temperature at 30°N paleolatitude in the southeastern United States. Estimates are based in part on percentage of entire-leafed margins in plants. [From Wolfe and Upchurch, 1987] *Reprinted with permission of Elsevier Science Publishers.*

also be found in this region (Brouwers et al., 1987), but there is some difference of opinion as to whether seasonal migration occurred for these organisms (Brouwers et al., 1987; Parrish et al., 1987). If they overwintered, it is not clear how the organisms spent their time in the dark winter months.

In the Southern Hemisphere, $\delta^{18}O$ records suggest temperatures 9–14°C on the Antarctic Peninsula in the Campanian (75 Ma; cf. Barrera et al., 1987 and Pirrie and Marshall, 1990). These records also indicate at least a 2°C cooling between the Campanian and Maastrichtian (~74–66 Ma). A similar cooling occurred in the Northern Hemisphere (Parrish and Spicer, 1988). Similar temperatures from some deep-water deposits suggest that high latitudes were at least a partial source of deep waters during this time of warmer climates (Saltzman and Barron, 1982; cf. Section 8.1.2). However, bottom-water temperatures were significantly warmer than present for much of the late Cretaceous and early Cenozoic (Fig. 10.3). There are also frequent abrupt transitions in the deep-water record, which suggest that the system crossed threshold values at numerous stages, a response that may reflect instabilities in the climate system (Section 10.2.4).

Warm temperatures depicted in the deep-sea $\delta^{18}O$ record (Fig. 10.3) are supported by records

of sea level change (Fig. 10.4). Although the seismic sea level record may reflect in part expansion of ocean ridges and cooling of aging lithosphere along continental margins (Watts, 1982; Kominz, 1984), correlations of horizons with isotope and faunal transitions imply a significant glacio-eustatic component (e.g., Loutit and Kennett, 1981; Loutit and Keigwin, 1982; Miller et al., 1987b). Higher sea levels in the late Cretaceous and early Cenozoic are consistent with decreased ice volume.

Relative warmth continued into the early Cenozoic (Paleocene and Eocene; ~65–40 Ma). Plant fossils reflect such conditions in the high latitudes of both the Northern and Southern Hemispheres (cf. Andrews et al., 1972; Wolfe, 1978; Axelrod, 1984; Creber and Chaloner, 1985). In the Southern Hemisphere, the distribution of warm floras in continental interiors (Fig. 10.5) suggests that warmth was not just seasonal (Axelrod, 1984). The latter author suggested that polar expansions of warm flora can be explained in terms of low seasonality, i.e., their limits are more dependent on minimum winter temperatures rather than maximum summer temperatures.

Early Cenozoic SSTs around Antarctica (Fig. 10.6) were at least 10°C warmer than present (Shackleton and Kennett, 1975a; Barrera et al., 1987). Results from the Ocean Drilling Program

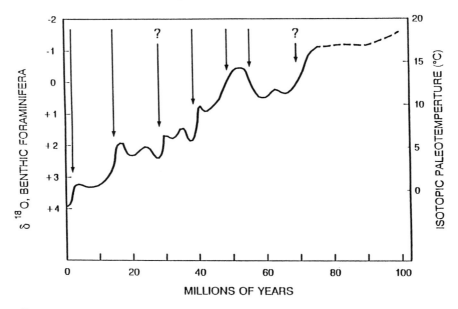

Fig. 10.3. $\delta^{18}O$ record of deep-water temperatures for the last 100 million years. Ice volume changes complicate the estimation of temperature over part of this record. Arrows indicate times of relatively abrupt transitions. [Modified from Douglas and Woodruff, 1981] *Reprinted with permission of R. G. Douglas and F. Woodruff, Deep sea benthic foraminifera, in "The Sea," v. 7, C. Emiliani (Ed.). Copyright 1981, John Wiley and Sons.*

(ODP) indicate planktonic assemblages suggestive of relative warmth near Antarctica from the late Cretaceous to the Eocene (Leg 113 Shipboard Party, 1987). This evidence is consistent with increased abundances of the clay minerals smectite and kaolinite, which are consistent with warm, humid conditions on Antarctica. Eocene pollen indicates a temperate beech forest, with ferns on the Antarctic peninsula (Leg 113 Shipboard Party, 1987).

There are some indications that Northern Hemisphere high-latitude warmth had a signifi-

cant seasonal signature. For much of the interval 40–80 Ma, a mixed conifer–hardwood forest, typical of seasonal vegetation, blanketed regions north of the Arctic Circle (Fig. 10.7; Azelrod, 1984). Similar evidence from the mid-Cretaceous suggests that this explanation may even apply in part to the thermal maximum of the last 100 million years (cf. Spicer and Parrish, 1986; Parrish and Spicer, 1988; Rich et al., 1988). Low winter temperatures are also indicated by scattered evidence for ice in Arctic rivers and coastal areas during parts of the late Cre-

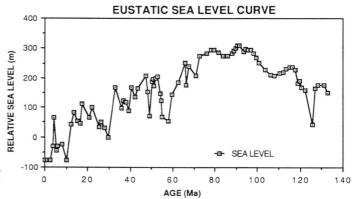

Fig. 10.4.   Estimated sea level changes over the last 140 million years. [Based on data from Haq et al., 1987]

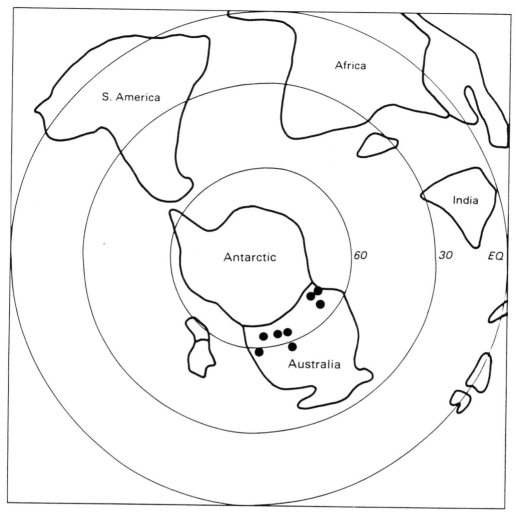

Fig. 10.5. Key early Cenozoic high-latitude plant sites in the Southern Hemisphere that support the interpretation of greater warmth during this time. [After Axelrod, 1984] *Reprinted with permission of Elsevier Science Publishers.*

taceous and early Cenozoic (Chumakov, 1981; Pickton, 1981).

The subject of seasonal versus year-round warmth is very important from a modeling perspective, since different mechanisms yield different signatures of warmth. For this reason we point out that even the presence of tropical marine deposits at high latitudes is an ambiguous indicator of year-round warmth at all longitudes. For example, mid- to late Cenozoic fauna from marine deposits east of Nova Scotia and from western Europe indicate very warm, tropical temperatures (e.g., Berggren and Hollister, 1974). However, low winter temperatures could

still have occurred on adjacent land masses. For comparison, the present day Gulf Stream flows within 100 km of the eastern North American shoreline as far north as Cape Hatteras. Yet winter temperatures inland can still be quite cold. A similar strong gradient can also be found in some high-latitude fossil sites. In Late Cretaceous (Maastrichtian, 70 Ma) deposits from the Labrador Sea, temperate latitude species in a nearshore site occur at about the same latitude as an offshore site dominated by tropical species (Gradstein and Srivastava, 1980).

Other evidence for post mid-Cretaceous climate change involves events at the Cenoman-

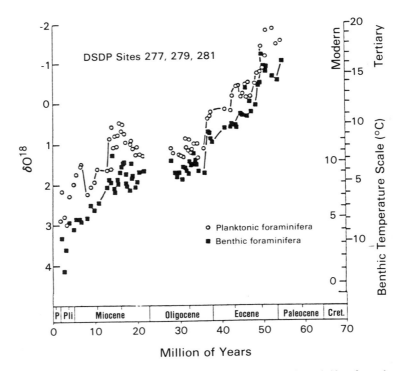

Fig. 10.6. Isotopic paleotemperature analyses of planktonic and benthonic foraminifera from the sub-Antarctic Pacific, indicating considerably warmer conditions in the Early Cenozoic. [From Shackleton and Kennett, 1985a]

Fig. 10.7. Key early Cenozoic high-latitude plant sites in the Northern Hemisphere that support the interpretation of greater warmth during this time. [After Axelrod, 1984] *Reprinted with permission of Elsevier Science Publishers.*

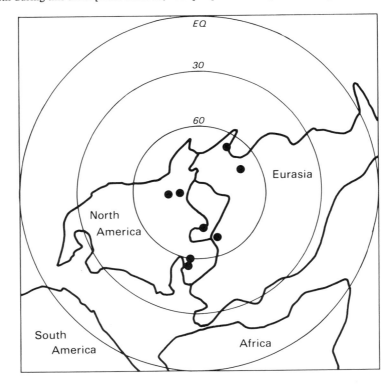

ian–Turonian (C-T) boundary (92 Ma). Arthur et al. (1987) summarize considerable evidence for a massive burial of organic carbon at the C-T boundary. These changes may indicate an enhanced level of oceanic productivity (Arthur et al., 1987). Additional measurements of the $\delta^{13}C$ of organic carbon show large excursions at the C-T. These changes have been interpreted as indicating changes in the partial pressure of dissolved CO_2 (Arthur et al., 1985a). The massive removal of carbon from the surface waters of the oceans may have caused a CO_2 drawdown (Arthur et al., 1988). Even with atmospheric CO_2 levels higher than the present, Arthur et al. (1988) calculate that the atmospheric reservoir could have been stripped of CO_2 in 120,000 years. This event could cause a significant climate cooling. There is some evidence for such a cooling (Arthur et al., 1985b; Arthur et al., 1988), but more is desirable. A modeling study suggests the high latitudes as the most likely site for the enhanced productivity (Sarmiento et al., 1988a).

10.1.3 Eocene Thermal Maximum

The Early Eocene (50–55 Ma) was the warmest period in the Cenozoic. Although Chap. 8 discussed at length the subject of an ice-free state, it is worthwhile to return to it because there are several advantages that Eocene deposits have over the Cretaceous deposits. Geographic control is better because there are many more sites from open oceans and continental interiors. Thus, there are more opportunities for isotopic inferences about the state of the open ocean. Deposits from continental interiors are also useful in discriminating seasonal from year-round warmth (deposits from coastal areas may indicate mild winters just due to the proximity of large bodies of water). Finally, Eocene organisms (both plants and animals) are much more similar to their present than their Cretaceous counterparts. Thus, paleoenvironmental interpretations are more reliable.

The Eocene warming is recorded in a number of climate indices. Bottom-water temperatures and sea level (Figs. 10.3 and 10.4) reached their highest level of the Cenozoic (Savin, 1977; Douglas and Woodruff, 1981; Haq et al., 1987). Rich tropical flora from the London Clay and

western North America indicate that tropical conditions extended to about 45°N paleolatitude (e.g., Reid and Chandler, 1933; Wolfe, 1980; Hubbard and Boulter, 1983; Axelrod and Raven, 1985). Laterite soil horizons indicative of warm temperatures with seasonal rainfall extend to 45°N in both hemispheres (Frakes, 1979). These and other climate indices suggest that tropical conditions extended 10–15° of latitude poleward of their present limits. Studies of North Atlantic plankton (foraminifer and coccoliths) indicate that the greatest poleward penetration of tropical assemblages occurred at this time (Haq et al., 1977). One of the most spectacular examples of high-latitude warmth comes from a rich fossil site from Ellesmere Island, west of Greenland (Dawson et al., 1976). Vertebrate fossils include alligators and flying lemurs, the latter occurring today only in southeastern Asia (Estes and Hutchinson, 1980; McKenna, 1980). Paleomagnetic results indicate that the paleolatitude of this site was about 78°N (McKenna, 1980).

The greater availability of open-ocean samples provides considerably more information about Eocene conditions than we have for the mid-Cretaceous (Section 8.1.3). Results suggest a low isotopic gradient in both hemispheres (Fig. 10.8) and isotopic SST estimates in tropical regions 4–7°C below present values (Shackleton and Boersma, 1981; cf. Boersma et al., 1987). This result could mean greater transport of heat out of the tropics into midlatitudes. However, calculations with an energy balance model (Horrell, 1990) suggest that the magnitude of required changes in poleward ocean heat transport are extremely large and very difficult to reconcile with theory. The $\delta^{18}O$ measurements could be affected by diagenesis (chemical alteration of sediment after deposition; Killingley, 1983) or salinity differences; increased tropical salinities and decreased poleward salinities could produce the same type of isotopic profiles (Poore and Matthews, 1984). We will return to the problem of Cenozoic tropical SSTs in Section 10.1.5.

The Early Eocene warming lasted for 2–5 million years and was associated with a significant change in the carbon reservoir (Fig. 10.9; Shackleton, 1985, 1986, 1987). There is an abrupt decrease in the bulk sediment deep-sea $\delta^{13}C$ record at the onset of the warming event (Paleocene–

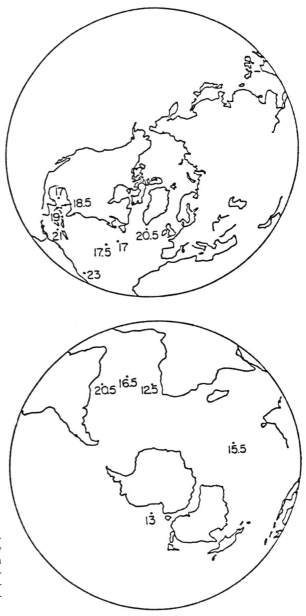

Fig. 10.8. Surface water $\delta^{18}O$ paleotemperature estimates for 50 Ma, which suggest an unusually low gradient between low and high latitudes. [From Shackleton and Boersma, 1981] *Reproduced by permission of the Geological Society from N. Shackleton and A. Boersma, 1981, Jour. Geol. Soc. (London) 138:153–157.*

Eocene boundary, about 55 Ma). A similar abrupt change occurs in mean size of eolian grains in Pacific sediments (Rea et al., 1985; Fig. 10.10), indicating a reduction in wind velocities. In neither record do values return to their previous levels after cessation of the warming event. This pattern suggests that a step function change in Cenozoic environments may have occurred at this time.

It has been suggested that the Early Eocene warming may have resulted from an increase in atmospheric carbon dioxide (Berner et al., 1983; Owen and Rea, 1985). A significant reorganization in plate motion occurred at this time and may have increased CO_2 through enhanced volcanism. The Norwegian–Greenland sea opened (Talwani and Eldholm, 1977) and volcanics were extruded throughout much of the northern

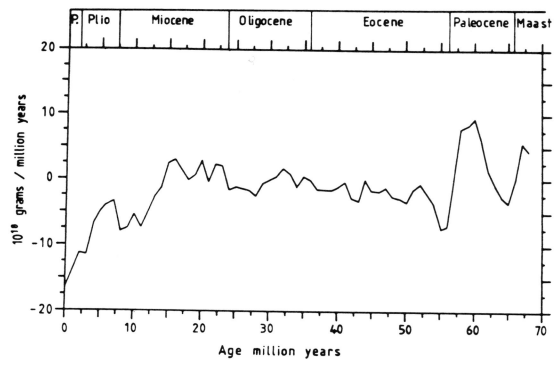

Fig. 10.9. Carbon isotope record of bulk sediment changes over the Cenozoic, scaled in terms of implied changes in the global organic carbon reservoir (10^{18} g/million years). Note the very large change at the Paleocene–Eocene boundary (~55 Ma) and relatively stable conditions for the next 30 million years. [From Shackleton, 1985] *Courtesy of American Geophysical Union.*

North Atlantic (the "Thulean" basalts; cf. Roberts et al., 1984). The latter authors estimated that the Thulean basaltic volume was similar to late Cretaceous Deccan basalts on India (cf. Duncan and Pyle, 1988; Courtillot et al., 1988). In addition, an increase in hydrothermal activity along spreading centers of midocean ridges released more calcium to seawater, which could have caused increased precipitation of calcium carbonate and release of CO_2 to the ocean and atmosphere (Owen and Rea, 1985). Although these changes could plausibly cause CO_2 increases, there is at present little proxy evidence for any significant changes during the Early Eocene.

10.1.4 Mid-Tertiary Transition

The Oligocene Epoch (~25–35 Ma) has long been considered transitional in nature between the warm periods of the early Cenozoic and the

Fig. 10.10. Grain size variations of both the total eolian component and of quartz grains extracted from a North Pacific core. Note the sharp reduction at the beginning of the Eocene (~55 Ma). [From Barron, 1987; based on data in Janecek and Rea, 1983] *Courtesy of American Geophysical Union.*

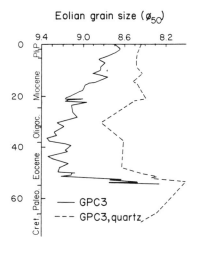

cold periods of the later Cenozoic. For a number of years there was a major debate about when ice first started to accumulate in polar regions—in the Oligocene or the Miocene (cf. Crowley, 1983b). This argument is of more than academic interest. The last 100 million years can be considered a yardstick against which to interpret climate change for as much as the last half of earth history. If one argument is valid, it suggests that glacial periods occur for perhaps only 15–20% of the last 2 billion years. If the other argument is valid, the number is much larger, perhaps greater than 50%, and previous conclusions about the general nature of climate through earth history need to be modified because they were based on a biased distribution of samples. In our opinion, the weight of evidence has now shifted to favor the latter model (cf. Matthews and Poore, 1980).

Following the Early Eocene thermal maximum, a significant cooling occurred over the interval 40–50 Ma (Fig. 10.3). This cooling may have contributed to the faunal turnover around 38–40 Ma—one of the largest extinction events in the Cenozoic (Corliss et al., 1984; Raup and Sepkowski, 1986; Hansen, 1987; Hut et al., 1987; Stanley, 1988). ODP drilling around Antarctica suggests that Antarctic glaciation may extend back to this time (Barron et al., 1989). Further changes occurred at about 34 Ma, with the appearance of ice-rafted debris in the Southern Ocean and glaciological evidence for some ice cover (e.g., Hayes and Frakes, 1975; Leckie and Webb, 1983; Barrett et al., 1987; Leg 113 Shipboard Party, 1987; Leg 119 Shipboard Party, 1988). A dramatic benthic $\delta^{18}O$ change at about 34 Ma (Fig. 10.3) has been interpreted in terms of development of a more vigorous, colder deep-water circulation (Kennett and Shackleton, 1976). Closely spaced core samples show that this cooling occurred in less than 100,000 years (Kennett and Shackleton, 1976) and was associated with a marked depression of the calcium carbonate compensation depth in the Pacific (van Andel, 1975). The deep-water interpretation is supported by evidence from some deep-sea fauna (Benson et al., 1984), which show a clear distinction between deep- and shallow-water faunas after the Eocene. The divergence may reflect development of a well-defined thermal stratification in the deep ocean.

Comparison of equatorial Pacific planktonic and benthonic $\delta^{18}O$ records indicates that about 0.3–0.4‰ of the $\delta^{18}O$ change at the 34-Ma transition can be attributed to ice volume changes (Keigwin, 1980; Keigwin and Corliss, 1986). [Covariance of planktonic and benthonic $\delta^{18}O$ records implies whole-ocean fluctuations (Shackleton, 1967; Matthews and Poore, 1980).] Utilizing a conversion factor of 0.1‰ $\delta^{18}O$ for a 10-m sea level change (Fairbanks and Matthews, 1978) yields a 30–40 m equivalent sea level change. This is about one-half the present volume of the East Antarctic Ice Sheet. The rest of the benthonic $\delta^{18}O$ change can be attributed to about a 3°C cooling of deep waters. This event therefore represents an important step in the evolution of Cenozoic climate and the development of the present deep-water circulation (Kennett and Shackleton, 1976).

Erosion on continental margins, plus deep-water and low-latitude surface water $\delta^{18}O$ records (Miller et al., 1987b), suggests a number of additional ice volume events in the Oligocene (Miller et al., 1987b). The most significant sea level event (Fig. 10.4) occurred at about 29–31 Ma and is associated with direct evidence for expanded Antarctic ice cover (Barrett et al., 1987), only a small change in benthic $\delta^{18}O$ (Fig. 10.3), but a larger change in planktonic $\delta^{18}O$. Prentice and Matthews (1988) argue that the latter record is a better measure of ice volume. There is widespread evidence for a significant cooling during the Oligocene. The pattern is recorded by reptiles, mammals, plants (e.g., Fig. 10.11), shallow-water invertebrates, and soil horizons (Wolfe, 1978; Hutchinson, 1982; Retallack, 1983; Zinsmeister, 1982; Axelrod and Raven, 1985; Prothero, 1985).

There were some peculiar features of the Oligocene ocean (e.g., Berger, 1981). Carbonate sedimentation rates were relatively low, as was planktonic diversity. There were also sporadic occurrences of "coccolith blooms" (*Braarudosphaera*) in Oligocene open-ocean sediments. Berger (1981) suggested that the Oligocene ocean had relatively low productivity. The relatively weak vertical thermal gradient (see below, Fig. 10.16) may have occasionally resulted in oceanic overturn and blooming of the conjectured "opportunistic species" in the suddenly fertility-enriched waters.

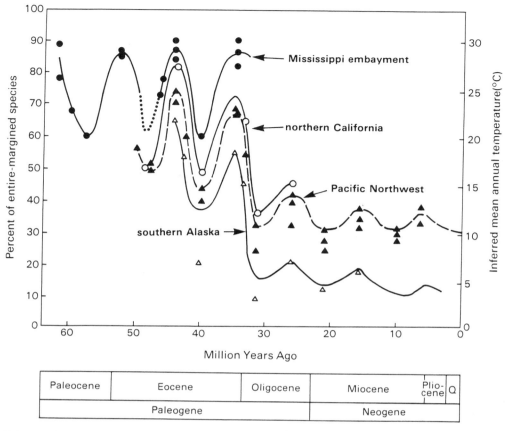

Fig. 10.11. Estimates of mean annual temperatures on North America, based on percentages of plant species with entire-margined leaves (an indication of warmth). It is inferred that a sharp drop in mean annual temperature took place in the Oligocene and has continued, at least at high latitudes, to the present day. At middle latitudes mean annual temperature has overall not changed since the Oligocene (Wolfe, 1978). Dotted intervals indicate that leaf margin data are either lacking or not considered reliable. [After Wolfe, 1978] *Reprinted by permission of American Scientist.*

10.1.5 Cenozoic Tropical SSTs Revisited

Changes in ocean heat transport have also been proposed for the Oligocene and younger intervals (cf. Section 10.1.3). This interpretation is based on altered gradients of $\delta^{18}O$ ratios in surface waters (e.g., Frakes and Kemp, 1973; Savin et al., 1975; Shackleton and Boersma, 1981; Keigwin and Corliss, 1986). However, the $\delta^{18}O$ data (e.g., Fig. 10.12) have been challenged on a number of grounds (see Section 10.1.3 and Matthews and Poore, 1980; Killingley, 1983; Poore and Matthews, 1984; Prentice and Matthews, 1988; Horrell, 1990). One objection involves the observation that SSTs did not change in the tropics at 18,000 BP (CLIMAP, 1981). Why should they have changed more during times of

presumably less ice volume (Matthews and Poore, 1980; Prentice and Matthews, 1988)? This argument certainly has merit; but it is not proof, and changes in the physical geometry of the ocean boundaries may affect ocean heat transport in unanticipated ways. Until we thoroughly understand the physical explanations for SST stability, we have to keep an open mind on the subject.

One way to investigate the possibility of larger tropical SST changes in the past is to examine other types of proxy information. For example, altered ocean heat transport should change gradients in both $\delta^{18}O$ (e.g., Fig. 10.12) and biotic distributions. Such a response is in fact consistent with changing abundances of tropical and high-latitude plankton in the South Pacific (e.g.,

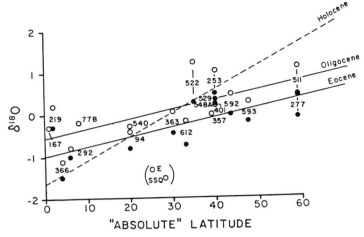

"ABSOLUTE" LATITUDE

Fig. 10.12. Average Late Eocene and Early Oligocene oxygen isotope results from surface-dwelling planktonic foraminifera, plotted against paleolatitude without respect for hemisphere ("absolute" latitude). Numbers refer to DSDP sites. The change in gradient may reflect changes in poleward ocean heat transport. The significant difference between the Eocene and Oligocene regression lines is the increase in the y intercept of 0.45‰, which probably reflects the effects of increased continental glaciation in earliest Oligocene time. [From Keigwin and Corliss, 1986] *Reprinted with permission of Geological Society of America. From "Stable isotopes in Late Middle Eocene to Oligocene foraminifers," L. Keigwin and B. Corliss, 1986, Geol. Soc. Amer. Bull. 97, 335–345.*

Lohmann and Carlson, 1981; Loutit et al., 1983; Kennett et al., 1985a). The distinctive out-of-phase relationship between low- and high-latitude groups is consistent with some changes in latitudinal temperature gradient (Fig. 10.13; Lohmann and Carlson, 1981).

There are additional types of biotic data that provide some support for the interpretation of lower tropical SSTs. Tropical shallow-marine mollusks and corals did not increase in diversity until the Late Cenozoic (Rosen, 1984; Valentine, 1984; Grigg, 1988). This pattern might be expected if diversity is primarily controlled by temperature (Valentine, 1984). In addition, pollen records from the Oligocene of Puerto Rico indicate an unusual level of temperate elements

Fig. 10.13. Temperature contrasts between Pacific low- and high-latitude sites over the interval 11–4 Ma. Note that when high-latitude assemblages cool, low-latitude assemblages tend to warm. These changes may indicate changes in ocean heat transport. Temperature estimates based on distributions of plankton (coccoliths) in Deep-Sea Drilling Project (DSDP) records. [After Lohmann and Carlson, 1981] *Reprinted with permission of Elsevier Science Publishers.*

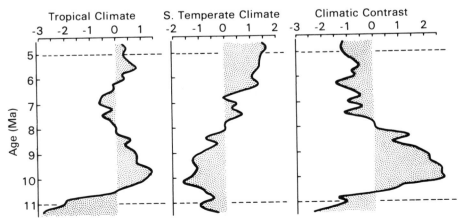

in the flora from this time period (Graham and Jarzen, 1969; see also Raven and Axelrod, 1974). We therefore believe that the argument on tropical SSTs should not be restricted to δ^{18}O; microfossils, plants, and shallow-water marine invertebrates may eventually help clarify the picture. [Note added in proof: further examination of the above groups provides additional support for the hypothesis that δ^{18}O-based SST estimates for the Cenozoic may be in error (Adams et al., 1990).]

10.1.6 Mid-Miocene Transition

Following a period of climate amelioration in the Early Miocene ($\sim$15–20 Ma), another significant change in climate occurred between 10 and 15 Ma (Figs. 10.3 and 10.4). A large δ^{18}O increase at 14–15 Ma, initially interpreted as the time of major Antarctic Ice Sheet growth (Shackleton and Kennett, 1975a), has also been described as primarily a deep-water cooling event (4–5°C) with some ice volume growth (Miller et al., 1987b; Moore et al., 1987). Expansion of Antarctic ice may have caused the sea level fall at 10–11 Ma (Fig. 10.14; Miller et al., 1987b; Moore et al., 1987). Greater variance in δ^{18}O records after this time is consistent with increasing ice volume fluctuations since the mid-

Miocene (Moore et al., 1981b). Decreases in low-latitude Pacific δ^{18}O have been interpreted as a warming of tropical regions as ice accumulated in high latitudes, i.e., the zonal temperature gradient in the ocean may have increased (e.g., Savin et al., 1985).

It therefore seems that there was extensive ice cover on Antarctica by $\sim$10 Ma, with significant ice being present back to at least 40 Ma. At present, ice cover on Antarctica consists of both East and West Antarctic ice sheets, separated from each other by the Transantarctic Mountains (Fig. 10.15). The East Antarctic Ice Sheet is an order of magnitude larger than the West Antarctic Ice Sheet (Denton et al., 1971). Together the ice masses compose a volume equivalent to about 55–60 m of sea level (Denton et al., 1971). Because the West Antarctic Ice Sheet is a marine-based ice sheet, it is considered to be considerably more unstable than the East Antarctic Ice Sheet (e.g., Stuiver et al., 1981).

Vincent and Berger (1985) proposed that major cooling in the Middle Miocene may be related to changes in atmospheric CO_2 levels. Significant changes in δ^{13}C occur in both surface- and bottom-dwelling organisms at this time (Vincent and Berger, 1985; Woodruff and Savin, 1985), and these changes precede δ^{18}O variations (Vincent and Berger, 1985). Large

Fig. 10.14. Sea level records for the Late Miocene, illustrating a significant drop at 10–11 Ma, plus 10^6-year scale oscillations. The two curves are similar, with the exception of two cycles, but the magnitude of the longer term fall on which the cycles are superimposed is much reduced in the curve of Moore et al. [After Moore et al., 1987; Haq et al. curve from 1987 paper.] *Courtesy of American Geophysical Union.*

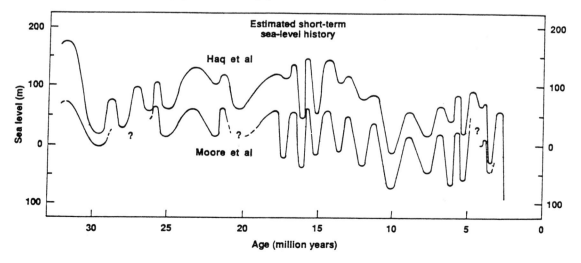

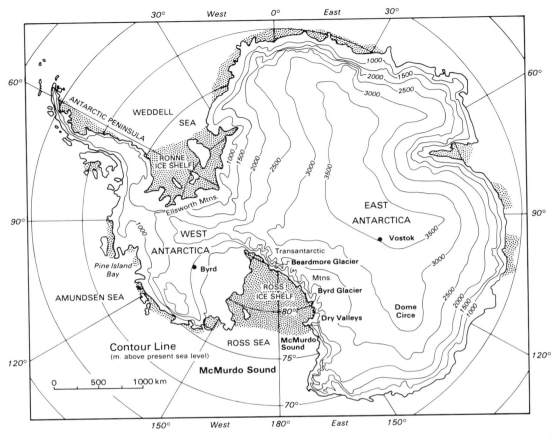

Fig. 10.15. Antarctic Ice Sheet, with ice shelves indicated by stippling. [Redrawn from Denton, 1985] *Reproduced with permission of South African Journal of Science.*

changes in upwelling may have caused a drawdown of atmospheric CO_2 levels and increased removal of carbon from the oceanic sink into sediments. Widespread occurrence of diatomaceous sediments around the rim of the Pacific basin (indicative of increased upwelling conditions) provide some support for this conjecture.

Another Miocene result of climate interest involves high-resolution $\delta^{18}O$ sampling of a deep-sea record at 16 ± 1 Ma (Fig. 10.16). This record indicates that the Miocene surface-bottom temperature difference in the Pacific was 12°C, one-half that at present (Shackleton, 1982). Furthermore, 100,000-year time scale fluctuations in this record are similar to those of the late Pleistocene, although the isotopic amplitude is only one-half of the later events. The 100,000-year amplitude occurs long before any conjectured Northern Hemisphere midlatitude glaciers, and has important implications for theories of the 100,000-year cycle (cf. Section 7.3.1 and 7.3.2).

10.1.7 Aridity Changes

There was a significant increase in aridity in the late Cenozoic. Wind-blown eolian deposits increase in the late Cenozoic (Rea et al., 1985). Vegetation shifts in high latitudes indicate larger seasonal swings of temperature, with vegetation types typical of drier conditions (Wolfe, 1985). Drying also occurred in midlatitude regions of western North America and Australia (Kemp, 1978; Axelrod and Raven, 1985). In Australia, there is a significant transition from widespread tropical forests in the Miocene to plant assemblages with an abundance of grasses (see Tedford, 1985). This transition coincided with an increased abundance of browsing and grazing herbivores, such as kangaroos (Tedford, 1985). Although some of this drying can be attributed to northward motions of Australia into the subtropical dry belt, that motion would only result in a horizontal translation of 200–300 km in 2–3 million years. Most of the drying therefore appears to reflect climate change (Tedford, 1985).

In tropical regions, plant fossils indicate a significant mid-Miocene expansion of savanna in eastern Africa (e.g., Axelrod and Raven, 1978; Yemane et al., 1985). Western India also became drier (Prakash, 1972). $\delta^{13}C$ analysis of soil horizons in Pakistan (Fig. 10.17) indicates a dramatic, abrupt change at 7.0–7.5 Ma (Quade et al., 1989). This has been interpreted as a change from forest to grassland and was probably associated with uplift of the Himalayas (cf. Section 10.2.1). Evolutionary radiation and dispersal of grazing herbivores (antelopes) at the same time in these areas support the interpretation of drying (Barry et al., 1985; Vrba, 1985). A further increase in African aridity occurred around 5–6 Ma and the Saharan flora, adapted to aridity, may date from this time or slightly later (van Zinderen Bakker and Mercer, 1986).

There is also some evidence for increased aridity in deep-sea sediments. Analysis of clay minerals in North Atlantic DSDP cores indicates a pronounced desert zone in the Miocene that does not appear in the Eocene (Chamley, 1979). (This conclusion was based on relative abundance of kaolinite and chlorite/illite; the former is indicative of moist, warm climates.) There are also increased concentrations of charcoal in North Pacific deep-sea sediments; drier conditions would be more conducive to fires (Herring, 1985). However, a South Pacific record does not show any net increase in eolian flux in the Late Cenozoic (Rea and Bloomstine, 1986). Overall, there is fairly widespread and coherent evidence for drier conditions in the late Cenozoic.

Fig. 10.16. Oxygen isotope records in planktonic and benthonic foraminifera, illustrating 100,000-year fluctuations in all records but decreased vertical temperature gradient in the Miocene (16 Ma). [After Shackleton, 1982] *Reprinted with permission from Prog. Oceanog. 11, 199–218, by N. Shackleton, copyright 1982, Pergamon Press plc.*

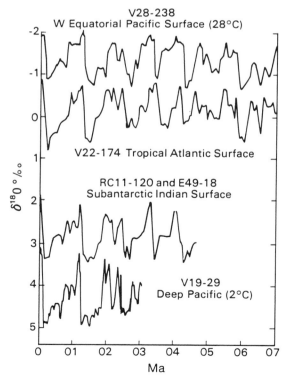

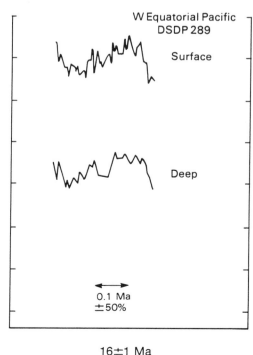

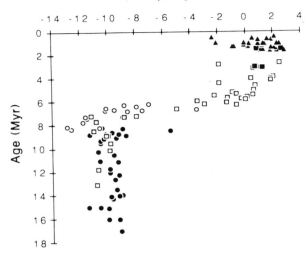

Fig. 10.17. $\delta^{13}C$ record of soil carbonate from Pakistan, illustrating a rapid change from 7.0–7.5 Ma that has been interpreted as indicating a transition to drier environments. [From Quade et al., 1989] *Reprinted by permission from Nature Vol. 342, pp. 163–166, Copyright (©) 1989 Macmillan Magazines Ltd.*

10.1.8   *Trends in Deep-Water Circulation*

At present there are two main sites of deep-water formation: the high latitudes of the North Atlantic and around Antarctica. In addition to the long-term cooling trend of Cenozoic deep waters (Fig. 10.3), there may also have been changes in the deep-sea circulation pattern. For example, it has long been suggested that the areas of expanded shallow seas in the tropics during the early Cenozoic may have been the site of warm, saline bottom-water formation (Chamberlin, 1906; Brass et al., 1982a). At present a small amount of very salty water forms in the Mediterranean. This water mass mixes with North Atlantic waters to form an intermediate water. If larger rates of the water mass were produced, they could have resulted in deep or bottom waters. Despite the attractiveness of this idea, there is at present relatively little firm evidence to actually support the existence of warm, saline bottom waters in the past.

A better documented long-term trend involves the history of North Atlantic Deep Water (NADW). This is one of the two main deep water masses in the world ocean. Various types of geologic records indicate that NADW production has been at least intermittent back to the early Oligocene ($\sim$34 Ma; Miller and Tucholke, 1983; Miller and Fairbanks, 1985). However, these records also indicate that in general NADW production was less than present until about 7–10 Ma, after which it switched to a pattern more typical of the present, but still with some tendencies for weaker flow (Keller and Barron, 1983; Miller and Fairbanks, 1985; Woodruff and Savin, 1989).

Some of the above conclusions are based on differences in the preservation of biogenous sediments between the Atlantic and the Pacific. At present the North Atlantic is a site of relatively young deep waters, which favor preservation of carbonate over silica in the sediments (Berger, 1970). The opposite preservation pattern holds for the North Pacific. However, if we examine the temporal patterns of biogenous sediment preservation for the late Cenozoic (Fig. 10.18), then the North Atlantic prior to about 10 Ma has a preservation pattern more typical of the North Pacific at present. This pattern holds for deep-water ^{13}C and cadmium/calcium trends, plus carbonate and silica deposition patterns (van Andel et al., 1975; Keller and Barron, 1983; Miller and Fairbanks, 1985; Woodruff and Savin, 1989; M. Delaney, personal communication, 1990).

10.1.9 Fluctuations of Climate in the Late Miocene and Early Pliocene

There are some climate variations of considerable interest that took place during the Late Miocene and Early Pliocene, with indications of significant changes in Antarctic Ice Sheet size. During the Late Miocene (5–7 Ma) the Antarctic Ice Sheet may have exceeded by perhaps 50% its glacial maximum dimensions (Fig. 10.19a). This may have happened two or three times. At another time it may have been reduced from its present dimensions by perhaps half (Fig. 10.19b). As the stability of the Antarctic Ice Sheet is of considerable climate interest, it is useful to examine the nature of the evidence.

Based on $\delta^{18}O$ analysis of DSDP samples, Shackleton and Kennett (1975b) first proposed a significant expansion of the Late Miocene Antarctic Ice Sheet. Denser sampling of this interval indicates that cold excursions about 5.0–5.5 Ma were followed by a warm event between

about 5.0 and 4.0 Ma (Hodell et al., 1986). During this phase of glacial growth, the Ross Ice Shelf expanded to the edge of the continental shelf (Hayes and Frakes, 1975) and there was a strong northward penetration of the Antarctic Circumpolar Current (Kemp et al., 1975). Significant equatorward penetrations of subarctic flora in the North Pacific (Keller, 1979) indicate that this cooling also affected the Northern Hemisphere. This interval was the driest of the Cenozoic in western North America (Axelrod and Raven, 1985), a conclusion consistent with evidence for an expansion in the range of grazing vertebrates at the same time (Repenning, 1967).

The postulated expansion of the Antarctic Ice Sheet at the Miocene–Pliocene boundary correlates with a sea level fall of about 50 m (Berggren and Haq, 1976). There is some evidence suggesting that the sea level fall isolated the Sea of Japan and turned it into a freshwater lake (Burckle and Akiba, 1978). The combination of sea

Fig. 10.18. Compilation of regional patterns of silica deposition in the world ocean from 5–23 Ma (based on data from Keller and Barron, 1983). Note that prior to about 10–12 Ma silica deposition was more common in the North Atlantic and after that time it was more common in the North Pacific. (However, silica deposition persisted in the North Atlantic as late as 2.4 Ma.) The different patterns of sediment accumulation can be interpreted in terms of changing deep-water circulation patterns. [From Woodruff and Savin, 1989] *Courtesy of American Geophysical Union.*

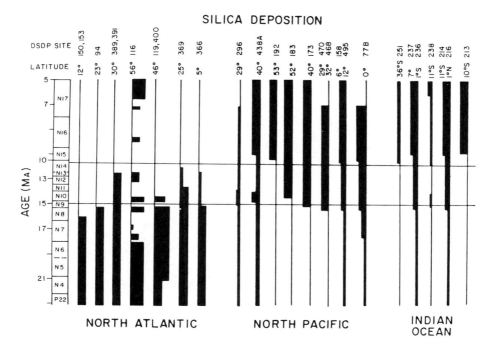

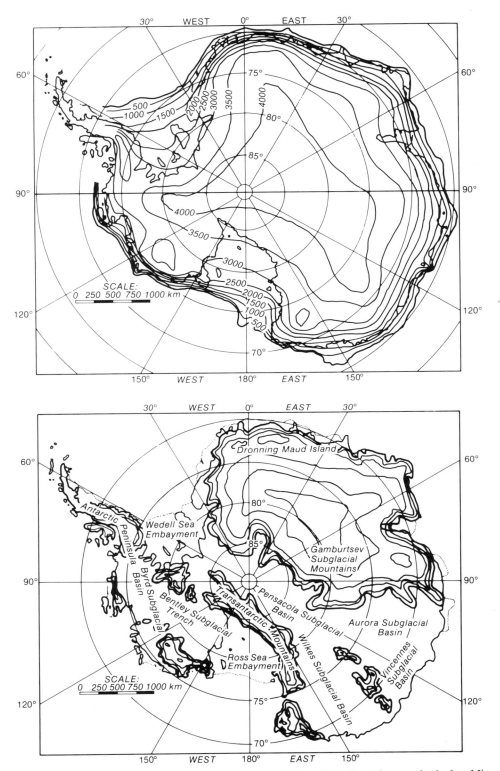

Fig. 10.19. *(Top)* Sketch of Antarctic Ice Sheet during proposed time of significant increase in the late Miocene or Pliocene; *(Bottom)* sketch of extent of Antarctic Ice Sheet during possible interval of extensive deglaciation 4–5 Ma. [From Denton, 1985] *Reproduced by permission of South African Journal Science.*

level fall and closure of the Straits of Gibraltar (due to tectonic impingement of Africa against Spain) also resulted in an isolation of the Mediterreanean Basin (e.g., Hsü et al., 1977). If the Straits of Gibraltar were closed today, the Mediterranean would dry up in about 1000 years because of excess evaporation over precipitation. Late Miocene salt deposits in the western Mediterranean indicate that the Mediterranean completely desiccated, forming a canyon 3000 m below sea level. Alteration of local base level resulted in a significant downcutting of the Nile and Rhone Rivers. Since the Mediterranean deposits are 2–3 km thick, and only 70 m of salt would be produced by isolation if it happened just once, the cycle of evaporation must have been repeated about 40 times in the latest Miocene, withdrawing about 6% of the salt from the world's ocean (Ryan, 1973). The net reduction in average ocean salinity by about 2.0‰ may have had a significant effect on ocean circulation.

There were also significant oscillations of ocean currents in the Northern Hemisphere (e.g., Fig. 10.20) prior to the postulated onset of midlatitude Northern Hemisphere glaciation (e.g., Ingle, 1973; Haq and Lohmann, 1976; Keller, 1979; Poore, 1981). There is a characteristic time scale of about 1 million years to these fluctuations (Poore, 1981), the same time scale as occurs in sea level fluctuations (Fig. 10.14;

Moore et al., 1987). The origin of the 1- to 2-Ma (Cenozoic) and 5- to 6-Ma (Cretaceous) fluctuations are not well understood (Schlanger, 1986).

$\delta^{18}O$ records suggest a significant reduction in size of the Antarctic Ice Sheet in the early Pliocene (about 5.0–4.0 Ma). There may have been a marine excursion into presently glaciated parts of East Antarctica (Webb et al., 1984). In addition to negative evidence for any tills of this age in either hemisphere (Mercer, 1984), SST in high southern latitudes may have been 5–8°C greater than present (Cieselski and Weaver, 1974). A speculative reconstruction of the dimensions of this reduced Antarctic Ice Sheet is illustrated in Fig. 10.19b (Denton, 1985).

The Northern Hemisphere was also warm in the early Pliocene (e.g., Zubakov and Borzenkova, 1988). Although at least seasonal Arctic Ocean ice cover and Alaskan glaciation date to perhaps 5.0–6.0 Ma (Clark et al., 1980; Lagoe and Eyles, 1988), Northern Hemisphere conditions were generally warmer until about 3.0 Ma. Miocene–Pliocene pollen from several areas in northern Canada (70–80°N paleolatitude) record warm-temperate conditions (Hills et al., 1974; Bujak and Davies, 1981; Norris, 1982, 1986; Omar et al., 1987). Terrestrial vegetation occurs as late as 4.0–5.0 Ma in shallow-water marine deposits from northern Greenland (82°N; Funder et al., 1985). An abundance of

Fig. 10.20. Evidence for 10^6-year climate oscillations in the Northern Hemisphere prior to the postulated time of midlatitude glacial inception. Figure is a schematic representation of major late Cenozoic oscillations of temperature-sensitive planktonic foraminiferal biofacies within the California Current system and adjacent Alaskan Gyre. Inferred temperature boundaries of the different biofacies are 10°C (subarctic), 15°C (temperate), and 20°C (tropical). [After Ingle, 1973]

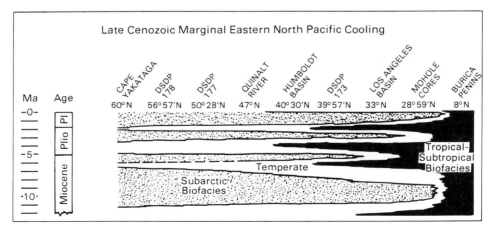

fauna and flora and well-sorted sediments at this site suggests open water (permanent sea ice cover, such as at present, prevents wave action on beaches).

There may have been another expansion of the Antarctic Ice Sheet in the later Pliocene (Mayewski, 1975; Denton et al., 1984; Webb et al., 1984; Prentice, 1985). Although there are some problems with establishing the timing of these events (Mercer, 1984; Prentice, 1985), the most likely intervals are between 2.5 and 4.0 Ma. Significant glaciation occurred on Patagonia during this interval (3.5 Ma; Mercer, 1976). Glacial drift in some of the dry valleys of West Antarctica indicate that the East Antarctic Ice Sheet may have overridden the divide between the two ice sheets (cf. Fig. 10.19a) and spilled into presently dry valleys (Denton et al., 1984; Prentice, 1985). A significant expansion of ice volume of the East Antarctic Ice Sheet is again implied. As tills associated with different evidence indicate different degrees of weathering, the glacial expansions may indicate two discrete events (Prentice, 1985).

10.1.10 Onset of Midlatitude Northern Hemisphere Glaciation

One of the best examples of a significant widespread climate change involves the glacial transition associated with the expansion of Northern Hemisphere ice in the Late Pliocene. As data from this interval are relevant to explanations for glacial growth and rapid climate change, we will examine it in some detail.

Before proceeding further, it is necessary to make a distinction between formation of the Greenland Ice Sheet and expansion of midlatitude Northern Hemisphere glaciers. Much evidence cited in discussions about Northern Hemisphere glacial initiation are based on ice-rafted debris in the subpolar North Atlantic or Norwegian Sea, $\delta^{18}O$ studies, or tills in Iceland and the Sierras (see text following). Some of this evidence is very good for assessing when midlatitude glaciation started; it is not very good for evaluating when the Greenland Ice Sheet formed. This feature cannot be detected by $\delta^{18}O$ studies—it has an isotopic signature of less than 0.1‰, too small to be resolvable from the noise

in isotopic records. Nor can occurrence of tills or ice-rafted debris in the Norwegian Sea or North Atlantic be used. The Greenland Ice Sheet is present today but is not associated with ice-rafted debris in these regions. Changes in sea ice through time presumably reflect conditions colder than those of the present. Probably the best place to assess when the Greenland Ice Sheet formed is the Labrador Sea. Presently, most of the outlet glaciers from Greenland are in this area. Seismic records in the Labrador Sea indicate that the Greenland Ice Sheet may have formed significantly earlier than 2.4–2.5 Ma (3–4 Ma?; Leg 105 Shipboard Party, 1986).

The exact timing of onset of this most recent phase of midlatitude glaciation has been a subject of some debate, but the matter now seems to be clarifying. Some variation occurs in $\delta^{18}O$ records by about 3.2 Ma (Shackleton and Opdyke, 1977; Prell, 1984a). There are well authenticated till dates of 3.15 Ma from Iceland and about 3.0 Ma from the Rocky Mountains (Curry, 1966; McDougall and Wensink, 1966). The most recent phase of active NADW formation may date from this time (Shackleton et al., 1984; Hodell et al., 1986).

Between 3.2 and 2.4 Ma the climate system appears to have evolved through another stage. Although glacial expansion was first interpreted as an abrupt step at 2.4 Ma (Shackleton et al., 1984), more recent work indicates significant, but smaller, $\delta^{18}O$ variations prior to 2.4 Ma (Raymo et al., 1989). $\delta^{18}O$ variations (Fig. 6.4) indicate ice sheet growth one-quarter to one-half as large as late Pleistocene ice volumes. Variations in plankton abundances indicate that large changes in sea surface temperature were occurring prior to 2.4 Ma (Raymo et al., 1986; Backman and Pestiaux, 1986; Loubere, 1988; Dowsett and Poore, 1990). However, these events were below the threshold needed to result in extensive ice rafting in the open North Atlantic.

Ice-rafted debris in the Norwegian Sea indicates the next phase of cooling occurred about 2.8–2.6 Ma (Jansen et al., 1988). Although ice rafting was restricted to this region and part of the Labrador Sea, significant cooling also occurred in the Southern Ocean at this time (Cieselski and Grinstead, 1986). At 2.4 Ma, the system underwent another step, with expanded ice rafting in the North Atlantic (Shackleton et al., 1984; Leg 105 Shipboard Party, 1986; Raymo et al, 1989). $\delta^{18}O$ values also increased (Fig. 6.4), although the original estimated changes (Shackleton et al., 1984) are somewhat larger than indicated by more recent measurements (Raymo et al., 1989).

Expansion of ice volume at 2.4 Ma (Fig. 6.4) was not unidirectional (Raymo et al., 1989), in that between 2.3 and 2.1 Ma the $\delta^{18}O$ amplitude again decreased. Although the long separation in these largest glacial stages may suggest a response to 400-KY (period in thousands of years) eccentricity fluctuations (see Sec. 7.1.2), overall the correlation is not very good, and the record is dominated by 41-KY power typical of the early Pleistocene (Raymo et al., 1989; cf. Fig. 7.15). These results suggest that climate variations at 100-KY periods may be decoupled from orbital forcing (see Oerlemans, 1980; Saltzman and Sutera, 1984; and Section 7.3.2).

There is unusually widespread evidence for a significant, abrupt climate change between 2.4 and 2.6 Ma. Increased abundance of ice-rafted debris in the Arctic Basin and expansion/initiation of ice rafting in the Arctic/North Pacific occurs at this level (Kent et al., 1971; Herman and Hopkins, 1980; Rea and Schrader, 1985; Sancetta and Silvesteri, 1986). Trees disappeared from northern Greenland around this time (Funder et al., 1985), and tundra expanded in Siberia (Wolfe, 1985). There was a significant increase in cooler vegetation in northwestern Europe (the Praetiglian time period; van der Hammen et al., 1971) and a corresponding turnover in vertebrate fauna marked by emergence of large vertebrates typical of Pleistocene deposits (Flint, 1971). The modern genus of horse *(Equus)* first dispersed from its North American home at this time. Loess deposition was initiated in China (Liu et al., 1985). Increased aridity in Asia is also reflected by an increase in eolian deposition in northwestern Pacific sediments (Rea et al., 1985).

The Late Pliocene climate change also affected tropical regions and the Southern Hemisphere. The aridity of northwestern Africa increased after 3.2 Ma (Stein and Sarnthein, 1984). Increased abundance of grazing vertebrates (e.g., *Equus,* elephants, antelopes, camels)

in the famous Siwalik beds in the foothills of the Himalayas in Pakistan reflects the onset of drier conditions at this time (Gaur and Chopra, 1984). The same trend toward drier conditions is indicated by the expansion of grazing vertebrates (antelopes) in eastern Africa and by increased abundances of small vertebrates adapted to dry conditions in the hominid-bearing Omo beds of Ethiopia (Wesselman, 1985; Vrba, 1985). The first appearance of the genus *Homo* *(Homo habilis;* cf. Fig. 6.3) is found near this level (Wesselman, 1985) and appears to mark a divergence in the family tree of hominids. In the Ethiopian highlands, pollen diagrams record increased grasses, implying cooler and drier climate between 2.3 and 2.5 Ma (Bonnefille, 1983).

In the western Atlantic and Caribbean, the 2.6-Ma cooling was associated with a 75% reduction in the number of species of tropical mollusks (Stanley and Campbell, 1981; Stanley, 1986; McNeil et al., 1988). Maximum exchange of vertebrates between North and South America—the Great American Faunal Interchange (Marshall et al., 1982)—occurred at this time. A long pollen record from Sabana de Bogota in Columbia also shows cooling about 2.4 Ma (Hooghiemstra, 1984). In the Southern Hemisphere, foraminifera and $\delta^{18}O$ analyses record significant cooling in New Zealand at 2.4 Ma (Kennett et al., 1971), although the most significant transition in Antarctic radiolaria occurred at 2.6 Ma (Hays and Opdyke, 1967; Cieselski and Grinstead, 1986). Antarctic diatoms show a similar strong change (Burckle, 1985).

10.2.0 Modeling Studies

10.2.1 Atmospheric Models

Models for the time interval 0–100 Ma have focused on factors responsible for the transition from a largely ice-free state to one with ice caps at both poles. These studies are an extension of modeling results from the thermal maximum of the mid-Cretaceous. As discussed in Chap. 8, there has been a modification of the time-honored view that changes in continental position were primarily responsible for the transition from an ice-free state to one with polar ice. This hypothesis (e.g., Donn and Shaw, 1977) was based on the observation that continental drift had resulted in a gradual northward drift of landmasses in the Cenozoic and that increased areal extent of snow in high latitudes would result in formation of polar ice caps.

Although reasonable, the above hypothesis could not explain apparent year-round warmth in high latitudes, nor could numerous sensitivity studies with more complicated models support that assumption. For example, GCM studies (Barron, 1985a) utilizing annually averaged insolation (cf. Sec. 2.3.2.1) indicate that plate tectonic-induced changes produced relatively little change in globally averaged temperatures over the interval 60–20 Ma (Fig. 10.21). Between 0 and 20 Ma cooling occurred due in part to formation of the Antarctic Ice Sheet (Barron,

1985a). The maximum effect of land–sea distribution on global annual-averaged temperatures over this interval is about 4–5°C (Barron et al., 1984; Barron, 1985a). It is not clear whether the non-ice sheet component of this number would be the same if seasonal cycle variations were considered (see text following).

Changing land–sea distributions in the Cenozoic should have resulted in significant modifications of the annual cycle (Watts and Hayder, 1984a; Crowley et al., 1986). Specifically, the breakup of large polar landmasses (North America/Europe and Australia/Antarctica) should have reduced continentality and therefore the magnitude of summer warming. Calculations employing a two-dimensional energy balance model (see Section 1.2.4) provide some support for this hypothesis (Crowley et al., 1986; Hyde et al., 1990). Northward movement of Greenland and opening of the Norwegian Sea decreased temperatures over central Greenland by ~15°C (Fig. 10.22). Land–sea changes in the Southern Hemisphere caused smaller decreases in summer warming over Antarctica, with the magnitude of the cooling being sensitive to the choice of paleogeographic reconstruction (Hyde et al., 1990)

Even though seasonal cycle changes may be important, models do not produce conditions warm enough in high latitudes to suggest that changes in seasonality represent the sole explanation for the long-term climate trend (Crowley, 1988). Temperatures were sufficiently warmer at high latitudes to require that other sources altered the energy balance. There are some other possible atmospheric effects that we will discuss in this section, plus ocean circulation changes and CO_2, which we will discuss in the following sections.

Kennett et al. (1985b; cf. Kennett and Thunell, 1977) note a significant increase in South Pacific volcanism at the Eocene–Oligocene boundary and suggested that this may have contributed to the cooling. However, the climate effect of volcanic eruptions over the last century has been relatively small and short-lived (Sear et al., 1987; Bradley, 1988), although there does seem to be some correlation between Little Ice Age fluctuations and volcanism (Sec. 5.2.1). By itself volcanism may not be enough to force the system to evolve so radically, but it could be an

Fig. 10.21. Global and hemispherically averaged surface temperatures computed at 20-million year intervals over the last 60 million years. The globally averaged annual surface temperature sensitivity for both the low-and high-albedo effects of the ice cap is indicated, as well as present global surface temperatures and the range of interpreted values for the late Cretaceous. [From Barron, 1985a] *Reprinted with permission of Elsevier Science Publishers.*

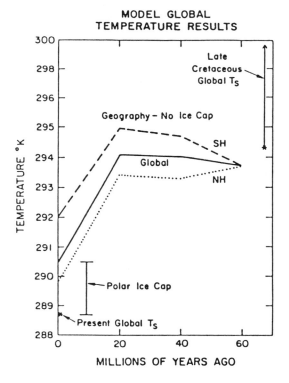

MODEL GLOBAL
TEMPERATURE RESULTS

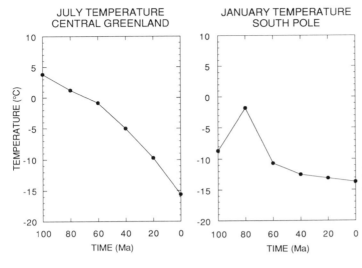

Fig. 10.22. Calculated summer temperature changes on Greenland and Antarctica due to changes in land-sea distribution during the last 100 million years. Simulations utilized a nonlinear EBM in order to evaluate the effect of snow-albedo feedback on summer temperatures. [Based on results presented in Hyde et al., 1990]

important trigger if the system were unstable and near a climate "bifurcation point" (see Section 10.2.4).

Changes in continental topography may have also affected the long-term evolution of climate (Ruddiman et al., 1986a; Ruddiman and Raymo, 1988). Substantial uplift has occurred in western North America, the Andes, and the Himalayas since the Pliocene (e.g., Izett, 1975; Hsü, 1978; Benjamin et al., 1987; Ruddiman et al., 1989a). GCM experiments indicate that increased elevations in these regions significantly affected the atmospheric circulation (Manabe and Terpstra, 1974; Kutzbach et al., 1989; Manabe and Broccoli, 1990; cf. Held, 1983). In January the amplitude of the midlatitude, upper tropospheric planetary waves increased with uplift (Fig. 10.23) and the low-level winds were progressively blocked or diverted around the topographic features. In July the progressive uplift caused monsoon-like circulations to develop in the vicinity of the Colorado and Tibetan plateaus (note especially the low pressure over Tibet in Fig. 10.23). The basic direction of most of the simulated changes is borne out by changes in the geologic record (Ruddiman and Kutzbach, 1989): winter cooling of North America, northern Europe, northern Asia, and the Arctic Ocean; summer drying of the North American west coast, the Eurasian interior, and the Mediterranean; winter drying of the North American northern plains and the interior of Asia; and changes over the ocean conducive to increased

formation of NADW (more southerly flow with mountains). Increased elevation would also presumably be associated with increased geochemical weathering, which may have affected atmospheric CO_2 levels (Raymo et al., 1988).

Modeling studies have also investigated the effect of changes in ocean boundary conditions on the atmosphere (cf. Kennett, 1977). For example, Oglesby (1989) calculated that, even if SSTs around Antarctica were 10°C warmer than the present, the effect of warmth was generally restricted to coastal regions; interior temperatures still dropped below zero in the winter (Fig. 10.24). Warmer SSTs lead to increased winter snow, which acts as a negative feedback to summer warming. In addition, transport of warm air from the continental margin cannot offset radiative cooling in the interior. Similar difficulties apply to the ocean as the source of mid-Cretaceous warmth (see Schneider et al., 1985; Covey and Barron, 1988; and Section 8.2.3).

In another study, Raymo et al. (1990a) examined the effect of an altered Arctic Ocean ice cover on an atmospheric GCM. Assuming that early Pliocene ice cover was absent in summer and reduced in winter, they found significant warming (~ 10°C) in the winter on land areas around the Arctic basin; summer changes were smaller. Most of the warming was restricted to the high latitudes of the Northern Hemisphere, a result in agreement with previous GCM simulations of the effect of a local heat source (see Fig. 4.8). Changes in ice cover had little effect on

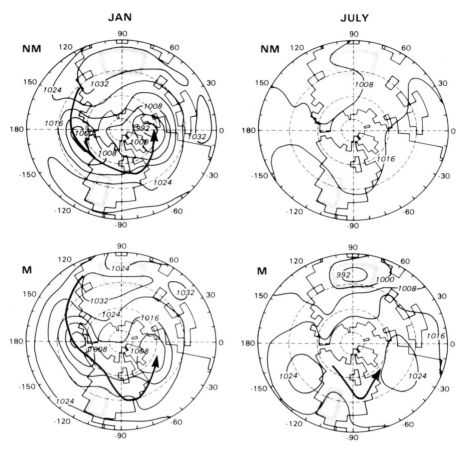

Fig. 10.23. Effect of changes in topography on seasonal surface pressure in the Northern Hemisphere. Results are from GCM simulations for mountain (M) and no-mountain (NM) cases (stippled regions in Asia and North America). Heavy arrow refers to path of jet stream. [After Ruddiman and Kutzbach, 1989] *Courtesy of American Geophysical Union.*

Fig. 10.24. Zonally averaged GCM calculated temperatures for the high-latitude Southern Hemisphere based on four different experiments: (1) with warm specified SSTs around Antarctica and present topographic relief for Antarctica; (2) with present SSTs and no topographic relief to Antarctica; (3) with warm SSTs and no relief; and (4) with present SSTs and topography. This figure illustrates that, regardless of choice of boundary conditions, it is not possible to get temperatures significantly above freezing in interior of Antarctica with the present continental configuration. [From Oglesby, 1989] *Reproduced with permission of Climate Dynamics and Springer-Verlag.*

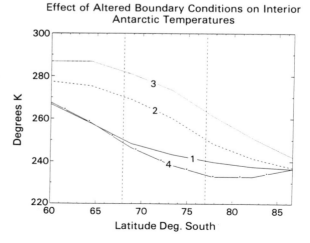

Effect of Altered Boundary Conditions on Interior Antarctic Temperatures

global temperatures, for there was an increase in cloudiness in the model (this effect offset the lower surface albedo).

10.2.2 Ocean Changes

Progress is finally being made in modeling changes in Cenozoic ocean circulation. There are a number of problems which can be addressed with ocean models: reality of inferred increases in ocean heat transport (Sections 10.1.3 and 10.1.5), effect of changes in land–sea distribution on ocean heat transport, and effect of altered ocean circulation on land records. For example, Horrell (1990) calculated that changes in ocean heat transport required to raise high-latitude SSTs and decrease low-latitude SSTs (Figs. 10.8 and 10.12) are so large as to appear unrealistic. It is conceivable that changes in warm, saline bottom-water (WSBW) production may have been responsible for the altered $\delta^{18}O$ trends (Brass et al., 1982a). Such changes should have led to warmer high-latitude temperatures (Oglesby and Saltzman, 1990). However, it is not clear whether WSBW production rates were large enough to maintain such a temperature profile.

Experiments with ocean circulation models (Seidov, 1986) indicate that changing the continental position can have a significant effect on poleward ocean heat transport. In fact, the Early Miocene warming (15–20 Ma) is simulated in the high latitudes of the Southern Hemisphere (Seidov, 1986) and may reflect a "transient" warming due to the fact that the Antarctic Circumpolar Current, although established, was still relatively weak. Heat stored in the subtropical gyres could be more easily transported to high latitudes (Seidov, 1986).

Geologists have long been interested in the effect of "ocean gateways" on past climates (Berggren and Hollister, 1974). For example, the Central American isthmus was open for much of the last 100 million years. Eastward motion of the Caribbean plate eventually blocked interocean transport. A strengthened North Atlantic circulation apparently resulted from this closure, since flow of the western boundary current in the Yucatan Straits and Blake Plateau (east of Florida) increased from about 3.0–4.0 Ma (Kaneps, 1979; Brunner, 1983/1984). This timing approximately coincides with divergence of planktonic trends on either side of the isthmus (Keigwin, 1982a,b) and with the first occurrence of North American vertebrates in South America—the Great American Faunal Interchange (Marshall et al., 1982).

The effect of an open Central American isthmus has been investigated with an ocean GCM (Maier-Reimer et al., 1990). Results indicate that the opening of the Central American isthmus significantly decreased poleward heat transport in the North Atlantic (Fig. 10.25; Maier-Reimer et al., 1990). Removal of the western boundary had two effects on the model. Physical removal of the barrier weakened the Atlantic western boundary current. Furthermore, the present 1.5‰ salinity difference between the North Pacific and North Atlantic was almost erased by mixing between the two regions (cf. Fig. 2.7). Warm water flowing poleward in the North Atlantic was less saline and less susceptible to vertical convection, which acts as a positive feedback at present (e.g., Broecker et al., 1985; see Fig. 4.13). Decreased poleward heat transport caused sea ice to form in the Norwegian Sea. The first result is in accord with geologic data for the western tropical Atlantic (see

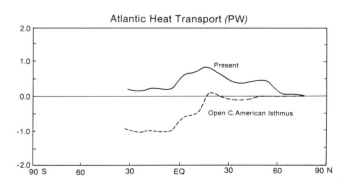

Fig. 10.25. Comparison of poleward ocean heat transport (1 PW = 10^{15}w) in the Atlantic for an ocean GCM simulation with present boundary conditions and with an open Central American isthmus. Based on results in Maier-Reimer et al., 1990] *Courtesy of American Geophysical Union.*

text preceding) but the second result exacerbates the problem of high-latitude warmth in the circum-North Atlantic basin because it implies that compensating effects (CO_2?) were required to account for warmer conditions in the high latitudes of the Northern Hemisphere.

The above results may have interesting implications for theories of past changes in ocean heat transport. It is evident that solid western boundaries and high salinities in tropical waters, coupled with a relatively strong meridional wind circulation (cf. Section 10.2.1), represent an optimal configuration for strong poleward ocean heat transport. This configuration exists at present. However, it is difficult to contemplate altered boundary conditions that would greatly enhance the role of ocean heat transport in the past. Viewed from this perspective, the Matthews–Poore criticisms of Oligocene $\delta^{18}O$ interpretations need to be recalled (Sec. 10.1.5). However, it should be evident from the above discussion that there are still a number of uncertainties, in both data and models, about past changes in ocean heat transport, and that it would be premature to pass any final judgment on the matter at this stage.

An additional insight gleaned from these initial ocean circulation studies involves the effect of the open Central American isthmus on the Pacific circulation. For example, Kennett et al. (1985a) suggested that increased North Pacific heat transport occurred in the late Miocene (8 Ma) as a result of closure of the Indonesian Seaway, which separates the Pacific and Indian ocean equatorial circulations. This interpretation is consistent with results from the Central American sensitivity experiment. The above results demonstrate the potential insights that further ocean-modeling studies may have on paleoceanography and paleoclimatology.

10.2.3   Carbon Dioxide Changes?

The inability of GCMs to simulate a largely ice-free state has led to the hypothesis that changes in atmospheric CO_2 levels may be responsible for such climates (Berner et al., 1983; Barron and Washington, 1984, 1985). Calculations with a geochemical model (see Section 8.2.4 and Fig. 10.26) provide support for this hypothesis for the mid-Cretaceous (Berner et al., 1983; Lasaga et al., 1985). CO_2 levels could have been a factor

Fig. 10.26.   Geochemical calculations of atmospheric CO_2 fluctuations over the last 100 million years. [After Lasaga et al., 1985] *Courtesy of American Geophysical Union.*

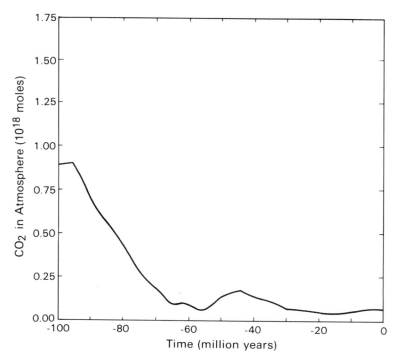

of 10 greater than those of the present (Fig. 10.26), although more recent efforts suggest this number is closer to seven (e.g., Berger and Spitz, 1988). However, estimated CO_2 levels do not exceed present levels by more than a factor of 2 for much of the last 70 million years (Fig. 10.26). These estimates do not take into account changes in weathering rates due to late Cenozoic uplift in many regions (Raymo et al., 1988). Sensitivity experiments (Bender, 1984) suggests that change in continental weathering may have a significant effect on CO_2 concentrations on long time scales.

Documenting the magnitude of past CO_2 fluctuations is an important problem in paleoclimatology. It is necessary to examine multiple lines of evidence to substantiate the changes. For example, mid-Cretaceous CO_2 concentrations should increase because plate rates and volcanism were significantly higher in the mid-Cretaceous (e.g., Alvarez et al., 1980b; Davis and Solomon, 1981; Kominz, 1984; Arthur et al., 1985b). The most recent estimate (Kominz, 1984) of mean ocean ridge volume (Fig. 10.27), which can be considered an integrated measure of the primary source for atmospheric CO_2, suggests that early Cenozoic values did not drop off as rapidly as the geochemical model suggests. However, ridge volume values (Fig. 10.27) for much of the last 40 million years are comparable to the present. It is interesting that this interval also represents approximately the time when significant ice started to accumulate on Antarctica.

As discussed in Section 8.2.5, sedimentological evidence for higher CO_2 levels yields an ambiguous picture of CO_2 changes over the last 100 million years. The evidence suggests that CO_2 levels during the mid-Cretaceous peak may have been 3–7 times present levels. However, CO_2 estimates for the Cenozoic are more uncertain. For example, whole-rock $\delta^{13}C$ and vertical profiles of $\delta^{13}C$ support the Lasaga et al. (1985) estimate of relatively low Cenozoic CO_2 levels (Arthur et al., 1985b; Shackleton, 1985; see Chap. 8). Although some $\delta^{13}C$ records of organic carbon suggest higher CO_2 values as late as 15 Ma (Arthur et al., 1985a; Dean et al., 1986), others do not (Lewan, 1986). Interpretation of the $\delta^{13}C_{org}$ profiles in terms of pCO_2 is based on observations that fractionation of $\delta^{13}C$ by plankton is dependent on the pCO_2 of surface waters (De-

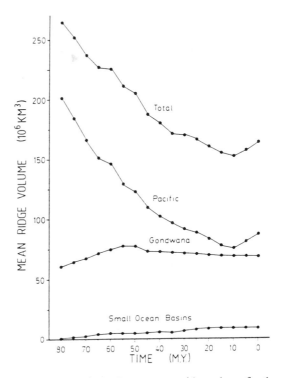

Fig. 10.27. Estimated mean ocean ridge volume for the last 80 million years, based on changes in spreading rates and ridge lengths. This index can be used to estimate changes in volcanism and CO_2 production through time. A decrease in spreading rates since the Cretaceous is the primary cause for changes in ridge volume through time. [From Kominz, 1984] *Reprinted by permission of the American Association of Petroleum Geologists.*

gens et al., 1968). The coexistence of different ^{13}C groups in amorphous kerogens suggests that past atmospheric pCO_2 levels were not high enough to cause consistent fractionation (Lewan, 1986). Similar conclusions have been drawn from $\delta^{13}C$ analyses of vascular land plants (Degens, 1969; Schoell, 1984), which also fractionate under different pCO_2 levels (Park and Epstein, 1960).

Other estimates of post-Eocene CO_2 concentrations based on carbon isotopes and carbonate preservation patterns suggest maximum CO_2 levels 2.5 times the present (Berger and Spitzy, 1988). These estimates are consistent with some geochemical measurements suggesting an increased rate of continental weathering in the late Cenozoic (Hodell et al., 1989; Shemesh et al., 1989; cf. Raymo et al., 1988)—increased weathering would increase the dissolved salts in the ocean, raising the alkalinity and lowering CO_2

levels (Bender, 1984; cf. Section 6.4.5). Although these results are promising, they conflict
with measurements of carbonate accumulation,
which suggest relatively constant levels of
weathering through time (Opdyke and Wilkinson, 1988). This maze of conflicting signals can
be further complicated by the necessity to consider whether carbonate is deposited on the continental shelfs or the deep sea (Volk, 1989b).
The most we can say at present is that there are
some indications Tertiary CO_2 levels may have
been at least a factor of two greater than the
present, but there is a great need for more quantitative constraints on the various assumptions
used to estimate CO_2 fluctuations (cf. Delaney
and Boyle, 1988).

Are CO_2 levels 2–2.5 times the present (Berger
and Spitzy, 1988) sufficient to account for all the
climate change that has occurred since 40 Ma?
A summary of results of GCM models examining the future impact of a doubling of CO_2 suggests that high-latitude temperatures might
change by $\sim 10°C$ in the winter and 3–4°C in the
summer, with the winter changes reflecting significant reductions in sea ice (Schlesinger and
Mitchell, 1987; Schlesinger, 1989; cf. Fig. 2.12).
There might also be a significant reduction in
Arctic ice cover in the summer (cf. Semtner,
1987), perhaps a melting of the West Antarctic
Ice Sheet, and northward movements of snow
lines by 5–10° latitude (e.g., Parkinson and Kellogg, 1979; Stuiver et al., 1981; Schlesinger and
Mitchell, 1987). The above changes are large
and, when coupled with changes discussed in
the previous sections, could be large enough to
explain inferred patterns for the late Cenozoic.

10.2.4 *Abrupt Transitions*

From the above discussions it is evident that at
least four processes are probably involved in the
long-term evolution of climate: changes in
land–sea distribution, ocean heat transport,
orography, and CO_2. The $\delta^{18}O$ record (Fig. 10.3)
also indicates that there are abrupt steps in climate. These steps could be due to either abrupt
changes in forcing or some type of climate instability (Berger, 1982; Brass, 1982b; North and
Crowley, 1985). Arthur et al. (1988) discussed
how rapid changes in ocean carbon accumulation at 90 Ma may have caused a rapid CO_2 and

climate change. The "ocean anoxic events" may
be due to some instability in the thermohaline
circulation. However, some of the Cenozoic
steps do not seem to be related to carbon events.

There are at least two additional types of climate instability that might explain the Cenozoic
steps. One type is called the "small ice cap instability" (e.g., North, 1984) and results from an
albedo discontinuity at the ice edge. This phenomenon has been extensively studied with simple climate models (see Section 1.3.4). Recent
experiments with a nonlinear two-dimensional
energy balance model (W. Hyde, pers. commun., 1989) indicate that slowly changing forcing over Antarctica can sometimes result in
rapid changes in ice cover.

Another type of instability involves the thermohaline circulation in the ocean (e.g.,
Broecker et al., 1985). Changes in surface salinity in high latitudes, or in areas of conjectured
warm saline bottom-water formation, may affect thermohaline overturn. There is growing evidence for multiple equilibria in ocean GCMs
(F. Bryan, 1986; Manabe and Stouffer, 1988).
Experiments with a simpler "plume" model for
warm, saline bottom-water formation also sug-

Fig. 10.28. Plot of streamfunction in a two-dimensional
(latitude-depth) ocean model as a function of time from
the moment a salinity perturbation is applied. Full line:
0.5‰ anomaly, dashed line 10^{-6}‰ anomaly. Units correspond to m^2/sec. Different signs for the streamfunction
represent different directions of flow. Note that an exceedingly small perturbation is still capable of getting the
system to flip into a different circulation mode. [From
Marotzke et al., 1988] *Reproduced with permission of
Munksgaard International Publishers.*

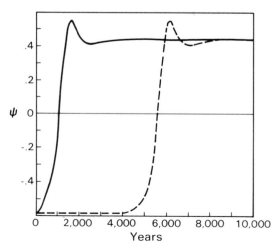

gest unstable behavior (Peterson, 1979). One of the advantages of using this simpler model is that multiple runs can be made to determine the effect of slowly changing behavior. Results indicate that slowly changing boundary conditions can indeed cause abrupt transitions. Another experiment with a simple ocean model (Marotzke et al., 1988) achieved a system transition (Fig. 10.28) with a salinity perturbation as small as 10^{-6}‰—a result very similar to the small ice cap instability simulations illustrated in Fig. 1.16. It is evident that the subject of abrupt transitions in the Cenozoic record warrants further investigation.

10.3 Summary

The long-term evolution of climate in the last 100 million years is characterized by a general trend from warmth to bipolar ice sheets, with increased aridity accompanying the cooling trend. There were significant 10^6-year oscillations superimposed on the long-term trend. Some of the transitions have been geologically abrupt.

Explanations for the Tertiary cooling trend are incomplete. Changes in continental location have significantly affected both annual and seasonal temperatures in high latitudes, but the temperature changes do not seem large enough to account for the entire evolution of Cenozoic temperatures. Although CO_2 fluctuations probably occurred, and may have been very important in the mid-Cretaceous (Chap. 8), calculated changes were much less in the Cenozoic. The smaller Cenozoic changes are consistent with some geologic evidence, but it is also possible that additional factors must be considered that may have increased CO_2 values and that were not included in the geochemical calculations. Changes in ocean heat transport and orography may also have significantly affected atmospheric circulation and high-latitude temperatures.

From the above discussion it should be clear that, although we have made significant progress in the analysis of the evolution of climate over the last 100 million years, there are still some major questions. We now have a number of pieces of the puzzle but are not sure how they fit together. Future progress will require more data (especially on CO_2 and tropical SSTs), more oceanographic and atmospheric simulations, and further geochemical calculations.

11. PALEOZOIC AND EARLY MESOZOIC CLIMATES (570–100 Ma)

As we travel farther back into the past, our ability to infer climate deteriorates. Statements are more generalized and climate simulations less definitive. Many of the problems encountered can be studied in greater detail by analyzing the climate of the last 100 million years. For this reason our discussion of the climate of the interval 570–100 Ma will be briefer (in terms of pages per million years). However, there are still a number of interesting climate questions that can be addressed. Many of them revolve around the subject of climates on supercontinents.

11.1 Major Features of Paleozoic–Early Mesozoic Climates

11.1.1 General Setting

Most textbooks on geology describe the interval from 570 to 100 Ma (see Appendix A for a geologic time scale) as generally mild, punctuated by two major glacial phases (Ordovician and Permo-Carboniferous). The impression of moderate climate was derived from both lack of evidence for frequent widespread glaciations and the environmental interpretations of sedimentary rocks, particularly limestone. Today, shallow-water lime sediments are accumulating primarily in the tropics (e.g., Matthews, 1974). Between 570 and 100 Ma, geologic evidence indicates that marine conditions were much more widespread. Large parts of continental interiors were covered by these shallow, epicontinental (epeiric) seas during marine "transgressions" (sea level rises). For example, more than two-thirds of North America was covered by a shallow sea 430 Ma ago (e.g., Dott and Batten, 1981). However, at least for the Mesozoic, most limestone deposits do not extend much farther poleward than the present (Ziegler et al., 1984), a fact possibly resulting from light limitations for carbonate-secreting organisms (Ziegler et al., 1984).

In general, Phanerozoic sea level variations are of broad geographic extent. In a landmark paper, Sloss (1963) showed that the sea level history of North America for the last 570 Ma (the Phanerozoic Eon) could be summarized as six sequences of transgressions and regressions (Fig. 11.1). Sea level patterns in Brazil, Africa, and Russia record the global scope of the fluctuations (Petters, 1979).

Although sea level changes of the last few million years have been dominated by fluctuations of continental ice volume, it is generally thought that the Phanerozoic transgression-regression sequences are not of glacio-eustatic origin. Rather, they may be related to temporal variations in plate tectonic processes. For example, times of increased upwelling of magma from the mantle are associated with a thermal expansion of ridges, displacing water onto the continents (e.g., Hays and Pitman, 1973; Pitman, 1978). The cooling of aging lithosphere may also result in subsidence and submergence along continental margins (e.g., Watts, 1982).

Historical geology textbooks may cite the expansion of seaways in the tropics as an indication of enhanced Paleozoic warmth, but there is some evidence (e.g., Frakes and Francis, 1988) that ice rafting was quite frequent in the Phanerozoic (Fig. 11.2). Caution is therefore required in equating the expansion of seas in equatorial regions with enhanced warmth in high latitudes, since most of the continents (especially well-studied North America and Europe) were in low latitudes in the Early Paleozoic (Fig. 11.3).

11.1.2 Earliest Paleozoic Warmth?

Our knowledge of earliest Paleozoic (570–460 Ma) climates is somewhat limited. Most geologic deposits are from low latitudes (Fig. 11.3) and naturally show this imprint (land plants had not yet evolved, so we have less information for

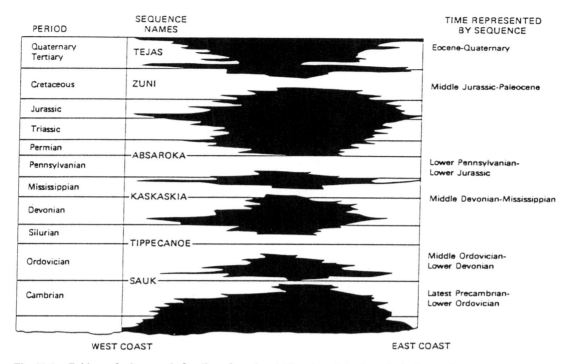

Fig. 11.1. Evidence for large-scale flooding of continental interiors through geologic time, as illustrated by sedimentary record of North America. Black areas represent large gaps in the rock record, which become smaller toward the continental margins. White areas (preserved strata) represent flooding of the continental interiors by sea level rises. See Fig. 8.1 for an example of the extent of flooded area during a maximum sea level rise. [From Sloss, 1963] *Reprinted with permission of Geological Society of America. From "Sequences in the cratonic interior of North America,"* *L. Sloss, 1963, Geol. Soc. Amer. Bull. 74, 93–114.*

Fig. 11.2 Occurrence by paleolatitude of ice-rafted deposits (shading) through the Phanerozoic. [Redrawn from Frakes and Francis (1988) by B. Opdyke.] *Reprinted with permission from Nature vol. 333, 547–549, Copyright (©) 1988 Macmillan Magazines Ltd.*

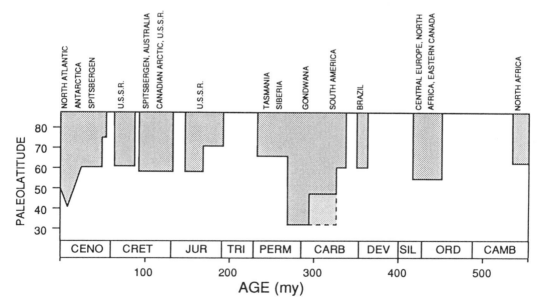

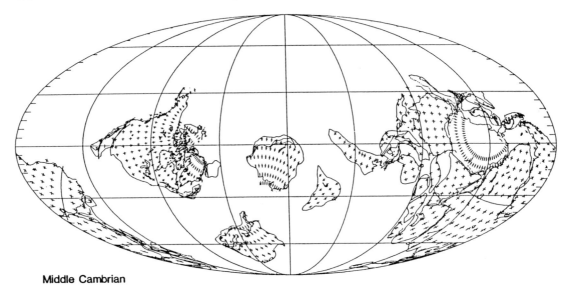

Middle Cambrian

Fig. 11.3. Middle Cambrian (530 Ma) paleogeography. [From Scotese, 1986; after Ziegler et al., 1979] *Reproduced with permission of C. Scotese and University of Texas Institute for Geophysics.*

the terrestrial realm). For example, the Northern Hemisphere was entirely oceanic north of about 30°N paleolatitude. Sea levels were at or near an all-time high for the Phanerozoic (Fig. 11.4), with high stands perhaps reflecting increased volcanic activity/ridge expansion after the breakup of a postulated late Precambrian supercontinent (e.g., Piper, 1976; Bond et al., 1984).

There are several lines of evidence suggesting that atmospheric CO_2 levels may have been significantly higher during the Early Paleozoic. For example, if we use the relationship between relative sea level and CO_2 that was discussed in Section 10.2.3, then we might expect CO_2 levels to be as high as any time in the Phanerozoic. This conclusion is supported by some geochemical evidence that links changes in preservation of various species of carbonate minerals with the partial pressure of CO_2 in surface waters. Since the latter is presumably in equilibrium with the atmosphere, changes in atmospheric pCO_2 can be inferred.

In general, variations in preservation of marine carbonate species support the sea level/CO_2 connection assumed for the Phanerozoic (Sandberg, 1983, 1985; Wilkinson and Given, 1986). For example, high-Mg carbonate shells (Fig. 11.5) are preferentially formed with lower pCO_2

and are more common in the record during times of low sea level (cf. Fig. 11.4). Aragonite (calcium carbonate with a different crystallographic orientation than calcite) follows the same trend (Fig. 11.5). Using data from modern oceans to calibrate a "carbon dioxide paleobarometer," Wilkinson and Given (1986) estimated that Phanerozoic pCO_2 has varied from about its present concentration to levels approximately 10 times higher. Presence of calcite cement throughout the Phanerozoic limits atmospheric pCO_2 concentrations to 13–14 times present levels (the calcite would dissolve at higher concentrations; Wilkinson and Givens, 1986). These higher levels are comparable to estimates based on other geochemical models (cf. Fig. 10.26) and may be required for climate models in order to reconcile data and models for the mid-Cretaceous (see Chap. 8).

11.1.3 Late Ordovician Glaciation

The "maritime" climates of the early Paleozoic were interrupted by an interval of ice growth at about 440 Ma (Frakes, 1979). Evidence for a major Ordovician ice sheet can be found in rocks from the Sahara Desert (Fairbridge, 1970, 1979; Crowell, 1983; Caputo and Crowell, 1985). This glaciation extended to the Early Silurian (Caputo and Crowell, 1985). Ice flow indicators for Late Ordovician glaciation suggest flow from the vicinity of the present equator (Fig. 11.6). However, paleomagnetic reconstructions of paleolatitudes indicate that northern Africa was in the vicinity of the South Pole during the Ordovician (Fig. 11.7). In fact, Fig. 11.7 illustrates a remarkable correspondence between occurrences of Paleozoic glaciations and paleopole positions. The late Paleozoic glaciations

(see text following) also show a polar position for a different part of the Gondwanaland supercontinent (which moved as one unit through most of the Paleozoic).

A very interesting feature of the Late Ordovician glaciation is that it occurred during a time of apparently high CO_2 (van Houton, 1985). Examination of the relative sea level record (Fig. 11.4) suggests that CO_2 levels could have been as much as 10 times present levels. We will discuss

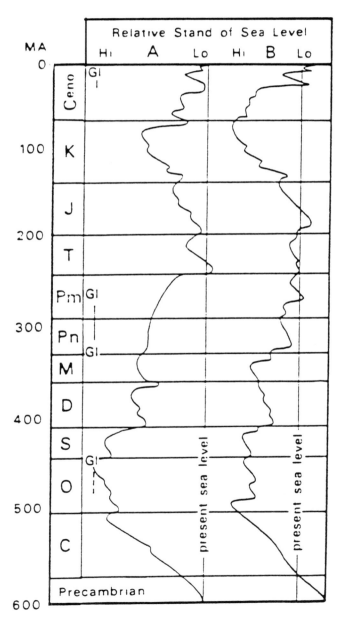

Fig. 11.4. Global sea level curves for the Phanerozoic, based on two different estimates. (A) after Hallam (1984); (B) after Vail et al. (1977). Gl = widespread glaciation. [Modified from van Houten, 1985] *Reprinted with permission of Geological Society of America. From "Oolitic ironstones and contrasting Ordovician and Jurassic paleogeography," F. B. Van Houten, 1985, Geology 13, 722–724.*

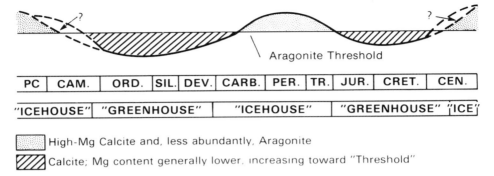

Fig. 11.5. Inferred first-order cycles in nonskeletal carbonate mineralogy, showing correlation to conjectured climate supercycles of Fischer (1982). Variations in carbonate species preservation may be related to total amount of CO_2 partial pressure in surface waters (and the atmosphere). [Modified from Sandberg, 1983] *Reprinted by permission from Nature vol. 305, pp. 19–22, Copyright (©) 1983 Macmillan Magazines Ltd.*

a possible explanation for this paradox in the modeling section (11.2.1).

11.1.4 Mid-Paleozoic Transition

Following the Late Ordovician/Early Silurian glaciation, there is no evidence for glaciation for a 60 million year time period (Caputo and Crowell, 1985). Although detailed climate information is somewhat limited for this time interval, there are several features of interest. For ex-

ample, the subsidence of the seas gradually exposed more land area, thereby restricting ocean circulation (Ziegler et al., 1981) and increasing the seasonal cycle of temperature on land (see Section 11.2.1).

Expansion of land plants also occurred for the first time in the Devonian. This evolutionary milepost would affect climate in several different ways. Vegetated surfaces can decrease surface albedo by as much as 10–15% (e.g., Posey and Clapp, 1964). In addition, the hydrologic cycle

Fig. 11.6. Evidence for an Ordovician glaciation in the Sahara. Striations and grooves in the bedrock immediately below Ordovician tillites reveal the direction of flow (arrows) of a great ice sheet. Darker areas are sand covered today. [From Eicher and McAlester, 1980; after Fairbridge, 1970] *From Don L. Eicher/A. Lee McAlester, History of the Earth, (©) 1980, p. 270. Reprinted by permission of Prentice Hall, Inc.*

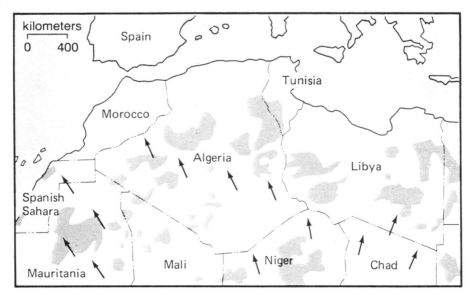

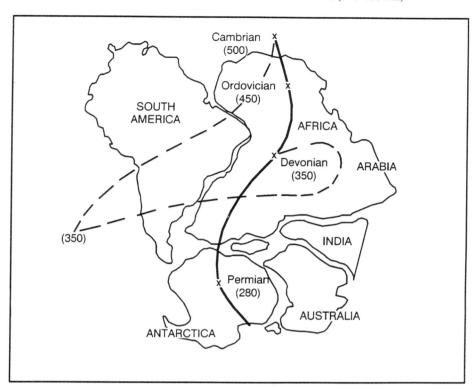

Fig. 11.7. Postulated Paleozoic apparent polar wander paths of the South Pole with respect to Gondwanaland. Dashed line indicates alternate path based on different data sources. Numbers in parentheses are ages in Ma. [From Crowley et al., 1987; adapted from Morel and Irving, 1978] *Reprinted by permission from Nature vol. 329, pp. 803– 807, Copyright (©) 1987 Macmillan Magazines Ltd.*

intensifies and the annual range of temperature decreases (e.g., Shukla and Mintz, 1982). Since the floral occupation of the land occurred in three stages of evolution—seedless (e.g., ferns), seeded (e.g., evergreens), and flowering plants (angiosperms)—the present features of the plant-related hydrologic cycle may not have developed until the origination of the last group at about 100–120 Ma.

Land plants may also have affected geochemical cycles, for they created a new reservoir in the global carbon cycle (Holland et al., 1986; Berner, 1987), with the net result being a decrease in atmospheric CO_2. Additionally, plants reduce the stability of soil minerals through a net export of ions from soil waters and through the release of complexing organic acids (Knoll and James, 1987). First-order increases in mineral weathering probably occurred in the middle Paleozoic and Late Cretaceous/Early Tertiary (Knoll and James, 1987). A long-term increase

in plant-induced weathering should also have decreased atmospheric CO_2 levels (Volk, 1989a).

There was a first-order change in $\delta^{18}O$ records at the end of the Devonian (~350 Ma; Fig. 11.8). More negative $\delta^{18}O$ levels in the early Paleozoic could be due to several factors. Early Paleozoic temperatures could have been 12°C warmer than later times; there were significant (and not well-understood) changes in the mean $\delta^{18}O$ levels of seawater; or ocean surface salinities were less than present (Veizer et al., 1986; Railsback et al., 1990; cf. Shemesh et al., 1983). None of these options are very satisfactory (Veizer et al., 1986)—the high temperatures would seem to exceed the tolerance limit of organisms and there seems to be relatively little justification for changing the mean $\delta^{18}O$ of seawater. Lower surface salinities (with warm, saline deep water) could explain some of the pattern (Railsback et al., 1990), but it is not clear whether the explanation could apply to the en-

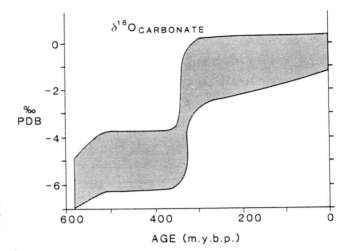

Fig. 11.8. Secular isotopic age curves for seawater as derived from the oxygen isotope record in carbonates, principally brachiopods. Note the mid-Paleozoic transition. [From Kump, 1989; after Veizer et al., 1986] *Reprinted with permission from Geochimica et Cosmochimica Acta 50:1679–1696, copyright 1986, Pergamon Press plc.*

tire 200-million year time interval (Fig. 11.8). The higher temperature scenario would be consistent with postulated higher CO_2 levels for the early Paleozoic (Section 11.1.2).

There were additional important geochemical changes in the mid-Paleozoic, some of which appear to be related to climate. Lower Paleozoic marine deposits are characterized by relatively high organic carbon content, which presumably reflects lower oxygen content in deep waters. A mid-Paleozoic transition has been interpreted as resulting from two processes: progressive ventilation of the ocean due to onset of Gondwanan glaciation (see text following) and increased oxygen content of downwelled water as a result of the evolution of land plants (Berry and Wilde, 1978; Wilde, 1987). If there was a sudden enhancement of oceanic overturn, it may have affected organisms through enhanced toxicity due to low-oxygen waters and sulphur release (Holser, 1977; Wilde and Berry, 1984; Holser and Magaritz, 1987).

11.1.5 Late Paleozoic Glaciation

The two most significant climate events that followed the mid-Paleozoic transition were the formation of massive coal deposits and glaciation on Gondwanaland. Berner (1987; Berner and Canfield, 1989) estimated that sequestering of organic carbon reached a Phanerozoic peak during these times of coal formation (Fig. 11.9). The distribution of coals falls into two main categories: a tropical belt of coal swamps extending

from the midwestern United States through Europe. Other major regions of coal formation are in Siberia and midlatitude Gondwanaland. Deposits on the eastward side of the land masses extend farther poleward than on the western side of the landmasses—an effect presumably reflecting transport of water by warm and cold ocean currents, respectively (Ziegler et al., 1981). Studies of well-preserved wood (Chaloner and Creber, 1973) indicate that the tropical coals had no or faint growth rings, suggesting that growth was not interrupted by seasonal dry spells. Midlatitude coal deposits had prominent growth rings.

Although there was some glaciation in the Late Devonian of South America (Caputo and Crowell, 1985), the main phase of Late Paleozoic glaciation started in the Carboniferous, and may have been triggered by uplift in Australia and South America (Powell and Veevers, 1987). The areal extent of glaciers (Fig. 11.10) was approximately the same as in the Pleistocene. Glaciation started in South America and Africa, with centers of activity later spreading to Antarctica and Australia. In fact, the Permo-Carboniferous Gondwanan glaciers were the first identified pre-Pleistocene glacial deposits (the most famous of them being the South African Dwyka Tillite). Study of Gondwanan glaciation contributed greatly to development of the theory of continental drift in the early twentieth century by such figures as Alfred Wegener (1929) in Germany and Alexander DuToit in South Africa (1937). For example, glacial stria-

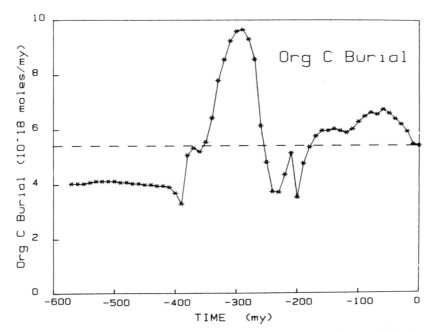

Fig. 11.9. Estimated organic carbon burial rates through the Phanerozoic. [After Berner, 1987] *Reprinted by permission of American Journal of Science.*

Fig. 11.10. Evidence for extensive Carboniferous glaciation in Gondwanaland. Figure illustrates Westphalian (Late Carboniferous; ⁓305 Ma) paleogeography and distribution of climatically controlled sediments. Circles, tillites; squares, coals; inverted "v," evaporites; dark shading, highland; medium shading, lowland; light shading, continental shelf. [From Parrish et al., 1986] *Reproduced by permission of the Smithsonian Institution Press from "The Ecology and Biology of Mammal-like Reptiles" edited by N. Hotton II, P. D. MacLean, J. J. Roth, and E.C. Roth. Smithsonian Institution, Washington, D.C. 1986, p. 101, fig. 1.*

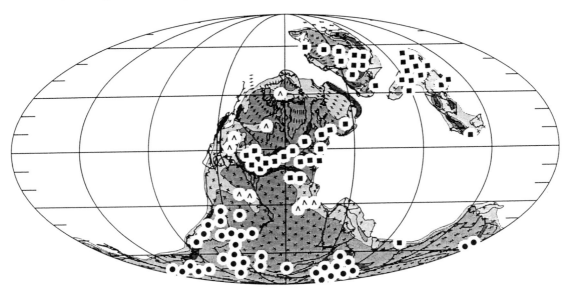

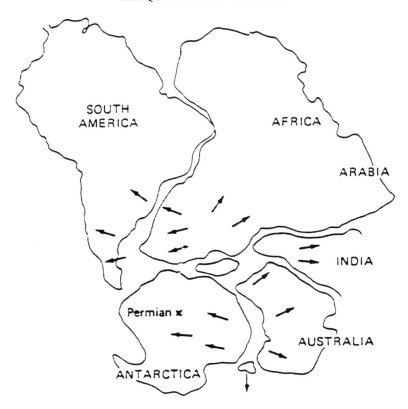

Fig. 11.11. Generalized diagram illustrating evidence for origination of Permo-Carboniferous glaciation on one large landmass. Arrows indicate direction of glacial flow. X = Permian pole position. [After Sullivan, 1974] *Reproduced by permission from W. Sullivan, "Continents in Motion: The New Earth Debate," copyright 1974, McGraw-Hill Publishing Co.*

tions in South America record ice movement out of the present South Atlantic (Fig. 11.11) but with a predrift reconstruction are almost perfectly consistent with ice directions in South Africa.

The Gondwanan glacial deposits occur in almost all the presently widely dispersed parts of Gondwanaland—South America, Africa, Antarctica, India, and Australia (the landmass name comes from an Indian sequence and means "Land of the Gonds"). The Gondwanan deposits are part of a characteristic sequence (Fig. 11.12) with Permo-Carboniferous glacial deposits near the base, sometimes interbedded with, but more often succeeded by, nonmarine sandstones and coal. The coals include the predominant *Glossopteris* plant fossils (a seed-fern-like plant which helped form the immense coal deposits). Coals are more common in the Permian and red beds more common in the Triassic. Fi-

nally, the overlying Jurassic sequence often has basaltic lavas and sills associated with the rifting and the breakup of Pangaea.

Permo-Carboniferous glaciation lasted over 100 million years, with peak extent being about 60 million years (Crowell, 1983; Caputo and Crowell, 1985; Fig. 11.2). Fluctuations of the ice sheet have long been linked with sea level fluctuations in North America, Europe, and Russia (Wanless and Shepard, 1936; Ross and Ross, 1985; Heckel, 1986; Veevers and Powell, 1987). There are significant cyclical variations of sediments in these regions. The "cyclothems" record fluctuations between relatively high sea level (marine carbonates) and nonmarine conditions (nonmarine sandstones and shale). The great North American coal fields usually developed during the nonmarine phase of the sea level oscillation. However, one of the major advances of the Gondwanan ice sheets may have been associated with enhanced tropical precipitation and coal formation in low latitudes (Raymond et al., 1989).

The cyclothem deposits (Fig. 11.13) were analyzed for quasi-periodic tendencies (Ross and Ross, 1985; Heckel, 1986). Intercontinental fluctuations have a time scale of about 2 million years (Ross and Ross, 1985). The North America sequence (Heckel, 1986) may record fluctuations with a characteristic time of approximately 400,000 years—the dominant period in

the eccentricity record of orbital insolation changes (see Chap. 7). Estimated sea level fluctuations were on the order of 100–200 m, but this value is a maximum glacio-eustatic effect and in North America may be significantly influenced by tectonic changes associated with uplift of the Appalacian Mountains (Klein and Willard, 1989). The estimated sea level changes are comparable to those which occurred in the Pleistocene (cf. Chap. 3). [Note added in proof: a significant revision of the Carboniferous time

Typical Gondwanan Sequence

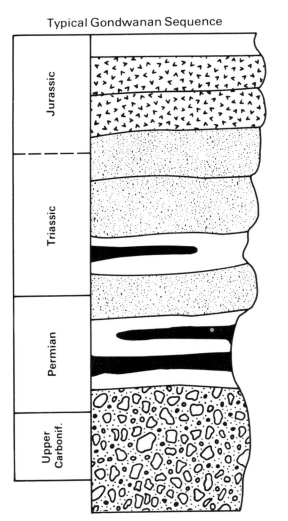

Basalt

Sandstone

Coal

Glacial tillites

G   Glossopteris flora

Fig. 11.12. Example of classic Gondwanan rock sequence, found on all major Gondwanan segments (Fig. 11.11) and indicative of deposition under one uniform regime. Vertical bar refers to range of *Glossopteris* flora, a common Gondwanan plant assemblage. [After Dott and Batten, 1981] *Reproduced by permission. Modified from Dott, R. H., and R. L. Batten, "Evolution of the Earth," copyright 1981, McGraw-Hill Publishing Co.*

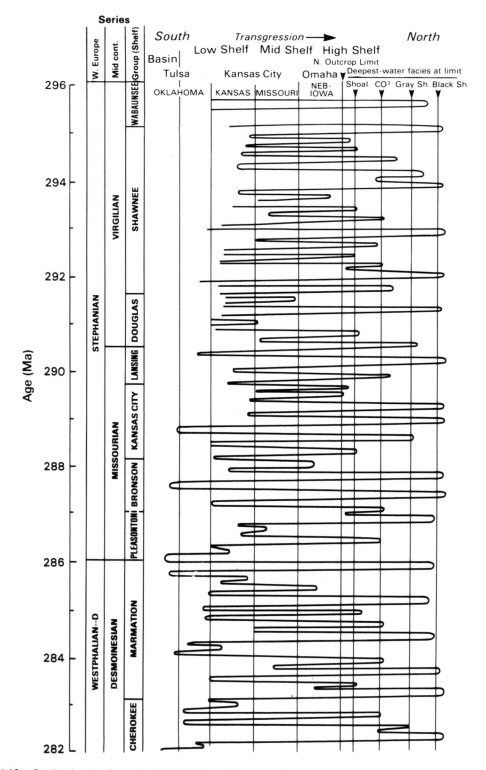

Fig. 11.13. Sea level curve for part of Middle-Upper Pennsylvanian sequence in the midcontinent North America outcrop belt, based on shoreline positions estimated from farthest basinward extent of exposure surfaces and fluvial–deltaic complexes. Correlations with time scale on left are only approximate and useful mainly as a frame of reference for estimating approximate periods of major sea level variations. [Modified from Heckel, 1986] *Reprinted with permission of Geological Society of America. Modified from P. Heckel, "Sea-level curve for Pennsylvanian eustatic marine transgressive-regressive depositional cycles along midcontinent outcrop belt, North America," 1986, Geology 14, 330–334.*

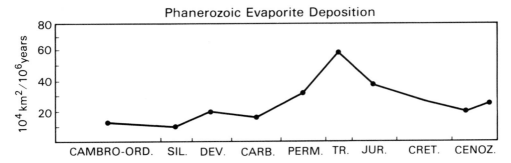

Fig. 11.14. Areas covered by evaporite deposits throughout the Phanerozoic. Evaporite deposits form in arid environments, and the Triassic peak in formation rate may be due to extreme aridity on a very large landmass. [After Gordon, 1975] *Reproduced by permission from W. Gordon, "Distribution by latitude of Phanerozoic evaporite deposits," Jour. Geol. 83, 671–684, copyright 1975, University of Chicago Press.*

scale raises serious questions as to whether the oscillations in Figure 11.13 correlate with Milankovitch periods—see Klein, 1990.]

11.1.6 Pangaea

The final suturing of the continents to form Pangaea was completed during the Triassic (~220 Ma). Since that time, continental fragmentation has proceeded to the present day. The combination of a gigantic landmass and emergent shorelines should have resulted in extremely continental climates. This conjecture is consistent with geologic data, inferences, and reconstructions (Robinson, 1973; Parrish et al., 1982) of very arid conditions in the Permian and Triassic (extensive red beds and salt deposits). Triassic evaporite (salt) deposits (Fig. 11.14), which form in arid environments, are more extensive than at any other time in the last 600 million years (Gordon, 1975). However, the geographic range of the sedimentary deposits does not seem significantly different than that of the present (Fig. 11.15). Increased evaporites in

Fig. 11.15. Histograms of the distribution of coals and evaporites by latitude. The length of the bars, which are marked off in increments of 5, indicates the number of deposits in each 10°-wide latitudinal band. Data from Ziegler et al. (1983). [After Parrish et al., 1982] *Reprinted with permission of Elsevier Science Publishers.*

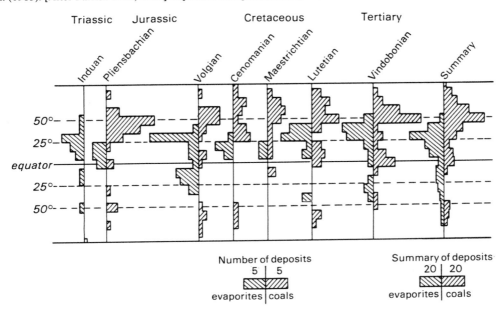

North America and western Europe may also reflect uplift of the Appalachian Mountains and enhanced aridity in the lee of the mountain chain (Bambach et al., 1981; Hay et al., 1982).

Extensive evaporite formation affected the mean ocean salinity. For example, the salt content of the evaporite deposits is equivalent to about 10% of the present ocean's salt (Stevens, 1977). Average oceanic salinities prior to 200 Ma should therefore have been 3.0‰ greater than the present average of about 35.0‰. This increase is more than twice that estimated for Pleistocene ocean enrichment due to removal of freshwater for ice sheet construction (Worthington, 1968).

We have relatively little information on changes of climates over the ~70 million year interval between disappearance of Gondwanan ice and large-scale breakup of the landmass. For much of this time interval, there is little evidence for glaciation (Fig. 11.2). When the initial breakup of Pangaea started in the Triassic, formation of narrow seas in eastern North America

and western Europe may have caused regional increases in precipitation inferred for this region (Simms and Ruffell, 1989). Studies of long sequences of sedimentary deposits in these lakes (Olsen, 1986) indicate variations typical of orbital periods (cf. Chap. 7). The variations are probably due to orbitally induced variations in monsoon intensity (cf. Section 4.5.1).

Later on in the Jurassic, warming is indicated by ~10° of latitude northward movement (Fig. 11.16) in the Indo-European/Siberian floral boundary (Vakhrameev, 1964). This warming trend more or less paralleled the rise in sea level (and CO_2?; cf. Fig. 11.4). Black shale intervals (ocean anoxic events; see Section 8.1.4) in the deep sea occur during the periods of high sea level (Jenkyns, 1988).

Pangaea existed from the Permian to the Jurassic. It is of interest to briefly review Jurassic deposits as a possible window into older intervals of supercontinent configurations, when plant and animal distributions are more incomplete. Examination of these distributions may

Fig. 11.16. Evidence for climate change in the Jurassic as illustrated by the shift in the boundary of Indo-European and Siberian floral province boundary in Eurasia during the early (1) and mid- (2) to late (3) Jurassic. [From Hallam, 1982; adapted from Vakhrameev, 1964] *Reprinted from "Climate in Earth History," with permission from the National Academy Press, Washington, DC.*

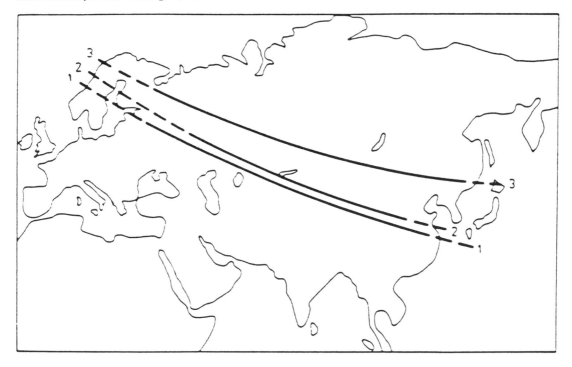

also provide insight into whether Pangaean climates were significantly warmer than the present—a point often assumed for the Jurassic (e.g., Hallam, 1982, 1985).

Despite lack of evidence for glacial deposits (e.g., Fig. 11.2), evidence for enhanced global warmth is ambiguous for the Triassic and Jurassic. For example, both Triassic and Jurassic reefs are restricted to about the same latitudes as the present (Frakes, 1979) with only a slight poleward extension in the area of the warm Tethys intrusion. Another marine group (ammonites and belemnites—invertebrate marine cephalopods) has been divided into Tethyan and boreal provinces (Hallam, 1982), but the interpretation of these distributions in terms of temperature variations is in dispute (Hallam, 1982). Jurassic laterites primarily occur in the Tethys region and probably reflect localized warmth from the Tethys Sea (van Houten, 1985; laterites are iron-rich, deeply weathered soils that form in warm, moist environments).

Vertebrate fossils extend farther poleward than reefs or laterites (Parrish et al., 1986), but it is not entirely clear whether such distributions could be accommodated by seasonal migrations. For example, tree ring studies show significant seasonality in high latitudes during the middle Mesozoic (Francis, 1984; cf. Jefferson, 1982). However, another plant group (ferns), whose present relatives are intolerant to cold, range to 50°N and 60°S (Barnard, 1973). Vakhrameev (1964) concluded from these distributions that winter temperatures never fell below freezing. This paleoecological extrapolation may be questionable because there have been significant evolutionary changes within the group and an entirely new class (Angiosperms, flowering plants) has evolved in the interim (100–120 Ma) and presumably occupied some of the niches inhabited by their more "primitive" colleagues.

To summarize, Pangaean climates were relatively dry, and there is some indication of a warming trend in the later Jurassic. It is not at all clear that Triassic/Early Jurassic climates were significantly warmer than the present, at least in terms of global mean annual temperatures. As we will see in the modeling section, there is good reason to believe that seasonal temperatures on the supercontinent may have been higher than present.

11.2 Modeling Studies

Although there have been various attempts to reconstruct climates of the Paleozoic and early Mesozoic, actual modeling studies are few in number. In this section we will discuss what modeling progress has been made for this time interval. One of the principal topics of modeling interest in Paleozoic and early Mesozoic climates involves the effects of supercontinents on climate. The discussions below primarily address this feature. Because studies are still at a relatively early stage, some of the topics may undergo revision as further work ensues.

11.2.1 "Gondwanaland" Experiments with Idealized Geography

As the supercontinent of Gondwanaland persisted as a large landmass for several hundred million years, it should have had a significant effect on climate, particularly on the seasonal cycle. As discussed in Section 1.2.4, two-dimensional energy balance models (EBMs) do a good job of resolving the seasonal cycle on landmasses and are particularly useful for examining the effects of supercontinents on the seasonal cycle.

EBM experiments with simplified geography (Fig. 11.17) help clarify the relationship between continental size, location, and seasonal cycle amplitude. For example, the area of an idealized "disk continent" is approximately equivalent to the area of Gondwanaland and extends from the South Pole to 45°S. When the South Pole was located at the edge of the supercontinent (Fig. 11.7), calculated summer temperatures are quite low (Crowley et al., 1987). As low summer temperatures favor preservation of permanent snow fields, such configurations might be conducive to glaciation (Fig. 11.17).

When the South Pole is more centrally located in the interior of a large landmass, intense heating results in summer temperatures that could have exceeded 20°C and possibly even 30°C (Fig. 11.17). Since this calculation was done without ice-albedo feedback, it is possible that summer warming in the continental interior would be less than simulated (Ledley, 1988). However, direct comparisons of EBMs and GCMs show quite good levels of agreement

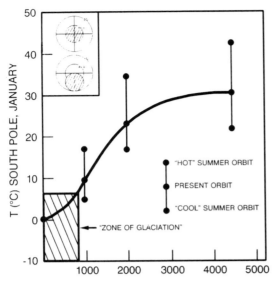

DISTANCE (km) OF POLE FROM EDGE OF CONTINENT

Fig. 11.17. Summer temperature at the South Pole for different positions of an idealized supercontinent (see inset) as calculated by a linear EBM. Heavy line represents temperatures for the present orbital configuration. Variations around this value represent changes in forcing compatible with the maximum changes in orbital configurations for the Pleistocene (highest points, "hot summer" orbit; lowest points, "cool summer" orbit). Diagonal striping marks the postulated region of parameter space affected by glaciation if ice-albedo feedback were included. [From Crowley et al., 1987] *Reprinted by permission from Nature, vol. 329, pp. 803–807, copyright © 1987 Macmillan Magazines Ltd.*

(cf. Fig. 1.10; Crowley et al., 1986; Hyde et al., 1989). Furthermore, precipitation in the interior of large landmasses is generally very low, and the amount of winter snow required for melting might be small enough to allow significant summer warming. Calculated high temperatures in continental interiors should be interpreted with some caution until further sensitivity tests have been conducted. Initial testing of this hypothesis with a nonlinear EBM supports the above conclusion (Hyde et al., 1990).

The above experiments suggest that the most favorable conditions for continental glaciation may occur when the pole is more coastally located, as it was in the Late Ordovician and Late Paleozoic (Fig. 11.7). But the Ordovician glaciation occurs during a time of postulated high CO_2 (van Houten, 1985; see Sections 11.1.2 and 11.1.3)—possibly as high as the mid-Cretaceous

(cf. Sections 8.2.4 & 8.2.5; compare relative sea levels in Fig. 11.4).

How can extensive ice sheets and high CO_2 coexist? One consideration involves the seasonal cycle—if the pole is located in coastal regions, there may be a small seasonal cycle because of the proximity of water. Summer temperatures might not get above freezing even with higher CO_2. The above hypothesis has been tested with EBMs by increasing the solar constant by 4 and 8% in order to mimic a CO_2 increase (Crowley et al., 1987). Increasing the solar constant has a greater effect on the mean annual temperature than on the seasonal cycle; glaciated conditions may still be possible for higher CO_2 levels if the pole is located in coastal regions (Crowley et al., 1987). Thus a perplexing paleoclimate paradox may be resolvable in an economic way and in a manner entirely consistent with present climate theory. However, this conclusion warrants considerably more testing before it can be accepted.

11.2.2 Gondwanaland Experiments with Realistic Geography

Simulations with realistic geography (Figs. 11.18 and 11.19) substantiate some of the conclusions derived from the idealized geography experiments and provide additional insight into regional variations in temperature. Flooding of continental margins by shallow seas effectively reduces the size of the supercontinent by almost 30% and thus also reduces the magnitude of summer warming. Simulated January temperatures for a coastally located position (Late Ordovician) are about 0°C (Fig. 11.18); they are ~20°C for a centrally located pole (Fig. 11.19), as may have occurred in the Devonian. These simulations are consistent with geologic data indicating glaciation in North Africa during the Late Ordovician (Caputo and Crowell, 1985), and for no ice for a 60 million year interval centered in the Early Devonian.

Devonian pole positions are sufficiently uncertain (e.g., Kent and May, 1987) that the Devonian experiment should be considered more as an illustration of an effect rather than a true simulation. For comparison we illustrate a different Devonian reconstruction (Bambach et al., 1981) and simulation in Fig. 11.19 (right). Summer temperatures are still high at the South

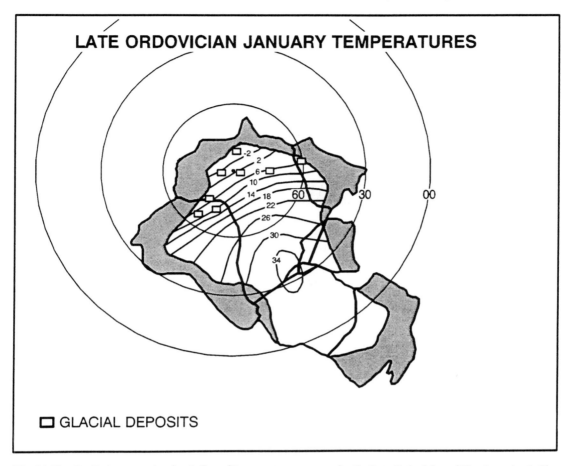

Fig. 11.18. Realistic geography simulation of January temperatures for the Late Ordovician of Gondwanaland. Circular contours, lines of latitude; other contours, January temperatures. Shaded areas represent continental margins flooded by shallow seas. Rectangles are glacial deposits from Caputo and Crowell (1985). [From Crowley et al., 1987] *Reprinted by permission from Nature, vol. 329, pp. 803–807, copyright (©) 1987 Macmillan Magazines Ltd.*

Pole, but temperatures are somewhat lower in the coastal zone than in Fig. 11.19 (left). Note the excellent correspondence between the regions of lowest temperatures on land and high-latitude biogeographic provinces based on marine invertebrates and plants.

Since the Gondwanan and Laurentide Ice Sheets covered comparable areas, it is of interest to inquire as to the effect of the Gondwanan supercontinent on ice sheet temperatures. Preliminary experiments (Fig. 11.20) suggest that summer warming over the Gondwanan Ice Sheet was considerably greater than over the Laurentide Ice Sheet (if an elevation correction is made, the ice sheet would still have been below freezing for most of its area). Winter cooling was correspondingly greater over the Gondwanan Ice Sheet (not shown). Given the expected extensive melting along the northern edge of the Gondwanan Ice Sheet, there may have been significant ponding of meltwater in the poor-drainage areas usually associated with such regimes. Perhaps some of the Gondwanan coal sequences formed in these environments.

The above calculation suggests that there may have been some fundamental differences between the Laurentide and Gondwanan Ice Sheets. The summer warming on the Gondwanan Ice Sheet was as large or greater than Milankovitch/CO_2 warming associated with Pleistocene deglaciations. How could the Gondwanan Ice Sheet have maintained its stability,

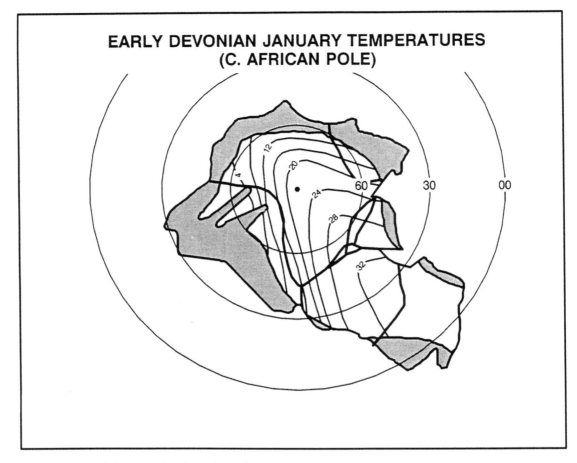

Fig. 11.19. Realistic geography simulations of January temperature for the Early Devonian (~390 Ma) of Gondwanaland. Because the pole position is not well known, two different reconstructions are figured. Left: simulation with central Africa (mid-Devonian) pole of Hurley and van der Voo (1987). [From Crowley et al., 1987] Right: simulation with South African pole of Bambach et al. (1981). Note that for the South African pole there is a very good agreement between regions of lowest temperatures and boreal biogeographic provinces as illustrated by shallow-water marine (Malvinokaffric realm) and plant distributions. [The latter two from Ziegler et al. (1981) and Boucot and Gray, (1983)] *Reprinted by permission from Nature vol. 329, pp. 803–807, copyright (©) 1987 Macmillan Magazines Ltd.*

especially when winter accumulation usually decreases on very large landmasses? Furthermore, if the ice sheet were to melt, summer temperatures could have reached 20°C even at the South Pole (cf. Figs. 11.17 and 11.19). How could the ice sheet have then been reinitiated? Is it possible that the only stable configuration for the Gondwanan Ice Sheet is one of very large dimensions, i.e., that it did not wax and wane like the Pleistocene ice sheets? The North American sea level history (Heckel, 1986) suggests that this was not the case (Fig. 11.13). It is evident that at this stage there are some significant problems

concerning the stability of the Gondwanan Ice Sheet.

11.2.3 Time Scale for Glacial Periods

The above EBM calculations may also yield information about the characteristic time scale of glacial periods. During the last 1 billion years major glacial periods have lasted on the order of a few tens of millions of years (e.g., Fig. 13.1). Geologists have sometimes tried to relate such recurrent events to some type of periodic forcing. But mere repetition does not automatically

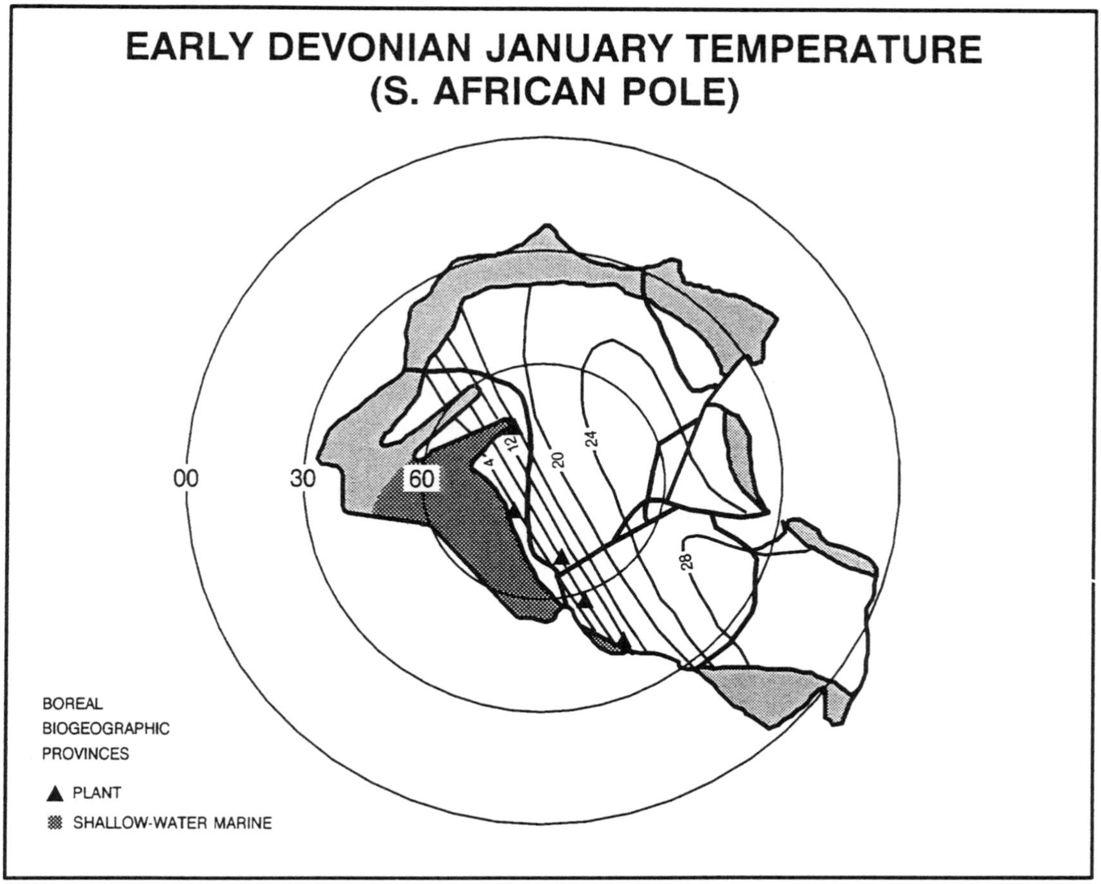
EARLY DEVONIAN JANUARY TEMPERATURE
(S. AFRICAN POLE)
00
30
60
4
12
20
24
28
BOREAL
BIOGEOGRAPHIC
PROVINCES
PLANT
SHALLOW-WATER MARINE

Fig. 11.19. *(continued)*

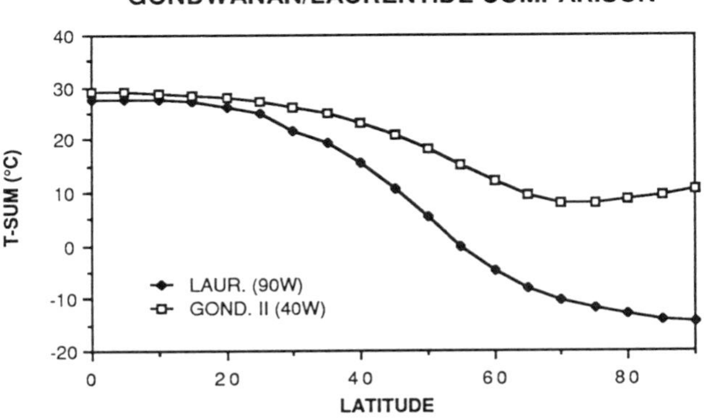
GONDWANAN/LAURENTIDE COMPARISON
T-SUM (°C)
40
30
20
10
0
-10
-20
LAUR. (90W)
GOND. II (40W)
0
20
40
60
80
LATITUDE

Fig. 11.20. Comparison of summer temperature profiles over the Gondwana and Laurentide Ice Sheets. EBM calculated temperatures are for sea level (i.e., no ice sheet elevation) at the longitude of maximum equatorward ice sheet extension. Although the ice sheets are approximately the same size, the much higher summer temperatures over the Gondwanan Ice Sheet are due to its location on a very large landmass, which heats up greatly in summer.

imply periodicity; there may be "earth-bound" events (e.g., tectonics) which may interject characteristic time scales into the geologic record.

Figure 11.17 may provide a clue as to how characteristic time scales can be due to tectonic processes. Note that the "zone of glaciation" is located in the region of parameter space less than about 6°C and within 800 km of the coastline. Glaciation could occur for more interior sites, but it is less likely. How long would it take a supercontinent to migrate from a coastal polar position to an interior position (see Fig. 11.7)? If Gondwanaland moved at the same rate as present plates with land masses (~2 cm/year), it would require on the order of 40 million years. The rates may have varied somewhat, and the path of migration may not have been normal to the coast, but the above simple calculation suggests that the characteristic time scale of glacial fluctuations of the last 1 billion years may have been set by the rate of plate motion and the interactions between geography and climate.

11.2.4   Temperature Variations on Pangaea

It is evident from the Gondwanaland modeling experiments that the supercontinent has a significant influence on climate. Similar patterns apply to the Permo-Triassic Pangaean supercontinent. In this section we will discuss modeling studies that explore the extreme effects of the supercontinent on temperature, precipitation, and the ocean circulation.

EBM experiments (Crowley et al., 1989) suggest that mean monthly summer temperatures in the interior of Pangaea (Fig. 11.21) exceeded 35°C—more than 6°C above maximum model temperatures for the present. Since daytime highs usually exceed mean monthly averages by 6–10°C or more, then daytime highs in the interior of Pangaea may have approached 45–50°C. Such high temperatures may have significantly influenced biotic distributions (Crowley et al., 1989). However, summer temperatures were relatively low in some polar regions (e.g., Fig. 11.21). These results are consistent with evidence for glacial deposits in eastern Australia and possibly Siberia (Epshteyn, 1981; Caputo and Crowell, 1985).

There was a very large seasonal cycle over most of southern Pangaea (Fig. 11.22); most of

the interior south of 30°S paleolatitude had a seasonal cycle greater than 30°C—a range found today only in northern Canada and Siberia. The very large seasonal cycle poses some dilemmas for paleoecological interpretations of vertebrate distributions. For example, a famous South African vertebrate locality (cf. Parrish et al., 1986) at about 60°S paleolatitude occurs in the core of the seasonal cycle maximum (Fig. 11.22). How could presumably warmth-loving vertebrates cope with such low winter temperatures?

Several factors could have contributed to high summer temperatures on southern Pangaea: the unification of Laurussia and Gondwanaland to form a larger supercontinent, 10° northward movement of the entire block between the Carboniferous and Permian, or waxing and waning of the Gondwanan Ice Sheet. Sensitivity experiments (Crowley et al., 1989) suggest that the collision with Laurussia had only a modest effect on maximum temperatures in the Southern Hemisphere—the two landmasses were in different hemispheres, and warming on each was 6 months out of phase. Presence of the Gondwanan Ice Sheet had a more significant effect on Southern Hemisphere temperatures. The disappearance of the late Paleozoic Gondwanan Ice Sheet therefore appears to have initiated the warmest phase of the supercontinent's climates.

Disappearance of the Early Permian ice sheets (e.g., Veevers and Powell, 1987) occurred at about the same time as the largest known terrestrial vertebrate extinction event (Benton, 1987), which was associated with a transition from an amphibian-dominated population to one characterized by advanced mammal-like reptiles (Olson, 1982). Since Pleistocene studies indicate that ice sheets can disappear very rapidly (5000–10,000 years), the disappearance of the Gondwanan Ice Sheet may have caused a relatively abrupt climate change to warmer and drier climates and possibly contributed to the extinction event (Crowley et al., 1989). The warm and dry climates (see text following) of the late Paleozoic also provide an interesting and perhaps important backdrop to the great Permian–Triassic extinctions (Fig. 9.1). This was the largest extinction in the geologic record and was associated with an estimated elimination of 96% of marine species (Raup, 1979). The calculated large drop in atmospheric pO_2 at this time (Berner and

KAZANIAN 255 MA (JANUARY)

Fig. 11.21. Simulated January temperatures for the Late Permian (Kazanian, 255 Ma). Base map, with topography (dark areas) for reference, is from Scotese (1986). Open circles refer to evidence for ice cover (after Parrish et al., 1986). [From Crowley et al., 1989] *Reprinted with permission of Geological Society of America. From T. Crowley, W. Hyde, and D. Short, "Seasonal cycle variations on the supercontinent of Pangaea," 1989, Geology 17, 457–460.*

Fig. 11.22. Annual range of temperatures for the Late Permian (Kazanian). Base map, with topography for reference, is from Scotese (1986). [From Crowley et al., 1989] *Reprinted with permission of Geological Society of America. From T. Crowley, W. Hyde, and D. Short, "Seasonal cycle variations on the supercontinent of Pangaea," 1989, Geology 17, 457–460.*

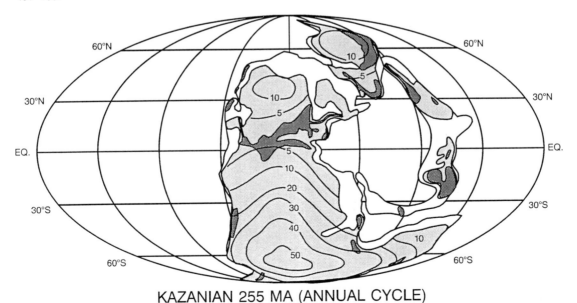

KAZANIAN 255 MA (ANNUAL CYCLE)

ANNUAL PRECIPITATION minus EVAPORATION (mm/day)

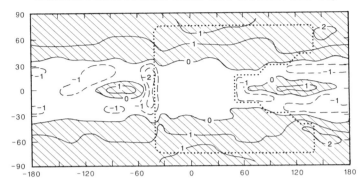

Fig. 11.23. Mean annual P-E for idealized Pangaea. Regions with positive P-E are hatched. [From Kutzbach and Gallimore, 1989] *Courtesy of American Geophysical Union.*

Canfield, 1989) may also have an important bearing on the extinctions (the O_2 trends parallel the organic carbon curve in Fig. 11.9).

11.2.5 Precipitation Variations on Pangaea

GCM experiments for an idealized Pangaea (approximate Triassic configuration) indicate that, despite the intense summer heating, precipitation, and net precipitation-minus-evaporation (P–E) was relatively low in the subtropics of the Pangaean landmass (cf. Fig. 11.23; Kutzbach and Gallimore, 1989). The very long "fetch" of land essentially squeezed out most of the moisture from the air before it reached Pangaean interiors. The net result (Fig. 11.23) is that most of the landmass equatorward of about 40° was relatively dry. Averaged over all land area, net P–E on Pangaea was only about 50% of the present level (Fig. 11.24). These results may be somewhat sensitive to specification of topographic boundary conditions, for the presence of a high plateau (e.g., the Tibetan Plateau) intensifies the monsoonal circulation (Flohn, 1968; Hahn and Manabe, 1975). A similar plateau may have existed in western Europe during the late Paleozoic (Menard and Molnar, 1988; location was approximately 15°N and 10°E in Fig. 11.21).

Average temperatures over land in the GCM experiment were almost 10°C warmer than present (17°C versus 7°C), a result attributable in part to the fact that no permanent ice was prescribed in the experiment (Kutzbach and Galli-

more, 1989). A very large seasonal cycle of temperatures is also simulated for the idealized Triassic landmass (see Fig. 11.22). The GCM simulated sea ice to 70° paleolatitude (see Hunt, 1984).

In another experiment Kutzbach and Gallimore (1989) mimicked a fivefold CO_2 increase by raising the solar constant (Fig. 11.25). Precipitation increased 8% and snowlines retreated. There was still a very large seasonal cycle and significant sea ice, and the area of very high summer temperatures (40°C) expanded—a result possibly approaching or exceeding the tolerance limits of organisms.

11.2.6 "Panthalassa" Ocean

The Pangaean supercontinent was surrounded by a superocean—"Panthalassa." Kutzbach et al. (1990) used output from the atmospheric

Fig. 11.24. Comparison of net P-E (land) as calculated for present geography (M), an idealized Pangaea (P), an idealized Pangaea with higher CO_2 forcing (P+), and an idealized Pangaea with topography elevated 1 km (arrows on "P" bar). [From Kutzbach and Gallimore, 1989] *Courtesy of American Geophysical Union.*

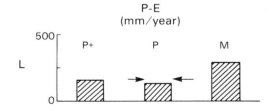

JJA SURFACE TEMPERATURE (°C)

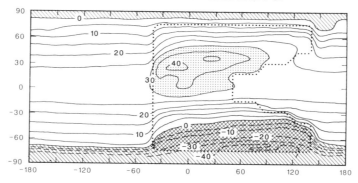

Fig. 11.25. Surface temperatures (°C) for June-July-August, for an idealized Pangaea with higher ($\sim 4\times$) CO_2 levels. Over the ocean the 0°C contour corresponds approximately with the sea ice limit. Note that even with higher CO_2 levels, there is significant cooling in the winter hemisphere, and that mean seasonal temperatures, which are daytime/ nighttime averages, exceed 40°C in low latitudes (daytime highs could have exceeded this value by at least 10–15°C). [From Kutzbach and Gallimore, 1989] *Courtesy of American Geophysical Union.*

GCM simulation to force an ocean model. Results indicate that ocean circulation feedbacks may have reinforced the atmosphere circulation patterns (Fig. 11.26). For example, there was a relatively strong western boundary current along the east coast of Pangaea. This current carried warm water into higher latitudes in these regions. The area of polar warming by the ocean occurs in the same region as the area affected by northward displacement of the atmospheric circulation by the Tethys Ocean. The combinations of atmosphere and ocean circulation patterns may provide the explanation for warm conditions on the eastern side of the Paleozoic continents (e.g., Ziegler et al., 1981).

Although there have not yet been any calculations for the Paleozoic with a fully coupled atmosphere–ocean model, experience with modern climate might provide insight into how warm penetration of ocean currents might have affected atmospheric temperatures in high latitudes. Of particular importance in this respect is the observation that the currents are moving poleward along the eastern boundary of Pangaea. Interior Pangaea temperatures at high latitudes may not have significantly ameliorated by the warm currents because they are downwind from the continents. A good analogy might be the east coast of the United States today, where the warm Gulf Stream follows the coast as far

Fig. 11.26. Simulated mean annual surface ocean currents (cm/sec) for idealized Panthalassa Ocean, utilizing atmospheric boundary conditions from simulation of Kutzbach and Gallimore (1989). [From Kutzbach et al., 1990] *Courtesy of American Geophysical Union.*

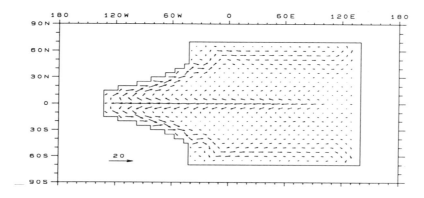

north as Cape Hatteras. Yet interior temperatures can frequently be quite cold in winter. Thus, the main effect of the warm currents would probably be restricted to coastal regions.

11.2.7 The Question of "Equable" Climates

In this final section we will summarize results from both modeling and observational studies to take another look at the term "equable" as it is often applied to past climates. This term connotes warm climate lacking in extreme variations. Is the term justified by our review of geologic data and modeling studies? It is difficult to say. For example, distributions of Jurassic reefs, laterites, coals, and evaporites are often not significantly different than those of the present, except in the Tethys Sea area, which may have been locally affected by a warm current (cf. Section 11.1.6). The widespread distribution of some marine invertebrates has been interpreted to mean greater warmth (Hallam, 1982). This interpretation receives some support from the distribution of fossil reptiles and plants. However, vertebrate distributions could be biased by seasonal migrations (cf. Parrish et al., 1987) or alternate interpretations in terms of "warm-blooded dinosaurs" (Bakker, 1975, 1980). Evolutionary changes may also not justify paleoclimatic inferences for land plants. Finally, there is occasional evidence for at least seasonal ice in parts of the Mesozoic.

Modeling studies indicate that a very large seasonal cycle, with probable winter cooling, seems to be almost an inescapable feature of Gondwanan and Pangaean climates. This conclusion is supported by a nonlinear two-dimensional energy balance model simulation for the Early Jurassic (Fig. 11.27)—the supercontinent interval for which there is the greatest amount of discussion about equable climates. Note that the annual cycle in both hemispheres is approximately the same as the modeled present annual cycle on Eurasia (Hyde et al., 1990). Such a large annual range with attendant low winter temperatures can hardly be considered "equable."

The above discussion suggests that it is difficult to reconcile large variations of temperatures on supercontinents with the term "equable." How would changes in ocean heat transport or CO_2 affect this conclusion? As discussed earlier, ocean heat transport would primarily affect Pangaean climates in coastal areas. CO_2 would certainly increase winter temperatures, but initial experiments indicate that temperatures would still be below freezing (Section 11.2.5). Furthermore, summer temperatures, already high in some areas, would become even higher. Experience with climate models (e.g., Manabe and Wetherald, 1986) indicates that once summer temperatures exceed a certain level, soil moisture is lost. With drying of the soil, temperatures are further amplified because the land heats up faster. To summarize, a large seasonal cycle might still result even if snow cover is reduced because soil moisture depletion might result in a positive temperature feedback.

The above experiments and arguments imply that it is very difficult to completely remove sea ice, and any increases in winter temperature are partially offset by increases in summer temperatures that reach very high levels. When considered in the light of "hierarchy of believability" of climate model results, the above results are fairly robust, for we have a good understanding of the linkage between land–sea distribution and the seasonal cycle (Fig. 1.9).

From a modeling viewpoint, the entire equability question reduces down to the observation that heat loss in continental interiors occurs at a greater rate than can be compensated for by transport from the oceans. The larger the landmass, the larger the discrepancy in the energy budget of continental interiors. Consequently, larger landmasses cool more than smaller landmasses. To emphasize this point, consider the climate of western Eurasia at present, which is washed by the warm Gulf Stream. At 57°N, the climate of northwestern coastal Scotland is sufficiently mild in winter to support an outdoors subtropical garden. Yet winter temperatures only one fourth of the distance into the interior of the landmass (Moscow, 56°N) can be exceedingly cold.

The above analysis calls into question widespread geologic usage of the term "equable climate." Do the data unambiguously indicate warm conditions with low variability? Are giant seasonal cycles and polar sea ice equable? This term originated in the misty past, when our

JURASSIC ANNUAL RANGE

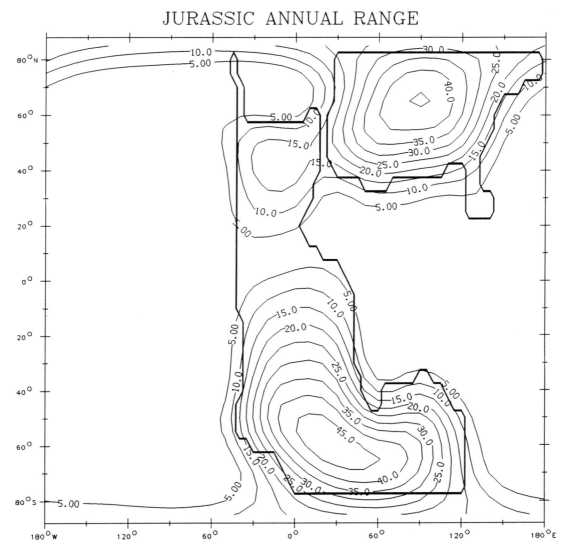

Fig. 11.27. Annual range of temperatures (°C) for the Early Jurassic (~195 Ma), as simulated by a nonlinear two-dimensional energy balance model (the nonlinear feature addresses the snow-albedo feedback; cf. Sec. 1.3.2). Geography from Parrish et al. (1982); model described in Hyde et al. (1990). *Figure provided courtesy of William Hyde.*

knowledge of climate and interpretation of biotic distributions was not nearly as great as now. For probably most of this century, the term, like a tradition, has been honorably passed down from geologist to geologist in Historical Geology classes. Yet from our own experience, we cannot recall a meteorologically trained climatologist or climate modeler ever using the term. The term does not even seem applicable to times of higher sea level and enhanced continental fragmenta-

tion (e.g., mid-Cretaceous or Early Eocene; cf. Sloan and Barron, 1990). We therefore suggest that geologists reconsider the widespread use of this venerable word (cf. Sloan and Barron, 1990) and ponder whether it should go the way of "geosynclines."

The above conclusion has considerable implications for interpretations of biotic distributions through time. The conclusion also represents a nice example of how results from one field (climate modeling) may challenge long-standing precepts in another field. In fact, what we are seeing is a movement beyond using paleoclimate data to validate climate models. Sometimes the model results may help us better understand the data. One example of such a spinoff involves the discussion of warm-blooded dinosaurs (Bakker, 1975, 1980). Viewed from a climate modeling perspective, a warm-blooded adaptation might be very advantageous to dinosaurs, as it would enable them to better survive low winter temperatures and also allow enhanced rates of seasonal migration. Future studies may provide many more examples of beneficial interactions between the climate and paleontology communities.

12. PRECAMBRIAN CLIMATES

Although comprising 85% of earth history, we can say much less about the climate of the Precambrian (4.6–0.57 Ga; cf. Appendix A) than we can for later periods because much of the record has been removed by erosion. Nevertheless, enough progress has been made to sketch the outline of the main features of climate and the environment over this time. Since climate scenarios depend on some familiarity with principal features of the evolution of the earth; these features will be included where appropriate.

12.1 Archean Climate (4.5–2.5 Ga)

12.1.1 Tectonic Setting

The first half of earth history is significantly different from later times. Because some of the tectonic changes are relevant to climate history, this section will briefly review some of the main differences between the Archean and later earth history. For example, the areal extent of continents during the first half of earth history may have been significantly less than at present. From 4.6 to 3.9 Ga (billion years ago) the crust was probably in a higher temperature state as a result of radioactive heat flux two to three times greater than present values (Goodwin, 1981) and intense meteorite bombardment during planetary accretion (the first 100 million years and also during the interval 4.2–3.9 Ga; Goodwin, 1976). The large impacts may have annihilated early ecosystems by producing globally lethal conditions following evaporation of large volumes of ocean water (Sleep et al., 1989). The gravitational energy released during the formation of the earth's core could have raised the average temperature of the planet by 1200°C (RamaMurthy, 1976). Virtual restriction of magnesium-rich basalts (komatiites) to Archean rocks supports this model (e.g., Nesbit, 1982). Such magmas are indicative of temperatures about 300–400°C greater than those of the present.

Radiometric dates from western Australia, northern Canada, southwestern Greenland, and eastern India indicate crustal formation by 3.8–4.0 Ga, with possible generation as early as 4.2–4.3 Ga (Moorbath et al., 1975; Basu et al., 1981; Compston and Pidgeon, 1986; Bowring et al., 1989). Rounded sedimentary particles in the Greenland rocks record the presence of a hydrosphere. Deuterium levels on Mars and Venus also suggest the presence of an early hydrosphere, which was subsequently lost—HDO/H_2O ratios are higher than expected, thereby indicating mass loss of the lighter H_2O (Donahue et al., 1982; Owen et al., 1988).

Between 3.8 and 2.5 Ga several models of crustal evolution suggest only microcontinent scale landmasses until the end of the Archean (e.g., Burke et al., 1976; Goodwin, 1981). The Archean earth may have been covered by many small plates (Hargraves, 1986). However, occurrences of high-pressure metamorphic minerals (diamonds, kyanite) in 3.5- to 3.8-Ga rocks also suggests that, at least locally, continental fragments had evolved very deep roots (Boak and Dymek, 1982; Boyd and Gurney, 1986).

The formation of extensive continental platforms may have been a consequence of the thermal evolution of the planet. The virtual absence of komatiites after the Archean suggests a transition to lower thermal regimes after 2.5 Ga (Goodwin, 1981). The weakening of the geotherm might have been associated with the segregation of lighter, silica-rich, granitic magmas (e.g., Lambert, 1976). This reaction may then have triggered the massive thermal event between 3.0 and 2.5 Ga, an event that produced 50–85% of the present granitic continental crust (Ronov, 1968; Condie, 1981; Dewey and Windley, 1981; Condie, 1989). Subsequent to 2.5 Ga there has been no shortage of continent-sized landmasses.

The change in tectonic regime was also associated with a significant change in atmospheric composition. Whereas the early atmosphere may have been slightly reducing, massive iron formations (banded iron formations) occur in the interval 2.5–2.0 Ga. These rocks are thought to reflect the neutralization of photosynthetic oxygen and presumably mark a threshold level in the expansion of photosynthetic organisms. There was also a marked increase in shelf (platform) carbonates after this time (Ronov, 1964), with most of the carbonates indicating stromatolitic formation (finely laminated rocks). The origin of this rock type seems dependent on the presence of bacteria or blue–green algae (e.g., Schopf, 1980).

12.1.2   Archean Climates

Perhaps the most challenging climate problem for the early earth involves the "Faint Young Sun" paradox (e.g., Ulrich, 1975; cf. Gilliland, 1989). Most models of solar evolution (e.g., Newman and Rood, 1977) indicate a solar luminosity increase of 20–30% during the past 4.7 Ga (Fig. 12.1). The luminosity estimates are not dependent on estimates of nuclear reaction rates (Endal, 1981), which are used to predict phenomena such as neutrino flux [see, for example, Davis et al., (1978) and Newman (1986)]. Scaling arguments identify luminosity as being primarily dependent on the mean molecular weight of the sun. This quantity increases with time because of the conversion of hydrogen to helium in the sun's core. [Note that Willson et al. (1987) suggested that the early sun may have undergone a significant mass loss and may have been considerably brighter than usually assumed.]

The above results are significant because climate models indicate that a 5–10% reduction of the solar constant from present values should result in an ice-covered planet (e.g., Wetherald and Manabe, 1975; North et al., 1981). Strong ice-albedo feedback effects might then maintain this condition even if the solar constant were to increase to values greater than the present (see Section 1.3.2). Feedback processes are therefore required in order to explain an apparently ice-

free earth for the first half of its history (cf. Cogley and Henderson-Sellers, 1984).

Despite climate model predictions, the oldest rocks of glacial origin are "only" 2.7 Ga (Frakes, 1979; Crowell, 1982; cf. Fig. 13.1), and there is no strong evidence for extreme climates in the Archean (Walker, 1982). A number of climate feedbacks are therefore required to prevent an ice-covered planet. Postulated early earth feedbacks are of several types. During the earliest stages of earth formation, impact-induced heating produced a large blanket of volatiles in the earth's proto-atmosphere (Matsui and Abe, 1986). The main source of heat for the proto-atmosphere would involve the impact energy. For such conditions, the atmosphere is heated from below, and the amount of heating is a function of the atmosphere's opacity, or optical thickness, which would have absorbed most of the outgoing radiation from the surface. The first rain on earth may have had temperatures as high as 600°K (Matsui and Abe, 1986). Impact heating should also have been important during the interval 4.2–3.9 Ga (see Section 12.1.1).

An increased atmospheric greenhouse effect is the most commonly cited factor for moderate temperatures during lower luminosity (e.g., Hart, 1978; Owen et al., 1979; Kasting, 1989). Sagan and Mullen (1972) proposed that ammonia may have been the agent responsible for the greenhouse effect. However, NH_3 photodissociates after only 40 years (Kuhn and Atreya, 1979). Increased concentrations of CO_2 and H_2O may be more likely candidates for a greenhouse effect. At present, the crust contains over 99.9% of the carbon in the atmosphere–ocean–crust system (Garrels and MacKenzie, 1972). In fact, the total amount of carbon in the earth's crust–ocean–atmosphere is comparable to that of Venus (Owen, 1978), except in the latter case almost all of it is in the atmosphere, with the net effect being a supergreenhouse effect of several hundred degrees centigrade.

If higher CO_2 levels were responsible for a warm early earth, it is necessary to devise explanations for maintaining higher atmospheric concentrations. Higher partial pressures of CO_2 may be attributed in part to increased early outgassing due to large geothermal energy sources (e.g., Henderson-Sellers and Cogley, 1982). It has long been suggested (Urey, 1952; Pollack

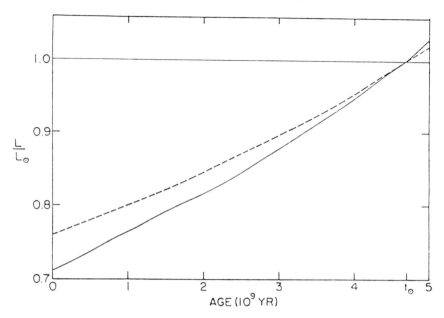

Fig. 12.1. Long-term evolution of the solar luminosity. Dashed curve shows evolution predicted by the idealized model of Endal (1981). The solid curve illustrates predictions from a detailed computer model (Endal and Sofia, 1981). [From Endal, 1981]

and Yung, 1980) that the CO_2 would be removed via the breakdown of silicate minerals and the precipitation of inorganic calcium carbonate. Atmospheric CO_2 levels would remain relatively high in the Archean because the area of the weathering "sink" (continents) was small and thus would retard removal (e.g., Kasting, 1989). Furthermore any drops in atmospheric temperatures should be accompanied by decreased weathering rates, which would decrease CO_2 removal rates and cause a negative feedback (Walker et al., 1981).

Eventually the CO_2 would be incorporated into the crust in the form of carbonate minerals. A negative feedback to CO_2 removal involves the observation that at present transfer of weathered carbonate to the crust mainly occurs with biological mediaries (e.g., Matthews, 1974). Precambrian carbonate formations may also have been controlled by organisms, for the carbonates commonly occur in the form of stromatolites. The earliest stromatolites are essentially coeval with these earliest carbonates (3.3–3.5 Ga; Walter et al., 1980; Byerly et al., 1986; Schopf and Parker, 1987). Carbon isotope studies show that organisms significantly influenced the carbon cycle as far back as the record exists

(Schidlowski, 1988). However, if the total biomass was small during the Archean, then the amount of CO_2 removal via the biological mediary may also have been small.

Early Precambrian CO_2 partial pressures estimated at 0.01–0.10 atm exceed present values by 30–300 times (Garrels and Perry, 1974; Pollack and Yung, 1980). The high Mg concentrations in Precambrian carbonates (e.g., Tucker, 1982) are consistent with such an increased partial pressure of CO_2. An enhanced CO_2–H_2O greenhouse effect could have produced a mean temperature of 57°C at 4.2 Ga (the present mean is 15°C) and could potentially have reached values of 85–110°C (Kasting and Ackerman, 1986). Although isotopic analyses of cherts indicate groundwater temperatures of 70°C as late as 2.8 Ga (Knauth and Epstein, 1976), these values may have been strongly affected by subsequent metamorphism. Occurrences of the evaporite mineral gypsum ($CaSO_4$), which is only stable at temperatures less than 58°C, provides an upper limit to some early Precambrian temperature estimates (Holland, 1978). The presence of life forms back almost to the beginning of the rock record also places an upper constraint on Archean temperatures, for it seems unlikely that life forms whose biochemistry is based on lipids and proteins could tolerate temperatures much in excess of 90°C before breakdown would occur (Kasting, 1989). It seems that most organisms die before this point, because procaryotes (organisms without a cell nucleus—the only kind present in the Archean) do not survive above about 70°C (Valentine, 1985). Combining the mineralogic and paleontologic results yield maximum estimated Archean temperatures <60°C. Although these numbers are useful constraints, it would be desirable to acquire other types of geochemical data to test more rigorously CO_2 models for the early earth.

Despite the focus on CO_2 as an Archean climate feedback, other climate mechanisms may also be important. A water-covered planet would also have a lower surface albedo, thereby increasing insolation receipt (Henderson-Sellers and Cogley, 1982). Lower initial surface temperatures would decrease cloudiness because this variable is dependent on the moisture content of the atmosphere, which in turn is strongly dependent on surface temperatures. Archean cloudiness could have decreased from its present level of about 50% to about 38%, resulting in a significant increase in the percentage of incoming energy absorbed. Calculated surface temperatures with an interactive-cloud climate model were 277°K, warm enough to support liquid water (Henderson-Sellers and Cogley, 1982).

Ocean circulation on a watery planet may also have helped maintain locally warm conditions. Calculations with an ocean general circulation model and decreased solar constant illustrate (Fig. 12.2) that reduced continental barriers diminish poleward ocean heat transport (Henderson-Sellers and Henderson-Sellers, 1988). High latitudes get significantly colder with ice to 58° latitude, but equatorial temperatures are still relatively warm (15°C).

12.2 Proterozoic Climates (2.5–0.57 Ga)

The late Precambrian (Proterozoic) contains evidence for two phases of continental glaciation, at 2.3 and 0.9–0.6 Ga (Frakes, 1979; cf. Fig. 13.1). The first evidence for glaciation occurs in 2.7- to 2.3-Ga-old rocks from North America, South Africa, and Australia (Frakes, 1979). The areal extent of the individual glaciers is not well

known. Their presence may mark a threshold temperature through which the atmosphere passed. The combination of low luminosity (Fig. 12.1) and a "thinned" greenhouse shield could explain the low temperatures. However, a puzzling dilemma arises when comparing the 2.3-Ga glacial event with the climate of the subsequent 1.4 Ga. There is little evidence for glaciation between 2.3 and 0.9 Ga (Frakes, 1979). As mentioned in the introduction, the gap may be due to erosion. Nevertheless, aluminous clay minerals (kaolinite) indicate intense chemical weathering, typical of tropical environments, in sediments overlying the Canadian glacial deposits (Young, 1973). Physical explanations for the conjectured moderate climates of the period 2.3–0.9 Ga have not been extensively explored.

One of the most unusual climatic events in earth history took place between about 0.9 and

Fig. 12.2. Archean total northward heat transport simulated by an ocean GCM with reduced land area (top) as compared with transport with present continental configuration (bottom). In the lower figure, solid line represents total transport and the dashed line diffusive transport (the comparison should be with the latter). [From Henderson-Sellers and Henderson-Sellers, 1988] *Reprinted by permission from Nature, vol. 336, pp. 117–118, copyright © 1988 Macmillan Magazines Ltd.*

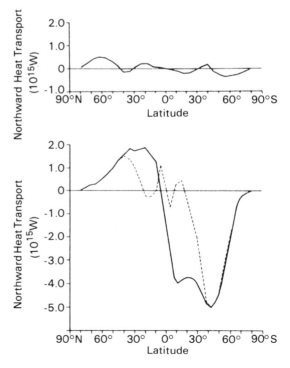

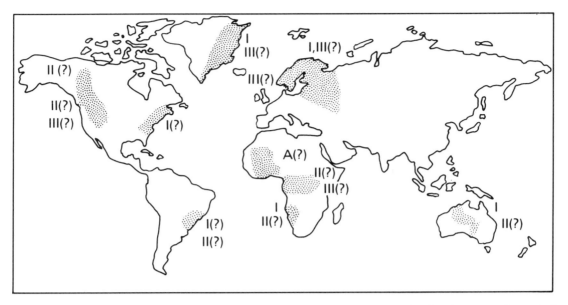

Fig. 12.3. Global distribution of major late Precambrian glacial centers on a map showing the present dispersal of continents. I, II, III refer to glaciations identified by Williams (1975) as centered on ~610 Ma, 750 Ma, and 950 Ma, respectively. A subsequent summary of late Precambrian glaciations (Hambrey and Harland, 1981a) suggests that these glaciations may not be as episodic as inferred by Williams. The letter A signifies that all three time intervals may be represented. [Modified from Frakes, 1979] *Reprinted by permission from L. Frakes, "Climates Throughout Geologic Time," copyright, 1979, Elsevier Scientific Publishers.*

0.6 Ga. At least three glaciations of continental scale occurred (Williams, 1975). Virtually all regions on the earth that contain Precambrian rocks show some evidence for glaciation during this time (Fig. 12.3). Furthermore, paleomagnetic evidence suggests that many of the glaciations may have occurred in low latitudes (McWilliams and McElhinny, 1980; Christie-Blick, 1982; Embleton and Williams, 1986). Land elevation may have been relatively low, for marine sediments are interlayered with some of the glacial deposits (Frakes, 1979). In fact, Frakes (1979, p. 88) stated that "if mixtites were not known from the late Precambrian, the proportions of shelf carbonate could be taken as evidence for widespread and continuously warm climates."

The causes of late Precambrian glaciations are poorly understood. That the glaciers penetrated into lower latitudes than their recent counterparts may not necessarily be very significant, for luminosity output was still 5–10% less than the present output (Fig. 12.1). Even a synchronous low-latitude glaciation does not imply an ice-covered earth. The planet was presumably still

covered by 70% water, and factors responsible for sea ice formation and preservation are quite complex. Nevertheless, the problem is intriguing and warrants some theoretical exploration.

In summary, the interval from 2.5 to 0.57 Ga records two main glacial events, one near the beginning and one near the end. Observations indicate two perplexing paleoclimatologic problems: (1) the occurrence of moderate climates during a period of presumably low luminosity and low greenhouse effect and (2) the possibility of extensive low-latitude glaciations. The onset of late Precambrian glaciations coincides with one of the major tectonic events in earth history (Goodwin, 1981). These tectonic events link the late Precambrian with the present tectonic regimes of seafloor spreading (Goodwin, 1981) and may reflect, in part, formation of a Precambrian supercontinent (Piper, 1976, 1983). The end of the major phase of late Precambrian glaciation is marked by the postulated breakup of the supercontinent, significant changes in the global carbon cycle (and CO_2?), and the first appearance of multicelled organisms (Cloud and Glaessner, 1982; Bond et al., 1984; Dyson,

1985; Knoll et al., 1986; Lambert et al., 1987). It would be intriguing to determine if any of these first-order changes are related to each other. Such changes provide us with a natural bridge leading to the Paleozoic, a topic covered separately in Chapter 11.

12.3 Summary

Despite the fact that the Precambrian covers a huge part of earth history, only a relatively small amount can be said about paleoclimates. The earliest glaciation occurred between ~2.7–2.3 Ga, and another period of possibly low-latitude glaciation occurred between 0.9–0.6 Ga. Most modeling studies have focused on the reason why the earth did not become ice-covered during the Archean as a result of lower solar luminosity. Explanations generally converge on CO_2 as the culprit, although as we have discussed elsewhere (Sections 8.2.5 and 10.2.3), geochemical and sedimentological evidence for past CO_2 variations are not as well constrained as we would like. Presence of liquid water, gypsum, and single-celled organisms back to almost the beginning of the sediment record (3.8 Ga) probably constrain mean earth temperatures to <58°C for the remainder of earth history. These values may be lower than can be obtained with some of the more extreme CO_2 scenarios.

The other major climate problem of the Precambrian involves the occurrence of apparently low-latitude glaciation in the late Precambrian (0.9–0.6 Ga). To our knowledge, this has not been explored at all theoretically. Perhaps one of the reasons for this is that evidence for low-latitude glaciation is rarely presented in a convincing manner easily grasped by climate modelers. Results are often shown as paleomagnetics plots. Interpretation of such plots do not tap one of the stronger points of a climatologist's training. It would be desirable to see some interaction between modelers and geologists on this interesting problem.

PART IV
SUMMARY AND SYNTHESIS

13. SUMMARY AND SYNTHESIS OF PAST CLIMATES

Now that we have taken the grand tour through the geologic record, it is instructive to step back and summarize some of the main trends in paleoclimatology. This chapter will address four main topics: (1) summary of the major findings for individual time periods; (2) synthesis across time scales; (3) comparisons of atmospheric and oceanic circulation features for warm and cold climates; and (4) unsolved problems. A separate aspect of synthesis (paleoclimate perspectives on a future greenhouse warming) will be examined in the next chapter.

13.1 Principal Findings

1. From 4.6 to $\sim$2.5 Ga the earth was apparently ice-free despite a substantially lower solar luminosity (Faint Sun Paradox). An enhanced atmospheric greenhouse effect may have compensated for the decreased insolation receipt.

2. At $\sim$2.5 Ga evidence for the first glaciation (Fig. 13.1) seems to make a threshold temperature below which the atmosphere dipped.

3. From $\sim$2.5 to 0.9 Ga the earth was apparently ice-free, despite low luminosity and a presumably depleted greenhouse effect. This phenomenon has not received much attention by modelers.

4. From 0.9 to 0.6 Ga at least three major phases of glaciations occurred, with paleomagnetic data suggesting ice in low latitudes. This phenomenon has also received little attention. It should be noted that because there is no evidence bearing on the presence of low-latitude sea ice, low-latitude continental glaciation should not be construed to imply an ice-covered earth.

5. From 600 to 100 Ma climates were thought to be generally mild but punctuated by two major phases of ice growth. However, there is some evidence that at least seasonal ice may have been much more extensive than previously thought. Seasonal cycle variations on the supercontinents of Gondwanaland and Pangaea were very large. Climate model simulations suggest that such conditions would be conducive to intense summer heating and dry continents, conditions in agreement with geologic data (extensive red beds and evaporites). Calculations also suggest that postulated coexistence of high CO_2 and extensive ice sheets in the late Ordovician may possibly be explained by a coastal location for the Gondwanan south pole.

6. From 100 to 50 Ma mild, generally nonglacial climate prevailed. The mid-Cretaceous (100 Ma) has served as a testing ground for atmospheric and oceanic circulation models of a presumed ice-free world. Changes in land–sea distribution cannot adequately account for inferred high polar temperatures at this time. Additional factors (CO_2) may be important for explaining the origin of nonglacial climates. There is a continued need to test the CO_2 model with new types of geochemical data.

7. From 50 to $\sim$3.0 Ma sequential cooling and drying of the globe occurred. The long-term evolution of climate appears to be driven by plate tectonic changes, CO_2, and ocean heat transport. Plate movements caused changes in seasonal and mean annual temperatures on land masses. Uplift of the Tibetan Plateau and western North America mountains also affected the planetary circulation in midlatitudes of the Northern Hemisphere. However, these changes do not seem to be large enough to account for the entire evolution of climate over the last 50 million years. Ocean heat transport may also have changed in the past, but initial calculations suggest that such changes may not ameliorate some

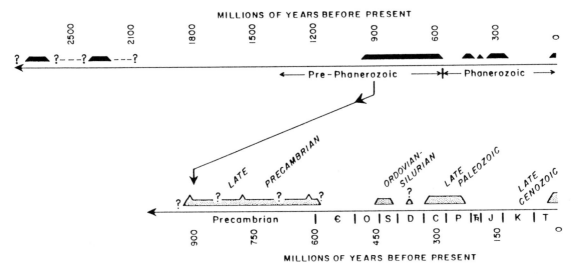

Fig. 13.1. Occurrence in time of ancient ice ages plotted linearly, to two scales (see Appendix A for time periods). [From Crowell, 1982] *Reprinted from "Climate in Earth History," with permission from the National Academy Press, Washington, DC.*

problems of warmth in high-latitude continental interiors. It is also very difficult to understand the origin of such postulated large oceanographic changes. There is some evidence that CO_2 levels may have been about twice as high between 50 Ma and 3 Ma. However, the CO_2 estimates require considerably more constraint from geochemical data before they can be considered robust. If CO_2 values were higher, then this effect, together with the plate-tectonic and ocean circulation changes, may be enough to account for climate change over this time interval. The relative importance of the different mechanisms has yet to be quantitatively evaluated.

8. There are numerous abrupt transitions in the geologic record. Such transitions suggest that the system has rapidly evolved at a number of different times to new equilibrium states, perhaps due to ice-albedo effects or thermohaline instabilities. Ice on Antarctica occurred at least as far back as 40 Ma, whereas the Greenland Ice Sheet apparently formed much later (3–4 Ma?) and midlatitude ice sheets later still (2.4–3.2 Ma).

9. From 3.0 to 0.0 Ma numerous oscillations of Northern Hemisphere ice sheets occurred, with intervals as warm as today occupying only 10% of the late Quaternary record. During the last glacial maximum (18,000 BP), sea ice and ice sheets migrated into midlatitudes. There were significant changes in the thermohaline circulation and atmospheric composition (CO_2, CH_4). SST changes were generally small in regions not directly affected by sea ice. Equatorward of the continental ice margins the surface of the planet dried out considerably. There were significant wet (pluvial) stages in the tropics that correlate with precession-induced summer warming, usually during the early stages of interglacials (e.g., 9000 BP).

10. Modeling studies indicate that we have a good first-order understanding of 18,000 BP changes in the northern high latitudes and monsoon increases at 9000 BP. There is also a convergence of explanations for Southern Hemisphere climate change, although some significant uncertainties still exist. However, there are large model–data discrepancies in the tropics at 18,000 BP (temperature, precipitation, trade wind strength) that are not understood.

11. Late Quaternary time series indicate a strong correlation between ice volume fluctuations and orbital perturbations; but the nature of the interaction is unclear.

Feedback processes in the land–sea–air–ice system appear to be required in order to explain phenomena such as rapid deglaciation and the magnitude of the 100,000-year cycle. The precise pathways of the interaction are a subject of debate, but there is growing evidence for the importance of climate instabilities in effecting climate transitions.

12. Significant climate fluctuations have occurred on the decadal to millennial time scale. Analysis of records over the last 1000 years indicates that a cool period (Little Ice Age) was apparently global in scope, but that there was a significant regional overprint on many of the climate signals. Various mechanisms have been proposed to explain such changes—solar variability, volcanism, or internal interactions in the ocean–atmosphere system. Although there is some evidence to support the first two processes, the relative importance of each of the mechanisms has still not been satisfactorily quantified.

13.2 Synthesis Across Time Scales

The perspective afforded by an overview of the entire geologic record allows for some additional syntheses. The inferred planetary temperature history during the last 4.5 billion years apparently reflects changes of both a secular and a fluctuating nature in the components of the climate system. The secular variation in solar luminosity has caused changes in global insolation receipt on a time scale of 10^9 years. An early greenhouse effect may have offset the lower luminosity. Fluctuations of temperature involve mechanisms with characteristic time scales of 10^0–10^8 years. On time scales of 10^7–10^8 years, paleogeographic factors (e.g., continental drift, ocean circulation changes) and atmospheric CO_2 changes appear to have played an important role in controlling fluctuations of global temperature.

There is growing evidence for significant variance on 10^6-year time scales, but the origin of such variance is not well understood; it could represent "high-frequency" tectonic fluctuations or perhaps very long period orbital insolation forcing. On a time scale of 10^3–10^5 years

the earth's climate appears to be sensitive to both external forcing by orbital perturbations and internal feedback interactions (including CO_2) within the land–sea–air–ice system. Finally, on a time scale of 10^0–10^2 years the earth's climate appears to be sensitive to solar variability, volcanism, and (possibly) internal feedback interactions within the ocean–atmosphere system.

The above results can be summarized with a plot of climatic fluctuations across a broad range of frequencies (Fig. 13.2). Kutzbach (1976) showed that a log-log plot of the variance spectrum in the North Atlantic sector indicates three main features: a generally white noise pattern at high frequencies that is characterized by relatively high variance, a midfrequency record of relatively low variance, and a low frequency record of increasing variance. The latter "red noise" shape (see Section 7.3.2) resembles that of a sluggish system being forced by white noise (cf. Appendix C).

13.3 Comparison of Circulation Statistics for Warm and Cold Climates

For many different warm and cold climate scenarios, some important zonally averaged circulation statistics do not change as much as might be expected. For example, NASA/GISS model results (Rind, 1986) show no significant difference in jet stream positions for different climates (Fig. 13.3a). The subtropical jet is located at 31°N in all cases. Even the high-resolution forecast model of the European Centre for Medium Range Weather Forecasts locates the subtropical jet at this latitude (L. Bengtsson, pers. commun., 1986). Similar indications of zonal stability have been found in NCAR GCM runs for the Cretaceous (Barron and Washington, 1982a) and GFDL runs for CO_2 doubling (Manabe and Wetherald, 1980). Latitudinal profiles of combined ocean/atmosphere poleward heat transport (Fig. 13.4) also have similar patterns for different levels of CO_2 forcing (Manabe and Bryan, 1985). The subtropical belt of maximum moisture depletion is located at about the same latitude for all of these CO_2 experiments. These results are consistent with geologic data indicating that the latitudes of the subtropical high-pressure systems (and associated evaporites) have

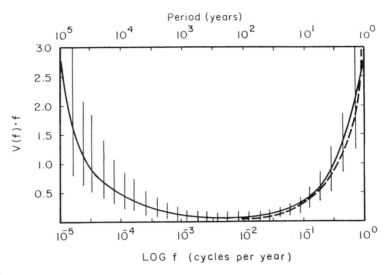

Fig. 13.2. Schematic version of the variance spectrum of temperature fluctuations in the North Atlantic sector on time scales of 10^0—10^5 years [after data from Kutzbach and Bryson (1974) and Hays et al. (1976b)]. The ordinate is $V(f)$, variance spectral density, times f, frequency, with units of $(°K)^2$; the abscissa is a logarithmetic frequency scale. These coordinates are chosen such that equal area under the curve represents equal variance. Vertical lines are used to indicate the relative degree of uncertainty in the shape of the curve. The dashed curve indicates a white noise spectrum fitted approximately to the high-frequency portion of the spectrum. [Modified from Kutzbach, 1976 by Crowley, 1983b] *Reprinted with permission from Quaternary Research.*

not varied much over the last 500 million years (Fig. 13.5; Gordon, 1975).

The above results imply that there are important compensating mechanisms in the climate system which maintain overall equator-to-pole gradients, even if some of the individual components of the climate system may change (cf. Stone, 1978; Covey and Barron, 1988). For an example from the present climate, Trenberth (1979) postulated that poleward ocean heat transport is greater in the Southern than the Northern Hemisphere in order to compensate for the low values of stationary eddy activity in the Southern Hemisphere. Similar adjustments can be found for some paleoclimatic simulations. For example, although eddy-generated sensible heat transport increases in cold climates, latent heat transport tends to decrease (Rind, 1986).

13.4 Some Significant Problems in Paleoclimatology

We conclude the synthesis section by listing what we consider to be some of the most im-

portant unsolved problems in paleoclimatology. The list below is not meant to be exhaustive but rather to highlight the state of the field, and the sequence is not meant to imply relative importance of the problem and more or less progresses from lower frequency to higher frequency climate oscillations. We refer to the relevant chapters that discuss this problem further.

1. How extensive were ice-free states in the Phanerozoic? Although tradition describes most of this interval as warm, some studies suggest more extensive ice conditions than usually assumed (Sections 8.1.3, 10.1.2, 11.1.1).

2. Were warmer periods during the last 100 million years indicative of seasonal or year-round warmth? Although the traditional view is year-round, and although this explanation may fit some coastal regions, existence of at least seasonal ice suggests seasonally cold temperatures for some time intervals. Are ecological interpretations of biota capable of revision? For example, dinosaurs, long assumed to be cold-blooded, have been postulated to be warm-blooded.

Likewise, can seasonal migrations (or plant dormancy) explain presence of some high-latitude assemblages (Sections 8.1.3 and 10.1.2)?

3. Can geochemical models for long-term CO_2 fluctuations be better validated? Although calculations suggest that atmospheric CO_2 levels may have fluctuated sig-

nificantly, there is a need for more testing of the geochemical models with other geochemical data. Also, climate models predict that SSTs should have been 4–5°C higher in the tropics with higher CO_2 levels. Why are these changes not detected in $\delta^{18}O$ records (Sections 8.2.4, 8.2.5, 10.2.3)?

4. How have changes in solid-earth

Fig. 13.3. Comparison of climate statistics for five different climate simulations with the NASA GISS GCM, indicating the relative stability of circulation statistics for different climates. Zonally averaged January values of (a) zonal wind at 200 mb and (b) 200 mb northward transport of angular momentum by eddies. Ice Age I (short dashes); Ice Age II (dotted line); current climate (long dashes); double CO_2 (thin solid line); 65 Ma (thick solid line). Ice Age II values based on GISS calculation with CLIMAP 18,000 BP SST values everywhere decreased by 2°C to conform better to estimates of tropical aridity and temperature decreases (see Sections 4.1.2 & 4.1.4). [After Rind, 1986] *Reprinted from Journal of the Atmospheric Sciences with permission of the American Meteorological Society.*

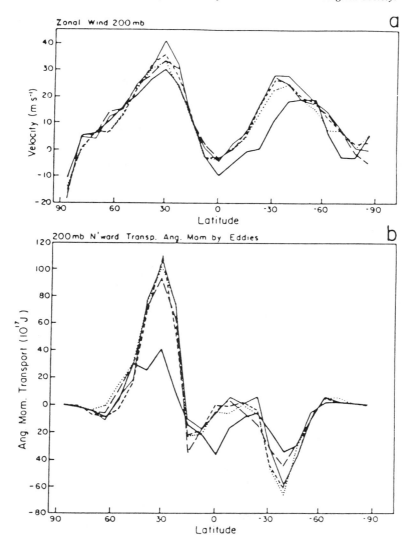

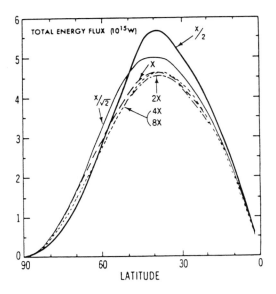

the model–data disagreements for the un-coupled mode.

6. What is the origin of the ice-age CO_2 fluctuations? At present we have only pieces of the puzzle (Section 6.4.5). Models have to be validated not only against the last glacial maximum, but against the record of CO_2 fluctuations for the last 150,000 years.

7. What is the origin of the abrupt warming at $\sim$13,000 BP (last deglaciation)? At present we do not have a good explanation for the near-global warmth (Sections 3.2.2 and 4.4.1).

8. What factors are responsible for ice sheet inception during the early part of a glacial cycle? Recent GCM experiments

Fig. 13.4. Poleward transport of total energy by the joint ocean–atmosphere system in a GFDL sector version of a coupled ocean–atmosphere model. Units are in 10^{15} W. Three sector oceans and three sector atmospheres, which cover an entire hemisphere, are involved in this transport. Different values of X refer to different atmospheric CO_2 levels ($X = 300$ ppm). [From Manabe and Bryan, 1985] This figure illustrates relative stability of zonal profiles of transport to different levels of total transport. *Courtesy of American Geophysical Union.*

Fig. 13.5. Distribution of ancient evaporites by latitude. The evaporites are plotted by era grouping with reference to paleolatitude. Note the tendency for evaporites to form in the same latitudes at different time periods, a result consistent with calculations illustrated in Figs. 13.3 and 13.4. [From Gordon, 1975] *Reproduced by permission from "Distribution by latitude of Phanerozoic evaporite deposits," W. A. Gordon, Jour. Geol. 83:671–684, copyright 1975, University of Chicago Press.*

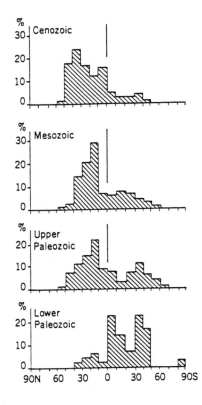

boundary conditions affected ocean heat transport in the past? Some data suggest that significant changes may have occurred, but preliminary results indicate it is difficult to understand the postulated magnitude of such changes or whether the changes alleviate the problem of high-latitude warmth in continental interiors (Sections 8.2.3 and 10.2.2).

5. What is the origin of the model–data discrepancies in the tropics at the glacial maximum (temperature, precipitation, trade wind velocities)? Each side of the argument (Sections 4.1.2 and 4.1.4) has good points, but something has to give. If tropical SSTs were warm, why were they warm? Are ocean circulation feedbacks involved in maintaining stable tropical SSTs (cf. Section 4.3.2)? Will decreases in ocean heat transport result in increased atmospheric transport when the two systems are coupled together? If so, then coupled ocean–atmosphere experiments may reconcile some of

suggest that growth under any condition is very difficult to explain. Is the system more sensitive than we have allowed for (Section 6.4.4)?

9. What is the origin of the ice-age 100,000-year cycles? Although many studies have focused on ice sheet/bedrock interactions on the explanation, there is some evidence for 100,000-year fluctuations in the pre-Pleistocene (i.e., before presumed significant ice volume fluctuations). Additionally, very rapid warming at the beginning of the last deglaciation is difficult to understand in terms of ice sheet/bedrock interactions. Instabilities in the ocean–atmosphere system represent one possible alternate explanation. Finally, there are many different types of models that produce quasi-100-KY oscillations. The models cannot all be right. We need more objective criteria to distinguish among them. Modelers have to be more explicit in defining their model differences, in defending

their model vis-à-vis other models, and in making predictions that geologists can test (Sections 7.3.1 and 7.3.2).

10. What is the explanation for 100,000-year fluctuations in the pre-Pleistocene (Sections 7.2.8 and 10.1.6)? Most models for late Pleistocene changes involve ice volume variations (Problem 9). But pre-Pleistocene changes occur during times of lower ice volume. Explanations for this response have been almost completely lacking.

11. What is the origin of decadal to millennial scale climate fluctuations (Section 5.2)? Three candidates have been proposed: solar variability, volcanism, and internal changes in the ocean circulation or heat storage. There is a need to quantify the relative importance of each of these mechanisms. We also need a global scale data base in order to extract a global signal from moderately noisy records. In addition, observations need to be tested against models for the different mechanisms.

14. PALEOCLIMATE PERSPECTIVES ON A GREENHOUSE WARMING

Some time in the next century the earth's climate is expected to undergo a major warming as a result of increasing concentrations of carbon dioxide and other trace gases. A doubling of CO_2 alone is estimated to cause an increase in global average temperatures of 1.9–5.2°C (e.g., Mitchell et al., 1989; Schlesinger, 1989), and a comparable increase may come from the contribution of other radiatively important trace gases (Dickinson and Cicerone, 1986). These estimates represent equilibrium values for a mixed layer ocean with no circulation (see Section 2.3.2.3), which may not apply in detail if there are significant response lags due to heat exchange with the deep ocean (e.g., Stouffer et al., 1989). Although these calculations have come under frequent criticism, we personally feel that none of the counterarguments have enough merit at this stage to seriously jeopardize the principal conclusion about a probable significant greenhouse warming. Indeed, the intrepid reader of our book will by now have realized that we cannot explain much of the climate variance over earth history unless the greenhouse effect is about as large as the models project.

Since warming of several degrees exceeds the ±0.4°C range (Fig. 14.1) of the past 100 years (e.g., Jones et al., 1986), we do not have any direct experience with a climate change of that magnitude. Furthermore, climate model simulations of the regional responses to a greenhouse warming, although instructive, are not so reliable that we can place great credence in their results. It is therefore appropriate to inquire how the projected large greenhouse warming compares with major changes of climate in the geologic record, and whether validation of climate models against past boundary conditions has helped us gain some level of confidence in climate models. The purpose of this chapter is to synthesize results from paleoclimate studies in order to establish a frame of reference for interpreting projections about the consequences of a future greenhouse warming (see Crowley, 1989).

14.1 Comparisons with Past Temperature Variations

14.1.1 Magnitude of Greenhouse Climate Perturbation

One of the first questions to ask about a greenhouse warming of several degrees is: "How significant is the projected warming?" The answer is "quite significant" (Fig. 14.2). A CO_2 doubling of 1.9–5.2°C is comparable in magnitude to the global average temperature decrease of about 4°C that occurred during the last glacial maximum (Hansen et al., 1984; Kutzbach and Guetter, 1986; Lautenschlager and Herterich, 1990a; cf. Section 4.2.1). An average increase of ~4°C represents almost one-half the estimated global temperature increase during the maximum warmth of the mid-Cretaceous (Thompson and Barron, 1981; cf. Chap. 8). CO_2 levels and other trace gas increases could exceed a factor of two by a significant amount. For example, utilizing all available fossil fuels could result in maximum atmospheric values perhaps eight times the present (see Keeling and Bacastow, 1977). A future greenhouse warming therefore represents a climate change of very significant levels. Unless there is a major shift in energy sources, atmospheric temperatures could approach levels assumed for an ice-free earth (global temperatures 6–10°C warmer than present). Furthermore, the rate of change of future global temperatures is comparable to or exceeds the greatest rates that occurred during the catastrophic deglaciations of the Pleistocene—the most extreme and abrupt well-documented climate change recorded in the geologic record (cf. Jones et al., 1987).

Even though future temperatures may in-

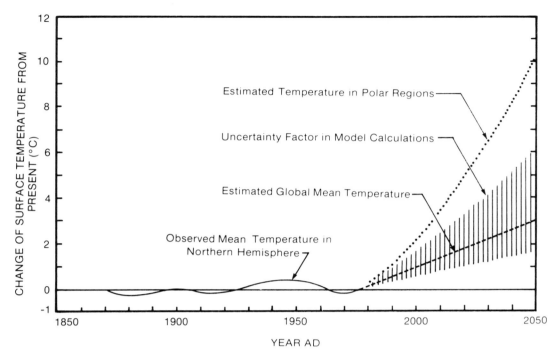

Fig. 14.1. Mean surface temperature for the Northern Hemisphere since 1860. Dashed line shows future global mean surface temperature change. Cross-hatching shows uncertainties in model calculations. [From Stuiver et al., 1981] *Reproduced with permission from M. Stuiver, G. H. Denton, T. J. Hughes, and J. L. Fastook, in "The Last Great Ice Sheets," G. H. Denton and T. J. Hughes (Eds.). Copyright 1981, John Wiley and Sons.*

crease greatly, perhaps to levels normally found for ice-free climates, it is uncertain whether an extreme greenhouse warming would completely destroy land-based ice sheets. The atmospheric CO₂ perturbation would eventually be neutralized by the deep-ocean carbonate reservoir after a few thousand years (Sundquist, 1985). About 6,000 years were required to melt the great Pleis-

Fig. 14.2. Schematic comparison of future greenhouse warming with past changes in temperature. Note that pre-Pleistocene changes are not well fixed in absolute magnitude, but the relative warmth of the intervals is approximately correct. [From Crowley, 1989] *Reproduced with permission of Kluwer Academic Publishers.*

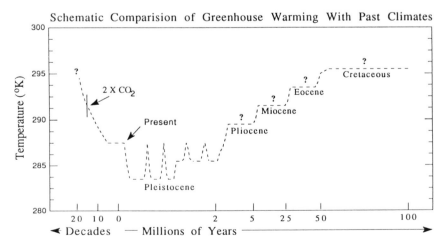

tocene ice sheets. A greenhouse warming could have repercussions even after this time. The time constants of the East Antarctic Ice Sheet are so long, perhaps tens of thousands of years (Whillans, 1981), that some climate fluctuations on Antarctica and elsewhere more than 10,000 years in the future may still bear the imprint of the greenhouse perturbation.

If CO_2 values reach several times the present level, the earth may therefore experience non-glacial level air temperatures and yet retain cold poles due to the inertial presence of the ice sheets. There is only one moderately well-documented example of this conjunction in the geologic record—the late Ordovician glaciation around 450 Ma (Section 11.1.3). Since this glaciation occurred after the atmosphere warmed, the sequence of events is opposite of that which will occur in the future. One may have to look to the late Precambrian (about 650 Ma) to find an example of a possible CO_2 increase following a major period of ice growth (e.g., Crowley and North, 1988). The conjectured CO_2 increase, driven by changes in seafloor spreading, would occur at a rate orders of magnitude slower than the anthropogenic CO_2 pulse.

The CO_2 perturbation therefore represents a major climate change, which will occur at a rate comparable to or exceeding any known to have occurred in earth history. The magnitude and duration of the event is significant even on geo-logic scales. Since there are no completely satisfactory geologic analogs for the warm atmosphere–cold pole combination, the future climate may represent a unique climate realization in earth history (Crowley, 1990).

14.1.2   Implications of Historical Climate Fluctuations for Detection and Modulation of a Greenhouse Warming

The end of the Little Ice Age (about 1880–1890) coincides approximately with expansion of the industrial revolution and the development of a relatively widespread instrumental network for monitoring climate. Studies of atmospheric CO_2 concentrations demonstrate that the "prean-thropogenic" CO_2 level was about 280 ppm (Fig. 14.3) during the Little Ice Age (Neftel et al., 1985). Atmospheric methane levels were about one-half of present values (Craig and Chou, 1982; Stauffer et al., 1985b). Changes in the latter value probably reflect increased numbers of herbivores and rice production in the last 200 years.

The increase in atmospheric CO_2 over the last 200 years has stimulated a great deal of interest in projecting the climate effects of higher CO_2 levels. This interest has been heightened by the observation that global average temperatures have been increasing over the past century (Fig.

Fig. 14.3. CO_2 concentration in air extracted from ice cores from Siple station (open squares) and from the South Pole station (black square), illustrating the rise in CO_2 since the seventeenth century. Also shown (crosses) are results from direct atmospheric samples at Mauna Loa. [From Friedli et al., 1986] *Reprinted by permission from Nature 324, 237–241 copyright (©) 1986 Macmillan Magazines Ltd.*

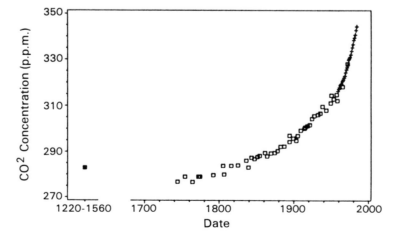

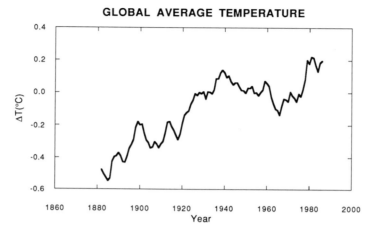

GLOBAL AVERAGE TEMPERATURE

Figure 14.4. Global average temperature record (5-yr. smoothing) from 1880–1988. Note the maximum values in the 1980's. The 1950–1970 cooling has provoked a number of discussions on the significance of the greenhouse effect and heightened the need to understand the nature of other sources of variability in the climate system. Based on data in Hansen and Lebedeff (1987, 1988).

14.4; note that much heat has been generated over the statistical significance of this rise). Are the two related? One manner in which our ideas can be tested involves comparisons of the instrumental record of the present century with longer time series, i.e., with historical climate fluctuations (cf. Chap. 5). For example, tree rings from the high latitudes of North America indicate that warming of the past century has exceeded the recent (post-1700) level of natural variability (Jacoby and D'Arrigo, 1989). A $\delta^{18}O$ record from the Tibetan Plateau (Fig. 14.5) suggests that the last fifty years may have been the warmest periods in the last 10,000 years (Thompson et al., 1989).

Despite the geologic evidence, some CO_2 doubling studies suggest that the atmosphere should already have warmed to levels greater than those of the present (e.g., Hansen et al., 1984; Schlesinger, 1986). Are the models wrong, or are other processes operating that are obscuring the trend? One possibility involves sequestering of excess heat in the intermediate and deep layers of the ocean (Hansen et al., 1984; Schlesinger, 1986), a possibility for which there is some empirical evidence, as intermediate waters in the North Atlantic appear to have warmed significantly over the last 30 years (Roemmich and Wunsch, 1984).

Another explanation for the CO_2 "delay" involves modulation of atmospheric signals by

volcanism and solar variability, i.e., by "natural" climate fluctuations. There have been significant decadal- and centennial-scale climate fluctuations over the last 1,000 years (e.g., Fig.

Fig. 14.5. $\delta^{18}O$ record for the last 35,000 years from the Qinghai-Tibetan Plateau. Note that $\delta^{18}O$ values for the twentieth century are heavier (more positive) than any values in the Holocene. [From Thompson et al., 1989] *Reproduced by permission from Science, 246, 474–477, 1989. Copyright 1989 by the AAAS.*

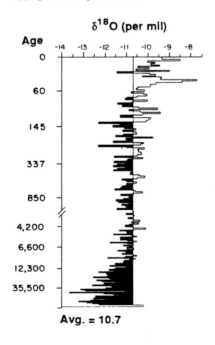

5.6). The characteristic temperature range of these fluctuations is 1.0–1.5°C, i.e., enough to modulate any uniform CO_2 warming trend by the amount observed in the instrumental record (Fig. 14.4).

Given the level of natural variability in the climate record and the possibility of different mechanisms contributing to climate change, how can we use the geologic record to sort out the relative importance of different mechanisms for interpreting the observed temperature trends of the 20th century? For example, calculations with a simple climate model suggest that the long-term CO_2 trend can be modified by an appropriate choice of parameters for volcano and solar forcing (Hansen et al., 1981; cf. Gilliland, 1982). However, these authors acknowledge that the choice of parameters, although not entirely arbitrary, is open to debate. This calculation suggests, but does not prove, that solar variability and volcanos may indeed be modifying the CO_2 signal but that more work is required to sort out the effects.

Internal interactions in a nonlinear coupled system may also generate low-frequency climate variance (Mitchell, 1966; Hasselmann, 1976;

Gaffin et al., 1986; cf. Sections 5.2.3 and 7.3.2). For example, a 100-year GCM run with a mixed-layer ocean (Fig. 14.6) generated internal climate variability with a range of about ±0.2°C (Hansen et al., 1988), i.e., about one-half of the fluctuations of the last 100 years.

By combining the results of the empirical and theoretical studies, it is possible to make an initial assessment of the relative importance of the different mechanisms proposed to account for climate change over the last few hundred years. Note that the following numbers reflect initial estimates that could be significantly modified by future work. Volcanism may account for at least 25% of the variance in the temperature record (Hammer et al., 1980; see caveats in Section 5.2.1) and solar variability perhaps a maximum of 10% of the variance (Cook and Jacoby, 1979; Mitchell et al., 1979; Hansen et al., 1981; Gilliland, 1982). A more significant solar–terrestrial connection may exist at periods of 140 years (Stuiver and Braziunas, 1989). The remaining unexplained variance reflects either a CO_2-induced warming or internal variations in the land–sea–air–ice system. The latter could be as much as 40% of the signal (cf. Fig. 14.6).

Fig. 14.6. Global mean annual surface air temperature in a 100-year GCM control run with a mixed-layer ocean. This figure suggests that a substantial amount of the observed variability of the last 100 years (Fig. 14.4) may be due to ocean–atmosphere interactions. [From Hansen et al., 1988]

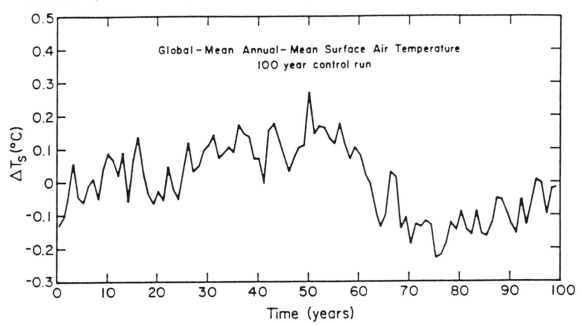

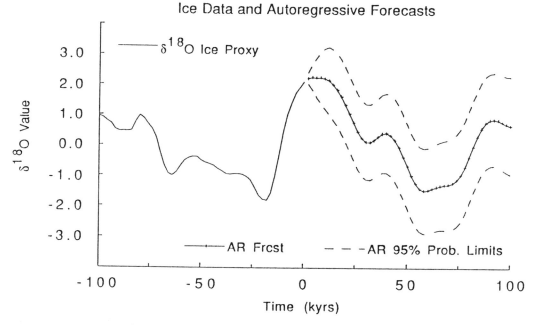

Fig. 14.7. Forecasts and 95% probability limits of future ice volume fluctuations using a linear autoregressive statistical model. Solid curve on left is observed proxy $\delta^{18}O$ record (from Imbrie et al., 1984). Middle curve on right represents ice volume forecast. Upper and lower curves represent 95% probability envelope. Positive values on vertical axis represent times of low global ice volume. [From Newton et al., 1989]

If the above estimates are at all correct, then a substantial fraction of atmospheric changes over the last century could be due to natural fluctuations. The amount of potential variance due to these natural fluctuations appears to be enough to modulate the course of any unidirectional warming due to CO_2 forcing (at least in the early stages of a warming). This conclusion is obviously open to debate. We include it here not because it may be right but because it may help focus the discussion as to the relative importance of volcanos, the sun, CO_2, and stochastic processes in determining the evolution of climate on decadal to centennial time scales.

14.1.3 *Potential for Orbital Insolation-Induced Ice Growth During the Next 10,000 Years*

Time series analyses of Pleistocene records indicate that Pleistocene ice volume changes were strongly influenced by orbital insolation changes—the Milankovitch effect (Hays et al., 1976b). As the characteristic time between ice

growth events is about 20,000 years and the last glacial maximum was 18,000 years ago, could the later stages of a future greenhouse warming be interrupted by a return to glacial growth? Several factors mitigate against such a possibility. Insolation changes over the next 10,000 years will not be nearly as large as they were for some postinterglacial intervals of the past because eccentricity levels, which modulate the 20,000-year precession signal, are very low [however, a small minimum in summer forcing is expected at 4000 AP (after present)].

There have been several attempts to forecast future ice volume fluctuations based on the pattern of past fluctuations in the Pleistocene (cf. Chap. 7). For example, a linear statistical model projects relatively small ice volume increases in the next 10,000 years (Fig. 14.7). These results are similar to those from a simple climate model that has been tuned to past orbital/ice volume relationships (Imbrie and Imbrie, 1980). These projections represent baseline forecasts for future climate change; however, a greenhouse warming should significantly alter the predictions of the Pleistocene model [the model assumed stationarity (see Appendix C), which will almost certainly not hold].

14.2 Response Scenarios

14.2.1 Regional Responses to a Greenhouse Warming

Paleoclimate studies also provide some information concerning possible regional consequences of a greenhouse warming. For example, a greenhouse warming might cause a considerable increase in summer aridity in central North America and Eurasia (e.g., Manabe and Wetherald, 1986; cf. Washington and Meehl, 1984). Similar drying occurred during the early Holocene warm period (6000–9000 BP), but since the forcing in the latter case was primarily due to seasonal changes in orbital insolation forcing (Kutzbach and Guetter, 1986), it is probably not wise to press the analogy too closely.

Another greenhouse effect may involve an increase of about 5 m in world sea level due to postulated melting of the marine-based West Antarctic Ice Sheet (Mercer, 1978; Stuiver et al., 1981). Bentley (1984) estimated that a minimum of 500 years would be required for this melting to occur. Since a similarly high sea level stand occurred during the last interglacial at 125,000 BP (Mesolella et al., 1969; Dodge et al., 1983), the sea level rise might be attainable with only a modest change in forcing. However, Robin (1985) suggested that the 5-m sea level increase may have been due to a thinner East Antarctic Ice Sheet at 120,000 BP. Alternatively, the West Antarctic Ice Sheet could have been "unhinged" by the very rapid sea level rise at the end of the penultimate glacial. The ~20 mm/yr. rise is an order of magnitude greater than the present 2.4 mm/yr. observed sea level rise (Peltier and Tushingham, 1989) and almost two orders of magnitude greater than the most recent projected sea level rise (0.6 mm/yr) for the next century (see Meier, 1990). The issue is even further complicated by the conjecture that melting of the Greenland Ice Sheet was actually responsible for the sea level rise during the last interglacial (Koerner, 1989).

Climate models also predict changes in high-latitude sea ice cover (Parkinson and Kellogg, 1979; Manabe and Stouffer, 1980; Parkinson

and Bindschadler, 1984; Washington and Meehl, 1984; Semtner, 1987). Geologic data appear to set an upper limit on the magnitude of this response during the early stage of greenhouse warming. Although Northern Hemisphere midlatitude glaciation was not initiated until about 2.4–3.2 Ma, the Arctic Ocean has apparently had some seasonal pack ice for at least the last 5 Ma. (Clark, 1982). Since the history of Arctic Ocean ice cover requires better documentation, this conclusion is subject to revision.

To date there has been relatively little attention given to the possible effects of a greenhouse warming on the East Antarctic Ice Sheet. Some geologic studies suggest that the East Antarctic Ice Sheet may have been much smaller at 4–5 Ma (Denton, 1985). The mechanisms responsible for such a climate change are not well understood. The mass balance of the East Antarctic Ice Sheet may actually be increasing at present (Bentley, 1984), perhaps due to higher temperatures resulting in greater moisture supply to the ice sheet (cf. Oerlemans, 1982b and Zwally et al., 1989).

Although the East Antarctic Ice Sheet could grow during the initial stages of a greenhouse warming, it is possible that melting could occur if CO_2 values reached very high levels. Since CO_2 doubling studies indicate winter warming around coastal Antarctica of 8–14°C (cf. Fig. 2.12), much higher CO_2 levels could tilt the mass balance of the ice sheet from accumulation to ablation. For example, such large forcing could affect portions of the East Antarctic Ice Sheet grounded below sea level, possibly resulting in glacial surges (cf. Wilson, 1969; Budd and McInnes, 1978; Hollin, 1980; Thomas, 1984). Hughes (1987a) suggested that ice streams that drain the large ice sheet may be unstable to sea level fluctuations and trigger a collapse of ice presently draining sub-sea level basins in East Antarctica. A reduction in East Antarctic ice to the levels of 4–5 Ma (Fig. 10.19b) could result, raising sea level by 25–30 m. A similar mechanism was proposed as an important component of ice sheet collapse at the last deglaciation (e.g., Hughes, et al., 1977; Hughes, 1987b). If part of the East Antarctic Ice Sheet were to collapse, the process could take thousands of years to transpire.

14.2.2 Possibility of Abrupt Transitions

Interest in the stability of climate states and the potential for abrupt climate transitions has grown steadily over the last few years (e.g., North et al., 1981; Broecker et al., 1985; Berger and Labeyrie, 1987). There are numerous examples of abrupt climate change in the geologic record (e.g., Crowley and North, 1988), with the long-term trend in climate during the last 100 million years involving the evolution from an ice-free state through several stages to one with polar ice caps. Transitions usually involved abrupt steps, with results suggesting that the Northern Hemisphere has most recently been glaciated. For example, areas in northern Greenland preserve fossil plants and animals indicative of significantly warmer climates as recently as 3–4 Ma (Funder et al., 1985).

If similar abrupt steps were to occur in the future, physical considerations and experience with the record suggest that the three most likely candidates for such a transition are the Arctic ice cap, the West Antarctic Ice Sheet, and the thermohaline circulation in the subpolar North Atlantic. The first two would change because of increased temperatures, the third because of possible lower surface salinities due to increased precipitation in the subpolar North Atlantic and increased melting of sea ice and Greenland. Lower salinities might significantly affect poleward heat transport in the North Atlantic (cf. Section 4.4.2).

14.2.3 Climate–Biosphere Interactions

A comparison of past abrupt transitions with the evolutionary record indicates that many transitions occur near intervals of extinction, with the largest biotic response occurring during the early stages of a climate change (Crowley and North, 1988). We therefore have the potential for a climate-induced extinction event in the next 200 years. At this stage, it is not possible to estimate the magnitude of such an event. In all likelihood it would be difficult to separate the climate-induced extinctions from those due to habitat destruction by man-induced changes that will likely have a much more devastating effect on the earth's ecosystem.

There may be a positive feedback between extinctions and climate. Changes in total biomass can result in a mass transfer of carbon from the biosphere to the atmosphere (e.g., Broecker, 1982a,b). Such a change may have occurred at the end of the Cretaceous and resulted in an estimated increase of atmospheric CO_2 levels by a factor of 2–3 (Hsü and McKenzie, 1985). Further feedbacks may involve ocean plankton. Partial collapse of the ocean's thermohaline circulation (due to warming of high-latitude surface waters) would affect the global carbon cycle because it would decrease the rate of nutrient cycling from deep waters to the surface (Bryan and Spelman, 1985). Carbon transfer from the ocean biosphere to the atmosphere could produce a climate feedback as strong as that caused by an enhanced uptake of heat from the atmosphere (Bryan and Spelman, 1985). This scenario has some geologic support. During the Cretaceous warm period, the thermohaline circulation was less intense, and there is fairly good geologic evidence indicating lower levels of ocean productivity at that time (Bralower and Thierstein, 1984).

14.3 Validation of Climate Models with Paleoclimate Data

A large uncertainty with respect to greenhouse predictions involves the reliability of model predictions. Although all models predict warming, there are significant differences in the regional response to warming. Paleoclimate data can be used as a check on climate models. Although we extensively discuss model-data comparisons in Chapter 4, we briefly review some of the major conclusions as they pertain to gauging the level of believability of GCM simulations for higher CO_2 levels.

14.3.1 High Latitudes

A summary of results from climate model experiments (Fig. 4.18) covering the last 20,000 years (Chap. 4) indicates that in northern mid-latitudes there is quite a good agreement between models and data. The large North American ice sheet had a significant effect on the entire circulation in the circum-North Atlantic (Fig. 4.1) and apparently caused the great extension of sea ice in this region, which in turn fed back on European temperatures and (possibly) NADW production. The perturbed flow pattern also seems to explain high lake levels in the southwestern United States at a time when the planet was generally dry.

Progress is being made in understanding the origin of high-latitude Southern Hemisphere climates. Changes in sea level due to formation of Northern Hemisphere ice sheets appears to play a key role in expansion of the West and East Antarctic ice sheets in the Pleistocene (cf. Denton et al., 1986). Ice-age changes in CO_2 seem most important for explaining Antarctic sea ice variations (Manabe and Broccoli, 1985b), with changes in interhemispheric heat transport (due to changes in NADW production) perhaps accounting for 20–30% of the variance (Crowley and Parkinson, 1988b). However, there is still some disagreement between models and data— models produce only about one third the measured temperature range on Antarctica (cf. Section 4.1.3). Overall, there is a fair understanding of processes occurring in the high latitudes of the Southern Hemisphere, but there is room for improvement.

14.3.2 Low Latitudes

In the tropics, there are some significant disagreements between models and data (Section 4.1.4). Models are too wet in tropical lowlands and too warm at upper levels—both factors attributable to warm glacial SSTs in the tropics. Furthermore, GCM tropical winds are generally too weak in ice-age simulations. The origin of these significant discrepancies are not well understood. They could indicate that CLIMAP SSTs are too warm (Webster and Streeten, 1978; Rind and Peteet, 1985), an alternative not in agreement with other geologic observations (Prell, 1985; Broecker, 1986; Brassell et al., 1986) or with calculations suggesting that CLIMAP SSTs are in energy balance with the earth–atmosphere system (Hyde et al., 1989). Perhaps some critical processes are not being included in climate models, e.g., increased dust content of the glacial atmosphere, lower CO_2 (which may also affect the stomatal response in plants; Wigley and Jones, 1985), or vegetation feedbacks (e.g., Sellers et al., 1986). As an example of the magnitude of these feedbacks, Coakley and Cess

(1985) calculated that increased tropospheric aerosols could reduce precipitation by 20% over central Africa. Such a change might resolve some model–data precipitation discrepancies in the tropics. Whatever the cause, the model–data discrepancies in the tropics represent a first-order disagreement and are a significant impediment to enhancing confidence in model results.

Subsequent changes in climate involving the evolution of the African–Asian monsoon also indicate good agreement between models and data (Kutzbach and Street-Perrott, 1985). During the early part of the present Holocene interglacial (about 6000–11,000 BP), lacustrine records indicate high levels throughout Africa and extending eastward to India. The primary forcing mechanism apparently involves significant orbitally induced insolation changes in the seasonal cycle (Kutzbach and Otto-Bliesner, 1982; Kutzbach and Guetter, 1984, 1986; Kutzbach and Street-Perrott, 1985; Mitchell et al., 1988). Experiments with GCMs of varying levels of complexity indicate that increased July insolation caused large changes in summer temperatures over Northern Hemisphere landmasses and small changes over the ocean (Section 4.5.1). Increased heating caused enhanced low-level inflow and decreased sea level pressure over the center of Eurasia. Stronger southwesterly winds resulted in enhanced moisture convergence in the northern summer intertropical convergence zone (ITCZ). A 10–20% increase in precipitation is simulated by the NCAR CCM (Kutzbach and Guetter, 1986). Comparisons of net precipitation minus evaporation with a hydrologic model for lakes show good agreement between model and data (Kutzbach and Street-Perrott, 1985; COHMAP, 1988).

14.3.3 Overall Evaluation of Model–Data Agreement

Based on the above comparisons, there is a fair-to-good agreement between models and observations of climate change over the last 20,000 years (Fig. 4.18). The level of agreement might tempt us to be moderately confident about GCM predictions for the future. This temptation should be resisted. The two cases where models agree best with the data (glacial maximum at northern latitudes and monsoon changes) may in some respects involve simple levels of forcing. In one case the North American ice sheet is so large that it may dominate any local climate change, with the type of response severely constrained by the enormity of the forcing. Likewise, the monsoon system is directly forced by seasonal solar heating of a large land mass, and changes in temperature response appear to be linearly related to changes in forcing (Crowley, 1988; cf. Fig. 1.12; the actual evolution of monsoon rain systems, however, clearly involves nonlinear processes). In the tropics, the one region where the system is not radically forced or linearly connected, the models and observations diverge quite severely at 18,000 BP. In a sense, this disagreement in the tropics may be a better measure of model reliability than the two successful cases, since the greenhouse climate will not involve locally forced or strong linear connections.

14.4 Summary and Conclusions

Geologic data provide a surprisingly rich set of results that enable us to peer into the future with some level of insight concerning the conse-quences of a greenhouse warming. The warming will probably be a very major climate change. Planetary temperatures could rise to levels normally associated with an ice-free state yet retain cold poles due to the inertial presence of ice sheets. The magnitude and rate of warming could be comparable to or exceed any that has occurred in earth history, with the resultant climate representing a perhaps unique occurrence in that history. The warming could significantly reduce Arctic and Antarctic sea ice cover, trigger instabilities in the marine-based West Antarctic Ice Sheet, and perhaps even affect the much larger East Antarctic Ice Sheet. The abrupt warming may also affect biotic extinction rates, but such extinctions will likely be dwarfed by anthropogenic-induced extinctions due to habitat destruction of ecosystems. Although there is moderately good agreement between climate models and inferred climate change over the last 20,000 years, there are a sufficient number of caveats to necessitate caution in utilization of existing paleoclimate results to establish confidence in climate model projections of a greenhouse warming.

Appendix A.   THE GEOLOGIC TIME SCALE

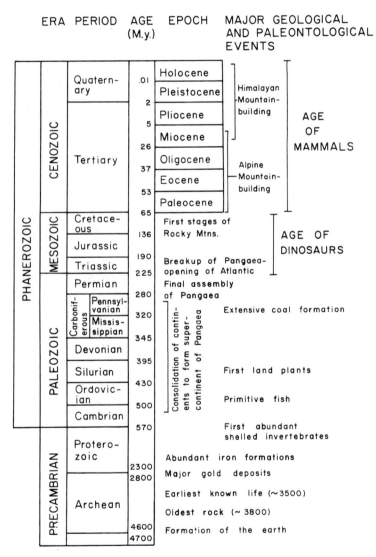

[From Crowley, 1983b] *Courtesy American Geophysical Union.*

Appendix B.   PALEOENVIRONMENTAL ESTIMATES FROM PROXY DATA

Although geologists have long interpreted past faunal and floral changes in terms of climate, only in the last 20 years have these estimates been quantified through the use of *transfer functions*. The basic idea behind this term involves relating data to some physical variable by means of regression. First the data from the present climate are used to find the regression coefficients; then the relationship can be used to infer past physical environments by using paleodata. In this appendix we will discuss two different methods for making that transference, using marine records as an example, but there is a growing body of literature on the subject and a wider applicability of the methods. Probably the two most influential papers that helped launch the field were Imbrie and Kipp (1971) and Webb and Bryson (1972). We refer the reader to the following papers for additional information (Sachs et al., 1977; Webb and Clark, 1977; Molfino et al., 1982; Howe and Webb, 1983; Arigo et al., 1984).

The basic approach underlying all proxy techniques involves the relatively high degree of differentiation of organisms according to their physical environment, and the assumption that physical variables can be estimated from biotic distributions once the degree of relationship has been objectively established. For example, there are about 30 species of planktonic foraminifera in the world ocean. Some live in tropical waters, others in polar regions. The different species tend to cluster into a smaller number of *assemblages* of species (also called factors or end members).

One method of estimating physical variables takes advantage of this natural clustering tendency in organisms by more objectively defining the clusters in terms of eigenvectors (any standard text on linear algebra should contain a discussion of eigenvectors). At any given fixed level of a core's depth (corresponding to a time in history) there will be population counts for the different species. We can form a vector $\vec{S}_d = (S_{d1}, S_{d2}, \ldots, S_{d30})$ corresponding to depth d of the core and whose components (1–30) are the counts for the various species. There will be a different vector $\vec{S}_d(\vec{x}_j)$ for each site $\vec{x}_j$, $j = 1, \ldots, 191$, where we have chosen the number of sites to be 191 corresponding to the number chosen by Kipp (1976).

We can form a 30 × 30 covariance matrix between species counts averaged across the cores (spatial sites):

$$C_{mn} = \overline{(\vec{S}_d)_m (\vec{S}_d)_n} \qquad (B.1)$$

where the overbar indicates averaging over the 191 sites at depth d (for the case cited $d = 0$, the "core top" or surface sediment sample). The 30 × 30 covariance matrix is symmetric and has 30 eigenvectors e_n^α, each labeled by $\alpha = 1, \ldots, 30$. They are formally defined by

$$\sum_{n=1}^{30} C_{mn} e_n^\alpha = \lambda_\alpha e_n^\alpha \qquad (B.2)$$

where λ_α is the *eigenvalue* corresponding to the eigenvector labeled by α. It is possible to show that the eigenvectors can be normalized to unit length and are orthogonal, which is expressed by

$$e^\alpha \cdot e^\beta = \sum_{n=1}^{30} e_n^\alpha e_n^\beta = \delta_{\alpha\beta} \qquad (B.3)$$

where $\delta_{\alpha\beta}$ is the Kronecker symbol, 1 for $\alpha = \beta$, 0 otherwise. The eigenvectors form a basis set so that any site sample can be represented in terms of

$$[\vec{S}_d(x_j)]_n = \sum_{\alpha=1}^{30} a_\alpha^d(x_j) e_n^\alpha \qquad (B.4)$$

The expansion coefficients $a_\alpha^d(x_j)$ are known as the *factor loadings* (cf. Imbrie and Kipp, 1971) for the depth d and the site at x_j. They can be found for each site by the inversion formula:

$$a_\alpha^d(x_j) = \sum_{n=1}^{30} [\vec{S}_d(x_j)]_n e_n^\alpha \qquad (B.5)$$

which makes use of the orthogonality property of the eigenvectors (B.3).

The total variance at a particular depth d across all the sites can be written as the sum of the diagonal elements of the covariance matrix. It can be shown that this is the sum of the eigenvalues and that the contribution of a particular eigenvector to the variance of a sum such as in (B.4) is proportional to the eigenvalue corresponding to the particular eigenvector. A very useful property encountered in these kinds of data analysis is that most of the variance is accounted for by only the first few eigenvectors. In fact, in the problem at hand the first six account for 92% of the total variance in the samples (Kipp, 1976). Hence, it will be convenient to truncate most of the sums over eigenvectors to only six terms. In most cases the other eigenvectors are not statistically different from those generated randomly (cf. North et al., 1982). In some cases it is convenient to perform a further linear transformation on the eigenvectors ("rotate them") to explain slightly more of the variance after the truncation (cf. Harmon, 1976; Walsh and Richman, 1981).

Analysis of 191 North Atlantic surface sediment samples, which are 1000–2000 year averages of the most recent oceanographic conditions, indicates that there is a remarkable correspondence between the geographic distribution of these factors (Fig. B.1) and major features of the North Atlantic circulation (Kipp, 1976). For example, consider the distribution of the Polar factor (i.e., assemblage of species). The area of greatest abundance is defined by the eigenvector with the highest factor loadings (i.e., accounts for the most variance). The Polar factor dominates in regions poleward of the oceanic polar front. The Transitional factor reflects the Gulf Stream/North Atlantic Current system, bounded on the north by the polar front and on the south by the subtropical convergence. The Gyre Margin assemblage reflects the higher productivity regions of the oceanic gyre margins, and its northward extension south of Newfoundland reflects the injection of warm Gulf

Stream waters into these regions. Similar relationships have been found for other planktonic groups in other oceans (e.g., Moore et al., 1981a; Molfino et al., 1982).

Now we want to consider how the eigenvectors can be used to estimate physically meaningful variables. To estimate surface temperature $T(x_j)$ at the various sites, we consider a linear form in terms of the factor loadings:

$$T(x_j) = \sum_{\alpha=1}^{6} T_\alpha a_\alpha^d(x_j) + E_j \qquad (B.6)$$

The unknown coefficients T_α are to be found from a least-squares fitting procedure (regression); E_j is the residual error term (note that the T_α are independent of site). We have measurements of $T(x_j)$ at core top and assume that the coefficients T_α can be found from the core top data alone and are independent of time (this implies that the species have not evolved appreciably in the time under consideration). Once these coefficients are found by the regression analysis, the factor loadings $a_\alpha^d(x_j)$ for any depth d and site x_j can be computed using (B.5) and the results inserted into (B.6) for that depth.

A scattergram of results for winter temperature (Fig. B.2) attests to the power of this method. The multiple correlation coefficient is 0.991, with a standard error of estimate of 1.2°C and an 80% confidence interval of 1.5°C (Kipp, 1976). These estimates are generally reproducible when different groups of organisms are used (Molfino et al., 1982).

As noted by Prell (1985), the Imbrie–Kipp method is especially useful when the number of samples in a surface calibration data set is relatively small. However, as the sample size increases, other techniques are also applicable and in some respects preferable. For example, the modern analog technique (MAT) is an alternative approach to estimating and evaluating SST patterns (Hutson, 1979; cf. Prentice, 1980; Overpeck et al., 1985; Prell, 1985). The MAT uses dissimilarity coefficients to measure the difference between multivariate (species) samples. This approach can be applied directly to core samples without an intervening step of computing eigenvectors, which may smooth or generalize the data. There are various ways of choosing dissimilarity coefficients (squared cosine of

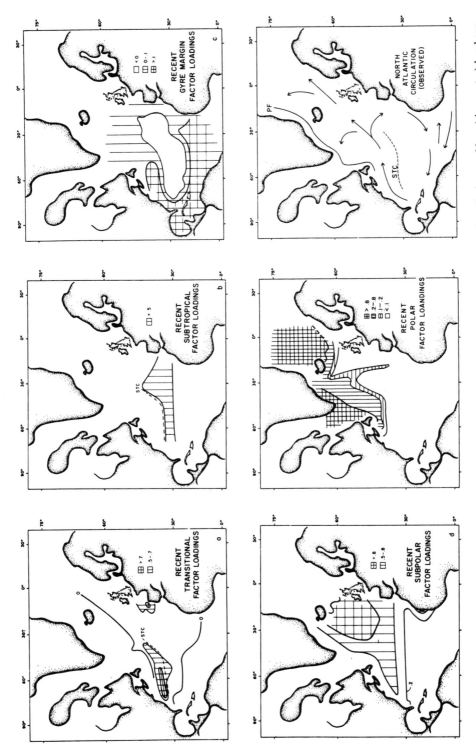

Fig. B.1. Comparison of five North Atlantic ecological assemblages with the observed North Atlantic circulation. Each assemblage (eigenvector) has been calculated from a factor analysis of surface sediment samples. Factor loadings are a measure of the abundance of a factor in a sample. Each map illustrates the geographic location where assemblage abundance reaches a maximum for a particular factor. Positions of the oceanic polar front (PF) and subtropical convergence (STC) are from Defant (1961). [From Crowley, 1981; based on data in Kipp, 1976. *Reproduced with permission of Elsevier Scientific Publishers.*]

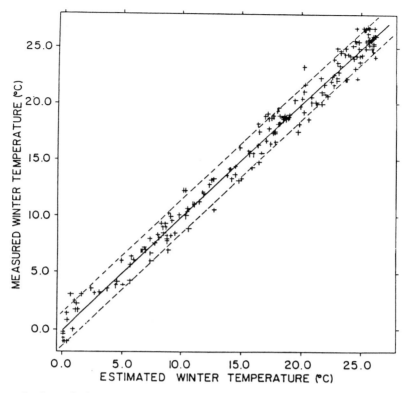

Fig. B.2. Observed values of winter temperature versus estimates calculated by a transfer function on 191 North Atlantic core top samples. Dashed lines represent the 80% confidence interval. [From Kipp, 1976] *Reprinted with permission of the Geological Society of America. From Geol. Soc. Amer. Mem. 145.*

the vector angle between modern and downcore samples, squared Euclidean distance, squared standardized Euclidean distance, and squared chord distance). Prell (1985) chose the squared chord distance to reevaluate CLIMAP SST estimates at 18,000 BP in order to determine whether the results were methodology-dependent, where

$$d_{ij} = \Sigma_k (P_{ik}^{1/2} - P_{jk}^{1/2})^2 \qquad (B.7)$$

where d is the dissimilarity measure, i is the subject sample, j is the analog sample, k is the species, and P is the proportion (fraction) of species k in the total sample. d_{ij} ranges from 0.0 to 2.0, with the lowest values indicating the most similarity.

After identifying the subset of most similar samples (i.e., lowest dissimilarity coefficients),

he used the chord distance similarity measure (S ranges from 1.0 to 0.0, with high values being most similar) to weight environmental estimates:

$$S_{ij} = \Sigma_k (P_{ik} \cdot P_{jk})^{1/2} \qquad (B.8)$$

where S is the similarity coefficient. The weighting approach gives approximately equal weighting to all samples if they are similar. However, it downweights contributions to the SST estimate of any sample that has low similarity.

When using the MAT, paleoenvironmental estimates (e.g., SST) are determined by taking simple averages or weighted averages of the SST data of the most similar subset rather than linear or curvilinear regression equations. Each SST estimate has its own standard deviation based on the range of values in the subset.

Appendix C.   A SHORT OUTLINE OF TIME SERIES ANALYSIS

In this appendix we present an outline of common terms used in time series analysis so that discussions such as that in Chap. 7 can be more easily digested. Interested readers can consult numerous texts on the subject (e.g., Jenkins and Watts, 1968; Bendat and Piersol, 1986; Newton, 1988). Here we simply give a series of heuristic definitions and illustrative examples.

Before discussing aspects of time series analysis, it is necessary to define some basic statistical terms. A *random variable x* can be thought of as a number drawn from a prescribed probability rule. For example, 1, 2, 3, 4, 5, and 6 are the equal probability values taken on by the toss of a die. In many earth science applications the variable is continuous and its *probability density function* (the probability that the random variable lies in a certain infinitesimal interval on a given evaluation) can be taken as a Gaussian or bell-shaped distribution. A particular drawing or evaluation of the random variable is called a *realization*.

The *mean* of a random variable is the arithmetic average over a large number of drawings of the random variable.

$$\langle x \rangle = \frac{1}{N} \sum_{1}^{N} x_i \qquad (C.1)$$

where N is the number of samples. The brackets indicate that the average has been taken over the whole population or *ensemble* (in other words, N is taken to be very large).

The *variance* is the mean squared differences of individual drawings from the overall mean.

$$Var(x) = \frac{1}{N} \sum_{1}^{N} (x_i - \langle x \rangle)^2$$
$$= \langle (x - \langle x \rangle)^2 \rangle \qquad (C.2)$$

where again it is understood that N is large. The *standard deviation* is the square root of the variance. It is a measure of the width or dispersion of a probability distribution.

A *time series* is a sequence of random variables labeled or indexed by time, which we take to be in unit increments. These might be a series of data entries, $x_1, x_2, \ldots, x_m, \ldots$. Examples of two time series *segments* are shown in Figs. C.1 and C.2. Each segment is for 256 time steps. The entry in Fig. C.1 is for a so-called *white noise* process and in Fig. C.2 for a *red noise* process (see text following).

The *covariance* is a measure of the tendency for two variables to vary together, defined as the ensemble average of the product of the deviation of one random variable from its mean with the deviation of the other from its mean:

$$Cov(x, y) = \frac{1}{N} \Sigma(x_i - \langle x \rangle)(y_i - \langle y \rangle)$$
$$= \langle (x - \langle x \rangle)(y - \langle y \rangle) \rangle \qquad (C.3)$$

The covariance between two random variables that tend to fluctuate together will be nonzero. *Statistically independent* random variables will have zero covariance (the converse is not always true).

The covariance of two variables normalized by the product of the standard deviations is called the *correlation* between the two variables. The correlation will lie between -1 and $+1$, with -1 for a perfect negative correlation and $+1$ for a perfect positive correlation.

A *stationary* time series is one for which the mean is independent of time and the covariance of the variable at one time and itself at a later time depends only on the time difference *(lag)*.

$$Cov(x_i, x_{i+n}) = fcn(n) \qquad (C.4)$$

That the lagged covariance depends only on the lag means that there is no preferred time origin in the series, i.e., it is translationally invariant along the time axis. An example of the opposite, a *nonstationary* time series, is shown in Fig. C.3a (note the abrupt shift in the mean).

It is often very convenient to analyze a time series into sinusoidal frequency components.

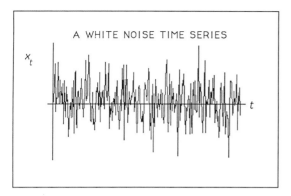

Fig. C.1. Example of a white noise time series for 256 time steps.

We can write the *discrete Fourier transform* of a time series segment $x_1, x_2, \ldots, x_N, \ldots$ as (Newton, 1988):

$$z_k = \sum_{t=1}^{N} x_t e^{2\pi i(t-1)\omega_k} \qquad (C.5)$$

where $\omega_k = (k-1)/N$, $k = 1, \ldots, N$. Note that there are N separate (complex) Fourier coefficients z_k, since $k = 1, \ldots, N$. The ω_k are frequencies running from $1/N$ (the lowest) to 1 (which is 1 over the interval between samples). The frequency is in units of 1/(length of time series segment). For example, for a series of measurements 400,000 years long with a resolution of 2000 years, we would have a coefficient at each 1/(400,000 years) out to a frequency of 1/(2000 years). [Later we will see that the highest useful frequency is actually only half of the latter.] We have used the complex exponential function to represent our sinusoids. The data can be recovered by the *inverse* of the discrete Fourier transform:

$$x_t = \frac{1}{N} \sum_{k=1}^{N} z_k e^{-2\pi i(t-1)\omega_k} \qquad (C.6)$$

which is the Fourier representation of the time series segment, i.e., a member x_t can be written as the sum of contributions from oscillations at different frequencies. The complex amplitudes z_k of these contributions are themselves random variables drawn from (complex) bell-shaped probability density distributions. For a long stationary time series the amplitudes from different frequencies are statistically independent.

The *variance* of a given member of a long stationary time series can be written as the sum of contributions from the variance of the amplitudes corresponding to the individual frequency terms in the Fourier representation.

$$Var(x) = \frac{1}{N^2} \sum_{k=1}^{N} \langle |z_k|^2 \rangle \qquad (C.7)$$

Study of the variance by looking at its decomposition into individual frequency components is a form of *analysis of variance.*

The amplitudes squared of the Fourier coefficients, multiplied by the length of the series $N|z_k|^2$, when plotted against the frequency ω_k is called the *periodogram* of the time series segment. Close examination shows that the periodogram is symmetric about $\omega_k = 1/2$ ($k = N/2$), and hence there are only $N/2$ independent members. The highest frequency that can be resolved is 2/(interval of sampling). The lowest frequency that can be resolved is $1/N$. The periodogram of the white noise series in Fig. C.1 is shown in Fig. C.4. Note how irregular it is, illustrating that the individual terms $N|z_k|^2$ are random and essentially independent.

If we take the periodogram from many different realizations of, say, length 256 of the white

Fig. C.2. Example of a red noise time series for 256 steps generated by the first-order autoregressive (AR1) process:

$$x_{n+1} = 0.9x_n + z_n$$

where the sequence z_n is a white noise sequence having the same properties as in Fig. C.1. Note that the AR1 process is much smoother because of the serial correlation.

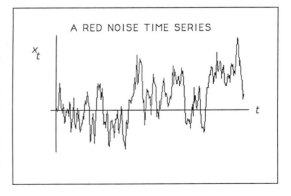

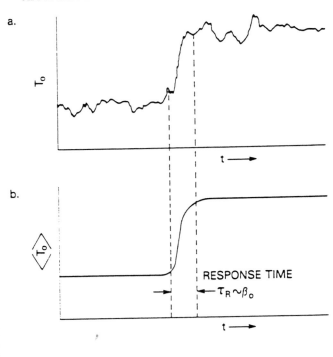

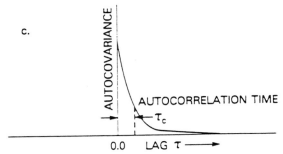

Fig. C.3. Schematic illustration of several statistical concepts. (a) A typical *nonstationary* time series of global temperature with a shift in the mean due to a step change in forcing at $t = 0$. (b) An *ensemble average* of a many sampled function similar to the function in Fig. C.3a. Averaging eliminates the fluctuations, leaving only the climatological mean, which changes from one stationary value to another during a time τ_R, the climate *response time*, proportional to the climate sensitivity β_0 (cf. Chap. 1). (c) *Autocovariance* of $\delta T_0 = T_0 - \langle T_0 \rangle$. This function decreases over a characteristic *lag time* T_c, the autocorrelation time for fluctuations in T_0. [From North et al., 1981]

noise process and average these together, we obtain the *spectral density* $f(\omega)$ (also called the *power spectrum*), which is typically much smoother than an individual periodogram. We can think of the $f(\omega)$ as $\langle N|z_k|^2 \rangle$. For example, for the white noise case $f(\omega)$ is a flat line. The theory of linear systems suggests that the spectral density can reveal information useful in identifying underlying mechanisms controlling the process responsible for the time series.

The observer never has the luxury of averaging over an ensemble of realizations and must inevitably resort to other means of estimating the spectral density. One way of approaching the problem is to smooth by averaging the terms in the periodogram over a *band* of frequencies. Basically, we are finding the average value of the

spectral density in the band by averaging over the neighboring spectral lines (values of the discrete variable ω_k). We can thereby improve our estimate of the spectral density over an interval by sacrificing frequency resolution. Figure C.4b and c illustrates this for the periodogram in panel a for two different bandwidths. Note that for a given sampling resolution, the number of spectral lines ω_k that can be packed in a given band increases with the length of the time series. Hence, increasing the length of the time series significantly improves the estimate of the spectral density for a given bandwidth. In other words, the confidence interval (sampling error bars) depends strictly on how many spectral lines are packed into the bandwidth and these error bars tighten as the time series segment is lengthened. In practice one uses other, more sophisticated methods of smoothing the spectrum (e.g., Newton, 1988), but the principle is the same as given here.

The *red noise spectrum* is more commonly encountered in geophysical applications (e.g., Båth, 1974). It is large at low frequencies and ta-

pers to zero at high frequencies roughly as the inverse second power. This spectrum arises in connection with time series in which there is serial correlation (*autocorrelation,* Fig. C.3c), i.e., in the case:

$$Cov(x_i, x_{i+1}) = \lambda \qquad -1 < \lambda < 1 \qquad (C.8)$$

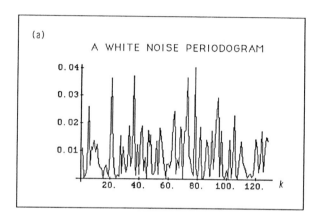

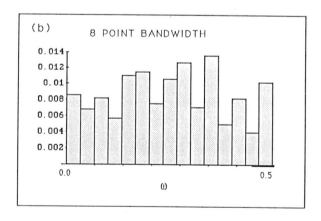

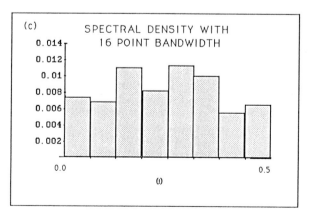

Fig. C.4. Several ways of showing the Fourier decomposition of the variance for the time series shown in Fig. C.1. (a) The periodogram. (b) The periodogram smoothed by an 8-point bandwidth. (c) Same as (b) but for a 16-point bandwidth. The theoretical spectral density is flat.

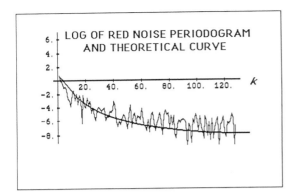

Fig. C.5. Natural log of the periodogram of the red noise time series segment (256 steps) shown in Fig. C.2 with the log of the theoretical spectrum superimposed. The abscissa is the Fourier index k, which runs from zero to 128.

This is the case of a first-order autoregressive process (AR1 process). The theoretical spectrum for a red noise (AR1) process is shown in Fig. C.5.

Many geophysical time series encountered in this book are red noise with *quasi-deterministic signals* at certain discrete frequencies superimposed (see Fig. 1.2). Time series analysis from such a process consists of computing the power spectrum and examining it for statistically significant peaks at the postulated frequencies. In this context *statistically significant* means that the peak rises appreciably above the level of sampling errors likely to occur in the red noise spectrum when no such signal is present. In some diagrams in Chap. 7 the abbreviation CI stands for *confidence interval,* which means that 95% of the analyses conducted with a time series generated by the same process and of the same record length and bandwidth should fall in this interval if the process were pure red noise.

Power spectra are displayed on a variety of different scales. For example, the periodograms shown in Fig. C.4 are in linear dimension on both abscissa and ordinate. The linear axes have the advantage that areas under the curve between values on the abscissa tell us directly how much variance is explained by that portion of the spectrum. There are certain advantages of looking at the logarithm of the periodogram or spectral density. In the red noise case this was used because the spectrum covers several orders of magnitude and our choice of log for the or-

dinate allows us to see over the entire range. Another advantage is that the CI for the process is proportional to the local value of the spectral density and this varies over several orders of magnitude. On the log scale, moreover, the CI or error bar is independent of the magnitude of the spectrum at the point in question; hence, a single error bar can be displayed that will serve for the entire graphic. Other methods of display such as log of the abscissa are also used from time to time mostly just to get the whole graph onto one picture (e.g., Fig. 1.2).

Another way of viewing power spectrum analysis is to consider a transformation of the original time series that filters out any portions of the variations that have frequency components outside the band of interest. A crude way of doing this would be to take the discrete Fourier transform of the time series segment, zero out the coefficients z_k corresponding to the unwanted frequencies, then transform back to obtain the modified or filtered series x'_t. A radio receiver uses this technique electronically as we *tune* to one station, filtering out all others. In spectral analysis we examine the variance of the filtered portion (band pass filtering) of the total variance one band at a time.

By *band pass filtering* we can compare two time series to see if they co-vary in certain frequency intervals. *Cross-spectrum analysis* is the study of the covariance across two time series at selected individual frequency bands. A peak in such a spectrum suggests that the two sequences are highly correlated in that band (it lies outside the CI for no covariance). This might be of importance in determining, for example, if the two time series are driven by the same periodic forcing agent.

The *coherency* function for a given frequency band for two time series is the analog of the correlation in that the covariance between the two bands is normalized by the standard deviations of the two individual bands. The coherency lies between 0 and $+1$, with the square being that fraction of the variance in one signal which is linearly related to the variance in the other.

We can also study the *phase coherence* between two band-passed time series to see if there is a clustering (peaking of the histogram) of the phase difference between the two sinusoidal variations at a given frequency. This might be important in identifying dynamic causes since in many cases theory will suggest that one time series lead the other by a certain fraction of a cycle. For example, a 0° phase relationship at a certain frequency indicates that there is no *lead* or *lag* between the two variables. Alternatively, a 90° phase lag for a 20,000-year period means that one variable lags the other by 5000 years (90° = ¼ of 360°).

A final word of caution concerning time series analysis. Taking a raw data sequence and seeing a peak above the CI interval for a red noise background does not necessarily mean that there is an underlying periodic mechanism at work. After all, if we have 20 individual frequency bandwidths, we might expect that one will lie outside the CI by pure chance! Hence, we should start our analysis of identifying periodicities in the data with firm physical hypotheses or, alternatively, if we insist on "going fishing," apply the appropriate more stringent statistical test that takes this into account.

Appendix D. ANNOTATED BIBLIOGRAPHY OF SELECTED REFERENCE WORKS

The following books provide the interested reader with additional information on various aspects of paleoclimatology and related disciplines.

Paleoclimatology

Abrupt Climate Change (Berger and Labeyrie, 1987) is an up-to-date assessment of evidence for various types of abrupt climate change in the geologic record—a topic of increasing importance in the field.

Milankovitch and Climate (Berger et al., 1984) is a two-volume series examining the orbital theory of glaciation, evidence for Milankovitch cycles in the geologic record, and models incorporating Milankovitch variations. The book is the most complete documentation of the role of orbital variations in the geologic record. After its publication, most remaining Milankovitch doubters were swept away by the flood of supporting documentation.

Quaternary Paleoclimatology (Bradley, 1985) is a good, balanced discussion of the various types of evidence used to reconstruct Quaternary climates. The book is particularly valuable for discussions on methodologies.

Investigations of Late Quaternary Paleoceanography and Paleoclimatology (Cline and Hays, 1976) is a showcase for the CLIMAP Project's contribution to Quaternary oceanography and climatology.

The Last Great Ice Sheets (Denton and Hughes, 1981b) is a reconstruction of global ice distributions at the last glacial maximum, supplemented with models of ice sheet configurations and mechanisms of ice decay.

Glacial and Quaternary Geology (Flint, 1971) is a comprehensive (but now somewhat dated) survey of all aspects of glacial geology by the late dean of the field; a classic reference text for glacial geologists.

Climates Throughout Geologic Time (Frakes, 1979) is a valuable book that discusses in some detail pre-Pleistocene glaciations, with an emphasis on evidence for the glaciations. The book has a useful bibliography.

The Little Ice Age (Grove, 1988) is a very detailed and highly praised examination of climate change during the Little Ice Age. This is an important reference work for those interested in the topic.

Paleoclimate, Paleomagnetism, and Continental Drift (Habicht, 1979) illustrates paleoclimatological world maps for each of the geologic periods of the last 570 million years.

Earth's Pre-Pleistocene Glacial Record (Hambrey and Harland, 1981a) is an encyclopedic listing and discussion of virtually all glacial deposits reported for the first 99.9% of earth history. Emphasis is on documentation rather than climatological interpretation. This is the type of compilation that occurs perhaps once every 20–25 years, so it is unlikely to be superseded soon.

Paleoclimate Analysis and Modeling (Hecht, 1985) examines principal results from various disciplines of paleoclimatology. The book has a number of valuable articles, and the discussions on historical climate fluctuations and tree rings are some of the best available.

Paleoecology of Beringia (Hopkins et al., 1982) is a survey of the various late Quaternary features of Beringia, the eastern Siberia–Bering Strait–western Alaska corridor used by man for his migration to the New World.

Ice Ages: Solving the Mystery (Imbrie and Imbrie, 1979) is a well-written introduction to the ice-age problem, complete with historical background.

Greenland Ice Core: Geophysics, Geochemistry, and the Environment (Langway et al., 1985) is a valuable discussion of various features of the Dye 3 (Greenland) ice core.

Causes of Climate Change (Mitchell, 1968) is a historic volume that contains a number of seminal papers in paleoclimatology. Although much has been done since 1968, the book is still important as perhaps the first significant attempt at quantitative paleoclimatology.

Climate in Earth History (National Academy of Sciences, 1982) is a summary by leading workers in the field on many important aspects of paleoclimatology.

Late Quaternary Environments of the United States, Volume 1, The Pleistocene (Porter, 1983) reviews a wide range of observations from this important area.

North America and Adjacent Oceans During the Last Deglaciation (Ruddiman and Wright,

1987) is a complete and thorough discussion of the many features related to the deglaciation of North America. The book includes discussions of land, marine, and ice records, plus modeling studies.

The Late Cenozoic Glacial Ages (Turekian, 1971) is a watershed book that separates "modern" from "classical" paleoclimatology. The book contains a number of classic papers that are still referenced. In addition to providing a summary of past accomplishments in paleoclimatology, the book set the tone for much of the research done in the subsequent decade.

Late Quaternary Environments of the Soviet Union (Velichko, 1983) includes 30 articles on Quaternary climate change in the Soviet Union.

Late Quaternary Climate Change in North America, Volume 2, The Holocene (Wright, 1983) is a companion volume to Porter (1983) and reviews a wide range of evidence for Holocene climate fluctuations.

Meteorology and Climate

Climate Processes and Climate Sensitivity (Hansen and Takahashi, 1984) includes important papers on various aspects of climate change (atmosphere and ocean dynamics, hydrologic cycle, albedo and radiation processes, cryospheric processes, and ocean chemistry). Many papers in this volume are frequently referenced.

An Introduction to Dynamic Meteorology (Holton, 1979) is one of the most frequently used texts on the subject.

The Global Climate (Houghton, 1984) contains a number of useful articles about various aspects of the climate system; the book is a nice step up from the introduction given in Chap. 2.

Climate: Present, Past and Future (Lamb, 1977) is an extensive summary of evidence for past climates, with a particularly good examination of records from historical times.

Physically Based Modelling and Simulation of Climate and Climate Change (Schlesinger, 1988) is a comprehensive discussion of the many different types of climate models, with particularly full discussions on various aspects of GCMs. The book includes a number of articles on applications of GCMs.

Atmospheric Science: An Introductory Survey (Wallace and Hobbs, 1977) is perhaps the most commonly consulted book on introductory aspects of atmospheric science.

An Introduction to Three-Dimensional Climate Modeling (Washington and Parkinson, 1986) is a very good, clearly written discussion of GCMs. The book is an ideal next place to turn for those who wish to move beyond what was discussed in Chap. 2 of our book.

Geology

History of the Earth (Eicher and McAlester, 1980) is a very readable and clearly illustrated introductory text on the principal physical and biological events in earth history.

Earth and Life Through Time (Stanley, 1989) is an excellent, lavishly illustrated textbook on earth history.

Oceanography

Marine Geology (Kennett, 1982) is a comprehensive survey of all aspects of marine geology, with much valuable information on the history of the ocean basins and ocean circulation.

Introductory Dynamic Oceanography (Pond and Pickard, 1983) is a fine introduction to modeling the ocean circulation.

Paleoceanography (Schopf, 1980) covers a broad spectrum of topics in paleoceanography, with some particularly useful sections on the evolution of the crust, atmosphere, and ocean (including ocean chemistry). The book has an extensive bibliography.

Evolution of Physical Oceanography (Warren and Wunsch, 1981) contains many valuable review papers on various aspects of physical oceanography.

Geochemistry

Tracers in the Sea (Broecker and Peng, 1982) is an excellent sourcebook for marine geochemistry, with particularly extensive discussions on the carbon cycle.

Natural Variations in the Carbon Cycle: Archean to Present (Sundquist and Broecker, 1985) is the bible for those who want to know about the carbon cycle and the geologic record. There are many important papers, both observational and modeling.

Solar Variability

The Ancient Sun (Pepin et al., 1980) contains the proceedings of a 1979 conference on solar variability, its nature, and fossil record in the earth, moon, and meteorites.

Variations of the Solar Constant (Sofia, 1981) is a wonderfully titled book illustrating that the solar constant is not constant, and discussing different processes responsible for the variations.

Secular Solar and Geomagnetic Variations in the Last 10,000 Years (Stephenson and Wolfendale, 1988) is a more recent examination of solar variability, containing a number of valuable articles.

REFERENCES

Absy, M. L., 1985. The palynology of Amazonia: The history of the forests as revealed by the palynological record. In: *Amazonia*. G. T. Prance and T. E. Lovejoy (Eds.). Pergamon Press, Oxford, England, pp. 72–82.

Adams, C. G., D. E. Lee, and B. R. Rosen, 1990. Conflicting isotopic and biotic evidence for tropical sea-surface temperatures during the Tertiary. *Palaeogeog., Palaeoclimatol., Palaeoecol.* 77:289–313.

Adamson, D. A., F. Gasse, F. A. Street, and M. A. J. Williams, 1980. Late Quaternary history of the Nile. *Nature* 288:50–55.

Adhémar, J. A., 1842. *Révolutions de la Mer.* Privately published, Paris.

Aharon, P., and J. Chappell, 1986. Oxygen isotopes, sea level changes and the temperature history of a coral reef environment in New Guinea over the last 10^5 years. *Palaeogeog., Palaeoclimatol., Palaeoecol.* 56:337–379.

Alexander, R. C., and R. L. Mobley, 1976. Monthly average sea surface temperatures and ice pack limits on a 1° global grid. *Mon. Wea. Rev.* 104:143–148.

Allison, I., and P. Kruss, 1977. Estimation of recent climate change in Irian Jaya by numerical modelling of its tropical glaciers. *Arct. Alp. Res.* 9:49–60.

Alvarez, L. W., 1987. Mass extinctions caused by large bolide impacts. *Phys. Today* 40:24–33 (July).

Alvarez, L. W., W. Alvarez, F. Asaro, and H. V. Michel, 1980a. Extraterrestrial cause for the Cretaceous-Tertiary extinction. *Science* 208:1095–1108.

Alvarez, W., D. V. Kent, I. P. Silva, R. A. Schweickert, and R. A. Larson, 1980b. Franciscan complex limestone deposited at 17° South paleolatitude. *Geol. Soc. Am. Bull.* 91:476–484.

Alvarez, W., L. W. Alvarez, F. Asaro, and H. V. Michel, 1984a. The end of the Cretaceous: Sharp boundary or gradual transition? *Science* 223:1183–1186.

Alvarez, W., et al., 1984b. Impact theory of mass extinctions and the invertebrate fossil record. *Science* 223:1135–1141.

Anderson, D. M., W. L. Prell, and N. J. Barratt, 1989. Estimates of sea surface temperature in the Coral Sea at the last glacial maximum. *Paleoceanog.* 4:615–627.

Anderson, T. L., and R. J. Charlson, 1990. Ice-age dust and sea salt. *Nature* 345:393.

Anderson, R. Y., 1984. Orbital forcing of evaporite sedimentation. In: *Milankovitch and Climate.* A. L. Berger, J. Imbrie, J. Hays, G. Kukla, and B. Saltzman (Eds.). D. Reidel, Dordrecht, Netherlands, pp. 147–162.

Andree, M. H., et al., 1986. Limits on the ventilation rate for the deep ocean over the last 12,000 years. *Clim. Dynam.* 1:53–62.

Andrews, J. T., 1973. The Wisconsin Laurentide ice sheet: Dispersal centers, problems of rates of retreat, and climatic implications. *Arct. Alp. Res.* 5:185–199.

Andrews, J. T., 1982. On the reconstruction of Pleistocene ice sheets: A review. *Quat. Sci. Rev.* 1:1–30.

Andrews, J. T., and R. G. Barry, 1978. Glacial inception and disintegration during the last glaciation. *Ann. Rev. Earth Plan. Sci.* 6:205–228.

Andrews, J. T., and M. A. W. Mahaffey, 1976. Growth rate of the Laurentide ice sheet and sea level lowering (with special emphasis on the 115,000 BP sea level low). *Quat. Res.* 6:167–183.

Andrews, J. T., G. K. Guennel, J. L. Wray, and J. D. Ives, 1972. An early Tertiary outcrop in north-central Baffin Island, Northwest Territories, Canada: Environment and significance. *Can. J. Earth Sci.* 9:233–238.

Andrews, J. T., P. T. Davis, and C. Wright, 1976. Little Ice Age permanent snowcover in the eastern Canadian Arctic: Extent mapped from Landsat-1 satellite imagery. *Geogr. Annir.* 58A:71–81.

Andrews, J. T., W. W. Shilts, and G. H. Miller, 1983. Multiple deglaciations of the Hudson Bay Lowlands, Canada, since deposition of the Missinaibi (last interglacial?) Formation. *Quat. Res.* 19:18–37.

Arakawa, A. 1988. Finite-difference methods in climate modeling. In: *Physically-based Modelling and Simulation of Climate and Climatic Change.* M. E. Schlesinger (Ed.). Kluwer, Dordrecht, Netherlands, pp. 79–168.

Arigo, R., S. E. Howe, and T. Webb III, 1984. Climatic calibration of pollen data. *U.S. Nuc. Reg. Comm. Cont. Rept.* NUREG/CR-3847.

Arthur, M. A., W. E. Dean, D. Bottjer, and P. A. Scholle, 1984. Rhythmic bedding in Mesozoic-Cenozoic pelagic carbonate sequences: The primary and diagenetic origin of Milankovitch-like cycles. In: *Milankovitch and Climate.* A. L. Berger, J. Imbrie, J. Hays, G. Kukla, and B. Saltzman (Eds.). D. Reidel, Dordrecht, Netherlands, pp. 191–222.

Arthur, M. A., W. E. Dean, and G. E. Claypool, 1985a. Anomalous ^{13}C enrichment in modern marine organic carbon. *Nature* 315:216–218.

Arthur, M. A., W. E. Dean, and S. O. Schlanger, 1985b. Variations in the global carbon cycle during the Cretaceous related to climate, volcanism, and changes in atmospheric CO_2. In: *The Carbon Cycle and Atmospheric CO_2: Natural Variations Archean to Present.* E. T. Sundquist and W. S. Broecker (Eds.). *Geophys. Mono.* **32,** Am. Geophys. Union, Washington, D.C., pp. 504–529.

Arthur, M. A., S. O. Schlanger, and H. C. Jenkyns, 1987. The Cenomanian-Turonian Oceanic Anoxic Event, II. Palaeoceanographic controls on organic-matter production and preservation. In: *Marine Petroleum Source Rocks.* J. Brooks and A. J. Fleet (Eds.). Geol. Soc. (London) Spec. Publ. 26:401–420.

Arthur, M. A., W. E. Dean, and L. M. Pratt, 1988. Geochemical and climatic effects of increased marine organic carbon burial at the Cenomanian/Turonian boundary. *Nature* 335:714–717.

Asaro, F., et al., 1988. Possible world-wide middle Miocene iridium anomaly and its relationship to periodicity of impacts and extinctions. In: *Global Catastrophes in Earth History: An Interdisciplinary Conference on Impacts, Volcanism, and Mass Mortality* (Abstracts). Lunar Plan. Inst., Houston, Texas, pp. 6–7.

Atkinson, T. C., K. R. Briffa, and G. R. Coope, 1987. Seasonal temperatures in Britain during the past 22,000 years reconstructed using beetle remains. *Nature* 325:587–592.

Axelrod, D. I., 1984. An interpretation of Cretaceous and Tertiary biota in polar regions. *Palaeogeog., Palaeoclimatol, Palaeoecol.* **45:**105–147.

Axelrod, D. I. and P. H. Raven, 1978. Late Cretaceous and Tertiary vegetation history of Africa. In: *Biogeography and Ecology of Southern Africa.* M. J. A. Werger (Ed.). Junk, The Hague, pp. 77–130.

Axelrod, D. I. and P. H. Raven, 1985. Origins of the Cordilleran flora. *J. Biogeog.* **12:**21–47.

Backman, J. and P. Pestiaux, 1986. Pliocene discoaster abundance variations, Deep Sea Drilling Project site 606: Biochronology and paleoenvironmental implications. *Init. Rep. Deep-Sea Drill. Proj.* **94:**903–909.

Bakker, R. T., 1975. Dinosaur renaissance. *Sci. Am.* 232:58–78.

Bakker, R. T., 1980. Dinosaur heresy—dinosaur renaissance: Why we need endothermic archosaurs for a comprehensive theory of bioenergetic evolution. In: *A Cold Look at the Warm-Blooded Dinosaurs.* R. D. K. Thomas and E. C. Olson (Eds.). Am. Assoc. Adv. Sci. Select Symp. **28:**351–462.

Bambach, R. K., C. R. Scotese, and A. M. Ziegler, 1981. Before Pangea: The geographies of the Paleozoic world. In: *Paleontology and Paleoenvironments, v.1.* B. J. Skinner (Ed.). William Kaufmann, Los Altos, Cal., pp. 116–128.

Barbetti, M., 1980. Geomagnetic strength over the last 50,000 years and changes in atmospheric ^{14}C concentration: Emerging trends. *Radiocarbon* 22:192–199.

Bard, E., M. Arnold, P. Maurice, J. Duprat, J. Moyes, and J.-C. Duplessy, 1987. Retreat velocity of the North Atlantic polar front during the last deglaciation determined by ^{14}C accelerator mass spectrometry. *Nature* 328:791–794.

Bard, E., B. Hamelin, R. G. Fairbanks, and A. Zindler, 1990. Calibration of the ^{14}C timescale over the past 30,000 years using mass spectrometric U-Th ages from Barbados corals. *Nature* 345:405–410.

Barker, P. F., and J. Burrell, 1977. The opening of Drake Passage. *Mar. Geol.* 25:15–34.

Barnard, P., 1973. Mesozoic floras. In: *Organisms and Continents Through Time.* N. Hughes (Ed.). Spec. Pap. Paleontol., Paleontol. Assoc., London, v. 12:175–188.

Barnett, T. P., L. Dümenil, U. Schlese, and E. Roeckner, 1988. The effect of Eurasian snow cover on global climate. *Science* 239:504–507.

Barnola, J. M., D. Raynaud, Y. S. Korotkevich, and C. Lorius, 1987. Vostok ice core provides 160,000-year record of atmospheric CO_2. *Nature* 329:408–414.

Baron, W. R., G. A. Gordon, H. W. Borus, and D. C. Smith, 1984. Frost-free record reconstruction for eastern Massachusetts, 1733–1980. *J. Clim. Appl. Met.* 23:317–319.

Barrera, E., B. T. Huber, S. M. Savin, and P.-N. Webb, 1987. Antarctic marine temperatures: Late Campanian through Early Paleocene. *Paleoceanog.* 2:21–47.

Barrett, P. J., D. P. Elston, D. M. Harwood, B. C. McKelvey, and P.-N. Webb, 1987. Mid-Cenozoic record of glaciation and sea-level change on the margin of the Victoria Land Basin, Antarctica. *Geology* 15:634–637.

Barron, E. J., 1983. A warm, equable Cretaceous: The nature of the problem. *Earth Sci. Rev.* **19:**305–338.

Barron, E. J., 1984. Climatic implications of the variable obliquity explanation of Cretaceous-Paleogene high-latitude floras. *Geology* 12:595–598.

Barron, E. J., 1985a. Explanations of the Tertiary global cooling trend. *Palaeogeog., Palaeoclim., Palaeoecol.* 50:45–61.

Barron, E. J., 1985b. Numerical climate modeling, a frontier in petroleum source rock prediction: Results based on Cretaceous simulations. *Am. Assoc. Petrol. Geol. Bull.* 69:448–459.

Barron, E. J., 1987. Eocene equator-to-pole surface

ocean temperatures: A significant climate problem? *Paleoceanog.* **2**:729–739.

Barron, E. J., 1989. Severe storms during earth history. *Geol. Soc. Am. Bull.* **101**:601–612.

Barron, E. J., and W. M. Washington, 1982a. Atmospheric circulation during warm geologic periods: Is the equator-to-pole surface-temperature gradient the controlling factor? *Geology* **10**:633–636.

Barron, E. J., and W. M. Washington, 1982b. Cretaceous climate: A comparison of atmospheric simulations with the geologic record. *Palaeogeog., Palaeoclim., Palaeoecol.* **40**:103–133.

Barron, E. J., and W. M. Washington, 1984. The role of geographic variables in explaining paleoclimates: Results from Cretaceous climate model sensitivity studies. *J. Geophys. Res.* **89**:1267–1279.

Barron, E. J., and W. M. Washington, 1985. Warm Cretaceous climates: High atmospheric CO_2 as a plausible mechanism. In: *The Carbon Cycle and Atmospheric CO_2: Natural Variations Archean to Present.* E. T. Sundquist and W. S. Broecker (Eds.). Geophys. Mono. **32**, Am. Geophys. Union, Washington, D.C., pp. 546–553.

Barron, E. J., and W. H. Peterson, 1989. Model simulation of the Cretaceous ocean circulation. *Science* **244**:684–686.

Barron, E. J., and W. H. Peterson, 1990. Mid-Cretaceous ocean circulation: Results from model sensitivity studies. *Paleoceanog.,* **5**:319–337.

Barron, E. J., J. L. Sloan, and C. G. A. Harrison, 1980. Potential significance of land-sea distribution and surface albedo variations as a climatic forcing factor: 180 M.y. to the present. *Palaeogeog., Palaeoclim., Palaeoecol.* **30**:17–40.

Barron, E. J., S. L. Thompson, and S. H. Schneider, 1981a. An ice-free Cretaceous? Results from climate model simulation. *Science* **212**:501–508.

Barron, E. J., C. G. A. Harrison, J. L. Sloan, and W. W. Hay, 1981b. Paleogeography, 180 million years ago to the present. *Ecol. Geol. Helv.* **74**:443–470.

Barron, E. J., S. L. Thompson, and W. W. Hay, 1984. Continental distribution as a forcing factor for global-scale temperature. *Nature* **310**:574–575.

Barron, E. J., M. A. Arthur, and E. G. Kauffman, 1985. Cretaceous rhythmic bedding sequences: A plausible link between orbital variations and climate. *Earth Plan. Sci. Lett.* **72**:327–340.

Barron, J. A., B. Larsen, and ODP 119 Shipboard Scientific Party, 1989. Preliminary results of drilling Kerguelen Plateau and Prydz Bay. *Eos (Trans. Am. Geophys. Union)* **70**:375.

Barry, J. C., N. M. Johnson, S. M. Raza, and L. L. Jacobs, 1985. Neogene mammalian faunal change in southern Asia: Correlations with climatic, tectonic, and eustatic events. *Geology* **13**:637–640.

Barry, R. G., 1967. Models in meteorology and climatology. In: *Models in Geography.* R. J. Chorley and P. Haggett (Eds.) Methuen, London, pp. 97–144.

Barry, R. G., 1983. Late-Pleistocene climatology. In: *Late-Quaternary Environments of the United States, The Late Pleistocene, v. 1.* H. E. Wright, and S. C. Porter (Eds.). Univ. Minn. Press, Minneapolis, pp. 390–407.

Bartlein, P. J., T. Webb III, and E. Fleri, 1984. Holocene climatic change in the northern Midwest: Pollen-derived estimates. *Quat. Res.* **22**:361–374.

Basu, A. R., S. L. Ray, A. K. Saha, and S. N. Sarkar, 1981. Eastern Indian 3800-million-year old crust and early mantle differentiation. *Science* **212**:1502–1506.

Bates, T. S., R. J. Charlson, and R. H. Gammon, 1987. Evidence for the climatic role of marine biogenic sulphur. *Nature* **329**:319–321.

Båth, M., 1974. *Spectral Analysis in Geophysics.* Dev. Solid Earth Geophys., v. 7. Elsevier, New York.

Beer, J., et al., 1985. ^{10}Be variations in polar ice cores. In: *Greenland Ice Core: Geophysics, Geochemistry, and the Environment.* C. C. Langway, H. Oeschger, and W. Dansgaard (Eds.). Geophys. Mono. **33**. Am. Geophys. Union, Washington, D.C., pp. 66–70.

Beer, J., et al., 1988. Information on past solar activity and geomagnetism from ^{10}Be in the Camp Century ice core. *Nature* **331**:675–680.

Begét, J. E., 1983. Radiocarbon-dated evidence of worldwide early Holocene climate change. *Geology* **11**:389–393.

Begét, J. E., and D. B. Hawkins, 1989. Influence of orbital parameters on Pleistocene loess deposition in central Alaska. *Nature* **337**:151–153.

Bekov, G. I., V. S. Letokhov, V. N. Radaev, D. D. Badyukov, and M. A. Nazarov, 1988. Rhodium distribution at the Cretaceous/Tertiary boundary analyzed by ultrasensitive laser photoionization. *Nature* **332**:146–148.

Belanger, P. E., 1982. Paleo-oceanography of the Norwegian Sea during the past 130,000 years: Coccolithophorid and foraminiferal data. *Boreas* **11**:29–36.

Bell, T. L., 1980. Climate sensitivity from fluctuation dissipation: Some simple model tests. *J. Atmos. Sci.* **37**:1700–1707.

Bendat, J. S. and A. G. Piersol, 1986. *Random Data, Analysis and Measurement Procedures.* John Wiley and Sons, New York.

Bender, M. L., 1984. On the relationship between ocean chemistry and atmospheric CO_2 during the Cenozoic. In: *Climate Processes and Climate Sensitivity.* J. E. Hansen and T. Takahashi (Eds.). Geophys. Mono **29**, Am. Geophys. Union, Washington, D.C., pp. 352–359.

Benjamin, M. T., N. M. Johnson, and C. W. Naeser, 1987. Recent rapid uplift in the Bolivian

Andes: Evidence from fission-track dating. *Geology* **15**:680–683.

Benson, L., and R. S. Thompson, 1987. The physical record of lakes in the Great Basin. In: *North America and Adjacent Oceans During the Last Deglaciation*. W. F. Ruddiman and H. E. Wright, Jr. (Eds.). The Geology of North America, v. **K-3**, Geol. Soc. Am., Boulder, Col., pp. 241–260.

Benson, R. H., R. E. Chapman, and L. T. Deck, 1984. Paleoceanographic events and deep-sea ostracodes. *Science* **224**:1334–1336.

Bentley, C. R., 1984. Some aspects of the cryosphere and its role in climatic change. In: *Climate Processes and Climate Sensitivity*. J. E. Hansen and T. Takahashi (Eds.). Geophys. Mono. **29**, Am. Geophys. Union, Washington, D.C., pp. 207–220.

Benton, M. J., 1987. Mass extinctions among families of non-marine tetrapods: The data: *Mem. Soc. Géol. France* **150**:21–32.

Berger, A. L., 1976. Obliquity and precession for the last 5,000,000 years. *Astron. Astrophys.* **51**:127–135.

Berger, A. L., 1978. Long-term variations of caloric insolation resulting from the earth's orbital elements. *Quat. Res.* **9**:139–167.

Berger, A. L., 1984. Accuracy and frequency stability of the earth's orbital elements during the Quaternary. In: *Milankovitch and Climate*. A. L. Berger, J. Imbrie, J. Hays, G. Kukla, and B. Saltzman (Eds.). D. Reidel, Dordrecht, Netherlands, pp. 3–39.

Berger, A. L., 1989. The spectral characteristics of pre-Quaternary climatic records, an example of the relationship between the astronomical theory and geo-sciences. In: *Climate and Geo-Sciences*. A. Berger, et al. (Eds.). Kluwer, Amsterdam, Netherlands, pp. 47–76.

Berger, A. L., and P. Pestiaux, 1984. Accuracy and stability of the Quaternary terrestrial insolation. In: *Milankovitch and Climate*. A. L. Berger, J. Imbrie, J. Hays, G. Kukla, and B. Saltzman (Eds.). D. Reidel, Dordrecht, Netherlands, pp. 83–111.

Berger, A. L., J. Imbrie, J. D. Hays, G. J. Kukla and B. Saltzman (Eds.). 1984. *Milankovitch and Climate*. D. Reidel, Dordrecht, Netherlands, 895 pp.

Berger, W. H., 1970. Biogenous deep-sea sediments: Fractionation by deep-sea circulation. *Deep-Sea Res.* **81**:31–43.

Berger, W. H., 1977. Deep-sea carbonate and the deglaciation preservation spike in pteropods and foraminifera. *Nature* **269**:301–304.

Berger, W. H., 1981. Paleoceanography: The deep-sea record. In: *The Sea, v. 7 The Oceanic Lithosphere*. C. Emiliani (Ed.). John Wiley and Sons, New York, pp. 1437–1519.

Berger, W. H., 1982. Deep-sea stratigraphy: Cenozoic climate steps and the search for chemo-climatic feedback. In: *Cyclic and Event Stratification*. G. Einsele and A. Seilacher (Eds.). Springer-Verlag, Berlin, pp. 121–157.

Berger, W. H., 1987. Ocean ventilation during the last 12,000 years: Hypothesis of counterpoint deep water production. *Mar. Geol.* **78**:1–10.

Berger, W. H., and R. S. Keir, 1984. Glacial-Holocene changes in atmospheric CO_2 and the deep-sea record. In: *Climate Processes and Climate Sensitivity*. J. E. Hansen and T. Takahashi (Eds.). Geophys. Mono. **29**, Am. Geophys. Union, Washington, D.C., pp. 337–351.

Berger, W. H., and E. Vincent, 1986. Sporadic shutdown of North Atlantic Deep Water production during the Glacial-Holocene transition? *Nature* **324**:53–55.

Berger, W. H., and L. D. Labeyrie (Eds.), 1987. *Abrupt Climatic Change: Evidence and Implications*. D. Reidel, Dordrecht, Netherlands.

Berger, W. H. and A. Spitzy, 1988. History of atmospheric CO_2: Constraints from the deep-sea record. *Paleoceanog.* **3**:401–411.

Berger, W. H., R. C. Finkel, J. S. Killingley, and V. Marchig, 1983. Glacial-Holocene transition in deep-sea sediments: Manganese-spike in the east-equatorial Pacific. *Nature* **303**:231–234.

Berger, W. H., J. S. Killingley, C. V. Metzler, and E. Vincent, 1985. Two-step deglaciation: ^{14}C-dated high-resolution $\delta^{18}O$ records from the tropical Atlantic Ocean. *Quat. Res.* **23**:258–271.

Berggren, W. A., and C. D. Hollister, 1974. Paleogeography, paleobiogeography, and the history of circulation in the Atlantic Ocean. In: *Studies in Paleo-Oceanography*. W. W. Hay (Ed.). Soc. Econ. Paleontol. Mineral. Spec. Pub. **20**, pp. 126–186.

Berggren, W. A., and B. Haq, 1976. The Andalusian stage (Late Miocene): Biostratigraphy, biochronology and paleoecology. *Palaeogeog., Palaeoclim., Palaeoecol.* **20**:67–129.

Bergthorsson, P., 1969. An estimate of drift ice and temperature in 1000 years. *Jökull* **19**:94–101.

Bernabo, J. C., 1981. Quantitative estimates of temperature changes over the last 2700 years in Michigan based on pollen data. *Quat. Res.* **15**:143–159.

Bernabo, J. C., and T. Webb III, 1977. Changing patterns in the Holocene pollen record of northeastern North America: A mapped summary. *Quat. Res.* **8**:64–96.

Berner, R. A., 1987. Models for carbon and sulfur cycles and atmospheric oxygen: Application to Paleozoic geologic history. *Am. J. Sci.* **287**:177–196.

Berner, R. A., 1989. Biogeochemical cycles of carbon and sulfur and their effect on atmospheric oxygen over Phanerozoic time. *Glob. Plan. Change* **1**:97–122.

Berner, R. A., and G. P. Landis, 1988. Gas bubbles

in fossil amber as possible indicators of the major gas composition of ancient air. *Science* 239:1406–1409.

Berner, R. A., and D. E. Canfield, 1989. A model for atmospheric oxygen over Phanerozoic time. *Am. J. Sci.,* 289:333–361.

Berner, R. A., A. C. Lasaga, and R. M. Garrels, 1983. The carbonate-silicate geochemical cycle and its effect on atmospheric carbon dioxide over the last 100 million years. *Am. J. Sci.* 283:641–683.

Berner, W., H. Oeschger, and B. Stauffer, 1980. Information on CO_2 cycle from ice core studies. *Radiocarbon* 22:227–235.

Berry, W. B. N., and P. Wilde, 1978. Progressive ventilation of the oceans—An explanation for the distribution of the Lower Paleozoic black shales. *Am. J. Sci.* 278:257–275.

Birchfield, G. E., J. Weertmann, and A. T. Lunde, 1981. A paleoclimate model of northern hemisphere ice sheets. *Quat. Res.* 15:126–142.

Birchfield, G. E., J. Weertmann, and A. T. Lunde, 1982. A model study of high-latitude topography in the climatic response to orbital insulation anomalies. *J. Atmos. Sci.* 39:71–87.

Birchfield, G. E., H. Wang, and M. Wyant, 1990. A bi-modal climate response controlled by water vapor transport in a coupled ocean-atmosphere box model. *Paleoceanog.* 5:383–395.

Bischoff, J. L., and R. J. Rosenbauer, 1981. Uranium series dating of human skeletal remains from the Del Mar and Sunnyvale sites, California. *Science* 213:1003–1005.

Bischoff, J. L., R. Julia, and R. Mora, 1988. Uranium-series dating of the Mousterian occupation at Abric Romani, Spain. *Nature* 332: 68–70.

Bloom, A. L., W. S. Broecker, J. M. A. Chappell, R. K. Matthews, and K. J. Mesolella, 1974. Quaternary sea-level fluctuations on a tectonic coast: New $^{230}Th/^{234}U$ dates from the Huon Peninsula, New Guinea. *Quat. Res.* 4:185–205.

Boak, J. L. and R. F. Dymek, 1982. Metamorphism of the ca. 3800 Ma supracrustal rocks at Isua, West Greenland: Implications for early Archaean crustal evolution. *Earth Plan. Sci. Lett.* 59:155–176.

Boersma, A., 1984. Campanian through Paleocene paleotemperature and carbon isotope sequence and the Cretaceous/Tertiary boundary in the Atlantic Ocean. In: *Catastrophism and Earth History, The New Uniformitarianism.* W. A. Berggren and J. A. Vancouvering (Eds.). Princeton Univ. Press, Princeton, N.J., pp. 247–278.

Boersma, A., and N. J. Shackleton, 1981. Oxygen and carbon isotope variations and planktonic foraminifer depth habitats, Late Cretaceous to Paleocene, Central Pacific, Deep Sea Drilling Project sites 463 and 465. *Init. Rep. Deep-Sea Drill. Proj.* 62:695–717.

Boersma, A., I. P. Silva, and N. J. Shackleton,

1987. Atlantic Eocene planktonic foraminiferal paleohydrographic indicators and stable isotope paleoceanography. *Paleoceanog.* 2:287–331.

Bohor, B. F., E. E. Foord, P. J. Modreski, and D. M. Triplehorn, 1984. Mineralogic evidence for an impact event at the Cretaceous-Tertiary boundary. *Science* 224:867–869.

Bohor, B. F., P. J. Modreski, and E. E. Foord, 1987. Shocked quartz in the Cretaceous-Tertiary boundary clays: Evidence for a global distribution. *Science* 236:705–709.

Bond, G. C., P. A. Nickeson, and M. A. Kominz, 1984. Breakup of a supercontinent between 625 Ma and 555 Ma: New evidence and implications for continental histories. *Earth Plan. Sci. Lett.* 70:325–345.

Bonnefille, R., 1983. Evidence for a cooler and drier climate in the Ethiopian uplands towards 2.5 Myr ago. *Nature* 303:487–491.

Bonnefille, R. and G. Riollet, 1988. The Kashiru pollen sequence (Burundi) palaeoclimatic implications for the last 40,000 yr. B.P. in tropical Africa. *Quat. Res.* 30:19–35.

Boucot, A. J., and J. Gray, 1983. A Paleozoic Pangaea. *Science* 222:571–581.

Boulton, G. S., G. D. Smith, A. S. Jones, and J. Newsome, 1985. Glacial geology and glaciology of the last mid-latitude ice sheets. *J. Geol. Soc. (London)* 142:447–474.

Bourgeois, J., T. A. Hansen, P. L. Wiberg, and E. G. Kauffman, 1988. A tsunami deposit at the Cretaceous-Tertiary boundary in Texas. *Science* 241:567–570.

Bourke, W., 1988. Spectral methods in global climate and weather prediction models. In: *Physically-Based Modelling and Simulation of Climate and Climatic Change.* M. E. Schlesinger (Ed.). Kluwer, Dordrecht, Netherlands, pp. 169–220.

Bowler, J. M., G. L. Hope, J. N. Jennings, G. Singh, and D. Walker, 1976. Late Quaternary climates of Australia and New Guinea. *Quat. Res.* 6:359–394.

Bowring, S. A., I. S. Williams, and W. Compston, 1989. 3.96 Ga gneisses from the Slave province, Northwest Territories, Canada. *Geology* 11:971–975.

Boyd, F. R., and J. J. Gurney, 1986. Diamonds and the African lithosphere. *Science* 232:472–477.

Boyle, E. A., 1983. Manganese carbonate overgrowths on foraminifera tests. *Geochim. Cosmochim. Acta* 47:1815–1819.

Boyle, E. A., 1984. Cadmium in benthic foraminifera and abyssal hydrography: Evidence for a 41 KYR obliquity cycle. In: *Climate Processes and Climate Sensitivity.* J. E. Hansen and T. Takahashi (Eds.). Geophys. Mono **29**, Am. Geophys. Union, Washington, D.C., pp. 360–368.

Boyle, E. A., 1988a. Cadmium: Chemical tracer of

deepwater paleoceanography. *Paleoceanog.* **3**:471–489.

Boyle, E. A., 1988b. The role of vertical chemical fractionation in controlling late Quaternary atmospheric carbon dioxide. *J. Geophys. Res.,* **93**:15701–15714.

Boyle, E. A., 1988c. Vertical oceanic nutrient fractionation and glacial/interglacial CO_2 cycles. *Nature* **331**:55–58.

Boyle, E. A., and L. D. Keigwin, 1982. Deep circulation of the North Atlantic over the last 200,000 years: Geochemical evidence. *Science* **218**:784–787.

Boyle, E. A., and L. D. Keigwin, 1985/1986. Comparison of Atlantic and Pacific paleochemical records for the last 215,000 years: Changes in deep ocean circulation and chemical inventories. *Earth Plan. Sci. Lett.* **76**:135–150.

Boyle, E. A., and L. D. Keigwin, 1987. North Atlantic thermohaline circulation during the past 20,000 years linked to high-latitude surface temperatures. *Nature* **330**:35–40.

Bradbury, J. P. et al., 1981. Late Quaternary climatic changes in northern South America. *Science* **214**:1299–1305.

Bradley, R. S., 1985. *Quaternary Paleoclimatology: Methods of Paleoclimatic Reconstruction.* Allen and Unwin, Boston, Mass.

Bradley, R. S., 1988. The explosive volcanic eruption signal in northern hemisphere continental temperature records. *Clim. Change* **12**:221–243.

Bradley, R. S., 1991. Pre-instrumental climate: How has climate varied during the past 500 years? In: *Greenhouse-Gas-Induced Climatic Change: A Critical Appraisal of Simulations and Observations.* M. E. Schlesinger (Ed.), Elsevier, Amsterdam, pp. 391–410.

Bradley, R. S., H. F. Diaz, P. D. Jones, and P. M. Kelly, 1987. Secular fluctuations of temperature over northern hemisphere land areas and mainland China by the mid-19th century. In: *The Climate of China and Global Climate.* D. Ye, C. Fu, J. Chao, and M. Yoshino (Eds.). 'Ocean Press/Springer-Verlag, Beijing, China, pp. 76–87.

Brakenridge, G. R., 1978. Evidence for a cold, dry full-glacial climate in the American Southwest. *Quat. Res.* **9**:22–40.

Bralower, T. J., and H. R. Thierstein, 1984. Low productivity and slow deep-water circulation in mid-Cretaceous oceans. *Geology* **12**:614–618.

Brandt, D. S., and R. J. Elias, 1989. Temporal variations in tempestite thickness may be a geologic record of atmospheric CO_2. *Geology* **17**:951–952.

Brass, G. W., J. R. Southam, and W. H. Peterson, 1982a. Warm saline bottom waters in the ancient ocean. *Nature* **296**:620–623.

Brass, G. W., et al., 1982b. Ocean circulation, plate tectonics, and climate. In: *Climate in Earth History.* Natl. Acad. Sci., Washington, D.C., pp. 83–89.

Brassell, S. C., E. Eglinton, I. T. Marlowe, U. Pflaumann, and M. Sarnthein, 1986. Molecular stratigraphy: A new tool for climatic assessment. *Nature* **320**:129–133.

Bretagnon, P., 1984. Accuracy of long term planetary theory. In *Milankovitch and Climate.* A. L. Berger et al. (Eds.) D. Reidel, Dordrecht, Netherlands, pp. 41–53.

Briskin, M., and W. A. Berggren, 1975. Pleistocene stratigraphy and quantitative paleo-oceanography of tropical North Atlantic core V16-205. In: *Late Neogene Epoch Boundaries,* Spec. Pub. **1.** T. Saito and L. H. Burckle (Eds.). Am. Mus. Nat. Hist., New York. pp. 167–198.

Briskin, M., and J. Harrell, 1980. Time-series analysis of the Pleistocene deep-sea paleoclimatic record. *Mar. Geol.* **36**:1–22.

Broccoli, A. J., and S. Manabe, 1987. The influence of continental ice, atmospheric CO_2, and land albedo on the climate of the last glacial maximum. *Clim. Dynam.* **1**:87–100.

Broecker, W. S., 1963. Radioisotopes and large-scale oceanic mixing. In: *The Sea, v. 2.* M. N. Hill (Ed.). Wiley-Interscience, New York, pp. 88–108.

Broecker, W. S., 1966. Absolute dating and the astronomical theory of glaciation. *Science* **151**:299–304.

Broecker, W. S., 1971. Calcite accumulation rates and glacial to interglacial changes in ocean mixing. In: *The Late Cenozoic Glacial Ages.* K. K. Turekian (Ed.). Yale Univ. Press, New Haven, Conn., pp. 239–265.

Broecker, W. S., 1974. *Chemical Oceanography.* Harcourt, Brace, Jovanovich, New York.

Broecker, W. S., 1975. Floating glacial ice caps in the Arctic Ocean. *Science* **188**:1116–1118.

Broecker, W. S., 1982a. Glacial to interglacial changes in ocean chemistry. *Prog. Oceanog.* **11**:151–197.

Broecker, W. S., 1982b. Ocean chemistry during glacial time. *Geochim. Cosmochim. Acta* **46**:1689–1705.

Broecker, W. S., 1984. Terminations. In: *Milankovitch and Climate.* A. L. Berger, J. Imbrie, J. Hays, G. Kukla, and B. Saltzman (Eds.). D. Reidel, Dordrecht, Netherlands, pp. 687–698.

Broecker, W. S., 1986. Oxygen isotope constraints on surface ocean temperatures. *Quat. Res.* **26**:121–134.

Broecker, W. S., 1989. The salinity contrast between the Atlantic and Pacific Oceans during glacial time. *Paleoceanog.* **4**:207–212.

Broecker, W. S., and J. van Donk, 1970. Insolation changes, ice volumes, and the ^{18}O record in deep-sea cores. *Rev. Geophys. Space Phys.* **8**:169–198.

Broecker, W. S., and T. Takahashi, 1977. Neutralization of fossil fuel CO_2 by marine calcium carbonate. In: *The Fate of Fossil Fuel CO_2 in the Oceans.* N. R. Andersen and A. Malahoff (Eds.). Plenum Press, New York, pp. 213–241.

Broecker, W. S., and T.-H. Peng, 1982. *Tracers in the Sea.* Eldigio Press, Palisades, New York.

Broecker, W. S., and T.-H. Peng, 1987. The role of $CaCO_3$ compensation in the glacial to interglacial atmospheric CO_2 change. *Global Biogeochem. Cycles* **1**:15–29.

Broecker, W. S. and G. H. Denton, 1989. The role of ocean-atmosphere reorganizations in glacial cycles. *Geochim. Cosmochim. Acta,* **53**:2465–2501.

Broecker, W. S., D. M. Peteet, and D. Rind, 1985. Does the ocean-atmosphere system have more than one stable mode of operation? *Nature* **315**:21–26.

Broecker, W. S., et al., 1988a. The chronology of the last deglaciation: Implications to the cause of the Younger Dryas event. *Paleoceanog.* **3**:1–19.

Broecker, W. S., et al., 1988b. Comparison between radiocarbon ages obtained on coexisting planktonic foraminifera. *Paleoceanog.* **3**:647–658.

Broecker, W. S., et al., 1988c. Preliminary estimates for the radiocarbon age of deep water in the glacial ocean. *Paleoceanog.* **3**:659–670.

Broecker, W. S., D. Oppo, T.-H. Peng, W. Curry, M. Andree, W. Wolfli, and G. Bonani, 1988d. Radiocarbon-based chronology for the $^{18}O/^{16}O$ record for the last deglaciation, *Paleoceanog.* **3**:509–515.

Broecker, W. S., et al., 1989. Routing of meltwater from the Laurentide Ice Sheet during the Younger Dryas cold episode. *Nature* **341**:318–321.

Brooks, C. E. P., 1949. *Climate Through the Ages.* London, Ernest Benn (republished by Dover, New York, 1970).

Brouwers, E. M., W. A. Clemens, R. A. Spicer, T. A. Ager, L. D. Carter, and W. V. Sliter, 1987. Dinosaurs on the North Slope, Alaska: High latitude, latest Cretaceous environments. *Science* **237**:1608–1610.

Brown, K. S., 1982. Paleoecology and regional patterns of evolution in neotropical forest butterflies. In: *Biological Diversification in the Tropics.* G. T. Prance (Ed.). Columbia Univ. Press, New York, pp. 255–308.

Brunner, C. A., 1983/1984. Evidence for increased volume transport of the Florida current in the Pliocene and Pleistocene. *Mar. Geol.* **54**:223–235.

Bryan, F., 1986. High-latitude salinity effects and interhemispheric thermohaline circulations. *Nature* **323**:301–304.

Bryan, F., and A. Oort, 1984. Seasonal variation of the global water balance based on aerological data. *J. Geophys. Res.* **89**:11717–11730.

Bryan, K., 1986. Poleward buoyancy transport in the ocean and mesoscale eddies. *J. Phys. Oceanog.* **16**:927–933.

Bryan, K., 1988. Efficient methods for finding the equilibrium climate of coupled ocean-atmosphere models. In: *Physically-Based Modelling and Simulation of Climate and Climatic Change.* M. E. Schlesinger (Ed.). Kluwer, Dordrecht, Netherlands, pp. 567–582.

Bryan, K., and M. J. Spelman, 1985. The ocean's response to a CO_2-induced warming. *J. Geophys. Res.* **90**:11679–11688.

Bryan, K., S. Manabe, and R. C. Pacanowski, 1975. A global ocean-atmosphere climate model. Part II. The oceanic circulation. *J. Phys. Oceanog.* **5**:30–46.

Bryson, R. A., 1989. Late Quaternary volcanic modulation of Milankovitch climate forcing. *Theor. Appl. Climatol.* **39**:115–125.

Bryson, R. A., and A. M. Swain, 1981. Holocene variation of monsoonal rainfall in Rajasthan. *Quat. Res.* **16**:135–145.

Bryson, R. A., W. M. Wendland, J. D. Ives, and J. T. Andrews, 1969. Radiocarbon isochrones on the disintegration of the Laurentide ice sheet. *Arct. Alp. Res.* **1**:1–14.

Budd, W. F., 1981. The importance of ice sheets in long-term changes of climate and sea level. *Int. Assoc. Hydrol. Sci., AISH Pub.* **131**:441–471.

Budd, W. F., and B. McInnes, 1978. Modelling surging glaciers and periodic surging of the Antarctic ice sheet. In: *Climate Change and Variability: A Southern Perspective.* A. B. Pittock, L. A. Frakes, D. Jenssen, J. A. Peterson, and J. W. Zillman (Eds.). Cambridge Univ. Press, London, pp. 228–234.

Büdel, J., 1949. Die räumliche und zeitliche Gliederung des Eiszeitklimas. *Naturwissenschaften* **36**:105–112; 133–149.

Budyko, M. I., 1968. On the origin of glacial epochs. *Meteorol. Gidrol.* **2**:3–8.

Budyko, M. I., 1969. The effect of solar radiation variations on the climate of the earth. *Tellus* **21**:611–619.

Budyko, M. I., 1978. The heat balance of the Earth. In: *Climate Change.* J. Gribbin (Ed.). Cambridge Univ. Press, London, pp. 85–113.

Budyko, M., and A. Ronov, 1979. Chemical evolution of the atmosphere in the Phanerozoic. *Geochem. Int.* **16**:1–9.

Buetz, J. H., 1923. The channeled scablands of the Columbia Plateau. *J. Geol.* **31**:617–649.

Bujak, J. P., and E. H. Davies, 1981. Neogene dinoflagellate cysts from the Hunt Dome Kopanoar M-13 well, Beaufort Sea, Canada. *Bull. Can. Petrol. Geol.* **29**:420–425.

Burckle, L. H., 1985. Diatom evidence for Neogene palaeoclimate and palaeoceanographic events in the world ocean. *S. Afr. J. Sci.* **81**:249.

Burckle, L. H., and F. Akiba, 1978. Implications

of late Neogene freshwater sediment in the Sea of Japan. *Geology* **6**:123–127.

Burckle, L. H., D. Robinson, and D. Cooke, 1982. Reappraisal of sea-ice distribution in Atlantic and Pacific sectors of the Southern Ocean at 18,000 yr BP. *Nature* **299**:435–437.

Burke, K., J. F. Dewey, and W. S. F. Kidd, 1976. Dominance of horizontal movements, arc and microcontinental collisions during the later permobile regime. In: *The Early History of the Earth*. B. F. Windley (Ed.). John Wiley and Sons, New York, pp. 113–129.

Buys, M., and M. Ghil, 1984. Mathematical methods of celestial mechanics illustrated by simple models of planetary motion. In: *Milankovitch and Climate*. A. L. Berger, J. Imbrie, J. Hays, G. Kukla, and B. Saltzman (Eds.). D. Reidel, Dordrecht, Netherlands, pp. 55–82.

Bye, B. A., F. H. Brown, T. E. Cerling, and I. McDougall, 1987. Increased age estimate for the lower Palaeolithic hominid site at Olorgesailie, Kenya. *Nature* **329**:237–239.

Byerly, G. R., D. R. Lower, and M. M. Walsh, 1986. Stromatolites from the 3,300–3,500-Myr Swaziland Supergroup, Barberton Mountain Land, South Africa. *Nature* **319**:489–491.

Cahalan, R. F., and G. R. North, 1979. A stability theorem for energy-balance climate models. *J. Atmos. Sci.* **36**:1205–1216.

Campbell, W. H., J. B. Blechman, and R. A. Bryson, 1983. Long-period tidal forcing of Indian monsoon rainfall: An hypothesis. *J. Clim. Appl. Meteorol.* **22**:287–296.

Caputo, M. V., and J. C. Crowell, 1985. Migration of glacial centers across Gondwana during the Paleozoic Era. *Geol. Soc. Am. Bull.* **96**:1020–1036.

Caratini, C., and C. Tissot, 1988. Paleogeographical evolution of the Mahakam Delta in Kalimantan, Indonesia during the Quaternary and late Pleistocene. *Rev. Palaeobot. Palynol.* **55**:217–228.

Catchpole, A. J. W., D. W. Moodie, and D. Milton, 1976. Freeze-up and break-up of estuaries on Hudson Bay in the eighteenth and nineteenth centuries. *Can. Geographer* **20**:279–297.

Causse, C., et al., 1989. Two high levels of continental waters in the southern Tunisian chotts at about 90 and 150 ka. *Geology* **17**: 922–925.

Cerling, T. E., 1989. Does the gas content of amber reveal the composition of palaeoatmospheres? *Nature* **339**:695–696.

Cess, R. D., and G. L. Potter, 1984. A commentary on the CO₂-climate controversy. *Clim. Change* **6**:365–376.

Cess, R. D., and G. L. Potter, 1988. A methodology for understanding and intercomparing atmospheric climate feedback processes in general circulation models. *J. Geophys. Res.* **93**:8305–8314.

Cess, R. D., et al., 1989. Interpretation of cloud-climate feedback as produced by 14 atmospheric general circulation models. *Science* **245**: 513–516.

Chaloner, W. G., and G. T. Creber, 1973. Growth rings in fossil woods as evidence of past climates. In: *Implications of Continental Drift to the Earth Sciences, 1.* D. H. Tarling and S. K. Runcorn (Eds.). Academic Press, New York, pp. 425–437.

Chamberlin, T. C., 1906. On a possible reversal of deep-sea circulation and its influence on geologic climates. *J. Geol.* **14**:363–373.

Chamley, H., 1979. North Atlantic clay sedimentation and paleoenvironment since the Late Jurassic. In: *Deep Drilling Results in the Atlantic Ocean: Continental Margins and Paleoenvironment*. M. Talwani, W. Hay, and W. B. F. Ryan (Eds.). Maurice Ewing Ser., 3, Am. Geophys. Union, Washington, D.C., pp. 342–361.

Chappell, J., 1983. A revised sea-level record for the last 300,000 years from Papua New Guinea. *Search* **14**:99–101.

Chappell, J., and N. J. Shackleton, 1986. Oxygen isotopes and sea level. *Nature* **324**:137–140.

Charlson, R. J., J. E. Lovelock, M. O. Andrea, and S. G. Warren, 1987. Oceanic phytoplankton, atmospheric sulphur, cloud albedo and climate. *Nature* **326**:655–661.

Christie-Blick, N., 1982. Pre-Pleistocene glaciation on earth: Implication for climatic history of Mars. *Icarus* **50**:423–443.

Chumakov, N. M., 1981. Scattered stones in Mesozoic deposits of north Siberia, U.S.S.R. In: *Earth's Pre-Pleistocene Glacial Record*. H. A. Hambrey and W. B. Harland (Eds.). Cambridge Univ. Press, Cambridge, p. 264.

Churchill, D. M., R. W. Galloway, and G. Singh, 1978. Closed lakes and the palaeoclimatic record. In: *Climatic Change and Variability, A Southern Perspective*. A. B. Pittock, L. A. Frakes, D. Jenssen, J. A. Peterson, and J. W. Zillman (Eds.). Cambridge Univ. Press, Cambridge, pp. 97–108.

Cieselski, P. F., and F. M. Weaver, 1974. Early Pliocene temperature changes in the Antarctic seas. *Geology* **2**:511–516.

Cieselski, P. F., and G. P. Grinstead, 1986. Pliocene variations in the position of the Antarctic convergence in the southwest Atlantic. *Paleoceanog.* **1**:197–232.

Cita, M. B., C. Vergnaud-Grazzini, C. Robert, H. Chamley, N. Ciaranfi, and S. d'Onofrio, 1977. Paleoclimatic record of a long deep sea core from the eastern Mediterranean. *Quat. Res.* **8**:205–235.

Clapperton, C. M., D. E. Sugden, J. Birnie, and M. J. Wilson, 1989. Late glacial and Holocene glacier fluctuations and environmental change on South Georgia, Southern Ocean. *Quat. Res.* **31**:210–228.

Clark, D. L., 1982. Origin, nature, and world climate effect of Arctic Ocean ice-cover. *Nature* **300**:321–325.

Clark, D. L., 1988. Early history of the Arctic Ocean. *Paleoceanog.* **3**:539–550.

Clark, D. L., R. R. Whitman, K. A. Morgan, and S. D. Mackey, 1980. Stratigraphy and glacial-marine sediments of the Amerasian Basin, central Arctic Ocean. *Geol. Soc. Am. Spec. Pap.* **181**.

Clark, D. L., C. W. Byers, and L. M. Pratt, 1986. Cretaceous black mud from the central Arctic Ocean. *Paleoceanog.* **1**:265–271.

Clark, G. M., 1968. Sorted patterned ground: New Appalachian localities south of the glacial border. *Science* **161**:355–356.

Clegg, S. L., and T. M. L. Wigley, 1984. Periodicities in precipitation in northeast China, 1470–1979. *Geophys. Res. Lett.* **11**:1219–1222.

CLIMAP Project Members, 1976. The surface of the ice-age earth. *Science* **191**:1131–1137.

CLIMAP Project Members, 1981. Seasonal reconstruction of the earth's surface at the last glacial maximum. *Geol. Soc. Am. Map Chart Ser.* **MC-36**.

CLIMAP Project Members, 1984. The last interglacial ocean. *Quat. Res.* **21**:123–224.

Cline, R. M., and J. D. Hays (Eds.), 1976. *Investigation of Late Quaternary Paleoceanography and Paleoclimatology.* Geol. Soc. Am. Mem. **145**, Geol. Soc. Am., Boulder, Col.

Cloud, P., and M. F. Glaessner, 1982. The Ediacarian period and system: Metazoa inherit the Earth. *Science* **217**:783–792.

Coakley, J. A., and R. D. Cess, 1985. Response of the NCAR Community Climate Model to the radiative forcing by the naturally occurring tropospheric aerosols. *J. Atmos. Sci.* **42**:1677–1692.

Cogley, J. G., and A. Henderson-Sellers, 1984. The origin and earliest state of the earth's hydrosphere. *Rev. Geophys. Space Phys.* **22**:131–175.

COHMAP Members, 1988. Climatic changes of the last 18,000 years: Observations and model simulations. *Science* **241**:1043–1052.

Colbert, E. H., 1973. Continental drift and the distribution of fossil reptiles. In: *Implications of Continental Drift to the Earth Sciences.* D. H. Tarling and S. K. Runcorn (Eds.). Academic Press, New York, pp. 395–412.

Colinvaux, P. 1972. Climate and the Galapagos Islands. *Nature* **240**:17–20.

Colinvaux, P. A., 1989. Ice-age Amazon revisited. *Nature* **340**:188–190.

Colinvaux, P. A., M. Frost, I. Frost, K.-B. Liu, and M. Steinitz-Kannan, 1988. Three pollen diagrams of forest disturbance in the western Amazon Basin. *Rev. Palaeobot. Palynol.* **55**:73–83.

Compston, W., and R. T. Pidgeon, 1986. Jack Hills, evidence of more very old detrital zircons in western Australia. *Nature* **321**:766–769.

Condie, K. C., 1981. *Archaean Greenstone Belts.* Elsevier, Amsterdam.

Condie, K. C., 1989. Origin of the earth's crust. *Glob. Plan. Change.* **1**:57–82.

Connell, J. H., 1978. Diversity in tropical rainforests and coral reefs. *Science* **199**:1302–1310.

Cook, E. R., and G. C. Jacoby, 1979. Evidence for quasi-periodic July drought in the Hudson Valley, New York. *Nature* **282**:390–392.

Cook, K. C., and I. M. Held, 1988. Stationary waves of the ice age climate. *J. Clim.* **1**:807–819.

Cooke, D. W., and J. D. Hays, 1982. Estimates of Antarctic Ocean seasonal sea-ice cover during glacial intervals. In: *Antarctic Geoscience.* C. Craddock et al. (Eds.). Univ. Wisconsin Press, Madison, pp. 1017–1025.

Coope, G. R., 1975. Climatic fluctuations in northwest Europe since the last interglacial, indicated by fossil assemblages of Coleoptera. In: *Ice Ages Ancient and Modern.* A. E. Wright and F. Moseley (Eds.). Geol. J. Spec. Issue No. **6**, Liverpool Univ. Press, Liverpool, pp. 153–168.

Corliss, B. H., et al., 1984. The Eocene/Oligocene boundary event in the deep sea. *Science* **226**:806–810.

Corliss, B. H., D. G. Martinson, and T. Keffer, 1986. Late Quaternary deep-ocean circulation. *Geol. Soc. Am. Bull.* **97**:1106–1121.

Costin, A. B., 1972. Carbon-14 dates from the Snowy Mountains area southeastern Australia and their interpretation. *Quat. Res.* **2**:579–590.

Courtillot, V., et al., 1988. Deccan flood basalts and the Cretaceous/Tertiary boundary. *Nature* **333**:843–846.

Covey, C., and P. L. Haagenson, 1984. A model of oxygen isotope composition of precipitation: Implications for paleoclimate data. *J. Geophys. Res.* **89**:4647–4655.

Covey, C., and E. Barron, 1988. The role of ocean heat transport in climatic change. *Earth Sci. Rev.* **24**:429–445.

Covey, C., and S. L. Thompson, 1989. Testing the effects of ocean heat transport on climate. *Glob. Plan. Change* **1**:331–341.

Covey, C., S. J. Ghan, J. J. Walton, and P. R. Weissman, 1990. Global environmental effects of impact-generated aerosols: Results from a general circulation model. In: *Global Catastrophes in Earth History.* V. L. Sharpton and P. Ward (Eds.). Geol. Soc. Am. Spec. Paper **247**, pp. 263–270.

Cox, M. D., 1985. An eddy resolving model of the ventilated thermocline. *J. Phys. Oceanog.* **15**:1312–1324.

Crafoord, C., and E. Källén, 1978. A note on the condition for existence of more than one steady state solution in Budyko-Sellers type models. *J. Atmos. Sci.* **35**:1123–1125.

Craig, H., and C. C. Chou. 1982. Methane: The record in polar ice cores. *Geophys. Res. Lett.* **9**:477–481.

Creber, G. T., and W. G. Chaloner, 1985. Tree growth in the Mesozoic and early Tertiary and the reconstruction of paleoclimates. *Palaeogeog., Palaeoclimatol., Palaeoecol.* **52**:35–60.

Croll, J., 1864. On the physical cause of the change of climate during geological epochs. *Phil. Mag.* **28**:121–137.

Croll, J., 1867a. On the eccentricity of the earth's orbit, and its physical relations to the glacial epoch. *Phil. Mag.* **33**:119–131.

Croll, J., 1867b. On the change in the obliquity of the ecliptic, its influence on the climate of the polar regions and on the level of the sea. *Phil. Mag.* **33**:426–455.

Cronin, T. M., 1985. Speciation and stasis in marine ostracoda: Climatic modulation of evolution. *Science* **227**:60–63.

Crowell, J. C., 1982. Continental glaciation through geologic times. In: *Climate in Earth History.* W. H. Berger and J. C. Crowell (Eds.). Natl. Acad. Press, Washington, D.C., pp. 77–82.

Crowell, J. C., 1983. Ice ages recorded on Gondwanan continents. *Trans. Geol. Soc. S. Afr.* **86**:237–262.

Crowley, T. J., 1981. Temperature and circulation changes in the eastern North Atlantic during the last 150,000 years: Evidence from the planktonic foraminiferal record. *Mar. Micropaleontol.* **6**:97–129.

Crowley, T. J., 1983a. Calcium-carbonate preservation patterns in the central North Atlantic during the last 150,000 years. *Mar. Geol.* **51**:1–14.

Crowley, T. J., 1983b. The geologic record of climatic change. *Rev. Geophys. Space Phys.* **21**:828–877.

Crowley, T. J., 1984. Atmospheric circulation patterns during glacial inception: A possible candidate. *Quat. Res.* **21**:105–110.

Crowley, T. J., 1985. Late Quaternary carbonate changes in the North Atlantic and Atlantic/Pacific comparisons. In: *The Carbon Cycle and Atmospheric CO₂: Natural Variations Archean to Present.* E. T. Sundquist and W. S. Broecker (Eds.). Geophys. Mono. **32**, Am. Geophys. Union, Washington, D.C., pp. 271–284.

Crowley, T. J., 1988. Paleoclimate modelling. In: *Physically-Based Modelling and Simulation of Climate and Climatic Change.* M. E. Schlesinger (Ed.). Kluwer, Amsterdam, pp. 883–949.

Crowley, T. J., 1989. Paleoclimate perspectives on a greenhouse warming. In: *Climate and Geosciences.* A. Berger, S. H. Schneider, and J.-C. Duplessy (Eds.). Kluwer, Dordrecht, Netherlands, pp. 179–207.

Crowley, T. J., 1990. Are there any satisfactory geologic analogs for a future greenhouse warming? *J. Clim.* **3**:1282–1292.

Crowley, T. J., and S. Häkkinen, 1988. A new mechanism for decreasing North Atlantic Deep Water production rates during the Pleistocene. *Paleoceanog.* **3**:249–258.

Crowley, T. J., and G. R. North, 1988. Abrupt climate change and extinction events in earth history. *Science* **240**:996–1002.

Crowley, T. J., and C. L. Parkinson, 1988a. Late Pleistocene variations in Antarctic sea ice I: Effect of orbital insolation changes. *Clim. Dynam.* **3**:85–91.

Crowley, T. J., and C. L. Parkinson, 1988b. Late Pleistocene variations in Antarctic sea ice II: Effect of interhemispheric deep ocean heat exchange. *Clim. Dynam.* **3**:93–105.

Crowley, T. J., D. A. Short, J. G. Mengel, and G. R. North, 1986. Role of seasonality in the evolution of climate over the last 100 million years. *Science* **231**:579–584.

Crowley, T. J., J. G. Mengel, and D. A. Short, 1987. Gondwanaland's seasonal cycle. *Nature* **329**:803–807.

Crowley, T. J., W. T. Hyde, and D. A. Short, 1989. Seasonal cycle variations on the supercontinent of Pangaea. *Geology* **17**:457–460.

Crutcher, H. L., and J. M. Meserve, 1970. Selected level heights, temperatures, and dew points for the northern hemisphere. *Rep NAV AIR 50-1C-52* (revised). Chief of Naval Operations, Washington, D.C.

Cullen, J. L., 1981. Microfossil evidence for changing salinity patterns in the Bay of Bengal over the last 20,000 years. *Palaeogeog., Palaeoclimatol., Palaeoecol.* **35**:315–356.

Currie, R. G., 1984. Evidence for 18.6 year lunar nodal drought in western North America during the past millenniun. *J. Geophys. Res.* **89**:1295–1308.

Curry, R. R. 1966. Glaciation about 3,000,000 years ago in the Sierra Nevada. *Science* **154**:770–771.

Curry, W. B., and G. P. Lohmann, 1982. Carbon isotopic changes in benthic foraminifera from the western South Atlantic: Reconstruction of glacial abyssal circulation patterns. *Quat. Res.* **18**:218–235.

Curry, W. B., and G. P. Lohmann, 1983. Reduced advection into Atlantic Ocean deep eastern basins during last glaciation maximum. *Nature* **306**:577–580.

Curry, W. B., and G. P. Lohmann, 1986. Late Quaternary carbonate sedimentation at the Sierra Leone Rise (eastern equatorial Atlantic Ocean). *Mar. Geol.* **70**:223–250.

Curry, W. B., and T. J. Crowley, 1987. The δ¹³C of equatorial Atlantic surface waters: Implications for ice-age pCO₂ levels. *Paleoceanog.* **2**:489–517.

Curry, W. B., J.-C. Duplessy, L. D. Labeyrie, and N. J. Shackleton, 1988. Changes in the distribution of δ¹³C of deep water ΣCO₂ between the last glaciation and the Holocene. *Paleoceanog.* **3**:317–341.

Damon, P. E., 1968. The relationship bewteen terrestrial factors and climate. *Met. Mono.* **8**:106–111.

Damon, P. E., J. C. Lerman, and A. Long, 1978.

Temporal fluctuations of atmospheric C-14: Causal factors and implications. *Ann. Rev. Earth Plan. Sci.* **6**:457–494.

Damuth, J. E., and R. W. Fairbridge, 1970. Equatorial Atlantic deep-sea arkosic sands and ice-age aridity in tropical South America. *Geol. Soc. Am. Bull.* **81**:189–206.

Dansgaard, W., and H. Tauber, 1969. Glacial oxygen-18 content and Pleistocene ocean temperatures. *Science* **166**:499–502.

Dansgaard, W., S. J. Johnsen, H. B. Clausen, and C. C. Langway, 1971. Climatic record revealed by the Camp Century ice core. In: *The Late Cenozoic Glacial Ages.* K. K. Turekian (Ed.). Yale Univ. Press, New Haven, Conn., pp. 37–56.

Dansgaard, W., S. J. Johnsen, N. Reeh, N. Gundestrup, H. B. Clausen, and C. U. Hammer, 1975. Climate changes, Norsemen, and modern man. *Nature* **255**:24–28.

Dansgaard, W., et al., 1984. North Atlantic climatic oscillations revealed by deep Greenland ice cores. In: *Climate Processes and Climate Sensitivity.* J. E. Hansen and T. Takahashi (Eds.). Geophys. Mono **29**, Am. Geophys. Union, Washington, D.C., pp. 288–298.

Dansgaard, W., J. W. C. White, and S. J. Johnsen, 1989. The abrupt termination of the Younger Dryas climate event. *Nature* **339**:532–534.

Davis, D. D., and S. C. Solomon, 1981. Variations in the velocities of the major plates since the Late Cretaceous. *Tectonophys.* **74**: 189–208.

Davis, M. B., R. W. Spear, and L. C. K. Shane, 1980. Holocene climate of New England. *Quat. Res.* **14**:240–250.

Davis, R., J. C. Evans, and B. T. Cleveland, 1978. The solar neutrino problem. In: *Long-Distance Neutrino Detection—1978.* Am. Inst. Phys. Conf. Proc. No. **52**. A. W. Saenz and H. Uberall (Eds.). Am. Inst. Phys., New York, pp. 17–27.

Dawson, M. R., R. M. West, W. Langston, and J. H. Hutchinson, 1976. Paleogene terrestrial vertebrates: Northernmost occurrence, Ellesmere Island, Canada. *Science* **192**:781–782.

Dean, W. E., M. A. Arthur, and G. E. Claypool, 1986. Depletion of ^{13}C in Cretaceous marine organic matter: Source, diagenetic, or environmental signal? *Mar. Geol.* **70**:119–157.

Dean, W. E., J. V. Gardner, and E. Hemphill-Haley, 1989. Changes in redox conditions in deep-sea sediments of the sub-Arctic north Pacific Ocean: Possible evidence for the presence of North Pacific Deep Water. *Paleoceanog.* **4**:639–653.

DeAngelis, M., N. I. Barkov, and V. N. Petrov, 1987. Aerosol concentrations over the last climatic cycle (160 kyr) from an Antarctic ice core. *Nature* **325**:318–321.

deBoer, P. L., and A. A. H. Winders, 1984. Astronomically induced rhythmic bedding in Cretaceous pelagic sediments near Moria (Italy). In: *Milankovitch and Climate.* A. L. Berger, J. Im-

brie, J. Hays, G. Kukla, and B. Saltzman (Eds.). D. Reidel, Dordrecht, Netherlands, pp. 177–190.

Defant, A., 1961. *Physical Oceanography,* v. **1.** Pergamon, New York.

Degens, E. T., 1969. Biochemistry of stable carbon isotopes. In: *Organic Geochemistry Methods and Results.* G. Eglinton and M. T. J. Murphy (Eds.). Springer-Verlag, New York, pp. 304–356.

Degens, E. T., R. R. L. Guillard, W. M. Sackett, and J. A. Hellebust, 1968. Metabolic fractionation of carbon isotopes in marine plankton—I. Temperature and respiration experiments. *Deep-Sea Res.* **15**:1–19.

Delaney, M. L., and E. A. Boyle, 1988. Tertiary paleoceanic chemical variability: Unintended consequences of simple geochemical models. *Paleoceanog.* **3**:137–156.

Delcourt, H. R., 1979. Late-Quaternary vegetation history of the eastern Highland Rim and adjacent Cumberland Plateau of Tennessee. *Ecol. Mono.* **49**:255–280.

Delcourt, H. R., and P. A. Delcourt, 1985. Quaternary palynology and vegetational history of the southeastern United States. In: *Pollen Records of Late-Quaternary North American Sediments.* V. M. Bryant and R. G. Holloway (Eds.). Am. Assoc. Strat. Palynol. Found., pp. 1–37.

Delcourt, P. A., and H. R. Delcourt, 1983. Late-Quaternary vegetational dynamics and community stability reconsidered. *Quat. Res.* **19**:265–271.

Delmas, R. J., J. M. Ascencio, and M. Legrand, 1980. Polar ice evidence that atmospheric CO_2 20,000 yr. BP was 50% of present. *Nature* **284**:155–157.

Delworth, T. L., and S. Manabe, 1988. The influence of potential evaporation on the variabilities of simulated soil wetness and climate. *J. Clim.* **1**:523–547.

Denton, G. H., 1985. Did the Antarctic Ice Sheet influence late Cainozoic climate and evolution in the Southern Hemisphere? *S. Afr. J. Sci.* **81**:224–229.

Denton, G. H., and W. Karlen, 1973. Holocene climatic variations—Their pattern and possible cause. *Quat. Res.* **3**:155–205.

Denton, G. H., and T. J. Hughes, 1981a. The Arctic ice sheet: An outrageous hypothesis. In: *The Last Great Ice Sheets.* G. H. Denton and T. J. Hughes (Eds.). John Wiley and Sons, New York, pp. 440–467.

Denton, G. H., and T. J. Hughes (Eds.), 1981b. *The Last Great Ice Sheets.* John Wiley and Sons, New York.

Denton, G. H., R. L. Armstrong, and M. Stuiver, 1971. The late Cenozoic glacial history of Antarctica. In: *The Late Cenozoic Glacial Ages.* K. K. Turekian (Ed.). Yale Univ. Press, New Haven, Conn., pp. 267–306.

Denton, G. H., et al., 1984. Late Tertiary history

of the Antarctic ice sheet: Evidence from the Dry Valleys. *Geology* **12**:263–267.

Denton, G. H., T. J. Hughes, and W. Karlén, 1986. Global ice-sheet system interlocked by sea level. *Quat. Res.* **26**:3–26.

Denton, G. H., J. G. Bockheim, S. C. Wilson, and M. Stuiver, 1989. Late Wisconsin and early Holocene glacial history, inner Ross Embayment, Antarctica. *Quat. Res.* **31**:151–182.

deVernal, A., C. Causse, C. Hillaire-Marcel, R. J. Mott, and S. Occhietti, 1986. Palynostratigraphy and Th/U ages of upper Pleistocene interglacial and interstadial deposits on Cape Breton Island, eastern Canada. *Geology* **14**:554–557.

Dewey, J. F., and B. F. Windley, 1981. Growth and differentiation of the continental crust. In: *The Origin and Evolution of the Earth's Continental Crust.* S. Moorbath and B. F. Windley (Eds.). Phil. Trans. Roy. Soc. London, Ser. **A. 301**:189–206.

Dickinson, R. E., 1984. Modeling evapotranspiration for three-dimensional global climate models. In: *Climate Processes and Climate Sensitivity.* J. E. Hansen and T. Takahashi (Eds.). Geophys. Mono **29**, Am. Geophys. Union, Washington, D.C., pp. 58–72.

Dickinson, R. E., and R. J. Cicerone, 1986. Future global warming from atmospheric trace gases. *Nature* **319**:109–115.

Dickinson, R. E., and A. Henderson-Sellers, 1988. Modelling tropical deforestation: A study of GCM-land parameterizations. *Quart. J. Roy. Meteorol. Soc.* **114**:439–462.

Dodge, R. E., R. G. Fairbanks, L. K. Benninger, and F. Maurrasse, 1983. Pleistocene sea levels from raised coral reefs of Haiti. *Science* **219**:1423–1425.

Dodson, J. R., and R. V. S. Wright, 1989. Humid to arid to subhumid vegetation shift on Pilliga Sandstone, Ulungra Springs, New South Wales. *Quat. Res.* **32**:182–192.

Donahue, T. M., J. H. Hoffman, R. R. Hodges, Jr., and A. J. Watson, 1982. Venus was wet: A measurement of the ratio of deuterium to hydrogen. *Science* **216**:630–633.

Donn, W. L., and D. M. Shaw, 1977. Model of climate evolution based on continental drift and polar wandering. *Geol. Soc. Am. Bull.* **88**:390–396.

Dott, R. H., and R. L. Batten, 1981. *Evolution of the Earth,* 3rd ed. McGraw-Hill, New York.

Douglas, R. G., and S. M. Savin, 1975. Oxygen and carbon isotope analyses of Tertiary and Cretaceous microfossils from Shatsky Rise and other sites in the North Pacific Ocean. In: *Init. Rep. Deep-Sea Drill. Proj.* **32**. R. L. Larson, R. Moberly et al. (Eds.). U.S. Gov. Print. Office, Washington, D.C., pp. 509–520.

Douglas, R. G., and F. Woodruff, 1981. Deep sea benthic foraminifera. In: *The Sea,* v. 7. C. Emiliani (Ed.). Wiley-Interscience, New York, pp. 1233–1327.

Dowsett, H. J., and R. Z. Poore, 1990. A new planktic foraminifer transfer function for estimating Pliocene-Holocene paleoceanographic conditions in the North Atlantic. *Mar. Micropaleontol.* **16**:1–23.

Dreimanis, A., and R. P. Goldthwait, 1973. Wisconsin glaciation in the Huron, Erie, and Ontario lobes. In: *The Wisconsinan Stage.* R. F. Black et al. (Eds.). Geol. Soc. Am. Mem. **136**, Geol. Soc. Am., Boulder, Col., pp. 71–106.

Dreimanis, A., and A. Raukas, 1975. Did middle Wisconsin, middle Weichselian, and their equivalents represent an interglacial or an interstadial complex in the Northern Hemisphere? In: *Quaternary Studies.* R. P. Suggate and M. M. Cresswell (Eds.). Roy. Soc. New Zealand, Wellington, pp. 109–120.

Drewry, G. E., A. T. S. Ramsay, and A. G. Smith, 1974. Climatically controlled sediments, the geomagnetic field and trade wind belts in Phanerozoic time. *J. Geol.* **82**:531–553.

Druffel, E. M., 1982. Banded corals: Changes in ocean Carbon-14 during the Little Ice Age. *Science* **218**:13–19.

Duncan, R. A., and D. G. Pyle, 1988. Rapid eruption of the Deccan flood basalts at the Cretaceous/Tertiary boundary. *Nature* **333**:841–843.

Duplessy, J.-C., 1978. Isotopes studies. In: *Climatic Change.* J. Gribbin (Ed.). Cambridge Univ. Press, New York, pp. 46–67.

Duplessy, J.-C., and N. J. Shackleton, 1985. Response of global deep-water circulation to earth's climatic change 135,000–107,000 years ago. *Nature* **316**:500–507.

Duplessy, J.-C., J. Moyes, and C. Pujol, 1980. Deep water formation in the North Atlantic Ocean during the last ice age. *Nature* **286**:479–482.

Duplessy, J.-C, et al., 1984. [13]C record of benthic foraminifera in the last interglacial ocean: Implications for the carbon cycle and the global deep water circulation. *Quat. Res.* **21**:225–243.

Duplessy, J.-C., et al., 1988. Deepwater source variations during the last climatic cycle and their impact on the global deepwater circulation. *Paleoceanog.* **3**:343–360.

Du Toit, A. L., 1937. *Our Wandering Continents.* Oliver & Boyd, Edinburgh.

Dymond, J., and M. Lyle, 1985. Flux comparisons between sediments and sediment traps in the eastern tropical Pacific: Implications for atmospheric CO_2 variations during the Pleistocene. *Limnol. Oceanog.* **30**:699–711.

Dyson, I. A., 1985. Frond-like fossils from the base of the late Precambrian Wilpena Group, South Australia. *Nature* **318**:283–285.

Eddy, J. A., 1976. The Maunder Minimum. *Science* **192**:1189–1202.

Edwards, R. L., J. H. Chen, T.-L. Ku, and G. J. Wasserburg, 1987. Precise timing of the last interglacial period from mass spectrometric determination of Thorium-230 in corals. *Science* **236**:1547–1553.

Eicher, D. L., and A. L. McAlester, 1980. *History of the Earth.* Prentice-Hall, Englewood Cliffs, N.J.

Embleton, B. J. J., and G. E. Williams, 1986. Low palaeolatitude of deposition for late Precambrian periglacial varvites in South Australia: Implications for palaeoclimatology. *Earth Plan. Sci. Lett.* **79**:419–430.

Embley, R. W., and J. J. Morley, 1980. Quaternary sedimentation and paleoenvironmental studies off Namibia (south-west Africa). *Mar. Geol.* **36**:183–204.

Emiliani, C., 1955. Pleistocene temperatures. *J. Geol.* **63**:538–578.

Emiliani, C., 1961. The temperature decrease of surface sea-water in high latitudes and of abyssal-hadal water in open oceanic basins during the past 75 million years. *Deep-Sea Res.* **8**:144–147.

Emiliani, C., 1966. Paleotemperature analysis of Caribbean cores P-6304-8 and P-6304-9 and a generalized temperature curve for the past 425,000 years. *J. Geol.* **74**:109–124.

Emiliani, C., 1972. Quaternary hypsithermals. *Quat. Res.* **2**:270–273.

Emiliani, C., C. Rooth, and J. J. Stipp, 1978. The late Wisconsin flood in the Gulf of Mexico. *Earth Plan. Sci. Lett.* **41**:159–162.

Emiliani, C., E. B. Kraus, and E. M. Shoemaker, 1981. Sudden death at the end of the Mesozoic. *Earth Plan. Sci. Lett.* **55**:317–334.

Endal, A. S., 1981. Evolutionary variations of solar luminosity. In: *Variations of the Solar Constant.* NASA Conf. Publ. **2191**:175–183.

Endal, A. S., and S. Sofia, 1981. Rotation in solar-type stars, I, Evolutionary models for the spindown of the sun. *Astrophys. Jour.* **243**:625–640.

Ennever, F. K., and M. B. McElroy, 1985. Changes in atmospheric CO_2: Factors regulating the glacial to interglacial transition. In: *The Carbon Cycle and Atmospheric CO_2: Natural Variations Archean to Present.* E. T. Sundquist and W. S. Broecker (Eds.). Geophys. Mono. **32**, Am. Geophys. Union, Washington, D.C., pp. 154–162.

Epshteyn, O. G., 1981. Late Permiam ice-marine deposits of the Atkan Formation in the Kolyma River Headwaters Region, U.S.S.R. In: *Earth's Pre-Pleistocene Glacial Record.* J. J. Hambrey and W. M. Harland (Eds.). Cambridge Univ. Press, Cambridge, pp. 270–273.

Epstein, S., R. Buchsbaum, H. A. Lowenstam, and H. C. Urey, 1953. Revised carbonate-water isotopic temperature scale. *Geol. Soc. Am. Bull.* **64**:1315–1326.

Erickson, D. J., and S. M. Dickson, 1987. Global trace-element biogeochemistry at the K/T boundary: Oceanic and biotic response to a hypothetical meteorite impact. *Geology* **15**:1014–1017.

Estes, R., and J. H. Hutchinson, 1980. Eocene lower vertebrates from Ellesmere Island, Canadian Arctic Archipelago. *Palaeogeog., Palaeoclimatol., Palaeoecol.* **30**:325–347.

Fabre, J., and N. Petit-Maire, 1988. Holocene climatic evolution of 22–23°N from two palaeolakes in the Taoudenni area (northern Mali). *Palaeogeog., Palaeoclimatol, Palaeoecol.* **65**:133–148.

Fairbanks, R. G., 1989. A 17,000-year glacio-eustatic sea level record: Influence of glacial melting rates on Younger Dryas event and deep-ocean circulation. *Nature* **342**:637–642.

Fairbanks, R. G., and R. K. Matthews, 1978. The marine oxygen isotype record in Pleistocene coral, Barbados, West Indies. *Quat. Res.* **10**:181–196.

Fairbridge, R. W., 1970. An ice-age in the Sahara. *Geotimes* **15**:18–20.

Fairbridge, R. W., 1979. Traces from the desert: Ordovician. In: *The Winters of the World.* B. S. John (Ed.). John Wiley and Sons, New York, pp. 131–153.

Feynman, J., and P. F. Fougere, 1984. Eighty-eight year periodicity in solar-terrestrial phenomena confirmed. *J. Geophys. Res.* **89**:3023–3027.

Fischer, A. G., 1982. Long-term climatic oscillations recorded in stratigraphy. In: *Climate in Earth History.* Natl. Acad. Press, Washington, D.C., pp. 97–104.

Fischer, A. G., and W. Schwarzacher, 1984. Cretaceous bedding rhythms under orbital control. In: *Milankovitch and Climate.* A. L. Berger, J. Imbrie, J. Hays, G. Kukla, and B. Saltzman (Eds.). D. Reidel, Dordrecht, Netherlands, pp. 163–175.

Flenley, J. R., 1979a. *The Equatorial Rain Forest, A Geological History.* Butterworths, London.

Flenley, J. R., 1979b. The Quaternary vegetational history of the equatorial mountains. *Prog. Phys. Geog.* **3**:488–509.

Flint, R. F., 1971. *Glacial and Quaternary Geology.* John Wiley and Sons, New York.

Flohn, H., 1968. Contributions to a meterology of the Tibetan Highlands. *Atmos. Sci. Pap.* **130**, Dept. Atmos. Sci., Col. State Univ., Fort Collins.

Flohn, H., 1979. On time scales and causes of abrupt paleoclimatic events. *Quat. Res.* **12**:135–149.

Flohn, H., 1986. Singular events and catastrophes now and in climatic history. *Naturwissenschaften* **73**:136–149.

Florschutz, F., J. Menendez-Amor, and T. A. Wijmstra, 1971. Palynology of a thick Quaternary succession in southern Spain. *Palaeogeog., Palaeoclimatol, Palaeoecol.* **10**:233–264.

Folland, C. K., T. N. Palmer, and D. E. Parker, 1986. Sahel rainfall and worldwide sea temperatures, 1901–85. *Nature* **320**:602–607.

Foster, T. D., and E. C. Carmack, 1976. Frontal zone mixing and Antarctic Bottom Water formation in the southern Weddell Sea. *Deep-Sea Res.* **23**:301–317.

Frakes, L. A., 1979. *Climates Throughout Geologic Time.* Elsevier, Amsterdam.

Frakes, L. A., and E. M. Kemp, 1972. Influence of continental positions on Early Tertiary climates. *Nature* **240**:97–100.

Frakes, L. A., and E. M. Kemp, 1973. Palaeogene continental positions and evolution of climate. In: *Implications of Continental Drift to the Earth Sciences.* D. H. Tarling and S. K. Runcorn (Eds.). Academic Press, London, pp. 539–559.

Frakes, L. A., and J. E. Francis, 1988. A guide to Phanerozoic cold polar climates from high-latitude ice-rafting in the Cretaceous. *Nature* **333**:547–549.

Francis, J. E., 1984. The seasonal environment of the Purbeck (Upper Jurassic) fossil forests. *Palaeogeog., Palaeoclimatol., Palaeoecol.* **48**:285–307.

Frankignoul, C., and K. Hasselmann, 1977. Stochastic climate models, II. Application to sea-surface temperature anomalies and thermocline variability. *Tellus* **29**:289–305.

Frenzel, B., and W. Bludau, 1987. On the duration of the interglacial to glacial transition at the end of the Eemian Interglacial (deep-sea Stage 5e): Botanical and sedimentological evidence. In: *Abrupt Climate Change: Evidence and Implications.* W. H. Berger and L. D. Labeyrie (Eds.). D. Reidel, Dordrecht, Netherlands, pp. 151–162.

Friedli, H., H. Lötscher, H. Oeschger, U. Siegenthaler, and B. Stauffer, 1986. Ice core record of the $^{13}C/^{12}C$ ratio of atmospheric CO_2 in the past two centuries. *Nature* **324**:237–241.

Fritts, H. C., G. R. Lofgren, and G. A. Gordon, 1979. Variations in climate since 1602 as reconstructed from tree rings. *Quat. Res.* **12**:18–46.

Funder, S., N. Abrahamsen, O. Bennike, and R. W. Feyling-Hanssen, 1985. Forested Arctic: Evidence from north Greenland. *Geology* **13**:542–546.

Gaffin, S. R., M. I. Hoffert, and T. Volk, 1986. Nonlinear coupling between surface temperature and ocean upwelling as an agent in historical climate variations. *J. Geophys. Res.* **91**:3944–3950.

Galloway, R. W., 1965. Late Quaternary climates of Australia. *J. Geol.* **73**:603–618.

Garrels, R. M., and F. T. MacKenzie, 1972. A quantitative model for the sedimentary rock cycle. *Mar. Chem.* **1**:27–41.

Garrels, R. M., and E. A. Perry, 1974. Cycling of carbon, sulphur, and oxygen through geologic time. In: *The Sea* v. 5. E. D. Goldberg (Ed.). John Wiley and Sons, New York, pp. 303–336.

Gascoyne, M., A. P. Currant, and T. C. Lord, 1981. Ipswichian fauna of Victoria Cave and the marine palaeoclimatic record. *Nature* **294**:652–654.

Gasse, F., B. Stabell, E. Fourtanier, and Y. van Iperen, 1989a. Freshwater diatom influx in intertropical Atlantic: Relationships with continental records from Africa. *Quat. Res.* **32**:229–243.

Gasse, F., V. Lédém, M. Massault, and J.-C. Fontes, 1989b. Water-level fluctuations of Lake Tanganyika in phase with oceanic changes during the last glaciation and deglaciation. *Nature* **342**:57–59.

Gates, W. L., 1976a. Modeling the ice-age climate. *Science* **191**:1138–1144.

Gates, W. L., 1976b. The numerical simulation of ice-age climate with a global general circulation model. *J. Atmos. Sci.* **33**:1844–1873.

Gaur, R., and S. R. K. Chopra, 1984. Taphonomy, fauna, environment and ecology of Upper Siwaliks (Plio-Pleistocene) near Chandigarh, Ir.dia. *Nature* **308**:353–355.

Gerstel, J., R. Thunell, and R. Ehrlich, 1987. Danian faunal succession: Planktonic foraminiferal response to a changing marine environment. *Geology* **15**:665–668.

Ghil, M., 1981. Internal climatic mechanisms participating in glaciation cycles. In: *Climatic Variations and Variability: Facts and Theories.* A. Berger (Ed.). D. Reidel, Dordrecht, Netherlands, pp. 539–557.

Ghil, M., and H. LeTreut, 1981. A climate model with cryodynamics and geodynamics. *J. Geophys. Res.* **86**:5262–5270.

Gillespie, R., F. A. Street-Perrott, and R. Switsur, 1983. Post-glacial arid episodes in Ethiopia have implications for climate prediction. *Nature* **306**:680–683.

Gilliland, R. L., 1981. Solar radius variations over the past 265 years. *Astrophys. J.* **248**:1144–1155.

Gilliland, R. L., 1982. Solar, volcanic, and CO_2 forcing of recent climatic changes. *Clim. Change* **4**:111–131.

Gilliland, R. L., 1989. Solar evolution. *Glob. Plan. Change* **1**:35–56.

Gilmore, J. S., J. D. Knight, C. J. Orth, C. L. Pillmore, and R. H. Tschudy, 1984. Trace element patterns at a non-marine Cretaceous-Tertiary boundary. *Nature* **307**:224–228.

Girard, C., and M. Jarraud, 1982. Short and medium range forecast differences between a spectral and a grid-point model. An extensive quasi-operational comparison. *European Center for Medium-Range Weather Forecasting, Tech. Rep.* 32.

Glancy, T. J., E. J. Barron, and M. A. Arthur, 1986. An initial study of the sensitivity of modeled Cretaceous climate to cyclical insolation forcing. *Paleoceanog.* **1**:523–537.

Gleissberg, W., 1966. Ascent and descent in the eighty-year cycle of solar activity. *J. Br. Astron. Assoc.* **76**:265–268.

Goodwin, A. M., 1976. Giant impacting and the development of continental crust. In: *The Early History of the Earth*. B. F. Windley (Ed.). John Wiley and Sons, New York, pp. 77–95.

Goodwin, A. M., 1981. Precambrian perspectives. *Science* **213**:55–61.

Goody, R. M., 1964. *Atmospheric Radiation I: Theoretical Basis*. Oxford University Press (Clarendon Press), London.

Goody, R. M. and J. C. G. Walker, 1972. *Atmospheres*. Prentice-Hall, Englewood Cliffs, N.J.

Gordon, A. L., 1981. Seasonality of Southern Ocean sea ice. *J. Geophys. Res.* **86**:4193–4197.

Gordon, W. A., 1973. Marine life and ocean surface currents in the Cretaceous. *J. Geol.* **81**:269–284.

Gordon, W. A., 1975. Distribution by latitude of Phanerozoic evaporite deposits. *J. Geol.* **83**:671–684.

Gradstein, F. M., and S. P. Srivastava, 1980. Aspects of Cenozoic stratigraphy and paleoceanography of the Labrador Sea and Baffin Bay. *Palaeogeog., Palaeoclimatol., Palaeoecol.* **30**:261–295.

Graham, A., and D. J. Jarzen, 1969. Studies in Neotropical paleobotany. I. The Oligocene communities of Puerto Rico. *Ann. Missouri Bot. Gard.* **56**:308–357.

Graham, N. E., and T. P. Barnett, 1987. Sea surface temperature, surface wind divergence, and convection over tropical oceans. *Science* **238**:657–659.

Gribbin, J., and H. H. Lamb, 1978. Climatic change in historical times. In: *Climatic Change*. J. Gribbin (Ed.). Cambridge Univ. Press, Cambridge, pp. 68–82.

Grigg, R. W., 1988. Paleoceanography of coral reefs in the Hawaiian-Emperor Chain. *Science* **240**:1737–1743.

Grootes, P. M., M. Stuiver, L. G. Thompson, and E. Mosley-Thompson, 1989. Oxygen isotope changes in tropical ice, Quelccaya, Peru. *J. Geophys. Res.* **94**:1187–1194.

Grosswald, M. G., 1980. Late Weichselian ice sheet of northern Eurasia. *Quat. Res.* **13**:1–32.

Grove, J. M., 1988. *The Little Ice Age*. Methuen, New York.

Groveman, B. S., and H. E. Landsberg, 1979. Simulated northern hemisphere temperature departures 1579–1880. *Geophys. Res. Lett.* **6**:767–769.

Guetter, P. J., and J. E. Kutzbach, 1990. A modified Köppen classification applied to model simulations of glacial and interglacial climates. *Clim. Change* **16**:193–215,

Guidon, N., and G. Delibrias, 1986. Carbon-14 dates point to man in the Americas 32,000 years ago. *Nature* **321**:769–771.

Guiot, J., A. Pons, J. L. de Beaulieu, and M. Reille, 1989. A 140,000-year continental climate reconstruction from two European pollen records. *Nature* **338**:309–313.

Guthrie, R. D., 1982. Mammals of the Mammouth Steppe as paleoenvironmental indicators. In: *Paleoecology of Beringia*. D. M. Hopkins, J. V. Matthews, C. E. Schweger, and S. B. Young (Eds.). Academic Press, New York, pp. 307–326.

Habicht, J. K. A., 1979. *Paleoclimate, Paleomagnetism, and Continental Drift*. Am. Assoc. Petrol. Geol. Stud. Geol., No. **9**, Am. Assoc. Petrol. Geol., Tulsa, Okla.

Haffer, J., 1969. Speciation in Amazonian forest birds. *Science* **165**:131–137.

Hahn, D. G., and S. Manabe, 1975. The role of mountains in the south Asian monsoon circulation. *J. Atmos. Sci.* **32**:1515–1541.

Hahn, D. G., and J. Shukla, 1976. An apparent relationship between Eurasian snow cover and Indian monsoon rainfall. *J. Atmos. Sci.* **33**:2461–2462.

Hallam, A., 1981. A revised sea level curve for the Early Jurassic. *J. Geol. Soc. (London)* **138**:735–743.

Hallam, A., 1982. The Jurassic climate. In: *Climate in Earth History*. National Academy Press, Washington, D.C., pp. 159–163.

Hallam, A., 1984. Pre-Quaternary sea-level changes. *Ann. Rev. Earth Plan. Sci.* **12**:205–243.

Hallam, A., 1985. A review of Mesozoic climates. *J. Geol. Soc. (London)* **142**: 433–445.

Hallam, A., 1986a. Origin of minor limestone-shale cycles: Climatically induced or diagenetic? *Geology* **14**:609–612.

Hallam, A., 1986b. The Pliensbachian and Tithonian extinction events. *Nature* **319**:765–768.

Hallam, A., 1987. End-Cretaceous mass extinction event: Argument for terrestrial causation. *Science* **238**:1237–1242.

Hambrey, H. A., and W. B. Harland (Eds.), 1981a. *Earth's Pre-Pleistocene Glacial Record*. Cambridge Univ. Press, Cambridge.

Hambrey, H. A., and W. B. Harland, 1981b. Summary of earth's pre-Pleistocene glacial record. In: *Earth's Pre-Pleistocene Glacial Record*. H. A. Hambrey and W. B. Harland (Eds.). Cambridge Univ. Press, Cambridge, pp. 943–969.

Hameed, S., W.-M. Yeh, M.-T. Li, R. D. Cess, and W.-C. Wang, 1983. An analysis of periodicities in the 1470–1974 Beijing precipitation record. *Geophys. Res. Lett.* **6**:436–439.

Hammer, C. U., H. B. Clausen, and W. Dansgaard, 1980. Greenland ice sheet evidence of post-glacial volcanism and its climatic impact. *Nature* **288**:230–235.

Hammer, C. U., H. B. Clausen, W. Dansgaard, A. Neftel, P. Kristinsdottir, and E. Johnson, 1985. Continuous impurity analysis along Dye 3 deep core. In: *Greenland Ice Core: Geophysics, Geo-*

chemistry, and the Environment. C. C. Langway, H. Oeschger, and W. Dansgaard (Eds.). Geophys. Mono. **33**, Am. Geophys. Union, Washington, D.C., pp. 90–94.

Han, Y.-J., 1988. Modelling and simulation of the general circulation of the ocean. In: *Physically-Based Modelling and Simulation of Climate and Climatic Change.* M. E. Schlesinger (Ed.). Kluwer, Dordrecht, Netherlands, pp. 465–508.

Hansen, J. E., and T. Takahashi (Eds.), 1984. *Climate Processes and Climate Sensitivity.* Geophys. Mono. **29**. Am. Geophys. Union, Washington, D.C.

Hansen, J. E., and S. Lebedeff, 1987. Global trends of measured surface air temperature. *J. Geophys. Res.* **92**: 13345–13372.

Hansen, J. E. and S. Lebedeff, 1988. Global surface air temperatures—update through 1987. *Geophys. Res. Lett.* **15**:323–326.

Hansen, J. E., et al., 1981. Climate impact of increasing atmospheric carbon dioxide. *Science* **213**:957–966.

Hansen, J. E., et al., 1983. Efficient three-dimensional global models for climate studies: Models I and II. *Mon. Wea. Rev.* **3**:609–662.

Hansen, J. E., et al., 1984. Climate sensitivity: Analysis of feedback mechanisms. In: *Climate Processes and Climate Sensitivity.* J. E. Hansen and T. Takahashi (Eds.). Geophys. Mono. **29**, Am. Geophys. Union, Washington, D.C., pp. 130–163.

Hansen, J. E., et al., 1988. Global climate changes as forecast by Goddard Institute for Space Studies three-dimensional model. *J. Geophys. Res.* **93**:9341–9364.

Hansen, T. A., 1987. Extinction of Late Eocene to Oligocene Molluscs: Relationship to shelf area, temperature changes, and impact events. *Palaios* **2**:69–75.

Haq, B. U., and G. P. Lohmann, 1976. Early Cenozoic calcareous nannoplankton biogeography of the Atlantic Ocean. *Mar. Micropaleo.* **1**:119–194.

Haq, B. U., I. Premoli-Silva, and G. P. Lohmann, 1977. Calcareous planktonic paleobiogeographic evidence for major climatic fluctuations in the Early Cenozoic Atlantic Ocean. *J. Geophys. Res.* **8**:427–431.

Haq, B. U., J. Hardenbol, and P. R. Vail, 1987. Chronology of fluctuating sea levels since the Triassic. *Science* **235**:1156–1167.

Hardie, L. A., A. Bosellini, and R. K. Goldhammer, 1986. Repeated subaerial exposure of subtidal carbonate platforms, Triassic, Northern Italy: Evidence for high frequency sea level oscillations on a 10^4 year scale. *Paleoceanog.* **1**:447–457.

Hargraves, R. B., 1986. Faster spreading or greater ridge length in the Archean? *Geology* **14**:750–752.

Harman, H. H., 1976. *Modern Factor Analysis.* Univ. Chicago Press, Chicago.

Harrison, S. P., 1989. Lake levels and climatic change in eastern North America. *Clim. Dynam.* **3**:157–167.

Hart, M. H., 1978. The evolution of the atmosphere of the earth. *Icarus* **33**:23–39.

Hartmann, D. L., 1984. On the role of global-scale waves in ice-albedo and vegetation-albedo feedback. In: *Climate Processes and Climate Sensitivity.* J. E. Hansen and T. Takahashi (Eds.). Geophys. Mono. **29**, Am. Geophys. Union, Washington, D.C., pp. 18–28.

Harvey, L. D. D., 1988. Climatic impact of ice-age aerosols. *Nature* **334**:333–335.

Hasselmann, K., 1976. Stochastic climate models, I, Theory. *Tellus* **28**:473–484.

Hasselmann, K., 1988. Some problems in the numerical simulation of climate variability using high-resolution coupled models. In: *Physically-Based Modelling and Simulation of Climate and Climatic Change.* M. E. Schlesinger (Ed.). Kluwer, Dordrecht, Netherlands, pp. 583–614.

Hastenrath, S., 1971. On the Pleistocene snowline depression in the arid regions of the South American Andes. *J. Glaciol.* **10**:255–267.

Hastenrath, S., 1980. Heat budget of tropical ocean and atmosphere. *J. Phys. Oceanog.* **10**:159–170.

Hastenrath, S., 1984. *The Glaciers of Equatorial East Africa.* Reidel, Dordrecht, Netherlands.

Hastenrath, S., 1985. *Climate and Circulation of the Tropics.* Reidel, Dordrecht, Netherlands.

Hay, W. W., J. F. Behensky, Jr., E. J. Barron, and J. L. Sloan II, 1982. Late Triassic-Liassic paleoclimatology of the proto-central North Atlantic rift system. *Palaeogeog., Palaeoclimatol., Palaeoecol.* **40**:13–30.

Hayes, D. E., and L. A. Frakes, 1975. General synthesis. In: *Init. Rep. Deep-Sea Drill. Proj.* v. **28**. U. S. Gov. Print. Office, Washington, D.C., pp. 919–942.

Hays, J. D., and N. D. Opdyke, 1967. Antarctic radiolaria, magnetic reversals, and climatic change. *Science* **158**:1001–1011.

Hays, J. D., and W. C. Pitman, 1973. Lithospheric plate motions, sea-level changes, and climatic and ecological consequences. *Nature* **246**:18–22.

Hays, J. D., T. Saito, N. D. Opdyke, and L. H. Burckle, 1969. Pliocene-Pleistocene sediments of the equatorial Pacific, their paleomagnetic, biostratigraphic and climatic record. *Geol. Soc. Am. Bull.* **80**:1481–1514.

Hays, J. D., J. A. Lozano, N. Shackleton, and G. Irving, 1976a. Reconstruction of the Atlantic and western Indian Ocean sectors of the 18,000 BP Antarctic Ocean. *Geol. Soc. Am. Mem.* **145**:337–372.

Hays, J. D., J. Imbrie, and N. J. Shackleton, 1976b. Variations in the earth's orbit: Pacemaker of the ice ages. *Science* **194**:1121–1132.

Hearty, P. J., G. H. Miller, C. E. Stearns, and B. J.

Szabo, 1986. Aminostratigraphy of Quaternary shorelines in the Mediterranean basin. *Geol. Soc. Am. Bull.* **97**:850–858.

Heath, G. R., T. C. Moore, and T. H. van Andel, 1977. Carbonate accumulation and dissolution in the equatorial Pacific during the past 45 million years. In: *The Fate of Fossil Fuel CO₂ in the Oceans.* N. R. Andersen and A. Malahoff (Eds.). Plenum Press, New York, pp. 627–639.

Hecht, A. D. (Ed.), 1985. *Paleoclimate Analysis and Modeling.* Wiley-Interscience, New York.

Heckel, P. H., 1986. Sea-level curve for Pennsylvanian eustatic marine transgressive-regressive depositional cycles along midcontinent outcrop belt, North America. *Geology* **14**:330–334.

Held, I. M., 1983. Stationary and quasi-stationary eddies in the extratropical troposphere: Theory. In: *Large-scale Dynamical Processes in the Atmosphere.* B. J. Hoskins and R. P. Pearce (Eds.). Academic Press, New York, pp. 127–167.

Held, I. M., D. I. Linder, and M. J. Suarez, 1981. Albedo feedback, the meridional structure of the effective heat diffusivity, and climatic sensitivity: Results from dynamic and diffusive models. *J. Atmos. Sci.* **38**:1911–1927.

Henderson-Sellers, A., and J. G. Cogley, 1982. The earth's early hydrosphere. *Nature* **298**:832–835.

Henderson-Sellers, A., and B. Henderson-Sellers, 1988. Equable climate in the early Archaean. *Nature* **336**:117–118.

Herbert, T. D., and A. G. Fischer, 1986. Milankovitch climatic origin of mid-Cretaceous black shale rhythms in central Italy. *Nature* **321**:739–743.

Herbert, T. D., S. J. Hills, and H. R. Thierstein, 1988. Earth obliquity control on stratification of lower Cretaceous sediments. *Eos (Trans. Am. Geophys. Union)* **69**:1254.

Herman, Y. and D. M. Hopkins, 1980. Arctic oceanic climate in late Cenozoic time. *Science* **209**:557–562.

Herring, J. R., 1985. Charcoal fluxes into sediments of the North Pacific Ocean: The Cenozoic record of burning. In: *The Carbon Cycle and Atmospheric CO₂: Natural Variations Archean to Present.* E. T. Sundquist and W. S. Broecker (Eds.). Geophys. Mono. **32,** Am. Geophys. Union, Washington, D.C., pp. 419–442.

Herron, M. M., and C. C. Langway, 1985. Chloride, nitrate, and sulfate in the Dye 3 and Camp Century, Greenland ice cores. In: *Greenland Ice Core: Geophysics, Geochemistry, and the Environment.* C. C. Langway, H. Oeschger, and W. Dansgaard (Eds.). Geophys. Mono. 33, Am. Geophys. Union, Washington, D.C., pp. 77–84.

Herterich, K., and K. Hasselmann, 1987. Extraction of mixed layer advection velocities, diffusion coefficients, feedback factors and atmospheric forcing parameters from the statistical analysis of north Pacific SST anomaly fields. *J. Phys. Oceanog.* **17**:2145–2156.

Hester, K., and E. Boyle, 1982. Water chemistry control of cadmium content in recent benthic foraminifera. *Nature* **298**:260–262.

Heusser, C. J., 1989. Polar perspective of late-Quaternary climates in the southern hemisphere. *Quat. Res.* **32**:60–71.

Heusser, C. J., and S. S. Streeter, 1980. A temperature and precipitation record of the past 16,000 years in southern Chile. *Science* **210**:1345–1347.

Heusser, C. J., and J. Rabassa, 1987. Cold climatic episode of Younger Dryas age in Tierra del Fuego. *Nature* **328**:609–611.

Heusser, C. J., L. E. Heusser, and D. M. Peteet, 1985. Late-Quaternary climatic change on the American North Pacific coast. *Nature* **315**:485–487.

Hickey, L. J., 1981. Land plant evidence compatible with gradual, not catastrophic, change at the end of the Cretaceous. *Nature* **292**:529–531.

Hildebrand, A. R., and W. V. Boynton, 1988. Impact wave deposits provide new constraints on the location of the K/T boundary impact. In: *Global Catastrophes in Earth History: An Interdisciplinary Conference on Impacts, Volcanism, and Mass Mortality* (Abstracts). Lunar Plan. Inst., Houston, Texas, pp. 76–77.

Hillaire-Marcel, C., and C. Causse, 1989. The late Pleistocene Laurentide glacier: Th/U dating of its major fluctuations and $\delta^{18}O$ range of the ice. *Quat. Res.* **32**:125–138.

Hills, L. V., J. E. Klovan, and A. R. Sweet, 1974. *Juglans eocinerea* n. sp., Beaufort Formation (Tertiary) southwestern Banks Island, Arctic Canada. *Can. J. Bot.* **52**:65–90.

Hobgood, J. S., and R. S. Cerveny, 1988. Ice-age hurricanes and tropical storms. *Nature* **333**:243–245.

Hodell, D. A., K. R. Elmstrom, and J. P. Kennett, 1986. Latest Miocene benthic ¹⁸O changes, global ice volume, sea level, and the "Messinian salinity crises." *Nature* **320**:411–414.

Hodell, D. A., P. A. Mueller, J. A. McKenzie, and G. A. Mead, 1989. Strontium isotope stratigraphy and geochemistry of the late Neogene ocean. *Earth Plan. Sci. Lett.* **92**:165–178.

Holland, H. D., 1978. *The Chemistry of the Atmosphere and Oceans.* Wiley-Interscience, New York.

Holland, H. D., 1984. *The Chemical Evolution of the Atmosphere and Oceans.* Princeton Press, Princeton, N.J.

Holland, H. D., B. Lazar, and M. McCaffrey, 1986. Evolution of the atmosphere and oceans. *Nature* **320**:27–33.

Hollin, J. T., 1980. Climate and sea level in isotope stage 5: An East Antarctic ice surge at 95,000 BP? *Nature* **283**:629–633.

Holser, W. T., 1977. Catastrophic chemical events in the history of the ocean. *Nature* **267**:403–408.

Holser, W. T., and M. Magaritz, 1987. Events near

the Permian-Triassic boundary. *Mod. Geol.* **11**:155–180.

Holser, W. T. et al., 1989. A unique geochemical record at the Permian/Triassic boundary. *Nature* **337**:39–44.

Holton, J. R. 1979. *An Introduction to Dynamic Meteorology*. Academic Press, New York.

Hooghiemstra, H, 1984. *Vegetational and Climatic History of the High Plain of Bogotá, Colombia: A Continuous Record of the last 3.5 million years*. Thesis, Univ. of Amsterdam. Cramer, Vaduz, Netherlands (also published as Diss. Bot. **79**).

Hooghiemstra, H., and C. O. C. Agwu, 1988. Changes in the vegetation and trade winds in equatorial northwest Africa 140,000–70,000 yr B.P. as deduced from two marine pollen records. *Palaeogeog., Palaeoclimatol., Palaeoecol.* **66**:173–213.

Hooghiemstra, H., A. Bechler, and H.-J. Beug, 1987. Isopollen maps for 18,000 years B.P. of the Atlantic offshore of Northwest Africa: Evidence for paleowind circulation. *Paleoceanog.* **2**:561–582.

Hope, G. S., J. A. Peterson, U. Radok, and I. Allison, 1976. *The Equatorial Glaciers of New Guinea*. Balkema, Rotterdam.

Hopkins, D. M., 1982. Aspects of the paleogeography of Beringia during the Late Pleistocene. In: *Paleoecology of Beringia*. D. M. Hopkins, J. V. Matthews, C. E. Schweger, and S. B. Young (Eds.). Academic Press, New York, pp. 3–28.

Hopkins, D. M., J. V. Matthews, C. E. Schweger, and S. B. Young (Eds.). 1982. *Paleoecology of Beringia*. Academic Press, New York.

Horrell, M. A., 1990. Energy balance constraints on ^{18}O based paleo-sea surface temperature estimates. *Paleoceanog.* **5**:339–348.

Houghton, J. T., 1977. *The Physics of Atmospheres*. Cambridge Univ. Press, Cambridge.

Houghton, J. T. (Ed.), 1984. *The Global Climate*. Cambridge Univ. Press, Cambridge.

Hovan, S. A., D. K. Rea, N. G. Pisias, and N. J. Shackleton, 1989. A direct link between the China loess and marine $\delta^{18}O$ records: Aeolian flux to the north Pacific. *Nature* **340**:296–298.

Howard, W. R., and W. L. Prell, 1984. A comparison of radiolarian and foraminiferal paleoecology in the southern Indian Ocean: New evidence for the interhemispheric timing of climatic change. *Quat. Res.* **21**:244–263.

Howe, S., and T. Webb III, 1983. Calibrating pollen data in climatic terms: Improving the methods. *Quat. Sci. Rev.* **2**:17–51.

Hsü, J. 1978. On the paleobotanical evidence for continental drift and Himalayan uplift. *Paleobot.* **25**:131–142.

Hsü, K. J., 1980. Terrestrial catastrophe caused by cometary impact at the end of Cretaceous. *Nature* **285**:201–203.

Hsü, K. J., and J. A. McKenzie, 1985. A "Strangelove" ocean in the earliest Tertiary. In: *The Carbon Cycle and Atmospheric CO_2: Natural Variations Archean to Present*. E. T. Sundquist and W. S. Broecker (Eds.). Geophys. Mono. **32**, Am. Geophys. Union, Washington, D.C., pp. 487–492.

Hsü, K. J., et al., 1977. History of the Mediterranean salinity crisis. *Nature* **267**:399–403.

Hubbard, R. N. L. B., and M. C. Boulter, 1983. Reconstruction of Palaeogene climate from palynological evidence. *Nature* **301**:147–150.

Hughes, T. J., 1987a. Deluge II and the continent of doom: Rising sea level and collapsing Antarctic ice. *Boreas* **16**:89–100.

Hughes, T. J., 1987b. Ice dynamics and deglaciation models when ice sheets collapsed. In: *North America and Adjacent Oceans During the Last Deglaciation*. W. F. Ruddiman and H. E. Wright Jr. (Eds.). The Geology of North America, v. K-3. Geol. Soc. Am., Boulder, Col., pp. 183–220.

Hughes, T. J., G. H. Denton, and M. G. Grosswald, 1977. Was there a late-Würm Arctic ice sheet? *Nature* **266**:596–602.

Hughes, T. J., et al., 1981. The last great ice sheets: A global view. In: *The Last Great Ice Sheets*. G. H. Denton and T. J. Hughes (Eds.). John Wiley and Sons, New York, pp. 263–317.

Hunt, B. G., 1984. Polar glaciation and the genesis of the ice ages. *Nature* **308**:48–51.

Huntley, B., and C. Prentice, 1988. July temperatures in Europe from pollen data 6000 years before present. *Science* **241**:687–690.

Hurley, N. F., and R. van der Voo, 1987. Paleomagnetism of upper Devonian reefal limestones, Canning Basin, Western Australia. *Geol. Soc. Am. Bull.* **98**:138–146.

Hut, P., W. Alvarez, W. P. Elder, T. Hansen, E. G. Kauffman, G. Keller, E. M. Shoemaker, and P. R. Weissman, 1987. Comet showers as a cause of mass extinctions. *Nature* **329**:117–126.

Hutchinson, J. H., 1982. Turtle, crocodilian, and champosaur diversity changes in the Cenozoic of the north-central region of the western United States. *Palaeogeog., Palaeoclimatol., Palaeoecol.* **37**:149–164.

Hutson, W. H., 1979. The Agulhas Current during the late Pleistocene: Analysis of modern faunal analogs. *Science* **207**:64–66.

Hyde, W. T., and W. R. Peltier, 1985. Sensitivity experiments with a model of the ice age cycle: The response to harmonic forcing. *J. Atmos. Sci.* **42**:2170–2188.

Hyde, W. T., and W. R. Peltier, 1987. Sensitivity experiments with a model of the ice age cycle: The response to Milankovitch forcing. *J. Atmos. Sci.* **44**:1351–1374.

Hyde, W. T., T. J. Crowley, K.-Y. Kim, and G. R. North, 1989. Comparison of GCM and energy balance model simulations of seasonal temperature changes over the past 18,000 years. *J. Clim.*, **2**:864–887.

Hyde, W. T., K.-Y. Kim, T. J. Crowley, and G. R. North, 1990. On the relation between polar continentality and climate: Studies with a non-linear energy balance model. *J. Geophys. Res.* **95**:18,653–18,668.

Imbrie, J., 1985. A theoretical framework for the Pleistocene ice ages. *J. Geol. Soc. (London)* **142**:417–432.

Imbrie, J., 1987. Abrupt terminations of late Pleistocene ice ages: A simple Milankovitch explanation. In: *Abrupt Climatic Change*. W. H. Berger and L. D. Labeyrie (Eds.). D. Reidel, Dordrecht, Netherlands, pp. 365–367.

Imbrie, J., and N. G. Kipp, 1971. A new micropaleontological method for quantitative paleoclimatology: Application to a late Pleistocene Caribbean core. In: *The Late Cenozoic Glacial Ages*. K. K. Turekian (Ed.). Yale Univ. Press, New Haven, Conn., pp. 71–179.

Imbrie, J., and K. P. Imbrie, 1979. *Ice Ages: Solving the Mystery*. Enslow, Hillside, N.J.

Imbrie, J., and J. Z. Imbrie, 1980. Modeling the climatic response to orbital variations. *Science* **207**:943–953.

Imbrie, J., J. van Donk, and N. Kipp, 1973. Paleoclimatic investigation of a late Pleistocene Caribbean deep-sea core: Comparison of faunal and isotopic methods. *Quat. Res.* **3**:10–38.

Imbrie, J., et al., 1984. The orbital theory of Pleistocene climate: Support from a revised chronology of the marine $\delta^{18}O$ record. In: *Milankovitch and Climate*. A. Berger et al. (Eds.). D. Reidel, Dordrecht, Netherlands, pp. 269–305.

Imbrie, J., A. McIntyre, and A. Mix, 1989. Oceanic response to orbital forcing in the late Quaternary: Observational and experimental strategies. In: *Climate and Geosciences*. A. Berger, S. H. Schneider, and J.-C. Duplessy (Eds.). D. Reidel, Dordrecht, Netherlands, pp. 121–164.

Ingle, J. C. 1973. Neogene foraminifera from the northeastern Pacific Ocean, Leg 18, Deep Sea Drilling Project. In: *Init. Rep. Deep-Sea Drill. Proj.* v. 18. L. D. Kulm, et al. (Eds.). U.S. Gov. Print. Office, Washington, D.C., pp. 517–567.

Irving, E., F. K. North, and R. Couillard, 1974. Oil, climate and tectonics. *Can. J. Earth Sci.* **11**:1–17.

Ives, J. D., J. T. Andrews, J. T. Barry, and R. G. Barry, 1975. Growth and decay of the Laurentide ice sheet and comparisons with Fenno-Scandinavia. *Naturwissenschaften* **62**:118–125.

Izett, G. A. 1975. Late Cenozoic sedimentation and deformation in northern Colorado and adjacent areas. *Geol. Soc. Am. Mem.* **144**:179–209.

Izett, G. A., 1987. Authigenic "spherules" in K-T boundary sediments at Caravaca, Spain, and Raton Basin, Colorado and New Mexico, may not be impact derived. *Geol. Soc. Am. Bull.* **99**:78–86.

Jacoby, G. C., and E. R. Cook, 1981. Past temperature variations inferred from a 400-year tree-ring chronology from Yukon territory, Canada. *Arc. Alp. Res.* **13**:409–418.

Jacoby, G. C., and R. D'Arrigo, 1989. Reconstructed northern hemisphere annual temperature since 1671 based on high-latitude tree-ring data from North America. *Clim. Change* **14**:39–59.

Jacoby, G. C., I. S. Ivanciu, and L. D. Ulan, 1988. 263 years of summer temperature for northern Quebec reconstructed from tree ring data and evidence of a major climatic shift in the early 1800's. *Palaeogeog., Palaeoclimatol., Palaeoecol.* **64**:69–78.

Jaeger, L., 1976. Monatskarten des Niederschlags für die ganze Erde: *Berichte des Deutschen Wetterdienstes,* No. **139**.

Janecek, T. R., and D. K. Rea, 1983. Eolian deposition in the northeast Pacific Ocean: Cenozoic history of atmospheric circulation. *Geol. Soc. Am. Bull.* **94**:730–738.

Janecek, T. R., and D. K. Rea, 1984. Pleistocene fluctuations in northern hemisphere tradewinds and westerlies. In: *Milankovitch and Climate*. A. L. Berger, J. Imbrie, J. Hays, G. Kukla, and B. Saltzman (Eds.). D. Reidel, Dordrecht, Netherlands, pp. 331–347.

Janecek, T. R., and D. K. Rea, 1985. Quaternary fluctuations in the northern hemisphere trade winds and westerlies. *Quat. Res.* **24**:150–163.

Jansen, E., U. Bleil, R. Henrich, L. Kringstad, and B. Slettemark, 1988. Paleoenvironmental changes in the Norwegian Sea and the northeast Atlantic during the last 2.8 m.y.: Deep Sea Drilling Project/Ocean Drilling Program sites 610, 642, 643, and 644. *Paleoceanog.* **3**:563–581.

Jansen, J. H. F., T. C. E. van Weering, R. Gieles, and J. van Iperen, 1984. Middle and late Quaternary oceanography and climatology of the Zaire-Congo fan and the adjacent eastern Angola Basin. *Neth. J. Sea Res.* **17**:201–249.

Jansen, J. H. F., A. Kuijpers, and S. R. Troelstra, 1986. A mid-Brunhes climatic event: Long-term changes in global atmosphere and ocean circulation. *Science* **232**:619–622.

Jarrett, R. D., and H. E. Malde, 1987. Paleodischarge of the late Pleistocene Bonneville Flood, Snake River, Idaho, computed from new evidence. *Geol. Soc. Am. Bull.* **99**:127–134.

Jefferson, T. H., 1982. Fossil forests from the lower Cretaceous of Alexander Island, Antarctica. *Palaeontol.* **25**:681–708.

Jenkins, G. M., and D. G. Watts, 1968. *Spectral Analysis and Its Applications*. Holden-Day, San Francisco, Cal.

Jenkyns, H. C., 1988. The early Toarcian (Jurassic) anoxic event: Stratigraphic, sedimentary, and geochemical evidence. *Am. J. Sci.* **288**:101–151.

Johnsen, S. J., W. Dansgaard, H. B. Clausen, and

C. C. Langway, 1970. Climatic oscillations 1200-2000 A.D. *Nature* 227:482-483.

Johnson, D. A., M. Ledbetter, and L. H. Burckle, 1977. Vema Channel paleo-oceanography: Pleistocene dissolution cycles and episodic bottom water flow. *Mar. Geol.* 23:1-33.

Johnson, K. R., and L. J. Hickey, 1988. Patterns of megafloral change across the Cretaceous-Tertiary boundary in the northern Great Plains and Rocky Mountains. In: *Global Catastrophes in Earth History: An Interdisciplinary Conference on Impacts, Volcanism, and Mass Mortality (Abstracts)*. Lun. Plan. Inst., Houston, Texas, p. 87.

Jones, G. A., and L. D. Keigwin, 1988. Evidence from Fram Strait (78°N) for early deglaciation. *Nature* 336:56-59.

Jones, P. D., T. M. L. Wigley, and P. B. Wright, 1986. Global temperature variations between 1861 and 1984. *Nature* 322:430-434.

Jones, P. D., T. M. L. Wigley, and S. C. B. Raper, 1987. The rapidity of CO_2-induced climatic change: Observations, model results and palaeoclimatic implications. In: *Abrupt Climatic Change: Evidence and Implications*. A. L. Berger and L. D. Labeyrie (Eds.). D. Reidel, Dordrecht, Netherlands, pp. 47-55.

Joussaume, S., and J. Jouzel, 1987. Simulation of paleoclimatic tracers using atmospheric general circulation models. In: *Abrupt Climatic Change*. W. H. Berger and L. D. Labeyrie (Eds.). D. Reidel, Dordrecht, Netherlands, pp. 369-381.

Jouzel, J., C. Lorius, L. Merlivat, and J.-R. Petit, 1987a. Abrupt climatic changes: The Antarctic ice record during the late Pleistocene. In: *Abrupt Climatic Change*. W. H. Berger and L. D. Labeyrie (Eds.). D. Reidel, Dordrecht, Netherlands, pp. 235-245.

Jouzel, J., et al., 1987b. Vostok ice core: A continuous isotope temperature record over the last climatic cycle (160,000 years). *Nature* 329:403-408.

Jouzel, J., et al., 1989. A comparison of deep Antarctic ice cores and their implications for climate between 65,000 and 15,000 years ago. *Quat. Res.* 31:135-150.

Juillet-Leclerc, A., and H. Schrader, 1987. Variations of upwelling intensity recorded in varved sediment from the Gulf of California during the past 3,000 years. *Nature* 329:146-149.

Kanari, S., N. Fuji, and S. Horie, 1984. The paleoclimatological constituents of paleotemperature in Lake Biwa. In: *Milankovitch and Climate*. A. L. Berger, J. Imbrie, J. Hays, G. Kukla, and B. Saltzman (Eds.). D. Reidel, Dordrecht, Netherlands, pp. 405-414.

Kaneps, A. G., 1979. Gulf Stream: Velocity fluctuations during the late Cenozoic. *Science* 204:297-301.

Kasting, J. F., 1989. Long-term stability of the earth's climate. *Glob. Plan. Change* 1:83-95.

Kasting, J. F., and T. P. Ackerman, 1986. Climatic consequences of very high carbon dioxide levels in the earth's early atmosphere. *Science* 234:1383-1385.

Kauffman, E. G., 1973. Cretaceous Bivalvia. In: *Atlas of Palaeobiogeography*. A. Hallam (Ed.). Elsevier, Amsterdam, pp. 353-383.

Kearney, M. S., and B. H. Luckman, 1983. Holocene timberline fluctuations in Jasper National Park, Alberta. *Science* 221:261-263.

Keeling, C. D., and R. B. Bacastow, 1977. Impact of industrial gases on climate. In: *Energy and Climate*. Nat. Acad. Sci. Washington, D.C., pp. 72-95.

Keen, D. H., R. S. Harmon, and J. T. Andrews, 1981. U-series and amino acid dates from Jersey. *Nature* 289:162-164.

Keffer, T., D. G. Martinson, and B. H. Corliss, 1988. The position of the Gulf Stream during Quaternary glaciations. *Science* 241:440-442.

Keigwin, L. D., 1980. Palaeoceanographic change in the Pacific at the Eocene-Oligocene boundary. *Nature* 287:722-725.

Keigwin, L. D., 1982a. Isotopic paleoceanography of the Caribbean and east Pacific: Role of Panama uplift in late Neogene time. *Science* 217:350-353.

Keigwin, L. D., 1982b. Neogene planktonic foraminifers from Deep Sea Drilling Project Sites 502 and 503. In: *Init. Rep. Deep-Sea Drill. Proj.* v. 68. W. L. Prell, and J. V. Gardner et al. (Eds.). U.S. Gov. Print. Office, Washington, D.C., pp. 269-288.

Keigwin, L. D., 1987. North Pacific deep water formation during the latest glaciation. *Nature* 330:362-364.

Keigwin, L. D., and G. Keller, 1984. Middle Oligocene cooling from equatorial Pacific DSDP Site 77B. *Geology* 12:16-19.

Keigwin, L. D., and E. A. Boyle, 1985. Carbon isotopes in deep-sea benthic foraminifera: Precession and changes in low-latitude biomass. In: *The Carbon Cycle and Atmospheric CO_2: Natural Variations Archean to Present*. E. T. Sundquist and W. S. Broecker (Eds.). Geophys. Mono. 32, Am. Geophys. Union, Washington, D.C., pp. 319-328.

Keigwin, L. D., and B. H. Corliss, 1986. Stable isotopes in late middle Eocene to Oligocene foraminifers. *Geol. Soc. Am. Bull.* 97:335-345.

Keigwin, L. D. and E. A. Boyle, 1989. Late Quaternary paleochemistry of high-latitude surface waters, *Palaeogeog., Palaeoclimatol., Palaeocol.* 73:77-84.

Keigwin, L. D., and G. A. Jones, 1989. Glacial-Holocene stratigraphy, chronology, and paleoceanographic observations on some North Atlantic sediment drifts. *Deep-Sea Research*, 36:845-867.

Keir, R. S., 1983. Reduction of thermohaline circulation during deglaciation: The effect on atmospheric radiocarbon and CO_2. *Earth Plan, Sci. Lett.* **64:**445–456.

Keir, R. S., 1988. On the late Pleistocene ocean geochemistry and circulation. *Paleoceanog.* **3:**413–445.

Keller, G., 1979. Late Neogene paleoceanography of the North Pacific DSDP sites 173, 310, and 296. *Mar. Micropaleontol.* **4:**159–172.

Keller, G., 1989. Extended Cretaceaous/Tertiary boundary extinctions and delayed population change in planktonic foraminifera from Brazos River, Texas. *Paleoceanog.* **4:**287–332.

Keller, G., and J. A. Barron, 1983. Paleoceanographic implications of Miocene deep-sea hiatuses. *Geol. Soc. Am. Bull.* **94:**590–613.

Kellogg, T. B., 1980. Paleoclimatology and paleooceanography of the Norwegian and Greenland seas: Glacial-interglacial contrasts. *Boreas* **9:**115–137.

Kellogg, W. W., and Z.-C. Zhao, 1988. Sensitivity of soil moisture to doubling of carbon dioxide in climate model experiments. Part I: North America. *J. Clim.* **1:**348–378.

Kelly, P. M., C. M. Goodess, and B. S. G. Cherry, 1987: The interpretation of the Icelandic sea ice records. *J. Geophys. Res.* **92:**10835–10843.

Kemp, E. M., 1978. Tertiary climatic evolution and vegetation history in the southeast Indian Ocean region. *Palaeogeog., Palaeoclimatol., Palaeoecol.* **24:**169–208.

Kemp, E. M., L. A. Frakes, and D. E. Hayes, 1975. Paleoclimatic significance of diachronous biogenic facies. In: *Init. Rep. Deep-Sea Drill. Proj.* **v. 28.** U.S. Gov. Print. Office, Washington, D.C., pp. 909–917.

Kennett, J. P., 1977. Cenozoic evolution of Antarctic glaciation, the circum-Antarctic Ocean, and their impact on global paleoceanography. *J. Geophys. Res.* **82:**3843–3860.

Kennett, J. P., 1982. *Marine Geology.* Prentice Hall, Englewood Cliffs, N.J.

Kennett, J. P., and N. J. Shackleton, 1975. Laurentide ice sheet meltwater recorded in Gulf of Mexico deep-sea cores. *Science* **188:**147–150.

Kennett, J. P., and N. J. Shackleton, 1976. Oxygen isotope evidence for the development of the psychrosphere 38 M. yr. ago. *Nature* **260:**513–515.

Kennett, J. P., and R. C. Thunell, 1977. On explosive Cenozoic volcanism and climatic implications. *Science* **196:**1231–1234.

Kennett, J. P., N. D. Watkins, and P. Vella, 1971. Paleomagnetic chronology of Pliocene-Early Pleistocene climates and the Plio-Pleistocene boundary in New Zealand. *Science* **171:**276–279.

Kennett, J. P., G. Keller, and M. S. Srinivasan, 1985a. Miocene planktonic foraminiferal biogeography and paleoceanographic development of the Indo-Pacific region. *Geol. Soc. Am. Mem.* **163:**197–236.

Kennett, J. P. et al., 1985b. Palaeotectonic impli-

cations of increased late Eocene–early Oligocene volcanism from South Pacific DSDP sites. *Nature* **316:**507–511.

Kent, D., and S. R. May, 1987. Polar wander and paleomagnetic reference pole controversies. *Rev. Geophys.* **25:**961–970.

Kent, D., N. D. Opdyke, and M. Ewing, 1971. Climate change in the North Pacific using ice-rafted detritus as a climatic indicator. *Geol. Soc. Am. Bull.* **82:**2741–2754.

Kershaw, A. P., 1986. Climatic change and Aboriginal burning in northeast Australia during the last two glacial/interglacial cycles. *Nature* **322:**47–49.

Killingley, J. S., 1983. Effects of diagenetic recrystallization on $^{18}O/^{16}O$ values of deep-sea sediments. *Nature* **301:**594–597.

Kipp, N. G., 1976. New transfer function for estimating sea-surface conditions from sea-bed distribution of planktonic foraminiferal assemblages in the North Atlantic. In: *Investigation of Late Quaternary Paleoceanography and Paleoclimatology.* R. M. Cline and J. D. Hays (Eds.). Geol. Soc. Am. Mem. **145,** Geol. Soc. Am., Boulder, Col., pp. 3–42.

Kitchell, J. A., and D. L. Clark, 1982. Late Cretaceous-Paleogene paleogeography and paleocirculation: Evidence of north polar upwelling. *Palaeogeog., Palaeoclimatol., Palaeoecol.* **40:**135–165.

Klein, G. de V., 1990. Pennsylvanian time scales and cycle periods. *Geology* **18:**455–457.

Klein, G. deV., and T. A. Ryer, 1978. Tidal circulation patterns in Precambrian, Paleozoic, and Cretaceous epeiric and mioclinal shelf seas. *Geol. Soc. Am. Bull.* **89:**1050–1058.

Klein, G. deV., and D. A. Willard, 1989. Origin of the coal-bearing Pennsylvanian cyclothems of North America. *Geology* **17:**152–155.

Knauth, L. P., and S. Epstein, 1976. Hydrogen and oxygen isotope ratios in nodular and bedded cherts. *Geochim. Cosmochim. Acta* **40:**1095–1108.

Knoll, A. H., J. M. Hayes, A. J. Kaufman, K. Swett, and I. B. Lambert, 1986. Secular variation in carbon isotope ratios from upper Proterozoic successions of Svalbard and east Greenland. *Nature* **321:**832–838.

Knoll, M. A., and W. C. James, 1987. Effect of the advent and diversification of vascular land plants on mineral weathering through geologic time. *Geology* **15:**1099–1102.

Knox, J. C., 1983. Responses of river systems to Holocene climates. In: *Late Quaternary Environments of the United States, v. 2.* H. E. Wright, Jr. (Ed.). Univ. Minnesota Press, Minneapolis, pp. 26–41.

Knox, J. C., 1985. Responses of floods to Holocene climatic change in the upper Mississippi Valley. *Quat. Res.* **23:**287–300.

Koeberl, C., V. L. Sharpton, A. V. Murali, and K.

Burke, 1990. Kara and Ust-Kara impact structures (USSR) and their relevance to the K/T boundary event. *Geology* **18**:50–53.

Koerner, R. M. 1977. Devon Island ice cap: Core stratigraphy and paleoclimate. *Science* **196**:15–18.

Koerner, R. M. 1989. Ice core evidence for extensive melting of the Greeland ice sheet in the last interglacial. *Science* **244**:964–968.

Kolla, V., P. E. Biscaye, and A. F. Hanley, 1979. Distribution of quartz in Late Quaternary Atlantic sediments in relation to climate. *Quat. Res.* **11**:261–277.

Kominz, M. A., 1984. Oceanic ridge volumes and sea-level change—An error analysis. In: *Interregional Unconformities and Hydrocarbon Accumulation.* J. S. Schlee (Ed.). Am. Assoc. Petrol. Geol., Tulsa, Okla., **Mem. 36**, pp. 108–123.

Kominz, M. A., and N. G. Pisias, 1979. Pleistocene climate: Deterministic or stochastic? *Science* **204**:171–173.

Kominz, M. A., G. R. Heath, T.-L. Ku, and N. G. Pisias, 1979. Brunhes time scales and the interpretation of climatic change. *Earth Plan. Sci. Lett.* **45**:394–410.

Köppen, W., and A. Wegener, 1924. *Die Klimate der Geologischen Vorzeit.* Gebrüder Borntraeger, Berlin.

Krassilov, V. A., 1973. Climatic changes in eastern Asia as indicated by fossil floras, I, early Cretaceous. *Palaeogeog., Palaeoclimatol., Palaeoecol.* **13**:261–273.

Krassilov, V. A., 1981. Changes of Mesozoic vegetation and the extinction of dinosaurs. *Palaeogeog., Palaeoclimatol., Palaeoecol.* **34**:207–224.

Kraus, E. B., 1973. Comparison between ice age and present general circulation. *Nature* **245**:129–133.

Kroopnick, P. M., 1985. The distribution of ^{13}C of ΣCO_2 in the world oceans. *Deep-Sea Res.* **32**:57–84.

Kroopnick, P., S. V. Margolis, and C. S. Wong. 1977. $\delta^{13}C$ variations in marine carbonate sediments as indicators of the CO_2 balance between the atmosphere and oceans. In *The Fate of Fossil Fuel CO_2 in the Oceans.* N. R. Anderson and A. Malahoff (Eds.). Plenum Press, New York, pp. 295–321.

Ku, T.-L., M. A. Kimmel, W. H. Easton, and T. J. O'Neil, 1974. Eustatic sea level 120,000 years ago on Oahu, Hawaii. *Science* **183**:959–962.

Kuhle, M., 1987. Subtropical mountain and highland glaciation as ice age triggers and the waning of the glacial periods in the Pleistocene. *Geo. J.* **14**:393–421.

Kuhn, W. R., and S. K. Atreya, 1979. Ammonia photolysis and the greenhouse effect in the primordial atmosphere of the earth. *Icarus* **37**:207–213.

Kukla, G. J., 1970. Correlation between loesses and deep-sea sediments. *Geologiska Föreningeas i Stockholm Förhandlingar* **92**:148–180.

Kukla, G. J., 1977. Pleistocene land-sea correlations, I, Europe. *Earth Sci. Rev.* **13**:307–374.

Kukla, G., 1987. Loess stratigraphy in central China. *Quat. Sci. Rev.* **6**:191–219.

Kukla, G., F. Heller, L.-X. Ming, X.-T. Chun, L.-T. Sheng, and A.-Z. Sheng, 1988. Pleistocene climates in China dated by magnetic susceptibility. *Geology* **16**:811–814.

Kump, L. R. 1989. Chemical stability of the atmosphere and ocean. *Palaeogeog., Palaeoclimatol., Palaeoecol.* **75**:123–136.

Kunk, M. J., G. A. Izett, R. A. Haugerud, and J. F. Sutter, 1989. ^{40}Ar-^{39}Ar dating of the Manson impact structure: A Cretaceous-Tertiary boundary crater candidate. *Science* **244**:1565–1568.

Kutzbach, J. E., 1976. The nature of climate and climatic variations. *Quat. Res.* **6**:471–480.

Kutzbach, J. E., 1987. Model simulations of the climatic patterns during the deglaciation of North America. In: *North America and Adjacent Oceans During the Last Deglaciation.* W. F. Ruddiman and H. E. Wright, Jr. (Eds.). The Geology of North America, v. K-3, Geol. Soc. Am., Boulder, Col., pp. 425–446.

Kutzbach, J. E., and R. A. Bryson, 1974. Variance spectrum of Holocene climatic fluctuations in the North Atlantic sector. *J. Atmos. Sci.* **31**:1958–1963.

Kutzbach, J. E., and B. L. Otto-Bliesner, 1982. The sensitivity of the African-Asian monsoonal climate to orbital parameter changes for 9000 years B.P. in a low-resolution general circulation model. *J. Atmos. Sci.* **39**:1177–1188.

Kutzbach, J. E., and P. J. Guetter, 1984. The sensitivity of monsoon climates to orbital parameter changes for 9000 years B.P: Experiments with the NCAR general circulation model. In: *Milankovitch and Climate.* A. Berger, J. Imbrie, J. D. Hays, G. J. Kukla, and B. Saltzman (Eds.). D. Reidel, Dordrecht, Netherlands, pp. 801–820.

Kutzbach, J. E., and F. A. Street-Perrott, 1985. Milankovitch forcing of fluctuations in the level of tropical lakes from 18–0 kyr BP. *Nature* **317**:130–134.

Kutzbach, J. E., and H. E. Wright, 1985. Simulation of the climate of 18,000 yr BP: Results for the North American/North Atlantic/European sector and comparison with the geologic record. *Quat. Sci. Rev.* **4**:147–187.

Kutzbach, J. E., and P. J. Guetter, 1986. The influence of changing orbital parameters and surface boundary conditions on climate simulations for the past 18,000 years. *J. Atmos. Sci.* **43**:1726–1759.

Kutzbach, J. E., and R. G. Gallimore, 1988. Sensitivity of a coupled atmosphere/mixed layer ocean model to changes in orbital forcing at 9000 years B.P. *J. Geophys. Res.* **93**:803–821.

Kutzbach, J. E., and R. G. Gallimore, 1989. Pangean climates: Megamonsoons of the megacontinent. *J. Geophys. Res.* **94**:3341–3357.

Kutzbach, J. E., P. J. Guetter, W. F. Ruddiman, and W. L. Prell, 1989. The sensitivity of climate to late Cenozoic uplift in southeast Asia and the American southwest: Numerical experiments. *J. Geophys. Res.* **94**:18393–18407.

Kutzbach, J. E., P. J. Guetter, and W. M. Washington, 1990. Simulated circulation of an idealized ocean for Pangaean time, *Paleoceanog.* **5**:299–317.

Kvale, E. P., A. W. Archer, and H. R. Johnson, 1989. Daily, monthly, and yearly tidal cycles within laminated siltstones of the Mansfield Formation (Pennsylvanian) of Indiana. *Geology* **17**:365–368.

Kyle, H. L., P. E. Ardanuy, and E. J. Hurley, 1985. The status of the Nimbus-7 earth-radiation-budget data set. *Bull. Am. Met. Soc.* **66**:1378–1388.

Kyte, F. T., and J. T. Wasson, 1986. Accretion rate of extraterrestrial matter: Iridium deposited 33 to 67 million years ago. *Science* **232**:1225–1230.

Kyte, F. T., L. Zhou, and J. T. Wasson, 1988. New evidence on the size and possible effects of a late Pliocene oceanic asteroid impact. *Science* **241**:63–65.

Labeyrie, L. D., et al., 1986. Melting history of Antarctica during the past 60,000 years. *Nature* **322**:701–706.

Labeyrie, L. D., J.-C. Duplessy, and P. L. Blanc, 1987. Variations in mode of formation and temperature of oceanic deep waters over the past 125,000 years. *Nature* **327**:477–482.

Labitzke, K., and H. van Loon, 1988. Associations between the 11-year solar cycle, the QBO and the atmosphere. Part I: The troposphere and stratosphere in the northern hemisphere in winter. *J. Atmos. Terr. Phys.* **50**:197–206.

Labracherie, M., et al., 1989. The last deglaciation in the Southern Ocean. *Paleoceanog.* **4**:629–638.

Lagoe, M. B., and N. Eyles, 1988. The late Miocene-Pleistocene Yakataga Fm., Gulf of Alaska: The depositional record of temperate glaciation in the northeast Pacific Ocean. *Eos (Trans. Am. Geophys. Union)* **69**:1253.

LaMarche, V. C., 1974. Paleoclimatic inferences from long tree-ring records. *Science.* **183**:1043–1048.

LaMarche, V. C., and K. K. Hirschboeck, 1984. Frost rings in trees as records of major volcanic eruptions. *Nature* **307**:121–126.

Lamb, H. H., 1969. Climatic fluctuations. In: *World Survey of Climatology, v. 2.* H. E. Landsberg (Ed.). Elsevier, Amsterdam, pp. 173–249.

Lamb, H. H., 1970. Volcanic dust in the atmosphere: With a chronology and assessment of its meteorological significance. *Phil. Trans. Roy. Soc. (London)* **266**:425–533.

Lamb, H. H., 1977. *Climate: Present, Past and Future, v. 2.* Barnes and Noble, New York.

Lamb, H. H., 1979. Climatic variation and changes in the wind and ocean circulation: The Little Ice Age in the North Atlantic. *Quat. Res.* **11**:1–20.

Lambert, I. B., M. R. Walter, Z. Wenlong, L. Songnian, and M. Guogans, 1987. Palaeoenvironment and carbon isotope stratigraphy of upper Proterozoic carbonates of the Yangtze Platform. *Nature* **325**:140–142.

Lambert, R. S. J., 1976. Archean thermal regimes, crustal and upper mantle temperatures, and a progressive evolutionary model for the earth. In: *The Early History of the Earth.* B. F. Windley (Ed.). John Wiley and Sons, New York, pp. 363–373.

Landsberg, H. E., 1985. Historic weather data and early meteorological observations. In: *Paleoclimate Analysis and Modeling.* A. D. Hecht (Ed.). Wiley-Interscience, New York, pp. 27–70.

Langway, C. C., H. Oeschger, and W. Dansgaard (Eds.). 1985. *Greeland Ice Core: Geophysics, Geochemistry, and the Environment.* Geophys. Mono. **33**, Am. Geophys. Union, Washington, D.C.

Lasaga, A. C., R. A. Berner, and R. M. Garrels, 1985. An improved geochemical model of atmospheric CO_2 fluctuations over the past 100 million years. In: *The Carbon Cycle and Atmospheric CO_2: Natural Variations Archean to Present.* E. T. Sundquist and W. S. Broecker (Eds.). Geophys. Mono. **32**, Am. Geophys. Union, Washington, D.C., pp. 397–411.

Laskar, J., 1989. A numerical experiment on the chaotic behaviour of the solar system. *Nature* **338**:237–238.

Lautenschlager, M. and K. Herterich, 1990a. Atmospheric response to ice-age conditions—climatology near the earth's surface. *Jour. Geophys. Res.* 95:22,547–22,557.

Lautenschlager, M. and K. Herterich, 1990b. Climatic response to ice-age conditions Part 1: The atmospheric circulation. *Clim. Dynam.* (in press).

LaViolette, P. A., 1985. Evidence of high cosmic dust concentrations in late Pleistocene polar ice (20,000–14,000 years BP). *Meteorites* **20**:545–558.

Lean, J., and D. A. Warrilow, 1989. Simulation of the regional climatic impact of Amazon deforestation. *Nature* **342**:411–413.

Leckie, R. M., and P.-N. Webb, 1983. Late Oligocene-early Miocene glacial record of the Ross Sea, Antarctica: Evidence from DSDP Site 270. *Geology* **11**:578–582.

Ledbetter, M. T., 1984. Bottom-current speed in the Vema Channel recorded by particle size of sediment fine-fraction. *Mar. Geol.* **58**:137–149.

Ledley, T. S., 1984. Sensitivities of cryospheric models to insolation and temperature variations using a surface energy balance. In: *Milankovitch and Climate.* A. L. Berger, J. Imbrie, J. Hays, G.

Kukla, and B. Saltzman (Eds.). D. Reidel, Dordrecht, Netherlands, pp. 581–597.

Ledley, T. S., 1988. A wandering Gondwanaland's impact on summer temperatures. *Geophys. Res. Lett.* **15**:1397–1400.

Lefield, J., 1971. Geology of the Djadokhta Formation at Bayn Dzak (Mongolia). *Palaeontol. Pol.* **25**:101–127.

Leg 105 Shipboard Scientific Party, 1986. High-latitude palaeoceanography. *Nature* **230**:17–18.

Leg 113 Shipboard Scientific Party, 1987. Glacial history of Antarctica. *Nature* **328**:115–116.

Leg 119 Shipboard Scientific Party, 1988. Early glaciation of Antarctica. *Nature* **333**:303–304.

Legrand, M., and R. J. Delmas, 1987. A 220-year continuous record of volcanic H_2SO_4 in the Antarctic ice sheet. *Nature* **327**:671–676.

Legrand, M. R., R. J. Delmas, and R. J. Charlson, 1988a. Climate forcing implications from Vostok ice-core sulphate data. *Nature* **334**:418–420.

Legrand, M. R., C. Lorius, N. I. Barkov, and V. N. Petrov, 1988b. Vostok (Antarctica) Ice Core: Atmospheric chemistry changes over the last climatic cycle (160,000 years). *Atmos. Envir.* **22**:317–331.

Leith, C. E., 1975. Climate response and fluctuation-dissipation. *J. Atmos. Sci.* **32**:2022–2026.

Leith, C. E., 1978. Predictability of climate. *Nature* **276**:352–256.

Lemke, P., 1977. Stochastic climate models, III. Application to zonally averaged energy models. *Tellus* **29**:385–392.

Lemke, P., 1987. A coupled one-dimensional sea ice-ocean model. *J. Geophys. Res.* **92**:13164–13172.

Leopold, A. C., 1964. *Plant Growth and Development.* McGraw-Hill, New York.

Le Roy Ladurie, E., 1971: *Times of Feast, Times of Famine: A history of climate since the year 1000.* New York, Doubleday.

Le Roy Ladurie, E., and M. Baulant, 1980. Grape harvests from the fifteenth through the nineteenth centuries. *Inderdisc. Hist.* **10**:839–849.

LeTreut, H., J. Portes, J. Jouzel, and M. Ghil, 1988. Isotopic modeling of climatic oscillations: Implications for a comparative study of marine and ice core records. *J. Geophys. Res.* **93**:9365–9383.

Leventer, A., and R. B. Dunbar, 1988. Recent diatom record of McMurdo Sound, Antarctica: Implications for history of sea ice extent. *Paleoceanog.* **3**:259–274.

Leventer, A., D. F. Williams, and J. P. Kennett, 1982. Dynamics of the Laurentide ice sheet during the last deglaciation: Evidence from the Gulf of Mexico. *Earth Plan. Sci. Lett.* **59**:11–17.

Levitus, S., 1982. *Climatological Atlas of the World Ocean.* NOAA Professional Paper No. 13, U.S. Gov. Print. Office, Washington, D.C.

Lewan, M. D., 1986. Stable carbon isotopes of amorphous kerogens from Phanerozoic sedimentary rocks. *Geochim. Cosmochim. Acta* **50**:1583–1591.

Lewis, J. S., G. H. Watkins, H. Hartman, and R. G. Prinn, 1982. Chemical consequences of major impact events on Earth. *Geol. Soc. Am. Spec. Pap.* **190**:215–222.

Leyden, M. S., 1984. Guatemalan forest synthesis after Pleistocene aridity. *Proc. Natl. Acad. Sci.* **81**:4856–4859.

Li, W.-X, et al., 1989. High-precision mass-spectrometric uranium-series dating of cave deposits and implications for palaeoclimate studies. *Nature* **339**:534–536.

Lian, M. S., and R. D. Cess, 1977. Energy balance climate models: A reappraisal of ice-albedo feedback. *J. Atmos. Sci.* **34**:1058–1062.

Lin, R.-Q., and G. R. North, 1990. A study of abrupt climate change in a simple nonlinear climate model. *Clim. Change* **4**:253–261.

Lindstrom, D. R., and D. R. MacAyeal, 1989. Scandinavian, Siberian, and Arctic Ocean glaciation: Effect of Holocene atmospheric CO_2 variations. *Science* **245**:628–631.

Lindzen, R. S., and B. Farrell, 1977. Some realistic modifications of simple climate models. *J. Atmos. Sci.* **34**:1487–1501.

Liu, K.-B., and P. A. Colinvaux, 1985. Forest change in the Amazon basin in the last glacial maximum. *Nature* **318**:556–557.

Liu, T. S., Z. An, B. Yuan, and J. Han, 1985. The loess-paleosol sequence in China and climatic history. *Episodes* **8**:21–41.

Lloyd, C. R., 1982. The mid-Cretaceous earth: Paleogeography, ocean circulation and temperature, atmospheric circulation. *J. Geol.* **90**:393–413.

Locke, C. W., and W. W. Locke, 1977. Little ice age snow-cover extent and paleoglaciation thresholds: North-central Baffin Island, NWT, Canada. *Arc. Alp. Res.* **9**:291–300.

Lohmann, G. P., 1978. Response of the deep sea to ice ages. *Oceanus* **21**:58–64.

Lohmann, G. P., and J. J. Carlson, 1981. Oceanographic significance of Pacific Late Miocene calcareous nannoplankton. *Mar. Micropaleontol.* **6**:553–579.

Lorenz, E. N., 1968. Climatic determinism. *Meteorol. Mono.* **8**:1–19.

Lorius, C., L. Merlivat, J. Jouzel, and M. Pourchet, 1979. A 30,000-yr isotope climatic record from Antarctic ice. *Nature* **280**:644–648.

Lorius, C., et al., 1985. A 150,000-year climatic record from Antarctic ice. *Nature* **316**:591–596.

Loubere, P., 1988. Gradual late Pliocene onset of glaciation: A deep-sea record from the northeast Atlantic. *Palaeogeog., Palaeoclimatol., Palaeoecol.* **63**:327–334.

Loutit, T. S., and J. P. Kennett, 1981. New Zealand and Australian Cenozoic sedimentary cycles and global sea-level changes. *Am. Assoc. Petrol. Geol. Bull.* **65**:1586–1601.

Loutit, T. S., and L. D. Keigwin, 1982. Stable isotopic evidence for latest Miocene sea-level fall in the Mediterranean region. *Nature* **300:**163–166.

Loutit, T. S., J. P. Kennett, and S. M. Savin, 1983. Miocene equatorial and southwest Pacific paleoceanography from stable isotope evidence. *Mar. Micropaleontol.* **8:**215–233.

Lowenstam, H., and S. Epstein, 1954. Paleotemperatures of the post-Aptian Cretaceous as determined by the oxygen isotope method. *J. Geol.* **62:**207–248.

Luck, J. M., and K. K. Turekian, 1983. Osmium-187/Osmium-186 in manganese nodules and the Cretaceous-Tertiary boundary. *Science* **222:**613–615.

Lundelius, E. L., et al., 1983. Terrestrial vertebrate faunas. In: *Late-Quaternary Environments of the United States, v.* **1.** H. E. Wright, Jr. (Ed.). Univ. Minnesota Press, Minneapolis, pp. 311–353.

Luther, M. E., J. J. O'Brien, and W. L. Prell, 1990. Variability in upwelling fields in the northwestern Indian Ocean Part 1: Model experiments for the past 18,000 years. *Paleoceanog.* **5:**433–445.

Luyendyk, B. P., D. Forsyth, and J. D. Phillips, 1972. Experimental approach to the paleocirculation of the oceanic surface waters. *Geol. Soc. Amer. Bull.* **83:**2649–2664.

Luz, B., 1973. Stratigraphic and paleoclimatic analysis of late Pleistocene tropical southeast Pacific cores (with an appendix by N. J. Shackleton). *Quat. Res.* **3:** 56–72.

Luz, B., and N. J., Shackleton, 1975. CaCO$_3$ solution in the tropical east Pacific during the past 130,000 years. In: *Dissolution of Deep-Sea Carbonates,* Cushman Found. Foram. Res. Spec. Publ. **13.** W. V. Sliter, A. W. H. Bé, and W. H. Berger (Eds.) U.S. Nat. Mus., Washington, D.C., pp. 142–150.

Lyle, M., 1988. Climatically forced organic carbon burial in equatorial Atlantic and Pacific Oceans. *Nature* **335:**529–532.

Maasch, K. A., 1989. Calculating climate attractor dimension from δ^{18}O records by the Grassberger-Procaccia algorithm. *Clim. Dynam.* **4:**45–55.

MacDougall, J. D., 1988. Seawater strontium isotopes, acid rain, and the Cretaceous-Tertiary boundary. *Science* **239:**485–487.

MacIntyre, R., 1971. Apparent periodicity of carbonatite emplacement in Canada. *Nature* **230:**79–81.

MacIntyre, R., 1973. Possible periodic pluming. *Eos (Trans. Am. Geophys. Union)* **54:**239.

Maejima, I., and Y. Tagami, 1984. Climate of Little Ice Age in Japan. In: *Historical weather records at Hirosaki, northern Japan, from 1661–1868.* I. Maejima, M. Nagami, S. Ika, and Y. Tagami (Eds.). *Geogr. Repts. Tokyo Metropolitan Univ.* **18:**113–152.

Magaritz, M., 1989. ^{13}C minima follow extinction events: A clue to faunal radiation. *Geology* **17:**337–340.

Maier-Reimer, E., and K. Hasselmann, 1987. Transport and storage of CO$_2$ in the ocean—an inorganic ocean-circulation carbon cycle model. *Clim. Dynam.* **2:**63–90.

Maier-Reimer, E. T., and U. Mikolajewicz, 1989. Experiments with an OGCM on the cause of the Younger Dryas. Rept. #39, Max-Planck-Institut für Meteorologie, Hamburg, Federal Republic of Germany.

Maier-Reimer, E., K. Hasselmann, D. Olbers, and J. Willebrand, 1982. *An Ocean Circulation Model for Climate Studies.* Tech. Report, Max Planck Institut für Meteorologie, Hamburg, Federal Republic of Germany.

Maier-Reimer, E., U. Mikolajewicz, and T. J. Crowley, 1990. Ocean GCM sensitivity experiment with an open central American isthmus. *Paleoceanog.* **5:**349–366.

Malone, R. C., L. A. Auer, G. A. Glatzmaier, M. C. Wood, and O. B. Toon, 1986. Nuclear Winter: Three-dimensional simulations including interactive transport, scavenging, and solar heating of smoke. *J. Geophys. Res.* **91:**1039–1053.

Manabe, S. and R. F. Strickler, 1964. Thermal equilibrium of the atmosphere with a convective adjustment. *J. Atmos. Sci.* **21:**361–385.

Manabe, S., and R. T. Wetherald, 1967. Thermal equilibrium of the atmosphere with a given distribution of relative humidity. *J. Atmos. Sci.* **24:**241–259.

Manabe, S., and T. B. Terpstra, 1974. The effects of mountains on the general circulation of the atmosphere as identified by numerical experiments. *J. Atmos. Sci.* **31:**3–42.

Manabe, S., and D. G. Hahn, 1977. Simulation of the tropical climate of an ice age. *J. Geophys. Res.* **82:**3889–3911.

Manabe, S., and R. J. Stouffer, 1980. Sensitivity of a global climate model to an increase of CO$_2$ concentration in the atmosphere. *J. Geophys. Res.* **85:**5529–5554.

Manabe, S., and R. T. Wetherald, 1980. On the distribution of climate change resulting from an increase of CO$_2$ content of the atmosphere. *J. Atmos. Sci.* **37:**99–118.

Manabe, S., and K. Bryan, 1985. CO$_2$-induced change in a coupled ocean-atmosphere model and its paleoclimatic implications. *J. Geophys. Res.* **90:**11689–11708.

Manabe, S., and A. J. Broccoli, 1985a. A comparison of climate model sensitivity with data from the last glacial maximum. *J. Atmos. Sci.* **42:**2643–2651.

Manabe, S., and A. J. Broccoli, 1985b. The influence of continental ice sheets on the climate of an ice age. *J. Geophys. Res.* **90:**2167–2190.

Manabe, S., and R. T. Wetherald, 1986. Reduction in summer soil wetness induced by an increase in atmospheric carbon dioxide. *Science* **232:**626–628.

Manabe, S., and R. J. Stouffer, 1988. Two stable

equilibria of a coupled ocean-atmosphere model. *J. Clim.* **1**:841–866.

Manabe, S. and A. J. Broccoli, 1990. Mountains and arid climates of middle latitudes. *Science* **247**:192–195.

Mangerud, J., E. Sonstegaard, H. P. Sejrup, and S. Haldorsen, 1981. A continuous Eemian—early Weichselian sequence containing pollen and marine fossils at Fjosanger, western Norway. *Boreas* **10**:137–208.

Mankinen, E. A., and G. B. Dalrymple, 1979. Revised geomagnetic polarity time scale for the interval 0–5 m.y. B.P. *J. Geophys. Res.* **84**:615–626.

Manley, G., 1974. Central England temperatures: Monthly means 1659–1973. *Quat. J. Roy. Meteor. Soc.* **100**:389–405.

Margolis, S. V., J. F. Mount, E. Doehne, W. Showers, and P. Ward, 1987. The Cretaceous/Tertiary boundary carbon and oxygen isotope stratigraphy, diagenesis, and paleoceanography at Zumaya, Spain. *Paleoceanog.* **2**:361–377.

Markgraf, V., 1989. Palaeoclimates in Central and South America since 18,000 BP based on pollen and lake-level records. *Quat. Sci. Rev.* **8**:1–24.

Marotzke, J., P. Welander, and J. Willebrand, 1988. Instability and multiple steady states in a meridional-plane model of the thermohaline circulation. *Tellus* **40A**: 162–172.

Marshall, L. G., S. D. Webb, J. J. Sepkoski, and D. M. Raup, 1982. Mammalian evolution and the great American interchange. *Science* **215**:1351–1357.

Martinson, D. G., P. D. Killworth, and A. L. Gordon, 1981. A convective model for the Weddell Polynya. *J. Phys. Oceanog.* **11**:466–488.

Martinson, D. G., N. G. Pisias, J. D. Hays, J. Imbrie, T. C. Moore, and N. J. Shackleton, 1987. Age dating and the orbital theory of the ice ages: Development of a high-resolution 0–300,000-year chronostratigraphy. *Quat. Res.* **27**:1–29.

Matsui, T., and Y. Abe, 1986. Impact-induced atmospheres and oceans on Earth and Venus. *Nature* **322**:526–528.

Matthes, F. E., 1939. Report of committee on glaciers, April 1939. *Trans. Am. Geophys. Union* **20**:518–523.

Matthews, E., and I. Fung, 1987. Methane emission from natural wetlands: Global distribution, area, and environmental characteristics of sources. *Global Biogeochem. Cycles* **1**:61–86.

Matthews, R. K., 1974. *Dynamic Stratigraphy,* Prentice-Hall, Englewood Cliffs, N.J.

Matthews, R. K., and R. Z. Poore, 1980. Tertiary ^{18}O record and glacioeustatic sea-level fluctuation. *Geology* **8**:501–504.

Maurasse, F. J.-M. R., 1988. Step-wise extinctions at the Cretaceous/Tertiary boundary and their climatic implications. In: *Global Catastrophes in Earth History: An Interdisciplinary Conference on Impacts, Volcanism, and Mass Mortality*

(Abstract). Lun. Plan. Inst. Pub., Houston, Texas, p. 117.

Mayewski, P. A., 1975. Glacial geology and Late Cenozoic history of Trans-Antarctic Mountains, Antarctica, Institute of Polar Studies, Report **56**, Ohio State University, Columbus, Ohio.

Mayewski, P. A., G. H. Denton, and T. J. Hughes, 1981. Late Wisconsin ice sheets in North America. In: *The Last Great Ice Sheets.* G. H. Denton and T. J. Hughes (Eds.). John Wiley and Sons, New York, pp. 67–178.

Maykut, G. A., and D. K. Perovich, 1987. The role of shortwave radiation in the summer decay of a sea ice cover. *J. Geophys. Res.* **92**:7032–7044.

McAlester, A. L., 1977. *The History of Life.* Prentice-Hall, Englewood Cliffs, N.J.

McCartney, K., A. R. Huffman, and M. Tredoux, 1990. A paradigm for endogenous causation of mass extinctions. In: *Catastrophes in Earth History.* V. L. Sharpton and P. Ward (Eds.). Geol. Soc. Amer. Spec. Paper **247**, pp. 125–138.

McDonald, B. C., and W. W. Shilts, 1971. Quaternary events and stratigraphy, southeastern Quebec. *Geol. Soc. Am. Bull.* **82**:683–698.

McDougall, I., and H. Wensink, 1966. Paleomagnetism and geochronology of the Pliocene-Pleistocene lavas in Iceland. *Earth Plan. Sci. Lett.* **1**:232–236.

McElhinny, M. W., and W. E. Senanayake, 1982. Variations in the geomagnetic dipole 1: The past 50,000 years. *J. Geomagnet. Geoelect.* **34**:39–51.

McGlone, M. S., 1988. New Zealand. In: *Vegetation History,* B. Huntley and T. Webb, III (Eds.). Kluwer, Dordrecht, Netherlands, pp. 557–599.

McHone, J. F., R. A. Nieman, C. F. Lewis, and A. M. Yates, 1989. Stishovite at the Cretaceous-Tertiary boundary, Raton, New Mexico. *Science* **243**:1182–1184.

McIntyre, A., et al., 1976. The glacial North Atlantic 18,000 years ago: A CLIMAP reconstruction. In: *Investigation of Late Quaternary Paleoceanography and Paleoclimatology.* R. M. Cline and J. D. Hays (Eds.). Geol. Soc. Am. Mem. **145,** Geol. Soc. Am., Boulder, Col., pp. 43–76.

McIntyre, A., W. F. Ruddiman, K. Karlin, and A. C. Mix, 1989. Surface water response of the equatorial Atlantic Ocean to orbital forcing. *Paleoceanog.* **4**:19–55.

McKenna, M. C., 1980. Eocene paleolatitude, climate, and mammals of Ellesmere Island. *Palaeogeog., Palaeoclimatol., Palaeoecol.* **30**:349–362.

McLean, D. M., 1985. Mantle degassing induced dead ocean in the Cretaceous-Tertiary transition. In: *The Carbon Cycle and Atmospheric CO_2: Natural Variations Archean to Present.* E. T. Sundquist and W. S. Broecker (Eds.). Am. Geophys. Union, Washington, D.C., pp. 493–503.

McNeill, D. F., R. N. Ginsburg, S.-B. R. Chang, and J. L. Kirschvink, 1988. Magnetostratigraphic dating of shallow-water carbonates from San Salvador, Bahamas. *Geology* **16**:8–12.

McWilliams, M. O., and M. W. McElhinny, 1980. Lake Precambrian paleomagnetism of Australia: The Adelaide geosyncline. *J. Geol.* **88**:1–26.

Meehl, G. A., W. M. Washington, and A. J. Semtner, 1982. Experiments with a global ocean model driven by observed atmospheric forcing. *J. Phys. Oceanog.* **12**:301–312.

Meier, M. F., 1990. Reduced rise in sea level. *Nature* **343**:115–116.

Melosh, H. J., N. M. Schneider, K. J. Zahnle, and D. Latham, 1990. Ignition of global wildfires at the Cretaceous/Tertiary boundary. *Nature* **343**:251–254.

Menard, G., and P. Molnar, 1988. Collapse of a Hercynian Tibetan Plateau into a late Palaeozoic European basin and range province. *Nature* **334**:235–237.

Mengel, J. G., D. A. Short, and G. R. North, 1988. Seasonal snowline instability in an energy balance model. *Clim. Dynam.* **2**:127–131.

Mercer, J. H., 1968. Antarctic ice and Sangamon sea level. *Int. Assoc. Hydrol. Soc., AISH* Pub. **79**:217–225.

Mercer, J. H., 1976. Glacial history of southernmost South America. *Quat. Res.* **6**:125–166.

Mercer, J. H., 1978. West Antarctic ice sheet and CO_2 greenhouse effect: A threat of disaster. *Nature* **271**:321–325.

Mercer, J. H., 1984. Simultaneous climatic change in both hemispheres and similar bipolar interglacial warming: Evidence and implications. In: *Climate Processes and Climate Sensitivity.* J. E. Hansen and T. Takahashi (Eds.). Geophys. Mono. **29**, Am. Geophys. Union, Washington, D.C., pp. 307–313.

Mesolella, K. J., R. K. Matthews, W. S. Broecker, and D. L. Thurber, 1969. The astronomical theory of climatic change: Barbados data. *J. Geol.* **77**:250–274.

Milankovitch, M., 1930. Mathematische klimalehre und astronomische theorie der Klimaschwankungen. In: *Handbuch der Klimatologie,* **I.** W. Köppen and R. Geiger (Eds.). Gebrüder Borntraeger, Berlin.

Milankovitch, M., 1941. Canon of insolation and the ice age problem (in Yugoslavian). *K. Serb. Acad. Beorg.* Spec. Publ. **132**. (English translation by Israel Program for Scientific Translations, Jerusalem, 1969).

Miller, G. H., H. P. Sejrup, J. Mangerud, and B. G. Andersen, 1983. Amino acid ratios in Quaternary molluscs and foraminifera from western Norway: Correlation, geochronology and paleotemperature estimates. *Boreas* **12**:107–124.

Miller, J. R., and G. L. Russell, 1989. Ocean heat transport during the last glacial maximum. *Paleoceanog.* **4**:141–155.

Miller, J. R., G. L. Russell, and L. C. Tsang, 1983. Annual oceanic heat transports computed from an atmospheric model. *Dyn. Atmos. Oc.* **7**:95–109.

Miller, K. G., and R. G. Tucholke, 1983. Development of Cenozoic abyssal circulation south of the Greenland-Scotland Ridge. In *Structure and Development of the Greenland-Scotland Ridge, NATO Conf. Ser.* M. H. P. Bott, S. Saxov, M. Talwani, and J. Thiede (Eds.). Plenum Press, New York, pp. 549–589.

Miller, K. G., and R. G. Fairbanks, 1985. Oligocene to Miocene carbon isotope cycles and abyssal circulation changes. In: *The Carbon Cycle and Atmospheric CO_2: Natural Variations Archean to Present.* E. T. Sundquist and W. S. Broecker (Eds.). *Geophys. Mono.* **32**, Am. Geophys. Union, Washington, D.C., pp. 469–486.

Miller, K. G., T. R. Janecek, M. E. Katz, and D. J. Keil, 1987a. Abyssal circulation and benthic foraminiferal changes near the Paleocene/Eocene boundary. *Paleoceanog.* **2**:741–761.

Miller, K. G., R. G. Fairbanks, and G. S. Mountain, 1987b. Tertiary oxygen isotope synthesis, sea level history, and continental margin erosion. *Paleoceanog.* **2**:1–19.

Mintz, Y., 1981. A brief review of the present status of global precipitation estimates. In: *Precipitation Measurements from Space.* D. Atlas and O. W. Thiele (Eds.). NASA Workshop Report, pp. D-1–D-4.

Mitchell, J. F. B., N. S. Grahame, and K. H. Needham, 1988. Climate simulation for 9000 years before present: Seasonal variations and the effect of the Laurentide Ice Sheet. *J. Geophys. Res.* **93**:8282–8303.

Mitchell, J. F. B., C. A. Senior, and W. J. Ingram, 1989. CO_2 and climate: A missing feedback? *Nature* **341**:132–134.

Mitchell, J. M., 1966. Stochastic models of air-sea interaction and climatic fluctuation. Symposium on the Arctic Heat Budget and Atmospheric Circulation, Lake Arrowhead, Cal., 1966, *Mem. Rn-5233-NSF,* Rand Corp., Santa Monica, Cal.

Mitchell, J. M. (Ed.), 1968. *Causes of Climatic Change.* Met. Mono. **8**, Am. Meteorol. Soc., Boston, Mass.

Mitchell, J. M., C. W. Stockton, and D. M. Meko, 1979. Evidence of a 22-year rhythm of drought in the western United States related to the Hale solar cycle since the 17th century. In: *Solar-Terrestrial Influences on Weather and Climate.* B. M. McCormac and T. A. Seliga (Eds.). D. Reidel, Hingham, Mass., pp. 125–144.

Mix, A. C., 1987. The oxygen-isotope record of glaciation. In: *North America and Adjacent Oceans During the Last Deglaciation.* W. F. Ruddiman and H. E. Wright, Jr. (Eds.). The Geology of North America, v. **K-3.** Geol. Soc. Am., Boulder, Col., pp. 111–135.

Mix, A. C., 1989. Influence of productivity variations on long-term atmospheric CO_2. *Nature* **337**:541–544.

Mix, A. C., and R. G. Fairbanks, 1985. North Atlantic surface-ocean control of Pleistocene deep-ocean circulation. *Earth Plan. Sci. Lett.* **73**:231–243.

Mix, A. C., and W. F. Ruddiman, 1985. Structure and timing of the last deglaciation: Oxygen-isotope evidence. *Quat. Sci. Rev.* **4**:59–108.

Mix, A. C., and N. G. Pisias, 1988. Oxygen isotope analyses and deep-sea temperature changes: Implications for rates of oceanic mixing. *Nature* **331**:249–251.

Mix, A. C., W. F. Ruddiman, and A. McIntyre, 1986. Late Quaternary paleoceanography of the tropical Atlantic, 1: Spatial variability of annual mean sea-surface temperatures, 0–20,000 years B.P. *Paleoceanog.* **1**:43–66.

Molfino, B., N. G. Kipp, and J. J. Morley, 1982. Comparison of foraminiferal, coccolithophorid, and radiolarian paleotemperature equations: Assemblage coherency and estimate concordance. *Quat. Res.* **17**:279–313.

Molfino, B., L. H. Heusser, and G. M. Woillard, 1984. Frequency components of a Grande Pile pollen record: Evidence of precessional orbital forcing. In: *Milankovitch and Climate.* A. L. Berger, J. Imbrie, J. Hays, G. Kukla, and B. Saltzman (Eds.). D. Reidel, Dordrecht, Netherlands, pp. 391–404.

Molina-Cruz, A., 1977. The relation of the southern trade winds to upwelling processes during the last 75,000 years. *Quat. Res.* **8**:324–338.

Monin, A. S., 1975. Role of oceans in climatic models. In: *Physical Basis of Climate and Climate Modelling.* Report No. 16, GARP Publications Series. World Meteorol. Organiz. Geneva, pp. 201–205.

Montanari, A., R. L. Hay, W. Alvarez, F. Asaro, H. V. Michel, L. W. Alvarez, and J. Smit, 1983. Spheroids at the Cretaceous-Tertiary boundary are altered impact droplets of basaltic composition. *Geology* **11**:668–671.

Moorbath, S., R. K. O'Nions, and R. J. Pankhurst, 1975. The evolution of early Precambrian crustal rocks at Isua, West Greenland—Geochemical and isotopic evidence. *Earth Plan. Sci. Lett.* **27**:229–239.

Moore, T. C., W. H. Hutson, N. Kipp, J. D. Hays, W. Prell, P. Thompson and G. Boden, 1981a. The biological record of the ice-age ocean. *Palaeogeog., Palaeoclimatol., Palaeoecol.* **35**:357–370.

Moore, T. C., N. G. Pisias, and L. D. Keigwin, 1981b. Ocean basin and depth variability of oxygen isotopes in Cenozoic benthic foraminifera. *Mar. Micropaleontol.* **6**:465–481.

Moore, T. C., N. G. Pisias, and D. A. Dunn, 1982. Carbonate time series of the Quaternary and late Miocene sediments in the Pacific Ocean: A spectral comparison. *Mar. Geol.* **46**:217–234.

Moore, T. C., T. S. Loutit, and S. M. Grenlee, 1987. Estimating short-term changes in eustatic sea level. *Paleoceanog.* **2**:625–637.

Morel, P., and E. Irving, 1978. Tentative paleocontinental maps for the early Phanerozoic and Proterozoic. *J. Geol.* **86**:535–561.

Morgan, V. I., 1985. An oxygen isotope climate record from the Law Dome, Antarctica. *Clim. Change* **7**:415–426.

Morlan, R. E., and J. Clinq-Mars, 1982. Ancient Beringians: Human occupation in the late Pleistocene of Alaska and the Yukon Territory. In: *Paleoecology of Beringia.* D. M. Hopkins, J. V. Matthews, C. E. Schweger, and S. B. Young (Eds.). Academic Press, New York, pp. 353–381.

Morley, J. J., 1989. Variations in high-latitude oceanographic fronts in the southern Indian Ocean: An estimation based on faunal changes. *Paleoceanog.* **4**:547–554.

Morley, J. J., and J. D. Hays, 1981. Towards a high-resolution, global, deep-sea chronology for the last 750,000 years. *Earth Plan. Sci. Lett.* **53**:279–295.

Morrison, L. V., F. R. Stephenson, and J. Parkinson, 1988. Diameter of the sun in AD 1715. *Nature* **331**:421–423.

Mosley-Thompson, E., and L. G. Thompson, 1982. Nine centuries of microparticle deposition at the South Pole. *Quat. Res.* **17**:1–13.

Mosley-Thompson, E., L. G. Thompson, P. M. Grootes, and N. Gundestrup, 1990. Little Ice Age (Neoglacial) paleoenvironmental conditions at Siple Station, Antarctica. *Ann. Glac.*, **14**:199–204.

Mudie, P. J., and A. E. Aksu, 1984. Palaeoclimate of Baffin Bay from 300,000-year record of foraminifera, dinoflagellates and pollen. *Nature* **312**:630–634.

Muhs, D. R., and T. K. Kyser, 1987. Stable isotope compositions of fossil mollusks from southern California: Evidence for a cool last interglacial ocean. *Geology* **15**:119–122.

Nairn, A., 1961. *Descriptive Paleoclimatology.* Interscience, New York.

Nathorst, A. G., 1911. On the value of fossil floras of the Arctic regions as evidence of geological climates. *Geol. Mag.* **8**:217–225.

National Academy of Sciences, 1982. *Climate in Earth History.* Washington, D.C.

National Research Council (NRC), 1975. *Understanding Climatic Change: A Program for Action.* U.S. Committee for GARP, National Academy of Sciences, Washington, D.C.

Neftel, A., H. Oeschger, and H. Suess, 1981. Secular non-random variations of cosmogenic carbon-14 in the terrestrial atmosphere. *Earth Plan. Sci. Lett.* **56**:127–147.

Neftel, A., E. Moor, H. Oeschger, and B. Stauffer, 1985. Evidence from polar ice cores for the increase in atmospheric CO_2 in the past two centuries. *Nature* **315**:45–47.

Neftel, A., H. Oeschger, T. Staffelbach, and B. Stauffer, 1988. CO₂ record in the Byrd ice core 50,000–5,000 years BP. *Nature* **331**:609–611.

Nesbit, E. G., 1982. The tectonic setting and petrogenesis of komatiites. In: *Komatiites*. N. T. Arndt and E. G. Nesbit (Eds.). Allen and Unwin, London, pp. 501–520.

Newell, R. E., 1974. Changes in poleward energy flux by the atmosphere and ocean as a possible cause for ice ages. *Quat. Res.* **4**:117–127.

Newman, M. J., 1986. The solar neutrino problem: Gadfly for solar evolution theory. In: *Physics of the Sun*. P. A. Sturrock (Ed.). Reidel, Dordrecht, pp. 33–46.

Newman, M. J., and R. T. Rood, 1977. Implications of solar evolution for the earth's early atmosphere. *Science* **194**:1413–1414.

Newton, H. J., 1988. *TIMESLAB: A Time Series Analysis Laboratory*, Wadsworth & Brooks/ Cole, Pacific Grove, Ca.

Newton, H. J., G. R. North, and T. J. Crowley. 1991. Forecasting global ice volume. In: *Regional Forecasting with Global Atmospheric Models* (Second Year Report). T. J. Crowley, G. R. North and N. R. Smith (Eds.). Battelle-Pacific Northwest Laboratory Contract Report 017113-A-B1, pp. H.1-H.9. *J. Time Ser. Anal.* **12**:255–265.

Nichols, D. J., D. M. Jarzen, C. J. Orth, and P. Q. Oliver, 1986. Palynological and iridium anomalies at Cretaceous-Tertiary boundary, south-central Saskatchewan. *Science* **231**:714–717.

Nicholson, S. E., 1978. Climatic variations in the Sahel and other African regions during the past five centuries. *J. Arid Envir.* **1**:3–24.

Nicolis, C., 1984. Self-oscillations, external forcings, and climate predictability. In: *Milankovitch and Climate*. A. L. Berger, J. Imbrie, J. Hays, G. Kukla, and B. Saltzman (Eds.). D. Reidel, Dordrecht, Netherlands, pp. 637–652.

Norris, G., 1982. Spore-pollen evidence for early Oligocene high-latitude cool climatic episode in northern Canada. *Nature* **297**:387–389.

Norris, G., 1986. Systematic and stratigraphic palynology of Eocene to Pliocene strata in the imperial Nuktak C-22 well, MacKenzie Delta region, District of MacKenzie, N.W.T. *Geol. Surv. Canada Bull.* **340**:1–89.

North, G. R., 1975a. Analytical solution of a simple climate model with diffusive heat transport. *J. Atmos. Sci.* **32**:1301–1307.

North, G. R., 1975b. Theory of energy-balance climate models. *J. Atmos. Sci.* **32**:2033–2043.

North, G. R., 1984. The small ice cap instability in diffusive climate models. *J. Atmos. Sci.* **41**:3390–3395.

North, G. R., 1988. Lessons from energy balance models. In: *Physically-Based Modelling and Simulation of Climate and Climatic Change*. M. E. Schlesinger (Ed.). Kluwer, Dordrecht, Netherlands, pp. 627–651.

North, G. R., and J. A. Coakley, 1979. Differences between seasonal and mean annual energy balance model calculations of climate and climate sensitivity. *J. Atmos. Sci.* **36**:1189–1204.

North, G. R., and R. F. Cahalan, 1981. Predictability in a solvable stochastic climate model. *J. Atmos. Sci.* **38**:504–513.

North, G. R., and T. J. Crowley, 1985. Application of a seasonal climate model to Cenozoic glaciation. *J. Geol. Soc. (London)* **142**:475–482.

North, G. R., R. F. Cahalan, and J. A. Coakley, 1981. Energy balance climate models. *Rev. Geophys. Space Phys.* **19**:91–121.

North, G. R., T. L. Bell, R. F. Cahalan, and F. J. Moeng, 1982. Sampling errors in the estimation of empirical orthogonal functions. *Mon. Wea. Rev.* **110**:75–82.

North, G. R., J. G. Mengel, and D. A. Short, 1983. Simple energy balance model resolving the seasons and the continents: Application to the astronomical theory of the ice ages. *J. Geophys. Res.* **88**:6576–6586.

North, G. R., J. G. Mengel, and D. A. Short, 1984. On the transient response patterns of climate to time dependent concentrations of atmospheric CO₂. In: *Climate Processes and Climate Sensitivity*. J. E. Hansen and T. Takahashi (Eds.). Geophys. Mono **29**, Am. Geophys. Union, Washington, D.C., pp.164–170.

Oerlemans, J., 1980. Model experiments on the 100,000-yr. glacial cycle. *Nature* **287**: 430–432.

Oerlemans, J., 1981. Some basic experiments with a vertically-integrated ice sheet model. *Tellus* **33**:1–11.

Oerlemans, J., 1982a. Glacial cycles and ice-sheet modelling. *Clim. Change* **4**:353–374.

Oerlemans. J. 1982b. Response of the Antarctic ice sheet to a climatic warming: A model study. *J. Climatol* **2**:1–11.

Oeschger, H., 1985. The contribution of ice core studies to the understanding of environmental processes. In: *Greenland Ice Core: Geophysics, Geochemistry, and the Environment*. C. C. Langway, H. Oeschger, and W. Dansgaard (Eds.). Geophys. Mono. **33**, Am. Geophys. Union, Washington, D.C., pp. 9–17.

Oeschger, H., B. Stauffer, R. Findel, and C. C. Langway, 1985. Variations of the CO₂ concentration of occluded air and of anions and dust in polar ice cores. In: *The Carbon Cycle and Atmospheric CO₂: Natural Variations Archean to Present*. E. T. Sundquist and W. S. Broecker (Eds.). Geophys. Mono. **32**, Am. Geophys. Union, Washington, D.C., pp. 132–142.

Officer, C. B., and C. L. Drake, 1983. The Cretaceous-Tertiary transition. *Science* **219**:1383–1390.

Officer, C. B., and C. L. Drake, 1985. Terminal Cretaceous environmental events. *Science* **227**:1161–1167.

Oglesby, R. J., 1989. A GCM study of Antarctic glaciation. *Clim. Dynam.* **3**:135–156.

Oglesby, R. J., and J. Park, 1989. The effect of precessional insolation changes on Cretaceous climate and cyclic sedimentation. *J. Geophys. Res.* **94**:14793–14816.

Oglesby, R. J., and B. Saltzman, 1990. Extending the EBM: The effect of deep ocean temperatures on climate with applications to the Cretaceous. *Glob. Plan. Change* **2**:237–259.

Oglesby, R. J., K. A. Maasch, and B. Saltzman, 1989. Glacial meltwater cooling of the Gulf of Mexico: GCM implications for Holocene and present-day climates. *Clim. Dynam.* **31**:115–133.

O'Keefe, J. D., and T. J. Ahrens, 1982. The interaction of the Cretaceous/Tertiary extinction bolide with the atmosphere, ocean, and solid Earth. *Geol. Soc. Am. Spec. Pap.* **190**:103–120.

O'Keefe, J. D., and T. J. Ahrens, 1989. Impact production of CO_2 by the Cretaceous/Tertiary extinction bolide and the resultant heating of the Earth. *Nature* **338**:247–249.

Olsen, P. E., 1986. A 40-million-year lake record of early Mesozoic orbital climatic forcing. *Science* **234**:842–848.

Olson, E. C., 1982. Extinctions of Permian and Triassic nonmarine vertebrates. *Geol. Soc. Am. Spec. Pap.* **190**:501–511.

Omar, G., K. R. Johnson, L. J. Hickey, P. B. Robertson, M. R. Dawson, and C. W. Barnosky, 1987. Fission-track dating of Houghton Astrobleme and included biota, Devon Island, Canada. *Science* **237**:1603–1605.

Oort, A. H., and T. H. von der Haar, 1976. On the observed annual cycle in the ocean-atmosphere heat balance over the northern hemisphere. *J. Phys. Oceanog.* **6**:781–800.

Opdyke, B. N., and B. H. Wilkinson, 1988. Surface area control of shallow cratonic to deep marine carbonate accumulation. *Paleoceanog.* **3**:685–704.

Oppo, D. W., and R. G. Fairbanks, 1987. Variability in the southern polar ocean during the past 25,000 years: Northern hemisphere modulation. *Earth Plan. Sci. Lett.* **86**:1–15.

Orth, C. J., and M. Attrep, 1988. Iridium abundance measurements across bio-event horizons in the geologic record. In: *Global Catastrophes in Earth History: An Interdisciplinary Conference on Impacts, Volcanism, and Mass Mortality* (Abstracts). Lun. Plan. Inst., Houston, Texas, pp. 139–140.

Otto-Bliesner, B. L., G. W. Branstator, and D. D. Houghton, 1982. A global low-order spectral general circulation model. Part I: Formulation and seasonal climatology. *J. Atmos. Sci.* **39**:929–948.

Overpeck, J. T., 1985. A pollen study of a late Quaternary peat bog, south-central Adirondack Mountains, New York. *Geol. Soc. Am. Bull.* **96**:145–154.

Overpeck, J. T., T. Webb III, and I. C. Prentice,

1985. Quantitative interpretion of fossil pollen spectra: Dissimilarity coefficients and the method of modern analogs. *Quat. Res.* **23**:87–108.

Overpeck, J. T., L. C. Peterson, N. Kipp, J. Imbrie, and D. Rind, 1989. Climate change in the circum-North Atlantic region during the last deglaciation. *Nature* **338**:553–557.

Owen, R. M., and D. K. Rea, 1985. Sea-floor hydrothermal activity links climate to tectonics: The Eocene carbon dioxide greenhouse. *Science* **227**:166–169.

Owen, M. R., and M. H. Anders, 1988. Evidence from cathodoluminescence for non-volcanic origin of shocked quartz at the Cretaceous/Tertiary boundary. *Nature* **334**:145–148.

Owen, T., 1978. The origins and early histories of planetary atmospheres. In: *Evolution of Planetary Atmospheres and Climatology of the Earth* (Proc. Intl. Conf., Nice). Centre National D'Etudes Spatiales, Toulouse, France, pp. 1–10.

Owen, T., R. D. Cess, and V. Ramanathan, 1979. Early earth: An enhanced carbon dioxide greenhouse to compensate for reduced solar luminosity. *Nature* **277**:640–642.

Owen, T., J. P. Maillard, C. DeBergh, and B. L. Lutz, 1988. Deuterium on Mars: The abundance of HDO and the value of D/H. *Science* **240**:1767–1770.

Park, R., and S. Epstein, 1960. Carbon isotope fractionation during photosynthesis. *Geochim. Cosmochim. Acta* **21**:110–126.

Parkin, D. W., and N. J. Shackleton, 1973. Trade wind and temperature correlations down a deep-sea core off the Saharan coast. *Nature* **245**:455–457.

Parkinson, C. L., 1990. Search for the Little Ice Age in Southern Ocean sea ice records. *Ann. Glac.* **14**:221–225.

Parkinson, C. L., and W. W. Kellogg, 1979. Arctic sea ice decay simulated for a CO_2 induced temperature rise. *Clim. Change* **2**:149–162.

Parkinson, C. L., and W. M. Washington, 1979. A large-scale numerical model of sea ice. *J. Geophys. Res.* **84**:311–337.

Parkinson, C. L., and R. A. Bindschadler, 1984. Response of Antarctic sea ice to uniform atmospheric temperature increases. In: *Climate Processes and Climate Sensitivity*. J. E. Hansen and T. Takahashi (Eds.). Geophys. Mono. **29**, Am. Geophys. Union, Washington, D.C., pp. 254–264.

Parkinson, J. H., 1983. New measurements of the solar diameter. *Nature* **304**:518–520.

Parrish, J. M., J. T. Parrish, and A. M. Ziegler, 1986. Permian-Triassic paleogeography and paleoclimatology and implications for Therapsid distribution. In: *The Ecology and Biology of Mammal-like Reptiles*. N. Hotton II, P. D. MacLean, J. J. Roth, and E. C. Roth (Eds.).

Smithsonian Institution Press, Washington, D.C., pp. 109–131.

Parrish, J. M., J. T. Parrish, J. H. Hutchison, and R. A. Spicer, 1987. Late Cretaceous vertebrate fossils from the North Slope of Alaska and implications for dinosaur ecology. *Palaios* 2:377–389.

Parrish, J. T., and R. A. Spicer, 1988. Late Cretaceous terrestrial vegetation: A near-polar temperature curve. *Geology* 16:22–25.

Parrish, J. T., A. M. Ziegler, and C. R. Scotese, 1982. Rainfall patterns and the distribution of coals and evaporites in the Mesozoic and Cenozoic. *Palaeogeog., Palaeoclimatol., Palaeoecol.* 40:67–101.

Parrish, J. T., G. C. Gaynor, and D. J. P. Swift, 1984. Circulations in the Cretaceous western interior seaway of North America, a review. In: *The Mesozoic of Middle North America.* D. F. Scott and D. J. Glass (Eds.). Can. Soc. Petrol. Geol. 9:221–231.

Paterson, W. S. B., and C. U. Hammer, 1987. Ice core and other glaciological data. In: *North America and Adjacent Oceans During the Last Deglaciation.* W. F. Ruddiman and H. R. Wright (Eds.). The Geology of North America, v. K-3. Geol. Soc. Am., Boulder, Col., pp. 91–109.

Patrusky, B., 1980. Pre-Clovis man: Sampling the evidence. *Mosaic* 11:2–10.

Pedersen, T. F., 1983. Increased productivity in the eastern equatorial Pacific during the last glacial maximum (19,000 to 14,000 yr B.P.). *Geology* 11:16–19.

Peixoto, J. P., and A. H. Oort, 1983. The atmospheric branch of the hydrological cycle and climate. In: *Variations in the Global Water Budget.* A. Street-Perrot, M. Beran, and R. Ratcliffe, (Eds.). Reidel, Dordrecht, Netherlands, pp. 5–65.

Peltier, W. R., 1987. Glacial isostasy, mantle viscosity, and Pleistocene climatic change. In: *North America and Adjacent Oceans During the Last Deglaciation.* W. F. Ruddiman and H. E. Wright (Eds.). The Geology of North America v. K-3, Geol. Soc. Am., Boulder, Col., pp. 155–182.

Peltier, W. R. and A. M. Tushingham, 1989. Global sea level rise and the greenhouse effect: Might they be connected? *Science* 244:806–810.

Pepin, R. O., J. A. Eddy, and R. B. Merrill (Eds.), 1980. *The Ancient Sun (Fossil Record in the Earth, Moon, and Meteorites).* Pergamon Press, New York.

Perch-Nielsen, K., J. McKenzie, and Q. He, 1982. Biostratigraphy and isotope stratigraphy and the "catastrophic" extinction of calcareous nannoplankton at the Cretaceous/Tertiary boundary. *Geol. Soc. Am. Spec. Pap.* 190:353–371.

Peterson, L. C., 1986. The response of the deep-sea carbonate system to climatic forcing: Late Pleistocene records from the Indian Ocean. *Eos (Trans. Am. Geophys. Union)* 67:868.

Peterson, L. C., and W. L. Prell, 1985. Carbonate preservation and rates of climatic change: An 800 KYR record from the Indian Ocean. In: *The Carbon Cycle and Atmospheric CO$_2$: Natural Variations Archean to Present.* E. T. Sundquist and W. S. Broecker (Eds.). Geophys. Mono. 32, Am. Geophys. Union, Washington, D.C., pp. 251–269.

Peterson, W. H., 1979. A steady thermohaline convection model. *Tech. Report* TR-79-4, Univ. Miami, Rosenstiel School of Marine and Atmospheric Science, Miami, Flor.

Petit, J. R., M. Briat, and A. Royer, 1981. Ice age aerosol content from East Antarctic ice core samples and past wind strength. *Nature* 293:391–394.

Petit, J. R., et al., 1990. Palaeoclimatological and chronological implications of the Vostok core dust record. *Nature* 343:56–58.

Petit-Maire, N. 1986. Palaeoclimates in the Sahara of Mali: A multidisciplinary study. *Episodes* 9:7–16.

Petters, S. W., 1979. West African cratonic stratigraphic sequences. *Geology* 7:528–531.

Péwé, T. L., 1983. The periglacial environment in North America during Wisconsin time. In: *Late-Quaternary Environments of the United States.* H. E. Wright, Jr. (Ed.). Univ. of Minn. Press, Minneapolis, pp. 157–189.

Pfister, C., 1985. *CLIMHIST: A weather data bank for Central Europe.* Meteostat, Bern, pp. 1525–1863.

Pickton, C. A. G., 1981. Palaeogene and Cretaceous dropstones in Spitsbergen. In: *Earth's Pre-Pleistocene Glacial Record.* H. A. Hambrey and W. B. Harland (Eds.). Cambridge Univ. Press, Cambridge, pp. 567–569.

Pillmore, C. L., R. H. Tschudy, C. J. Orth, J. S. Gilmore, and J. D. Knight, 1984. Geologic framework of nonmarine Cretaceous-Tertiary boundary sites, Raton Basin, New Mexico and Colorado. *Science* 223:1180–1183.

Piper, J. D. A., 1976. Palaeomagnetic evidence for a Proterozoic supercontinent. *Phil. Trans. Roy. Soc. (London)* A 280:469–490.

Piper, J. D. A., 1983. Proterozoic paleomagnetism and single continent plate tectonics. *Geophys. Roy. Astron. Soc.* 74:163–197.

Pirrie, D., and J. D. Marshall, 1990. High-paleolatitude Late Cretaceous paleotemperatures: New data from James Ross Island, Antarctica. *Geology* 18:31–34.

Pisias, N. G., 1979. Model for paleoceanographic reconstructions of the California Current during the last 8000 years. *Quat. Res.* 11:373–386.

Pisias, N. G., and M. Leinen, 1984. Milankovitch forcing of the oceanic system: Evidence from the northwest Pacific. In: *Milankovitch and Climate*. A. L. Berger, J. Imbrie, J. Hays, G. Kukla, and B. Saltzman (Eds.). D. Reidel, Dordrecht, Netherlands, pp. 307–330.

Pisias, N. G., and W. L. Prell, 1985. High resolution carbonate records from the hydraulic piston cored section of site 572. *Init. Rep. Deep-Sea Drill Proj.* **85**:711–722.

Pisias, N. G., and J. Imbrie, 1986/87. Orbital geometry, CO_2, and Pleistocene climate. *Oceanus* **29**:43–49.

Pisias, N. G., and D. K. Rea, 1988. Late Pleistocene paleoclimatology of the central equatorial Pacific: Sea surface response to the southeast trade winds. *Paleoceanog.* **3**:21–37.

Pisias, N. G., J. P. Dauphin, and C. Sancetta, 1973. Spectral analysis of late Pleistocene-Holocene sediments. *Quat. Res.* **3**:3–9.

Pitman, W. C., 1978. Relationship between eustacy and stratigraphic sequences of passive margins. *Geol. Soc. Am. Bull.* **89**:1389–1403.

Pittock, A. B., 1983. Solar variability, weather and climate: An update. *Quart. J. R. Met. Soc.* **109**:23–55.

Pokras, E. M., 1987. Diatom record of late Quaternary climate change in the eastern equatorial Atlantic and tropical Africa. *Paleoceanog.* **2**:273–286.

Pokras, E. M., and A. C. Mix, 1985. Eolian evidence for spacial variability of late Quaternary climates in tropical Africa. *Quat. Res.* **24**:137–149.

Pokras, E. M., and A. C. Mix, 1987. Earth's precession cycle and Quaternary climatic changes in tropical Africa. *Nature* **326**:486–487.

Pollack, J. B., and Y. L. Yung, 1980. Origin and evolution of planetary atmospheres. *Ann. Rev. Earth Plan. Sci.* **8**:425–488.

Pollack, J. B., O. B. Toon, T. P. Ackerman, C. P. McKay, and R. P. Turco, 1983. Environmental effects of an impact-generated dust cloud: Implications for the Cretaceous-Tertiary extinctions. *Science* **219**:287–289.

Pollard, D., 1978. An investigation of the astronomical theory of the ice ages using a simple climate-ice sheet model. *Nature* **272**:233–235.

Pollard, D., 1982. A simple ice sheet model yields realistic 100 kyr glacial cycles. *Nature* **296**:334–338.

Pollard, D., 1984. Some ice-age aspects of a calving ice-sheet model. In: *Milankovitch and Climate*. A. L. Berger, J. Imbrie, J. Hays, G. Kukla, and B. Saltzman (Eds.). D. Reidel, Dordrecht, Netherlands, pp. 541–564.

Pollard, D., A. P. Ingersoll, and J. G. Lockwood, 1980. Response of a zonal climate-ice sheet model to the orbital perturbations during the Quaternary ice ages. *Tellus* **32**:301–319.

Pond, S., and G. L. Pickard, 1983. *Introductory Dynamical Oceanography*. Pergamon Press, Oxford, England.

Poore, R. Z., 1981. Late Miocene biogeography and paleoclimatology of the central North Atlantic. *Mar. Micropaleo.* **6**:599–616.

Poore, R. Z., and R. K. Matthews, 1984. Oxygen isotope ranking of Late Eocene and Oligocene planktonic foraminifers: Implications for Oligocene sea-surface temperatures and global ice volume. *Mar. Micropaleo.* **9**:111–134.

Porter, S. C., 1975. Equilibrium-line altitudes of late Quaternary glaciers in the Southern Alps, New Zealand. *Quat. Res.* **5**:27–47.

Porter, S. C., 1979. Hawaiian glacial ages. *Quat. Res.* **12**:161–187.

Porter, S. C., 1981. Pleistocene glaciation in the southern Lake District of Chile. *Quat. Res.* **16**:263–321.

Porter, S. C. (Ed.), 1983. *Late Quaternary Environments of the United States*, **v. 1, *The Pleistocene***. Univ. Minnesota Press, Minneapolis.

Porter, S. C., 1986. Pattern and forcing of northern hemisphere glacier variations during the last millennium. *Quat. Res.* **26**:27–48.

Porter, S. C., and G. Orombelli, 1985. Glacier contraction during the middle Holocene in the western Italian Alps: Evidence and implications. *Geology* **13**:296–298.

Posey, J. W., and P. F. Clapp, 1964. Global distribution of normal surface albedo. *Geofis. Int.* **4**:33–48.

Powell, C. McA., and J. J. Veevers, 1987. Namurian uplift in Australia and South America triggered the main Gondwanan glaciation. *Nature* **326**:177–179.

Prakash, U., 1972. Paleoenvironmental analysis of Indian Tertiary floras. *Geophyt.* **2**:178–205.

Prance, G. T., 1982. Forest refuges: Evidence from woody angiosperms. In: *Biological Diversification in the Tropics*. G. T. Prance (Ed.). Columbia Univ. Press, New York, pp. 137–158.

Preisinger, A., et al., 1986. The Cretaceous/Tertiary boundary in the Gosau Basin, Austria. *Nature* **322**:794–800.

Prell, W. L., 1982. Oxygen and carbon isotope stratigraphy for the Quaternary of Hole 502B: Evidence for two modes of isotopic variability. *Init. Rep. Deep-Sea Drill. Proj.* **68**:455–464.

Prell, W. L., 1984a. Covariance patterns of foraminiferal ^{18}O: An evaluation of Pliocene ice volume changes near 3.2 million years ago. *Science* **226**:692–694.

Prell, W. L., 1984b. Monsoonal climate of the Arabian Sea during the late Quaternary: A response to changing solar radiation. In *Milankovitch and Climate*, A. Berger et al. (Eds.). Reidel, Dordrecht, Netherlands, pp. 349–366.

Prell, W. L., 1985. The stability of low-latitude sea-surface temperatures: An evaluation of the CLIMAP reconstruction with emphasis on the positive SST anomalies. *U.S. Dept. Energy Cont. Rep.*, Cont. No. DE-ACO2-83ER60167.

Prell, W. L., and J. D. Hays, 1976. Late Pleistocene faunal and temperature patterns of the Colombia Basin, Caribbean Sea. In: *Investigation of Late Quaternary Paleoceanography and Paleoclimatology.* R. M. Cline and J. D. Hays (Eds.). Geol Soc. Am. Mem. **145**:201–220.

Prell, W. L., and J. E. Kutzbach, 1987. Monsoon variability over the past 150,000 years. *J. Geophys. Res.* **92**:8411–8425.

Prentice, I. C., 1980. Multidimensional scaling as a research tool in Quaternary palynology: A review of theory and methods. *Rev. Palaeobot. Palynol.* **31**:71–104.

Prentice, M. L., 1985. Peleus glaciation of Wright Valley, Antarctica. *S. Afr. J. Sci.* **81**:241–243.

Prentice, M. L., and R. K. Matthews, 1988. Cenozoic ice-volume history: Development of a composite oxygen isotope record. *Geology* **16**:963–966.

Prinn, R. G., and B. Fegley, 1987. Bolide impacts, acid rain, and biospheric traumas at the Cretaceous-Tertiary boundary. *Earth Plan. Sci. Lett.* **83**:1–15.

Prothero, D. R., 1985. Mid-Oligocene extinction event in North American land mammals. *Science* **229**:550–551.

Quade, J., T. E. Cerling, and J. R. Bowman, 1989. Development of Asian monsoon revealed by marked ecological shift during the latest Miocene in northern Pakistan. *Nature* **342**:163–166.

Railsback, L. B., S. C. Ackerly, T. F. Anderson, and J. L. Cisne, 1990. Palaeontological and isotope evidence for warm saline deep waters in Ordovician oceans. *Nature* **343**:156–159.

Raisbeck, G. M, et al., 1987. Evidence for two intervals of enhanced ^{10}Be deposition in Antarctic ice during the last glacial period. *Nature* **326**:273–277.

RamaMurthy, V., 1976. Composition of the core and early chemical history of the earth. In: *The Early History of the Earth.* B. F. Windley (Ed.). John Wiley and Sons, New York, pp. 21–31.

Ramanathan, V., 1988. The greenhouse theory of climate change: A test by an inadvertent global experiment. *Science* **240**:293–299.

Ramanathan, V., and J. A. Coakley, 1978. Climate modeling through radiative-convective models. *Rev. Geophys. Space Phys.* **16**:465–489.

Ramanathan, V., et al., 1989. Cloud-radiative forcing and climate: Results from the earth radiation budget experiment. *Science* **243**:57–63.

Rampino, M. R., and R. B. Stothers, 1988. Flood basalt volcanism during the past 250 million years. *Science* **241**:663–668.

Rampino, M. R., and T. Volk, 1988. Mass extinctions, atmospheric sulphur and climatic warming at the K/T boundary. *Nature* **332**:63–65.

Rasmusson, E. M., and T. H. Carpenter, 1983. The relationship between eastern equatorial Pacific sea surface temperatures and rainfall over India and Sri Lanka. *Mon. Wea. Rev.* **111**:517–528.

Rau, G. H., T. Takahashi, and D. J. Des Marais, 1989. Latitudinal variations in plankton δ^{13}C: Implications for CO_2 and productivity in past oceans. *Nature* **341**:516–518.

Raup, D. M., 1979. Size of the Permo-Triassic bottleneck and its evolutionary implications. *Science* **206**:217–218.

Raup, D. M., and J. J. Sepkoski, 1984. Periodicity of extinctions in the geologic past. *Proc. Natl. Acad. Sci. USA* **81**:801–805.

Raup, D. M., and J. J. Sepkoski, 1986. Periodic extinction of families and genera. *Science* **231**:833–836.

Raup, D. M., and G. E. Boyajian, 1988. Patterns of generic extinction in the fossil record. *Paleobiol.* **14**:109–125.

Raval, A., and V. Ramanathan, 1989. Observational determination of the greenhouse effect. *Nature* **342**:758–761.

Raven, P. H., and D. I. Axelrod, 1974. Angiosperm biogeography and past continental movements. *Ann. Missouri Bot. Gard.* **61**:539–673.

Raymo, M. E., W. F. Ruddiman, and B. M. Clement, 1986. Pliocene-Pleistocene paleoceanography of the North Atlantic at Deep Sea Drilling Project Site 609. *Init. Rep. Deep-Sea Drill. Proj.* **94**:895–901.

Raymo, M. E., W. F. Ruddiman, and P. N. Froelich, 1988. Influence of late Cenozoic mountain building on ocean geochemical cycles. *Geology* **16**:649–653.

Raymo, M. E., W. F. Ruddiman, J. Backman, B. M. Clement, and D. G. Martinson, 1989. Late Pliocene variations in northern hemisphere ice sheets and North Atlantic Deep Water circulation. *Paleoceanog.* **4**:413–446.

Raymo, M. E., W. Ruddiman, and D. Rind, 1990a. Climatic effects of variable Arctic sea ice in the GISS-II GCM: A possible analogy for the Pliocene. *Paleoceanog.* **5**:367–382.

Raymo M. E., W. F. Ruddiman, N. J. Shackleton, and D. W. Oppo, 1990b. Evolution of Atlantic-Pacific δ^{13}C gradients over the last 2.5 m.y. *Earth Plan. Sci. Lett.,* **97**:353–368.

Raymond, A., P. H. Kelley, and C. B. Lutken, 1989. Polar glaciers and life at the equator: The history of Dinantian and Namurian (Carboniferous) climate. *Geology* **17**:408–411.

Rea, D. K., 1989. Geologic record of atmospheric circulation on tectonic time scales. In: *Paleoclimatology and Paleometerorology: Modern and Past Patterns of Global Atmospheric Transport.* M. Leinen and M. Sarnthein (Eds.). Kluwer, Dordrecht, Netherlands, pp. 841–857.

Rea, D. K., and H. Schrader, 1985. Late Pliocene onset of glaciation: Ice-rafting and diatom stratigraphy of North Pacific DSDP cores. *Palaeogeog., Palaeoclim., Palaeoecol.* **49**:313–325.

Rea, D. K., and M. K. Bloomstein, 1986. Neogene

history of the South Pacific trade winds: Evidence for hemispherical asymmetry of atmospheric circulation. *Palaeogeog., Palaeoclim., Palaeoecol.* **55**:55–64.

Rea, D. K., M. Leinen, and T. R. Janecek, 1985. Geologic approach to the long-term history of atmospheric circulation. *Science* **227**:721–725.

Rea, D. K., et al., 1986. A 420,000-year record of cyclicity in oceanic and atmospheric processes from the eastern equatorial Pacific. *Paleoceanog.* **1**:577–586.

Rea, D. K., S. A. Hovan, and T. R. Janecek, 1990. Late Quaternary flux of eolian dust to the pelagic ocean. In: *Global Geoflux Study.* Nat. Res. Council, in press.

Reid, E. M., and M. E. J. Chandler, 1933. *The London Clay Flora.* British Museum, London.

Reid, G. C., 1987. Influence of solar variability on global sea surface temperatures. *Nature* **329**:142–143.

Ren, Z., 1987. The abnormal periods of climate in China over the past 5000 years and their causes. *Adv. Atmos. Sci.* **4**:210–217.

Rensberger, B., 1980. The emergence of *Homo sapiens. Mosaic* **11**:2–12.

Repenning, C. A., 1967. Subfamilies and genera of the Soricidae. *U.S. Geol. Surv. Prof. Paper* **565**.

Retalleck, G. J., 1983. Late Eocene and Oligocene paleosols from Badlands National Park, South Dakota. *Geol. Soc. Am. Spec. Paper* **193**.

Ribe, N. M., and A. B. Watts, 1982. The distribution of intraplate volcanism in the Pacific Ocean basin: A spectral approach. *Geophys. J. Roy. Astron. Soc.* **71**:333–362.

Ribes, E., J. C. Ribes, and R. Barthalot, 1987. Evidence for a larger sun with a slower rotation during the seventeenth century. *Nature* **326**:42–55.

Rich, P. V., et al., 1988. Evidence for low temperatures and biologic diversity in Cretaceous high latitudes of Australia. *Science* **242**:1403–1406.

Riehl, H. 1979. *Climate and Weather in the Tropics.* Academic Press, London.

Rind, D., 1984. The influence of vegetation on the hydrologic cycle in a global climate model. In: *Climate Processes and Climate Sensitivity.* J. E. Hansen and T. Takahashi (Eds.). Geophys. Mono **29**, Am. Geophys. Union, Washington, D.C., pp. 73–91.

Rind, D., 1986. The dynamics of warm and cold climates. *J. Atmos. Sci.* **43**:3–24.

Rind, D., 1987. Components of the ice age circulation. *J. Geophys. Res.* **92**:4241–4281.

Rind, D., 1988. Dependence of warm and cold climate depiction on climate model resolution. *J. Clim.* **1**:965–997.

Rind, D., and D. Peteet, 1985. Terrestrial conditions at the last glacial maximum and CLIMAP sea-surface temperature estimates: Are they consistent? *Quat. Res.* **24**:1–22.

Rind, D., D. Peteet, W. Broecker, A. McIntyre, and W. Ruddiman, 1986. The impact of cold North Atlantic sea surface temperatures on climate: Implications for the Younger Dryas cooling (11–10k). *Clim. Dynam.* **1**:3–33.

Rind, D., G. Kukla, and D. Peteet, 1989. Can Milankovitch orbital variations initiate the growth of ice sheets in a general circulation model? *J. Geophys. Res.* **94**:12851–12871.

Ritchie, J. C., and C. V. Haynes, 1987. Holocene vegetation zonation in the eastern Sahara. *Nature* **330**:645–647.

Ritchie, J. C., L. C. Cwynar, and R. W. Spear, 1983. Evidence from north-west Canada for an early Holocene Milankovitch thermal maximum. *Nature* **305**:126–128.

Roberts, D. G., A. C. Morton, and J. Backmann, 1984. Late Paleocene-Eocene volcanic events in the northern North Atlantic Ocean. *Init. Rep. Deep-Sea Drill. Proj.* **81**:913–923.

Robin, G. DeQ., 1985. Contrasts in Vostok ice core—changes in climate or ice volume? *Nature* **316**:578–579.

Robinson, P. L., 1973. Palaeoclimatology and continental drift. In: *Implications of Continental Drift to Earth Sciences,* v. 1. D. H.Tarling and S. K. Runcorn (Eds.). Academic Press, New York, pp. 451–476.

Robock, A., 1978 Internally and externally caused climate change. *J. Atmos. Sci.* **35**:1111–1122.

Roemmich, D., and Wunsch, C., 1984. Apparent changes in the climatic state of the deep North Atlantic Ocean. *Nature* **307**:447–450.

Ronov, A. B., 1964. Common tendencies in the chemical evolution of the earth's crust, ocean and atmosphere. *Geochem.* **8**:715–743.

Ronov, A. B., 1968. Probable changes in the composition of seawater during the course of geological time. *Sedimentol.* **10**:25–43.

Rooth, C., 1982. Hydrology and ocean circulation. *Prog. Oceanog.* **11**:131–149.

Rosen, B. R., 1984. Reef coral biogeography and climate through the Late Cainozoic: Just islands in the sun or a critical pattern of islands? In: *Fossils and Climate.* P. Brenchley (Ed.). John Wiley and Sons, New York, pp. 201–262.

Ross, C. A., and J. R. P. Ross, 1985. Late Paleozoic depositional sequences are synchronous and worldwide. *Geology* **13**:194–197.

Rossignol-Strick, M., 1983. African monsoons, an immediate climate response to orbital insolation. *Nature* **304**:46–49.

Rossignol-Strick, M., 1987. Rainy periods and bottom water stagnation initiating brine accumulation and metal concentrations: 1. The late Quaternary. *Paleoceanog.* **2**:333–360.

Rossignol-Strick, M., W. Nesteroff, P. Olive, and C. Vergnaud-Grazzini, 1982. After the deluge: Mediterranean stagnation and sapropel formation. *Nature* **295**:105–110.

Röthlisberger, F., 1986. *10,000 Jahre Gletschergeschichte der Erde.* Aarau, Verlag, Sauerländer.

Royer, A., M. De Angelis, and J. R. Petit, 1983. A 30,000 yr record of physical and optical properties of microparticles from an East Antarctic ice core and implications for climatic reconstruction models. *Clim. Change* **5**:381–412.

Royer, J. F., M. Deque, and P. Pestiaux, 1983. Orbital forcing of the inception of the Laurentide ice sheet? *Nature* **304**:43–46.

Royer, J. F., M. Deque, and P. Pestiaux, 1984. A sensitivity experiment to astronomical forcing with a spectral GCM: Simulation of the annual cycle at 125,000 BP and 115,000 BP. In: *Milankovitch and Climate.* A. L. Berger, J. Imbrie, J. Hays, G. Kukla, and B. Saltzman (Eds.). Reidel, Dordrecht, Netherlands, pp. 733–763.

Rubey, W. W., 1951. Geologic history of seawater. *Geol. Soc. Am. Bull.* **62**:1111–1148.

Ruddiman, W. F., 1968. Historical stability of the Gulf Stream meander belt: Foraminiferal evidence. *Deep-Sea Res.* **15**:137–148.

Ruddiman, W. F., 1971. Pleistocene sedimentation in the equatorial Atlantic: Stratigraphy and faunal paleoclimatology. *Geol. Soc. Am. Bull.* **82**:283–302.

Ruddiman, W. F., 1977. Late Quaternary deposition of ice-rafted sand in the sub-polar North Atlantic (lat. 40–65°N). *Geol. Soc. Am. Bull.* **88**:1813–1827.

Ruddiman, W. F., 1985. Climate studies in ocean cores. In: *Paleoclimate Analysis and Modeling.* A. D. Hecht (Ed.). John Wiley and Sons, New York, pp. 197–257.

Ruddiman, W. F., 1987. Northern oceans, In: *North America and Adjacent Oceans During the Last Deglaciation.* W. F. Ruddiman and H. E. Wright, Jr. (Eds.). The Geology of North America, v. **K-3**. Geol. Soc. of Am., Boulder, Col., pp. 137–154.

Ruddiman, W. F., and A. McIntyre, 1976. Northeast Atlantic paleoclimatic changes over the past 600,000 years. In: *Investigation of Late Quaternary Paleoceanography and Paleoclimatology.* R. M. Cline and J. D. Hays (Eds.). Geol Soc. Am. Mem. **145**:111–146.

Ruddiman, W. F., and A. McIntyre, 1979. Warmth of the subpolar North Atlantic Ocean during northern hemisphere ice-sheet growth. *Science* **204**:173–175.

Ruddiman, W. F., and A. McIntyre, 1981a. The mode and mechanism of the last deglaciation: Oceanic evidence. *Quat. Res.* **16**:125–134.

Ruddiman, W. F., and A. McIntyre, 1981b. The North Atlantic Ocean during the last deglaciation. *Palaeogeog., Palaeoclimatol., Palaeoecol.* **35**:145–214.

Ruddiman, W. F., and A. McIntyre, 1981c. Oceanic mechanisms for amplification of the 23,000-year ice-volume cycle. *Science* **212**:617–627.

Ruddiman, W. F., and A. McIntyre, 1982. Severity and speed of northern hemisphere glaciation pulses: The limiting case? *Geol. Soc. Am. Bull.* **93**:1273–1279.

Ruddiman, W. F., and A. McIntyre, 1984. Ice-age thermal response and climatic role of the surface Atlantic Ocean, 40°N to 63°N. *Geol. Soc. Am. Bull.* **95**:381–396.

Ruddiman, W. F. and J.-C. Duplessy, 1985. Conference on the last deglaciation: Timing and mechanism. *Quat. Res.* **23**:1–17.

Ruddiman, W. F. and H. E. Wright (Eds.), 1987. *North America and Adjacent Oceans During the Last Deglaciation.* The Geology of North America, v. **K-3**. Geol. Soc. Am., Boulder, Col.

Ruddiman, W. F., and M. E. Raymo. 1988. Northern hemisphere climate regimes during the past 3 Ma: Possible tectonic connections. *Phil Trans. Roy. Soc. (London)* **B 318**:411–430.

Ruddiman, W. F., and J. E. Kutzbach, 1989. Forcing of late Cenozoic northern hemisphere climate by plateau uplift in southeast Asia and the American southwest. *J. Geophy. Res.* **94**:18409–18427.

Ruddiman, W. F., B. Molfino, A. Esmay, and E. Pokras, 1980. Evidence bearing on the mechanism of rapid deglaciation. *Clim. Change* **3**:65–87.

Ruddiman, W. F., M. Raymo, and A. McIntyre, 1986a. Matuyama 41,000-year cycles: North Atlantic Ocean and northern hemisphere ice sheets. *Earth Plan. Sci. Lett.* **80**:117–129.

Ruddiman, W. F., N. J. Shackleton, and A. McIntyre, 1986b. North Atlantic sea-surface temperatures for the last 1.1 million years. *Geol. Soc. (London) Spec. Pub.* **21**:155–173.

Ruddiman, W. F., W. L. Prell, and M. E. Raymo, 1989a. History of late Cenozoic uplift in southeast Asia and the American southwest: Rationale for general circulation modeling experiments. *J. Geophys. Res.* **94**:18379–18391.

Ruddiman, W. F., M. E. Raymo, D. G. Martinson, B. M. Clement, and J. Backman, 1989b. Pleistocene evolution: Northern hemisphere ice sheets and North Atlantic Ocean. *Paleoceanog.* **4**:353–412.

Russell, D. A., 1979. The enigma of the extinction of the dinosaurs. *Ann. Rev. Earth Plan. Sci.* **7**:163–182.

Russell, D. A., 1982. The mass extinctions of the Late Mesozoic. *Sci. Am.* **246**:48–55.

Ryan, W. B. F., 1973. Geodynamic implications of the Messinian crisis of salinity. In: *Messinian Events in the Mediterranean.* C. W. Drooger (Ed.). Koninklijke Nederlands Akademie van Wetenschappen, North-Holland, Amsterdam, pp. 26–38.

Ryan, W. B. F., and M. B. Cita, 1977. Ignorance concerning episodes of ocean-wide stagnation. *Mar. Geol.* **23**:197–215.

Sachs, H. M., 1973. Late Pleistocene history of the North Pacific: Evidence from a quantitative study of radiolaria in core V21–173. *Quat. Res.* **3**:89–98.

Sachs, H. M., T. Webb, and D. R. Clark, 1977. Paleoecological transfer functions. *Ann. Rev. Earth Plan. Sci.* **5**:159–178.

Sadourny, R., and K. Laval, 1984. January and July performance of the LMD general circulation model. In: *New Perspectives in Climate Modelling.* Developments in Atmospheric Science **16**. A. Berger and G. Nicolis (Eds.). Elsevier, Amsterdam, pp. 173–198.

Sagan, C., and G. Mullen, 1972. Earth and Mars: Evolution of atmospheres and surface temperature. *Science* **177**:52–56.

Saigne, C., and M. Legand, 1987. Measurements of methanesulphonic acid in Antarctic ice. *Nature* **330**:240–242.

Saito, T., T. Yamanonoi, and K. Kaiho, 1986. Cretaceous devastation of terrestrial flora in the boreal Far East. *Nature* **323**:253–255.

Salati, E., 1985. The climatology and hydrology of Amazonia. In: *Amazonia.* G. T. Prance and T. E. Lovejoy (Eds.). Pergamon Press, Oxford, England, pp. 18–48.

Saltzman, B., 1985. Paleoclimate modeling. In: *Paleoclimate Analysis and Modeling.* A. D. Hecht (Ed.). Wiley-Interscience, New York, pp. 341–396.

Saltzman, B., 1987. Carbon dioxide and the $\delta^{18}O$ record of late Quaternary climatic change. *Clim. Dynam.* **1**:77–86.

Saltzman, B., 1988. Modelling the slow climatic attractor. In: *Physically-based Modelling and Simulation of Climate and Climatic Change.* M. E. Schlesinger (Ed.). Kluwer, Dordrecht, Netherlands, pp. 737–754.

Saltzman, B., and A. Sutera, 1984. A model of the internal feedback system involved in late Quaternary climatic variations. *J. Atmos. Sci.* **41**:736–745.

Saltzman, B., and A. Sutera, 1987. The mid-Quaternary climatic transition as the free response of a three-variable dynamical model. *J. Atmos. Sci.* **44**:236–241.

Saltzman, B., A. Sutera, and A. Evenson, 1981. Structural stochastic stability of a simple autooscillatory climatic feedback system. *J. Atmos. Sci.* **38**:494–503.

Saltzman, E. S., and E. J. Barron, 1982. Deep circulation in the Late Cretaceous: Oxygen isotope paleotemperatures from *Inoceramus* remains in DSDP cores. *Palaegeog., Palaeoclimatol., Palaeoecol.* **40**:167–181.

Sancetta, C., and S. M. Silvestri, 1986. Pliocene-Pleistocene evolution of the North Pacific ocean-atmosphere system, interpreted from fossil diatoms. *Paleoceanog.* **1**:163–180.

Sanchez, W. A., and J. E. Kutzbach, 1974. Climate of the American tropics and subtropics in the 1960s and possible comparisons with climatic variations of the last millenium. *Quat. Res.* **4**:128–135.

Sandberg, P. A., 1983. An oscillating trend in Phanerozoic nonskeletal carbonate mineralogy. *Nature* **305**:19–22.

Sandberg, P. A., 1985. Nonskeletal aragonite and pCO_2 in the Phanerozoic and Proterozoic. In: *The Carbon Cycle and Atmospheric CO_2: Natural Variations Archean to Present.* E. T. Sundquist and W. S. Broecker (Eds.). Geophys. Mono. **32**, Am. Geophys. Union, Washington D.C., pp. 585–594.

Sanford, R. L., J. Saldarriaga, K. E. Clark, C. Uhl, and R. Herrera, 1985. Amazon rain-forest fires. *Science* **227**:53–55.

Sarmiento, J. L., T. D. Herbert, and J. R. Toggweiler, 1988a. Causes of anoxia in the world oceans. *Glob. Biogeochem. Cycles* **2**:115–128.

Sarmiento, J. L., J. R. Toggweiler, and R. Najjar, 1988b. Ocean carbon-cycle dynamics and atmospheric pCO_2. *Phil Trans. Roy. Soc. (London)* Series A **325**:3–21.

Sarnthein, M., 1978. Sand deserts during glacial maximum and climatic optimum. *Nature* **272**:43–45.

Sarnthein, M., G. Tetzlaff, B. Koopman, K. Wolter, and U. Pflaumann, 1981. Glacial and interglacial wind regimes over the eastern subtropical Atlantic and north-west Africa. *Nature* **293**:193–196.

Sarnthein, M., K. Winn, J.-C. Duplessy, and M. R. Fontugne, 1988. Global variations of surface ocean productivity in low and mid latitudes: Influence on CO_2 reservoirs of the deep ocean and atmosphere during the last 21,000 years. *Paleoceanog.* **3**:361–399.

Sausen, R., K. Barthel, and K. Hasselmann, 1988. Coupled ocean-atmosphere models with flux correction. *Clim. Dynam.* **2**:145–163.

Savin, S. M., 1977. The history of the earth's surface temperature during the past 100 million years. *Ann. Rev. Earth Plan. Sci.* **5**:319–355.

Savin, S. M., R. G. Douglas, and F. G. Stehli, 1975. Tertiary marine paleotemperatures. *Geol. Soc. Am. Bull.* **86**:1499–1510.

Savin, S. M., et al., 1985. The evolution of Miocene surface and near-surface marine temperatures: Oxygen isotopic evidence. *Geol. Soc. Am. Mem.* **163**:49–82.

Schatten, K. H., 1988. A model for solar constant secular changes. *Geophys. Res. Lett.* **15**:121–124.

Schidlowski, M., 1988. A 3,800-million-year isotopic record of life from carbon in sedimentary rocks. *Nature* **333**:313–318.

Schlanger, S. O., 1986. High frequency sea-level fluctuations in Cretaceous time: An emerging geophysical problem. In: *Mesozoic and Cenozoic Oceans.* K. J. Hsü (Ed.). Am. Geophys. Union, Washington, D.C., pp. 61–74.

Schlesinger, M. E., 1984. Mathematical modeling and simulation of climate and climatic change. Report, Institut D'Astronomie et de Geophysique, Univ. Catholique de Louvain, Georges Lemaitre, Belgium.

Schlesinger, M. E., 1986. Equilibrium and transient climatic warming induced by increased atmospheric CO_2. *Clim. Dynam.* **1**:35–51.

Schlesinger, M. E. (Ed.), 1988. *Physically-Based Modelling and Simulation of Climate and Climatic Change.* Kluwer, Dordrecht, Netherlands.

Schlesinger, M. E., 1989. Model projections of the climatic changes induced by increased atmospheric CO_2. In: *Climate and Geo-Sciences.* A. Berger, S. H. Schneider, and J.-C. Duplessy (Eds.). D. Reidel, Dordrecht, Netherlands, pp. 375–415.

Schlesinger, M. E., and W. L. Gates, 1980. The January and July performance of the OSU two-level atmospheric general circulation model. *J. Atmos Sci.* **37**:1914–1943.

Schlesinger, M. E., and J. F. B. Mitchell, 1985. Model projections of the equilibrium climatic response to increased carbon dioxide. In: *Projecting the Climatic Effects of Increasing Carbon Dioxide.* M. C. MacCracken and F. M. Luther (Eds.). U.S. Dept. Energy Report, DOE/ER-0237, Washington, D.C., pp. 83–147.

Schlesinger, M. E., and J. F. B. Mitchell, 1987. Climate model simulations of the equilibrium climatic response to increased carbon dioxide. *Rev. Geophys.* **25**:760–798.

Schlesinger, M. E., and Z.-C. Zhao, 1989. Seasonal climatic changes induced by doubled CO_2 as simulated by the OSU atmospheric GCM/mixed-layer ocean model. *J. Clim.* **2**:459–495.

Schneider, S. H., and T. Gal-Chen, 1973. Numerical experiments in climate stability. *J. Geophys. Res.* **78**:6182–8194.

Schneider, S. H., and R. E. Dickinson, 1974. Climate modeling. *Rev. Geophys. Space Phys.* **12**:447–493.

Schneider, S. H., and S. L. Thompson, 1988. Simulating the climatic effects of nuclear war. *Nature* **333**:221–227.

Schneider, S. H., S. L. Thompson, and E. J. Barron, 1985. Mid-Cretaceous continental surface temperatures: Are high CO_2 concentrations needed to simulate above freezing winter conditions? In: *The Carbon Cycle and Atmospheric CO_2: Natural Variations Archean to Present.* E. T. Sundquist and W. S. Broecker (Eds.). Geophys. Mono. **32**. Am. Geophys. Union, Washington, D.C., pp. 554–560.

Schneider, S. H., D. M. Peteet, and G. R. North, 1987. A climate model intercomparison for the Younger Dryas and its implications for paleoclimatic data collection. In: *Abrupt Climatic Change.* W. H. Berger and L. D. Labeyrie (Eds.). D. Reidel, Dordrecht, Netherlands, pp. 399–417.

Schoell, M. 1984. Stable isotopes in petroleum research. In: *Advances in Petroleum Geochemistry,* **v.I.** J. Brooks and D. Welte (Eds.). Academic Press, London, pp. 215–245.

Scholle, P. A., and M. A. Arthur, 1980. Carbon isotope fluctuations in Cretaceous pelagic limestones: Potential stratigraphic and petroleum exploration tool. *Am. Assoc. Petrol. Geol. Bull.* **64**:67–87.

Scholz, C. A., and B. R. Rosendahl, 1988. Low lake stands in Lakes Malawi and Tanganyika, East Africa, delineated with multifold seismic data. *Science* **240**:1645–1648.

Schopf, J. W., and B. M. Parker, 1987. Early Archean (3.3 billion to 3.5-billion-year-old) microfossils from Warrawoona Group, Australia. *Science* **237**:70–73.

Schopf, T. J. M., 1980. *Paleoceanography.* Harvard Univ. Press, Cambridge, Mass.

Schramm, C. T., 1985. Implications of radiolarian assemblages for the late Quaternary paleoceanography of the eastern equatorial Pacific. *Quat. Res.* **24**:204–218.

Schutz, C., and W. L. Gates, 1972. *Global Climatic Data for Surface, 800 mb, 400 mb. (July).* Report R-1029-ARPA. Rand Corporation, Santa Monica, Cal.

Schwartz, S. E., 1988. Are global cloud albedo and climate controlled by marine phytoplankton? *Nature* **336**:441–445.

Scotese, C. R., 1986. Phanerozoic reconstructions: A new look at the assembly of Asia. *Univ. Texas Inst. for Geophys. Tech. Rep.* **66**.

Scotese, C. R., R. K. Bambach, C. Barton, R. Van der Voo, and A. M. Ziegler, 1979. Paleozoic base maps. *J. Geol.* **87**:217–277.

Scott, D. B., P. J. Mudie, V. Baki, K. D. MacKinnon, and F. E. Cole, 1989. Biostratigraphy and late Cenozoic paleoceanography of the Arctic Ocean: Foraminiferal, lithostratigraphic, and isotopic evidence. *Geol. Soc. Am. Bull.* **101**:260–277.

Scuderi, L. A., 1987. Late-Holocene upper timberline variation in the southern Sierra Nevada. *Nature* **325**:242–244.

Sear, C. B., P. M. Kelly, P. D. Jones, and C. M. Goodess, 1987. Global surface-temperature responses to major volcanic eruptions. *Nature* **330**:365–367.

Seidov, D. G., 1986. Numerical modelling of the ocean circulation and paleocirculation. In: *Mesozoic and Cenozoic Oceans.* K. Hsü (Ed.). Geodynamic Series, **v. 15.** Am. Geophys. Union, Washington, D.C., pp. 11–26.

Self, S., M. Rampino, and J. J. Barbera 1981. The possible effects of large 19th and 20th century volcanic eruptions on zonal and hemispheric surface temperatures. *J. Volcanol. Geothermal. Res.* **11**:41–60.

Sellers, P. J., Y. Mintz, Y. C. Sud, and A. Dalcher, 1986. A simple biosphere model (SiB) for use within general circulation models. *J. Atmos. Sci.* **43**:505–531.

Sellers, W. D., 1969. A climate model based on the energy balance of the earth-atmosphere system. *J. Appl. Meteorol.* **8**:392–400.

Semtner, A. J., 1987. A numerical study of sea ice and ocean circulation in the Arctic. *J. Phys. Oceanog.* **17**:1077–1099.

Semtner, A. J., and R. M. Chervin, 1988. A simulation of the global ocean circulation with resolved eddies. *J. Geophys. Res.* **93**:15502–15222.

Sergin, V. Y., 1980. Origin and mechanism of large-scale climatic oscillation. *Science* **209**:1477–1482.

SethuRaman, S., and A. J. Riordan, 1988. The Genesis of Atlantic Lows Experiment: The planetary-boundary-layer subprogram of GALE. *Bull. Am. Meteor. Soc.* **2**:161–712.

SethuRaman, S., A. J. Riordan, T. Holt, M. Stunder, and J. Hinman, 1986. Observations of the marine boundary layer thermal structure over the Gulf Stream during a cold air outbreak. *J. Clim. Appl. Meteor.* **25**:14–21.

Shackleton, N. J., 1967. Oxygen isotope analyses and Pleistocene temperatures re-assessed. *Nature* **215**:15–17.

Shackleton, N. J., 1969. The last interglacial in the marine and terrestrial records. *Proc. Roy. Soc. (London)* Ser. B **174**:135–154.

Shackleton, N. J., 1977. Carbon-13 in *Uvigerina*: Tropical rainforest history and the equatorial Pacific carbonate dissolution cycles. In: *The Fate of Fossil Fuel CO_2 in the Oceans.* N. R. Andersen and A. Malahoff (Eds.). Plenum Press, New York, pp. 401–428.

Shackleton, N. J., 1982. The deep-sea sediment record of climate variability. *Prog. Oceanog.* **11**:199–218.

Shackleton, N. J., 1985. Oceanic carbon isotope constraints on oxygen and carbon dioxide in the Cenozoic atmosphere. In: *The Carbon Cycle and Atmospheric CO_2: Natural Variations Archean to Present.* E. T. Sundquist and W. S. Broecker (Eds.). Geophys. Mono. **32**. Am. Geophys. Union, Washington, D.C., pp. 412–418.

Shackleton, N. J., 1986. Paleogene stable isotope events. *Palaeogeog., Palaeoclimatol., Palaeoecol.* **57**:91–102.

Shackleton, N. J., 1987. The carbon isotope record of the Cenozoic: History of organic carbon burial and of oxygen in the ocean and atmosphere. In: *Marine Petroleum Source Rocks.* J. Brooks and A. J. Fleet (Eds.). Geol. Soc. (London) Spec. Pub. **26**:423–434.

Shackleton, N. J., 1988a. Ocean carbon isotope data and the Vostok CO_2 record. *Eos (Trans. Am. Geophys. Union)* **69**:1235.

Shackleton, N. J., 1988b. Oxygen isotopes, ice volume, and sea level. *Quat. Sci. Rev.* **6**:183–190.

Shackleton, N. J., and N. D. Opdyke, 1973. Oxygen isotope and paleomagnetic stratigraphy of equatorial Pacific core V28-238: Oxygen isotope temperatures and ice volume on a 100,000 and 1,000,000 year scale. *Quat. Res.* **3**:39–55.

Shackleton, N. J., and J. P. Kennett, 1975a. Paleo-

temperature history of the Cenozoic and the initiation of Antarctic glaciation: Oxygen and carbon isotope analysis in DSDP sites 277, 279, and 281. In: *Init. Rep. Deep-Sea Drill. Proj.* v. **29**. J. P. Kennett et al. (Eds.). U.S. Gov. Print. Office, Washington, D.C., pp. 743–755.

Shackleton, N. J., and J. P. Kennett, 1975b. Late Cenozoic oxygen and carbon isotope changes at DSDP site 284: Implications for glacial history of the northern hemisphere and Antarctica. In: *Init. Rep. Deep-Sea Drill. Proj.* v. **29**. J. P. Kennett et al. (Eds.). U.S. Gov. Print. Office, Washington, D.C., pp. 801–807.

Shackleton, N. J., and N. D. Opdyke, 1976. Oxygen isotope and paleomagnetic stratigraphy of Pacific core V28-239 late Pliocene to latest Pleistocene. In: *Investigation of Late Quaternary Paleoceanography and Paleoclimatology.* R. M. Cline and J. D. Hays (Eds.). Geol. Soc. Am. Mem., **145**. Geol. Soc. Am., Boulder, Col., pp. 449–464.

Shackleton, N. J., and N. D. Opdyke, 1977. Oxygen isotope and paleomagnetic evidence for early northern hemisphere glaciation. *Nature* **270**:216–219.

Shackleton, N. J., and A. Boersma, 1981. The climate of the Eocene ocean. *J. Geol. Soc. (London)* **138**:153–157.

Shackleton, N. J., and N. G. Pisias, 1985. Atmospheric carbon dioxide, orbital forcing, and climate. In: *The Carbon Cycle and Atmospheric CO_2: Natural Variations Archean to Present.* E. T. Sundquist and W. S. Broecker (Eds.). Geophys. Mono. **32**. Am. Geophys. Union, Washington, D.C., pp. 303–317.

Shackleton, N. J., M. A. Hall, J. Line, and C. Shuxi, 1983. Carbon isotope data in core V19-30 confirm reduced carbon dioxide concentration of the ice age atmosphere. *Nature* **306**:319–322.

Shackleton, N. J., et al. 1984. Oxygen isotope calibration of the onset of ice-rafting and history of glaciation in the North Atlantic region. *Nature* **307**:620–623.

Shackleton, N. J., J.-C. Duplessy, M. Arnold, P. Maurice, M. A. Hall, and J. Cartlidge, 1988. Radiocarbon age of last glacial Pacific deep water. *Nature* **335**:708–711.

Shemesh, A., Y. Kolodny, and B. Luz, 1983. Oxygen isotope variations in phosphate of biogenic apatites, II. Phosphorite rocks. *Earth Plan. Sci. Lett.* **64**:405–416.

Shemesh, A., R. A. Mortlock, and P. N. Froelich, 1989. Late Cenozoic Ge/Si record of marine biogenic opal: Implications for variations of riverine fluxes to the ocean. *Paleoceanog.* **4**:221–234.

Shinn, R. A., and E. J. Barron, 1989. Climate sensitivity to continental ice sheet size and configuration. *J. Clim.* **2**:1517–1537.

Shoemaker, E. M., C. S. Shoemaker, and R. F.

Wolfe, 1988. Asteroid and comet flux in the neighborhood of the earth. In: *Global Catastrophes in Earth History: An Interdisciplinary Conference on Impacts, Volcanism, and Mass Mortality* (Abstracts). Lunar Plan. Inst., Houston, Texas, pp. 174–176.

Short, D. A., and J. G. Mengel, 1986. Tropical climatic phase lags and earth's precession cycle. *Nature* 323:48–50.

Short, D. A., G. R. North, T. D. Bess, and G. L. Smith, 1984. Infrared parameterization and simple climate models. *J. Clim. Appl. Meteorol.* 23:1222–1233.

Short, D. A., J. G. Mengel, T. J. Crowley, W. T. Hyde, and G. R. North, 1991. Filtering of Milankovitch cycles by earth's geography. *Quat. Res.* 35:157–173.

Shukla, J., and Y. Mintz, 1982. Influence of land-surface evapotranspiration on the earth's climate. *Science* 215:1498–1501.

Sikes, S. K., 1971. *The Natural History of the African Elephant.* Elsevier, New York.

Simms, M. J., and A. H. Ruffell, 1989. Synchroneity of climatic change and extinctions in the Late Triassic. *Geology* 17:265–168.

Slater, R. D., 1981. A numerical model of tides in the Cretaceous seaway of North America. MS Thesis, University of Chicago.

Sleep, N. H., K. J. Zahnle, J. F. Kasting and H. J. Morowitz, 1989. Annihilation of ecosystems by large asteroid impacts on the early Earth. *Nature* 342:139–142.

Sliter, W. V., 1989. Aptian anoxia in the Pacific Basin. *Geology* 17:909–912.

Sloan, L. C., and E. J. Barron, 1990. "Equable" climates during earth history? *Geology* 18:489–492.

Sloss, L. L., 1963. Sequences in the cratonic interior of North America. *Geol. Soc. Am. Bull.* 74:93–114.

Smiley, C. J., 1967. Paleoclimatic interpretations of some Mesozoic floral sequences. *Am. Assoc. Pet. Geol. Bull.* 51:849–863.

Smit, J., and J. Hertogen, 1980. An extraterrestrial event at the Cretaceous-Tertiary boundary. *Nature* 285:198–200.

Smit, J., and G. Klaver, 1981. Sanidine spherules at the Cretaceous-Tertiary boundary indicate a large impact event: *Nature* 292:47–49.

Smith, G. I., 1984. Paleohydrologic regimes in the southwestern great basin, 0-3.2 my ago, compared with other long records of "Global" climate. *Quat. Res.* 22:1–17.

Smith, G. I., and F. A. Street-Perrott, 1983. Pluvial lakes of the western United States. In: *Late-Quaternary Environments of the United States.* H. E. Wright, Jr. (Ed.). Univ. of Minn. Press, Minneapolis, pp. 190–212.

Smith, I. N., and W. F. Budd, 1981: The derivation of past climatic changes from observed changes of glaciers. In: *Sea Level, Ice and Climatic Change.* I. Allison (Ed.). Int. Assoc. Hydrol. Sci., Pub. 131:31–52.

Smith, N. D., A. C. Phillips, and R. D. Powell, 1990. Tidal drawdown: A mechanism for producing cyclic sediment laminations in glaciomarine deltas. *Geology* 18:10–13.

Sofia, S. (Ed.), 1981. *Variations of the Solar Constant.* NASA Conf. Pub. 2191, NASA/Goddard Space Flight Center, Greenbelt, Maryland.

Sofia, S., 1984. Solar variability as a source of climate change. In: *Climate Processes and Climate Sensitivity.* J. E. Hansen and T. Takahashi (Eds.). Geophys. Mono. 29. Am. Geophys. Union, Washington, D.C., pp. 202–206.

Sofia, S., D. W. Dunham, J. B. Dunham, and A. D. Fiala, 1983. Solar radius change between 1925 and 1979. *Nature* 304:522–526.

Sohl, N. F., 1969. North American Cretaceous biotic provinces delineated by gastropods. In: *Proceedings North American Paleontologic Convention II.* Allen Press, Lawrence, Kansas, pp. 1610–1637.

Sonett, C. P., and H. E. Suess, 1984. Correlation of bristlecone pine ring widths with atmospheric ^{14}C variations: A climate-sun relation. *Nature* 307:141–143.

Sonntag, C., et al., 1980. Isotopic identification of Saharian groundwaters, groundwater formation in the past. In: *Palaeoecology of Africa, v. 12.* E. M. van Zinderen Bakker and J. A. Coetzee (Eds.). A. A. Balkema, Rotterdam, pp. 159–187.

Spaulding, W. G., and L. J. Graumlich, 1986. The last pluvial climatic episodes in the deserts of southwestern North America. *Nature* 320:441–444.

Spicer, R. A., and J. T. Parrish, 1986. Paleobotanical evidence for cool north polar climates in middle Cretaceous (Albian-Cenomanian) time. *Geology* 14:703–706.

Stabell, B., 1986. Variations of diatom flux in the eastern equatorial Atlantic during the last 400,000 years (Meteor cores 13519 and 13521). *Mar. Geol.* 72:305–323.

Stahle, D. W., and M. K. Cleaveland, 1988. Texas drought history reconstructed and analyzed from 1698 to 1980. *J. Clim.* 1:59–74.

Stahle, D. W., M. K. Cleaveland, and J. G. Hehr, 1985. A 450-year drought reconstruction for Arkansas, United States. *Nature* 316:530–532.

Stahle, D. W., M. K. Cleaveland, and J. G. Hehr, 1988. North Carolina climate changes reconstructed from tree rings: A.D. 372 to 1985. *Science* 240:1517–1519.

Stanley, S. M., 1986. Anatomy of a regional mass extinction: Plio-Pleistocene decimation of the western Atlantic bivalve fauna. *Palaios* 1:17–36.

Stanley, S. M., 1988. Paleozoic mass extinctions: Shared patterns suggest global cooling as a common cause. *Am. J. Sci.* 288:334–352.

Stanley, S. M., 1989. *Earth and Life Through Time,* 2nd ed. W. H. Freeman, New York.

Stanley, S. M., and L. D. Campbell, 1981. Neogene mass extinction of western Atlantic molluscs. *Nature* **293**:457–459.

State Meteorological Administration (SMA), 1981. *Annals of 510 years' precipitation record in China* (in Chinese). The Meteorological Research Institute, Beijing.

Stauffer, B., A. Neftel, H. Oeschger, and J. Schwander, 1985a. CO$_2$ concentration in air extracted from Greenland ice samples. In: *Greenland Ice Core: Geophysics, Geochemistry, and the Environment.* C. C. Langway, H. Oeschger, and W. Dansgaard (Eds.). Geophys. Mono. **33**. Am. Geophys. Union, Washington, D.C., pp. 85–90.

Stauffer, B., G. Fischer, A. Neftel, and H. Oeschger, 1985b. Increase of atmospheric methane recorded in Antarctic ice core. *Science* **229**:1386–1388.

Stauffer, B., E. Lochbronner, H. Oeschger, and J. Schwander, 1988. Methane concentration in the glacial atmosphere was only half that of the preindustrial Holocene. *Nature* **332**:812–814.

Stein, R., and M. Sarnthein, 1984. Late Neogene events of atmospheric and oceanic circulation offshore northwest Africa: High resolution record from deep-sea sediments. In: *Palaeoecology of Africa*, v. **16**. J. A. Coetzee and E. M. van Zinderen Bakker (Eds.). Balkema, Rotterdam, pp. 9–36.

Stephens, G. L., G. G. Campell, and T. H. von der Haar, 1981. Earth radiation budgets. *J. Geophys. Res.* **86**:9739–9760.

Stephenson, F. R., and A. W. Wolfendale, 1988. *Secular Solar and Geomagnetic Variations in the last 10,000 years.* Kluwer, Dordrecht, Netherlands.

Stevens, C. H., 1977. Was development of brackish oceans a factor in Permian extinctions? *Geol. Soc. Am. Bull.* **88**:133–138.

Stigler, S. M., and M. J. Wagner, 1987. A substantial bias in nonparametric tests for periodicity in geophysical data. *Science* **238**:940–945.

Stockton, C. W., W. R. Boggess, and D. M. Meko, 1985. Climate and tree rings. In: *Paleoclimate Analysis and Modeling.* A. D. Hecht (Ed.). Wiley-Interscience, New York, pp. 71–150.

Stommel, H., and E. Stommel, 1981. The year without a summer. *Sci. Am.* **240**:176–186.

Stommel, H., and J. C. Swallow, 1983. Do late grape harvests follow large volcanic eruptions? *Bull. Am. Meteor. Soc.* **64**:794–795.

Stone, P. H., 1978. Constraints of dynamical transports of energy on a spherical planet. *Dyn. Atmos. Oc.* **2**:123–139.

Stothers, R. B., 1984. The great Tambora eruption in 1815 and its aftermath. *Science* **224**:1191–1198.

Stott, L. D., and J. P. Kennett, 1988. Cretaceous/Tertiary boundary in the Antarctic: Climatic cooling precedes biotic crisis. In: *Global Catastrophes in Earth History: An Interdisciplinary Conference on Impacts, Volcanism, and Mass Mortality* (Abstracts). Lun. Plan. Inst., Houston, Texas, pp. 184–185.

Stott, L. D. and J. P. Kennett, 1989. New constraints on early Tertiary palaeoproductivity from carbon isotopes in foraminifera. *Nature* **342**:526–529.

Stott, L. D., and J. P. Kennett, 1990. The paleoceanographic and paleoclimatic signature of the Cretaceous/Paleogene boundary in the Antarctic: Stable isotope results from ODP Leg 113. *Proc. Ocean Drill Prog.* **113**:829–848.

Stouffer, R. J., S. Manabe, and K. Bryan, 1989. Interhemispheric asymmetry in climate response to a gradual increase of atmospheric CO$_2$. *Nature* **342**:660–662.

Street, F. A., and A. T. Grove, 1979. Global maps of lake-level fluctuations since 30,000 BP. *Quat. Res.* **12**:83–118.

Street-Perrott, F. A., and S. P. Harrison, 1985. Lake levels and climate reconstruction. In: *Paleoclimate Analysis and Modeling.* A. D. Hecht (Ed.). John Wiley and Sons, New York, pp. 291–340.

Street-Perrott, F. A., and R. A. Perrott, 1990. Abrupt climatic fluctuations in the tropics—the influence of Atlantic Ocean circulation. *Nature* **343**:607–612.

Stringer, C. B., A. P. Currant, H. P. Schwarcz, and S. N. Collcutt, 1986. Age of Pleistocene faunas from Bacon Hole, Wales. *Nature* **320**:59–62.

Stuijts, I., J. C. Newsome, and J. R. Flenley, 1988. Evidence for late Quaternary vegetational change in the Sumatran and Javan highlands. *Rev. Palaeobot. Palynol.* **55**:207–216.

Stuiver, M., 1980. Solar variability and climatic change during the current millennium. *Nature* **286**:868–871.

Stuiver, M., and P. D. Quay, 1980. Changes in atmospheric Carbon-14 attributed to a variable sun. *Science* **207**:11–19.

Stuiver, M. and T. F. Braziunas, 1988. The solar component of the atmospheric ^{14}C record. In: *Secular Solar and Geomagnetic Variations in the Last 10,000 Years.* F. R. Stephenson and A. W. Wolfendale (Eds.). Kluwer, Dordrecht, Netherlands, pp. 245–266.

Stuiver, M., and T. F. Braziunas, 1989. Atmospheric ^{14}C and century-scale solar oscillations. *Nature* **338**:405–408.

Stuiver, M., G. H. Denton, T. J. Hughes, and J. L. Fastook, 1981. History of marine ice sheet in West Antarctica during the last glaciation: A working hypothesis. In: *The Last Great Ice Sheets.* G. H. Denton and T. J. Hughes (Eds.) Wiley-Interscience, New York, pp. 319–436.

Stuiver, M., P. D. Quay, and H. G. Ostlund, 1983. Abyssal water Carbon-14 distribution and the age of the world oceans. *Science* **219**:849–851.

Suarez, M. J., and I. M. Held, 1976. Note on modeling climate response to orbital parameter variations. *Nature* 263:46–47.

Suarez, M. J., and I. M. Held, 1979. The sensitivity of an energy balance climate model to variations in the orbital parameters. *J. Geophys. Res.* 84:4825–4836.

Suess, H. E., 1980. The radiocarbon record in tree rings of the last 8000 years. *Radiocarbon* 22:200–209.

Sullivan, W., 1974. *Continents in Motion: The New Earth Debate.* McGraw Hill, New York.

Sundquist, E. T., 1985. Geological perspectives on carbon dioxide and the carbon cycle. In: *The Carbon Cycle and Atmospheric CO₂: Natural Variations Archean to Present.* E. T. Sundquist and W. S. Broecker (Eds.). Geophys. Mono. 32. Am. Geophys. Union, Washington, D.C., pp. 5–60.

Sundquist, E. T., and W. S. Broecker (Eds.), 1985. *The Carbon Cycle and Atmospheric CO₂: Natural Variations Archean to Present.* Geophys. Mono. 32. Am. Geophys. Union, Washington, D.C.

Surlyk, F., and M. B. Johansen, 1984. End-Cretaceous brachiopod extinctions in the chalk of Denmark. *Science* 223:1174–1177.

Sutcliffe, A. J., T. C. Lord, R. S. Harmon, M. Ivanovich, A. Rae, and J. W. Hess, 1985. Wolverine in northern England at about 83,000 yr B.P.: Faunal evidence for climatic change during isotope Stage 5. *Quat. Res.* 24:73–86.

Taljaard, J. J., H. van Loon, H. L. Crutcher, and R. L. Jenne, 1969. *Climate of the Upper Air: Southern Hemisphere, Vol 1. Temperatures, Dew Points and Heights at Selected Pressure Levels* (NAVAIR 50-1C-55). Naval Weather Service Command, Washington, D.C.

Talwani, M., and O. Eldholm, 1977. Evolution of the Norwegian-Greenland Sea. *Geol. Soc. Am. Bull.* 88:969–999.

Tarduno, J. A., M. McWilliams, W. V. Sliter, H. E. Cook, M. C. Balke, and I. Premoli-Silva, 1986. Southern Hemisphere origin of the Cretaceous Laytonville limestone of California. *Science* 231:1425–1428.

Taylor, B. J., 1972. Stratigraphical correlation in southeast Alexander Island. In: *Antarctic Geology and Geophysics.* R. Adie (Ed.). Universitetsforlaget, Oslo, pp. 149–153.

Tedford, R. H., 1985. Late Miocene turnover of the Australian mammal fauna. *S. Afr. J. Sci.* 81:262–263.

Teller, J. T., 1987. Proglacial lakes and the southern margin of the Laurentide Ice Sheet. In: *North America and Adjacent Oceans During the Last Deglaciation.* W. F. Ruddiman and H. E. Wright, Jr. (Eds.) The Geology of North America, v. K-3. Geol. Soc. Am., Boulder, Col., pp. 39–69.

Thierstein, H. R., K. R. Geitzenauer, B. Molfino, and N. J. Shackleton, 1977. Global synchroneity of late Quaternary coccolith datum levels: Validation by oxygen isotopes. *Geology* 5:400–404.

Thomas, E., 1988. Mass extinctions in the deep sea. In: *Global Catastrophes in Earth History: An Interdisciplinary Conference on Impacts, Volcanism, and Mass Mortality* (Abstracts). Lunar Plan. Inst. Pub., Houston, Texas, pp. 192–193.

Thomas, R. H., 1984. Ice sheet margins and ice shelves. In: *Climate Processes and Climate Sensitivity.* J. E. Hansen and T. Takahashi (Eds.). Geophys. Mono. 29. Am. Geophys. Union, Washington, D.C., pp. 265–274.

Thompson, L. G., E. Mosley-Thompson, J. F. Bolzan, and B. R. Koci, 1985. A 1500-year record of tropical precipitation in ice cores from the Quelccaya Ice Cap, Peru. *Science* 229:971–973.

Thompson, L. G., E. Mosley-Thompson, W. Dansgaard, and P. M. Grootes, 1986. The Little Ice Age as recorded in the stratigraphy of the tropical Quelccaya Ice Cap. *Science* 234:361–364.

Thompson, L. G., et al., 1989. Holocene-Late Pleistocene climatic ice core records from Qinghai-Tibetan Plateau. *Science* 246:474–477.

Thompson, P. R., 1981. Planktonic foraminifera in the western North Pacific during the past 150,000 years: Comparison of modern and fossil assemblages. *Palaeogeog., Palaeoclimatol., Palaeoecol.* 35:241–179.

Thompson, P. R., and N. J. Shackleton, 1980. North Pacific paleoceanography: Late Quaternary coiling variations of planktonic foraminifer *Neogloboquadrina pachyderma. Nature* 287:829–833.

Thompson, S. L., and E. J. Barron, 1981. Comparison of Cretaceous and present earth albedos: Implications for the causes of paleoclimates. *J. Geol.* 89:143–167.

Thompson, S. L., and C. Covey, 1990. Global climatic effects of dust from large bolide impacts. In: *Global Catastrophes In Earth History.* (V. L. Sharpton and P. Ward, Eds.). Geol. Soc. Am. Spec. Paper, in press.

Thompson, S. L., and P. J. Crutzen, 1990. Acute effects of a large bolide impact simulated by a global atmospheric general circulation model. In: *Global Catastrophes In Earth History.* V. L. Sharpton and P. Ward (Eds.). Geol. Soc. Am. Spec. Paper, in press.

Thunell, R. C., and G. P. Lohmann, 1979. Planktonic foraminiferal fauna associated with eastern Mediterranean Quaternary stagnations. *Nature* 281:211–213.

Thunell, R. C., and D. F. Williams, 1989. Glacial-Holocene salinity changes in the Mediterranean Sea: Hydrographic and depositional effects. *Nature* 338:493–496.

Thunell, R. C., D. F. Williams, and M. B. Cita, 1983. Glacial anoxia in the eastern Mediterranean. *J. Foram. Res.* 13:283–290.

Thunell, R. C., D. F. Williams, and P. R. Belyea, 1984. Anoxic events in the Mediterranean Sea in relation to the evolution of late Neogene climates. *Mar. Geol.* **59**:105–134.

Thunell, R. C., D. F. Williams, and M. Howell, 1987. Atlantic-Mediteranean water exchange during the late Neogene. *Paleoceanog.* **2**:661–678.

Thunell, R. C., S. M. Locke, and D. F. Williams, 1988. Glacio-eustatic sea-level control on Red Sea salinity. *Nature* **334**:601–604.

Toggweiler, J. R., and J. L. Sarmiento, 1985. Glacial to interglacial changes in atmospheric carbon dioxide: The critical role of ocean surface water in high latitudes. In: *The Carbon Cycle and Atmospheric CO₂: Natural Variations Archean to Present.* E. T. Sundquist and W. S. Broecker (Eds.). Geophys. Mono. **32**. Am. Geophys. Union, Washington, D.C., pp. 163–184.

Toon, O. B., J. B. Pollack, T. P. Ackerman, R. D. Turco, C. P. McKay, and M. S. Liu, 1982. Evolution of an impact-generated dust cloud and its effects on the atmosphere. *Geol. Soc. Am. Spec. Pap.* **190**:187–200.

Trenberth, K. E., 1979. Mean annual poleward energy transports by the ocean in the Southern Hemisphere. *Dyn. Atmos. Oc.* **4**:57–64.

Tschudy, R. H., C. L. Pillmore, C. J. Orth, J. S. Gilmore, and J. D. Knight, 1984. Disruption of the terrestrial plant ecosystem at the Cretaceous-Tertiary boundary, western interior. *Science* **225**:1030–1032.

Tucker, M. E., 1982. Precambrian dolomites: Petrographic and isotopic evidence that they differ from Phanerozoic dolomites. *Geology* **10**:7–12.

Turco, R. P., O. B. Toon, T. P. Ackerman, J. B. Pollack, and C. Sagan, 1983. Nuclear Winter: Global consequences of multiple nuclear explosions. *Science* **222**:1283–1292.

Turco, R. P., O. B. Toon, T. P. Ackerman, J. B. Pollack, and C. Sagan, 1990. Climate and smoke: An appraisal of nuclear winter. *Science* **247**:166–176.

Turekian, K. K. (Ed.), 1971. *The Late Cenozoic Glacial Ages.* Yale Univ. Press, New Haven, Conn.

Tyson, P. D., 1986. *Climatic Change and Variability in Southern Africa.* Oxford University Press, Cape Town.

U.S. Committee for GARP (Global Atmospheric Research Program), 1975. *Understanding Climatic Change.* National Academy of Sciences, Washington, D.C.

Ulrich, R. K., 1975. Solar neutrinos and variations in the solar luminosity. *Science* **190**:619–624.

Urey, H. C., 1947. The thermodynamic properties of isotopic substances. *J. Chem. Soc. (London)*, pp. 562–581.

Urey, H. C., 1952. *The Planets, Their Origin and Development.* Yale Univ. Press, New Haven, Conn.

Vail, P. R., R. M. Mitchum, and S. Thompson, 1977. Seismic stratigraphy and global changes in sea level. In: *Seismic Stratigraphy: Application to Hydrocarbon Exploration.* Am. Assoc. Pet. Geol. Mem. **26**. Am. Assoc. Petrol. Geol., Tulsa, Okla., pp. 83–97.

Vakhrameev, V. A. 1964. Jurassic and Early Cretaceous floras of Eurasia and the paleofloristic provinces of this period. *Trans. Geol Inst. Moscow* **102**:1–263 (in Russian).

Vakhrameev, V. A., 1975. Main features of phytogeography of the globe in Jurassic and Early Cretaceous time. *Paleontol. J.* **2**:247–255.

Valentine, J. W., 1984. Neogene marine climate trends: Implications for biogeography and evolution of the shallow-sea biota. *Geology* **12**:647–650.

Valentine, J. W., 1985. Are interpretations of ancient marine temperatures constrained by the presence of ancient marine organisms? In: *The Carbon Cycle and Atmospheric CO₂: Natural Variations Archean to Present* E. T. Sundquist and W. S. Broecker (Eds.). Geophys. Mon. **32**. Am. Geophys. Union, Washington, D.C. pp. 623–627.

Valladas, H., et al., 1988. Thermoluminescence dating of Mousterian "Proto-Cro-Magnon" remains from Israel and the origin of modern man. *Nature* **331**:614–616.

van Andel, T. H., 1975. Mesozoic/Cenozoic calcite compensation depth and the global distribution of calcareous sediments. *Earth Plan. Sci. Lett.* **26**:187–195.

van Andel, T., G. R. Heath, and T. C. Moore, Jr., 1975. Cenozoic history of the central equatorial Pacific Ocean. *Geol. Soc. Am. Mem.* **143**:1–134.

van Campo, E., J.-C. Duplessy and M. Rossignol-Strick, 1982. Climatic conditions deduced from a 150-kyr oxygen isotope-pollen record from the Arabian Sea. *Nature* **296**:56–59.

van den Bogaard, C. M. Hall, H.-U. Schmincke and D. York, 1989. Precise single-grain ⁴⁰Ar/³⁹Ar dating of a cold to warm climate transition in central Europe. *Nature* **342**:523–525.

van den Dool, H. M., H. J. Krijnen, and C. J. E. Schuurmans, 1978. Average winter temperatures at De Bilt (the Netherlands): 1634–1977. *Clim. Change* **1**:319–330.

van der Hammen, T., 1974. The Pleistocene changes of vegetation and climate in tropical south America. *J. Biogeog.* **I**:3–26.

van der Hammen, T., 1985. The Plio-Pleistocene climatic record of the tropical Andes. *J. Geol. Soc. (London)* **142**:483–490.

van der Hammen, T., T. A. Wijmstra, and W. H. Zagwijn, 1971. The floral record of the late Cenozoic of Europe. In: *The Late Cenozoic Glacial Ages.* K. K. Turekian (Ed.). Yale Univ. Press, New Haven, Conn., pp. 391–424.

van Houten, F. B., 1985. Oolitic ironstones and

contrasting Ordovician and Jurassic paleogeography. *Geology* **13**:722–724.

van Loon, H., and J. C. Rogers, 1978. The seesaw in winter temperatures between Greenland and northern Europe. Part I, General description. *Mon. Wea. Rev.* **106**:296–310.

van Zinderen Bakker, E. M., and J. H. Mercer, 1986. Major late Cainozoic climatic events and palaeoenvironmental changes in Africa viewed in a world wide context. *Palaeogeog., Palaeoclimatol., Palaeoecol.* **56**:217–235.

Veevers, J. M., and C. McA. Powell, 1987. Late Paleozoic glacial episodes in Gondwanaland reflected in transgressive-regressive depositional sequences in Euramerica. *Geol. Soc. Am. Bull.* **98**:475–487.

Veizer, J., P. Fritz, and B. Jones, 1986. Geochemistry of brachiopods: Oxygen and carbon isotopic records of Paleozoic oceans. *Geochim. Cosmochim. Acta* **50**:1679–1696.

Velichko, A. A. (Ed.), 1983. *Late Quaternary Environments of the Soviet Union.* Univ. Minnesota Press, Minneapolis.

Venkatesan, M. I., and J. Dahl, 1989. Organic geochemical evidence for global fires at the Cretaceous/Tertiary boundary. *Nature* **338**:57–60.

Vernekar, A. D., 1968. Long-period global variations of incoming solar radiation. In: *Research on the Theory of Climate, v. 2.* Travelers Research Center. Hartford, Conn.

Vincent, E., and W. H. Berger, 1985. Carbon dioxide and polar cooling in the Miocene: The Monterey hypothesis. In: *The Carbon Cycle and Atmospheric CO2: Natural Variations Archean to Present.* E. T. Sundquist and W. S. Broecker (Eds.). Geophys. Mono. 32. Am. Geophys. Union, Washington, D.C., pp. 455–468.

Volk, T., 1989a. Rise of angiosperms as a factor in long-term climatic cooling. *Geology* **17**:107–110.

Volk, T., 1989b. Sensitivity of climate and atmospheric CO2 to deep-ocean and shallow-ocean carbonate burial. *Nature* **337**:637–640.

von Arx, W. S., 1952. A laboratory study of the wind-driven ocean circulation. *Tellus* **4**:311–318.

von Arx, W. S., 1957. An experimental approach to problems in physical oceanography. In: *Progress in Physics and Chemistry of the Earth, v. 2.* Pergamon Press, New York, pp. 1–29.

von der Haar, T. H., and A. H. Oort, 1973. New estimate of annual poleward energy transport by northern hemisphere oceans. *J. Phys. Oceanog.* **3**:169–172.

von Storch, H. V. (Ed.), 1988. *Climate simulations with the ECMWF T21-model in Hamburg. Part II: Climatology and Sensitivity Experiments.* Meteorologisches Institut der Universität Hamburg, Fed. Rep. Germany.

Vrba, E. S., 1985. African Bovidae: Evolutionary events since the Miocene. *S. Afr. J. Sci.* **81**:263–266.

Waitt, R. B., 1985. Case for periodic, colossal jökulhlaups from Pleistocene glacial Lake Missoula. *Geol. Soc. Am. Bull.* **96**:1271–1286.

Walker, J. C. G., 1982. Climatic factors on the Archean earth. *Palaeogeog., Palaeoclimatol., Palaeoecol.* **40**:1–11.

Walker, J. C. G., P. B. Hays, and J. F. Kasting, 1981. A negative feedback mechanism for the long-term stabilization of Earth's surface temperature. *J. Geophys. Res.,* **86**:9776–9782.

Wallace, J. M., and P. V. Hobbs (Eds.), 1977. *Atmospheric Science: An Introductory Survey.* Academic Press, New York.

Walsh, J. E., and M. B. Richman, 1981. Seasonality in the association between surface temperatures over the United States and the north Pacific Ocean. *Mon. Wea. Rev.* **109**:767–783.

Walter, M. R., R. Buick, and J. S. R. Dunlop, 1980. Stromatolites 3,400-3,500 M. yr. old from the North Pole area, western Australia. *Nature* **284**:443–445.

Wang, P. K., and D. Zhang, 1988. An introduction to some historical governmental weather records of China. *Bull. Am. Meteorol. Soc.* **69**:753–758.

Wang, Y., L. Guangyuan, X. Zhang, and C. Li, 1983. The relationships between tree rings of Qilianshan juniper and climatic change and glacial activity during the past 1000 years in China. *Kexue Tongbao* **28**:1647–1652.

Wanless, H. R., and F. P. Shepard, 1936. Sea level and climatic changes related to Late Paleozoic cycles. *Geol. Soc. Am. Bull.* **47**:1177–1206.

Ward, P. D., and K. MacLeod, 1988. Macrofossil extinction patterns at Bay of Biscay Cretaceous-Tertiary boundary sections. In: *Global Catastrophes in Earth History: An Interdisciplinary Conference on Impacts, Volcanism, and Mass Mortality* (Abstracts). Lun. Plan. Inst., Houston, Texas, pp. 206–207.

Ward, W. C., 1973. Influence of climate on the early diagenesis of carbonate eolianites. *Geology* **1**:171–174.

Ward, W. R., 1982. Comments on the long-term stability of the earth's obliquity. *Icarus* **50**:444–448.

Warren, B. A., and C. Wunsch (Eds.), 1981. *Evolution of Physical Oceanography.* MIT Press, Cambridge, Mass.

Washington, W., and G. Meehl, 1984. Seasonal cycle experiment on the climate sensitivity due to a doubling of CO2 with an atmospheric general circulation model coupled to a simple mixed layer ocean. *J. Geophys. Res.* **89**:9475–9503.

Washington, W. M., and C. L. Parkinson, 1986. *An Introduction to Three-Dimensional Climate Modeling.* Univ. Science Books, Mill Valley, Cal.

Washington, W. M., and G. A. Meehl, 1989. Climate sensitivity due to increased CO_2: Experiments with a coupled atmosphere and ocean general circulation model. *Clim. Dynam.* **4**:1–38.

Watts, A. B., 1982. Tectonic subsidence, flexure, and global changes of sea level. *Nature* **297**:469–474.

Watts, R. G., and M. E. Hayder, 1983. Climatic fluctuations due to deep ocean circulation. *Science* **219**:387–388.

Watts, R. G., and M. Hayder, 1984a. The effect of land-sea distribution on ice sheet formation. *Ann. Glac.* **5**:234–236.

Watts, R. G., and M. Hayder, 1984b. A possible explanation of differences between pre- and post-Jaramillo ice sheet growth. In: *Milankovitch and Climate.* A. L. Berger, J. Imbrie, J. D. Hays, G. J. Kukla, and B. Saltzman (Eds.). D. Reidel, Dordrecht, Netherlands, pp. 599–604.

Watts, W. A., 1980. Late-Quaternary vegetation history at White Pond on the inner coastal plain of South Carolina. *Quat. Res.* **13**:187–199.

Watts, W. A., 1983. Vegetational history of the eastern United States 25,000 to 10,000 years ago. In: *Late-Quaternary Environments of the United States.* H. E. Wright, Jr. (Ed.). Univ. of Minn. Press, Minneapolis, pp. 294–310.

Webb, P. N., D. M. Harwood, B. C. McKelvey, J. H. Mercer, and L. D. Stott, 1984. Cenozoic marine sedimentation and ice-volume variation on the east Antarctic craton. *Geology* **12**:287–291.

Webb, T., 1985. Holocene palynology and climate. In: *Paleoclimate Analysis and Modeling.* A. D. Hecht (Ed.). Wiley-Interscience, New York, pp. 163–196.

Webb, T., and R. A. Bryson, 1972. Late-and postglacial climatic change in the northern Midwest, USA: Quantitative estimates derived from fossil pollen spectra by multivariate statistical analysis. *Quat. Res.* **2**:70–115.

Webb, T., and D. R. Clark, 1977. Calibrating micropaleontological data in climatic terms: A critical review. *Ann. N.Y. Acad. Sci.* **288**:93–118.

Webb, T., P. J. Bartlein, and J. E. Kutzbach, 1987. Climatic change in eastern North America during the past 18,000 years: Comparisons of pollen data with model results. In: *North America and Adjacent Oceans During the Last Deglaciation.* W. F. Ruddiman and H. E. Wright, Jr. (Eds.) The Geology of North America, v. **K-3.** Geol. Soc. Am., Boulder, Col., pp. 447–462.

Webster, P. N., and N. Streeten, 1978. Late Quaternary ice age climates of tropical Australia, interpretation and reconstruction. *Quat. Res.* **10**:279–309.

Weertman, J., 1964. Rate of growth or shrinkage of non-equilibrium ice sheet. *J. Glaciol.* **5**:145–158.

Weertman, J., 1976. Milankovitch solar radiation variations and ice age ice sheet sizes. *Nature* **261**:17–20.

Wegener, A., 1929 (1966 translation). *The Origin of the Continents and Oceans.* Dover Publications, New York.

Wells, G. L., 1983. Late-glacial circulation over central North America revealed by aeolian features. In: *Variations in the Global Water Budget.* A. Street-Perrott et al. (Eds.). D. Reidel, Dordrecht, Netherlands, pp. 317–330.

Wenk, T., and U. Siegenthaler, 1985. The high-latitude ocean as a control of atmospheric CO_2: In: *The Carbon Cycle and Atmospheric CO_2: Natural Variations Archean to Present.* E. T. Sundquist and W. S. Broecker (Eds.). Geophys. Mon. **32.** Am. Geophys. Union, Washington, D.C., pp. 185–194.

Wesselman, H. B., 1985. Fossil micromammals as indicators of climatic change about 2.4 Myr ago in the Amo Valley, Ethiopia. *S. Afr. J. Sci.* **81**:260–261.

Wetherald, R. T., and S. Manabe, 1975. The effects of changing the solar constant on the climate of a general circulation model. *J. Atmos. Sci* **3**:2044–2059.

Wetherald, R. T., and S. Manabe, 1986. An investigation of cloud cover change in response to thermal forcing. *Clim. Change* **8**:5–23.

Weyl, P. K., 1968. The role of the ocean in climatic change: A theory of the ice ages. *Meterorol. Monog.* **8**:37–62.

Whillans, I. M., 1981. Reaction of the accumulation zone portions of glaciers to climatic change. *J. Geophys. Res.* **86**:4274–4282.

Wigley, T. M. L., 1976. Spectral analysis and the astronomical theory of climatic change. *Nature* **264**:629–631.

Wigley, T. M. L., 1988. The climate of the past 10,000 years and the role of the sun. In: *Secular Solar and Geomagnetic Variations in the Last 10,000 years.* F. R. Stephenson and A. W. Wolfendale (Eds.). Kluwer, Dordrecht, Netherlands, pp. 209–224.

Wigley, T. M. L., and P. D. Jones, 1985. Influence of precipitation changes and direct CO_2 effects on streamflow. *Nature* **314**:149–152.

Wigley, T. M. L., and B. D. Santer, 1988. Validation of general circulation climate models. In: *Physically-based Modelling and Simulation of Climate and Climatic Change.* M. E. Schlesinger (Ed.). Kluwer, Dordrecht, Netherlands, pp. 841–879.

Wijmstra, T. A., 1978. Paleobotany and climatic change. In: *Climatic Change.* J. Gribbin (Ed.). Cambridge Univ. Press, New York, pp. 25–45.

Wilde, P., 1987. Model of progressive ventilation of the late Precambrian-early Paleozoic ocean. *Am. J. Sci.* **287**:442–459.

Wilde, P., and W. B. N. Berry, 1984. Destabilization of the oceanic density structure and its significance to marine "extinction" events. *Palaeogeog., Palaeoclimatol., Palaeoecol.* **48**:143–162.

Wilkinson, B. H., and R. K. Given, 1986. Secular variation in abiotic marine carbonates: Constraints on Phanerozoic atmospheric carbon dioxide contents and oceanic Mg/Ca ratios. *J. Geol.* **94**:321–333.

Williams, G. E., 1975. Late Precambrian glacial climate and the earth's obliquity. *Geol. Mag.* **112**:441–465.

Williams, G. E., 1981. Sunspot periods in the late Precambrian glacial climate and solar-planetary relations. *Nature* **291**:624–628.

Williams, G. E., 1988. Cyclicity in the late Precambrian Elatina Formation, South Australia: Solar or tidal signature? *Clim. Change,* **13**:117–128.

Williams, J., R. G. Barry, and W. M. Washington, 1974. Simulation of the atmospheric circulation using the NCAR global circulation model with ice age boundary conditions. *J. Appl. Meteorol.* **13**:305–317.

Willson, L. A., G. H. Bowen, and C. Struck-Marcell, 1987. Mass loss on the main sequence. *Comments Astrophys.* **12**:17–34.

Wilson, A. T., 1969. The climatic effects of large-scale surges of ice sheets. *Can. J. Earth Sci.* **6**:911–918.

Wilson, A. T., C. H. Hendy, and C. P. Reynolds, 1979. Short-term climatic changes and New Zealand temperatures during the last millennium. *Nature* **279**:315–317.

Wilson, C. A., and J. F. B. Mitchell, 1987. A doubled CO_2 climate sensitivity experiment with a global climate model including a simple ocean. *J. Geophys. Res.* **92**:315–343.

Winograd, I. J., 1990. Dating sea level in caves. *Nature* **343**:217.

Winograd, I. J., T. B. Coplen, B. J. Szabo, and A. C. Riggs, 1988. A 250,000-year climatic record from Great Basin vein calcite: Implications for the Milankovitch theory. *Science* **242**:1275–1280.

Woillard, G. M., 1978. Grande Pile peat bog: A continuous pollen record for the last 140,000 years. *Quat. Res.* **9**:1–21.

Woillard, G. M., and W. G. Mook, 1982. Carbon-14 dates at Grande Pile: Correlation of land and sea chronologies. *Science* **215**:159–161.

Wolbach, W. S., R. S. Lewis, and E. Anders, 1985. Cretaceous extinctions: Evidence for wildfires and search for meteoritic material. *Science* **230**:167–170.

Wolbach, W. S., I. Gilmour, E. Anders, C. J. Orth, and R. R. Brooks, 1988. Global fire at the Cretaceous-Tertiary boundary. *Nature* **334**:665–669.

Wolfe, J. A., 1978. A paleobotanical interpretation of Tertiary climates in the Northern Hemisphere. *Am. Sci.* **66**:694–703.

Wolfe, J. A., 1980. Tertiary climates and floristic relationships at high latitudes in the northern hemisphere. *Palaeogeog., Palaeoclimatol., Palaeoecol.* **30**:313–323.

Wolfe, J. A., 1985. Distribution of major vegetational types during the Tertiary. In: *The Carbon Cycle and Atmospheric CO_2: Natural Variations Archean to Present.* E. T. Sundquist and W. S.

Broecker (Eds.). Geophys. Mono. **32**. Am. Geophys. Union, Washington, D.C., pp. 357–375.

Wolfe, J. A., 1990. Palaeobotanical evidence for a marked temperature increase following the Cretaceous/Tertiary boundary. *Nature* **343**:153–156.

Wolfe, J. A., and G. R. Upchurch, 1986. Vegetation, climatic and floral changes at the Cretaceous-Tertiary boundary. *Nature* **324**:148–152.

Wolfe, J. A., and G. R. Upchurch, 1987. North American nonmarine climates and vegetation during the Late Cretaceous. *Palaeogeog., Palaeoclimatol., Palaeoecol.* **61**:33–77.

Woodruff, F., and S. M. Savin, 1985. $\delta^{13}C$ values of Miocene Pacific benthic foraminifera: Correlations with sea level and biological productivity. *Geology* **13**:119–122.

Woodruff, F., and S. M. Savin, 1989. Miocene deepwater oceanography. *Paleoceanog.* **4**:87–140.

Worthington, L. V., 1968. Genesis and evolution of water masses. *Met. Mono.* **8**:63–67.

Wright, H. E. (Ed.), 1983. *Late Quaternary Environments of the United States,* v. 2, *The Holocene.* Univ. Minnesota Press, Minneapolis.

Wright, H. E., 1987. Synthesis: The land south of the ice sheets. In: *North America and Adjacent Oceans During the Last Deglaciation.* W. F. Ruddiman and H. E. Wright, Jr. (Eds.). The Geology of North America, v. K-3. Geol. Soc. Am., Boulder, Col., pp. 479–488.

Wu, P., and W. R. Peltier, 1983. Glacial isostatic adjustment and the free air gravity anomaly as a constraint on deep mantle viscosity. *Geophys. J. Roy. Astronom. Soc.* **76**:753–791.

Yamamoto, T., 1971. On the nature of the climatic change in Japan since the "Little Ice Age" around 1800 AD. *J. Meteorol. Soc. Japan* **49**:798–812.

Yemane, K., R. Bonnefille, and H. Faure, 1985. Palaeoclimatic and tectonic implications of Neogene microflora from the northwestern Ethiopian highlands. *Nature* **318**:653–656.

Young, G. M., 1973. Tillites and aluminous quartzites as possible time markers for middle Precambrian (Aphebian) rocks of North America. In: *Huronian Stratigraphy and Sedimentation.* G. M. Young (Ed.). Geol. Assoc. Can. Spec. Pap., **12**. Geological Association of Canada, Waterloo, Ont., pp. 97–127.

Young, M. A., and R. S. Bradley, 1984. Insolation gradients and the paleoclimatic record. In: *Milankovitch and Climate.* A. L. Berger, J. Imbrie, J. Hays, G. Kukla, and B. Saltzman (Eds.). D. Reidel, Dordrecht, Netherlands, pp. 707–713. Zhang (Ed.). Science Press, Beijing, pp. 18–31.

Zachos, J. C., and M. A. Arthur, 1986. Paleoceanography of the Cretaceous/Tertiary boundary event: Inferences from stable isotopic and other data. *Paleoceanog.* **1**:5–26.

Zachos, J. C., M. A. Arthur, and W. E. Dean, 1989. Geochemical evidence for suppression of pelagic marine productivity at the Cretaceous/ Tertiary boundary. *Nature* **337**:61–64.

Zahn, R., M. Sarnthein, and H. Erlenkeuser, 1987. Benthic isotope evidence for changes of the Mediterranean outflow during the late Quaternary. *Paleoceanog.* **2**:543–559.

Zhang, D., 1980. Winter temperature variation during the last 500 years in southern China. *Kexue Tongbao* **25**:497–500.

Zhang, D., 1984. Synoptic-climatic studies of dust fall in China since historic times. *Scientia Sinica (Ser. B)* **27**:825–836.

Zhang, D., 1988. The method for reconstruction of the dryness/wetness series in China for the last 500 years and its reliability. In: *The Reconstruction of Climate in China in Historical Time*. J. Zhang (Ed.). Science Press, Beijing, pp. 18–31.

Zhang, J., 1988. *The Reconstruction of Climate in China in Historical Time*. Science Press, Beijing.

Zhang, J., and T. J. Crowley, 1989. Historical climate records in China and reconstruction of past climates. *J. Clim.*, **2**:833–849.

Zhao, M., and J. L. Bada, 1989. Extraterrestrial amino acids in Cretaceous/Tertiary boundary sediments at Stevns, Klint, Denmark. *Nature* **339**:463–465.

Zheng, B., 1989. Controversy regarding the existence of a large ice sheet on the Qinghai-Xizang (Tibetan) Plateau during the Quaternary period. *Quat. Res.* **32**:121–123.

Zheng, S., and L. Feng. 1986. Historical evidence of climatic instability above normal in cool periods in China. *Scientia Sinica Ser. (B)* **29**:441–448.

Zhou, L., and F. T. Kyte, 1988. The Permian-Triassic boundary event: A geochemical study of three Chinese sections. *Earth Plan. Sci. Lett.* **90**:411–421.

Zhu, K. (Chu, K.), 1973. A preliminary study on the climatic fluctuations during the last 5,000 years in China. *Scientia Sinica* **16**:226–256.

Ziegler, A. M., C. R. Scotese, W. S. McKerrow, M. E. Johnson, and R. K. Bambach, 1979. Paleo-

zoic paleogeography. *Ann. Rev. Earth Plan. Sci.* **7**:473–502.

Ziegler, A. M., R. K. Bambach, J. T. Parrish, S. F. Barrett, E. H. Gierlowski, W. C. Parker, A. Raymond, and J. J. Sepkoski, 1981. Paleozoic biogeography and climatology. In: *Paleobotany, Paleoecology, and Evolution*. K. J. Niklas (Ed.). Praeger, New York, pp. 231–266.

Ziegler, A. M., C. R. Scotese, and S. F. Barrett, 1983. Mesozoic and Cenozoic paleogeographic maps. In: *Tidal Friction and the Earth's Rotation, II*. P. Brosche and J. Sundermann (Eds.). Springer-Verlag, Berlin, pp. 240–252.

Ziegler, A. M., M. L. Hulver, A. L. Lottes, and W. F. Schmachtenberg, 1984. Uniformitarianism and palaeoclimates: Inferences from the distribution of carbonate rocks. In: *Fossils and Climate*. P. Brenchley (Ed.). John Wiley and Sons, New York, pp. 3–25.

Zinsmeister, W. J., 1982. Late Cretaceous—Early Tertiary molluscan biogeography of the southern circum Pacific. *J. Paleontol.* **56**:84–102.

Zoller, W. H., J. R. Parrington, and J. M. Phelan Kotra, 1983. Iridium enrichment in airborne particles from Kilauea Volcano: January 1983. *Science* **222**:1118–1121.

Zubakov, V. A., and I. I. Borzenkova, 1988. Pliocene palaeoclimates: Past climates as possible analogues of mid-twenty-first century climate. *Palaeogeog., Palaeoclim., Palaeoecol.* **65**:35–49.

Zwally, H. J., et al., 1983. *Antarctic Sea Ice, 1973–1976: Satellite Passive-Microwave Observations*. NASA Spec. Pub. SP-459, National Aeronautics and Space Administration, Washington, D.C.

Zwally, H. J., A. C. Brenner, J. A. Major, R. A. Bindschadler, and J. G. Marsh, 1989. Growth of Greenland Ice Sheet: Measurement. *Science* **246**:1587–1591.

Zwiers, F. W., and G. J. Boer, 1987. A comparison of climates simulated by a general circulation model when run in the annual cycle and perpetual modes. *Mon. Wea. Rev.* **115**:2626–2644.

ADDENDUM: RECENT ADVANCES IN PALEOCLIMATOLOGY

Since the publication of the hardbound version of this book, there has been a flood of papers on past climates. At some point we plan to integrate these new findings into a comprehensive update of the subject. But for the purposes of this softbound version, we will restrict ourselves only to highlights of some of the findings. Because of space, our coverage of these developments is somewhat restricted and based on the significant problems identified in Sec. 13.4. We refer the reader to that section as a starting point to the following update. We conclude by adding a discussion of new findings on abrupt climate change and an editorial comment on what we consider a disturbing trend involving lack of "closure" in the field. But first we start with the original questions raised in Sec. 13.4.

13.4.1 *How extensive were ice-free states in the Phanerozoic?* Although uncertainties still remain, results continue to suggest that major glacial periods have occurred only about 20% of the time during the last 540 million years.

13.4.2 *Were warmer periods during the last 100 million years indicative of seasonal or year-round warmth?* There is growing evidence from the Eocene (55 Ma) that the winter freezing line was at least 15–20° of latitude north of its present position (Markwick, 1994; Wing and Greenwood, 1993). Permian studies also suggest that seasonal cooling was less in observations than in models (Yemane, 1993; Taylor et al., 1992). The Yemane paper noted the importance of large lakes and inland seas that were not specified in the paleo modeling runs. Until the full effect of these altered boundary conditions has been explored, it will not be possible to tell for sure that models disagree with observations. But the overall impression of the observational studies is that a significant model-data disagreement remains.

The above disagreement represents a troubling problem regarding other studies that purport to show agreements between models and data on tectonic time scales; most of these studies are based on comparisons of model-generated fields with observations on land. Yet the seasonal cycle of temperature should primarily control much of the circulation response on land. The question then arises: How is it possible to obtain good model-data agreements in some areas if the primary factor responsible for the circulation response involves a large disagreement between models and data? Until this question is answered, significant uncertainties remain about model-data comparisons for pre-Pleistocene climates.

13.4.3 *Can geochemical models for long-term CO_2 fluctuations be better validated? Is there any evidence for warmer tropical SSTs due to these CO_2 increases?* With respect to the first question, there have been some encouraging developments in the proxy CO_2 approach. Several different methods are now used, among them: (a) $\delta^{13}C$ analyses of carbonate profiles in soils (Cerling, 1991); (b) $\delta^{13}C$ analyses of marine organic carbon (e.g., Freeman and Hayes, 1992); (c) counts of stomatal (pore) densities in plant leaves (Van der Burgh et al., 1993); (d) $\delta^{13}C$ ratios of the carbonate component of the mineral goethite mineral goethite in soils (Yapp and Poths, 1992); and (e) vertical water column differences in the $\delta^{13}C$ ratios of marine carbonates (Broecker, 1982). The first four methods involve an observed relationship between measured changes and atmospheric CO_2 levels. The latter method utilizes differences in the water column $\delta^{13}C$ ratio as a measure of ocean productivity, which should affect atmospheric CO_2 levels (Broecker, 1982).

Although comparisons of the proxy CO_2 methods with geochemical models for Phanerozoic CO_2 variations show some encouraging first-order agreement (Berner, 1992), there is still probably a 50% level of uncertainty in the agreement between models and observations over the Phanerozoic. A more complete set of comparisons for the last 100 million years suggests that the uncertainty level may be also a factor of three between different CO_2 proxies (e.g., Arthur et al., 1991).

Sometimes these differences are less. For example, for the mid-Pliocene warm period (about 3 Ma), three different proxy approaches suggest a CO_2 increase of 100 ppm or less (Van der Burgh et al., 1993; Chen et al., 1994; Raymo and Rau, 1995).

A further development in the Phanerozoic CO_2 story involves changing emphasis of factors responsible for higher atmospheric CO_2 levels in the Phanerozoic. Although the early emphasis (Sec. 8.2.4) was on seafloor spreading changes and effects of sea level on area of land available for weathering (Berner et al., 1983), the most recent CO_2 model suggests that changing terrestrial carbon storage may be more important for regulating long-term changes in CO_2 (Berner, 1994). But complications still arise when predicting temperature/weathering feedbacks. For example, Crowley and Baum (1995) conducted a GCM study for the Late Ordovician glaciation, in which CO_2 levels were increased to 14X the preanthropogenic level and solar luminosity was reduced 4.5%. These changes where equivalent to about a net fourfold increase of CO_2 forcing. Results indicate that although average precipitation and precipitation-minus-evaporation increased per grid box, as predicted by other GCMs and assumed in the geochemical calculations, the much smaller land area of the lower Paleozoic (due to high sea levels—cf. Fig. 11.4) resulted in a decrease in runoff to the oceans. This response is opposite to that incorporated in the geochemical models and suggests a need for further revamping of the model formulations.

The overall impression from the above comparison is that although there have been encouraging developments documenting the CO_2 model, uncertainties still exist within the range of 50%. Thus, any attempt to provide more quantitative constraints on past CO_2 fluctuations will be hampered by this uncertainty. Before further progress can be made on this problem, it is critically important to better validate and compare different proxy CO_2 approaches.

A second major result discussed in this section involves conclusions about stability of tropical SSTs during past warm time periods. Again, some significant progress has occurred in this area. Compilations for the Pliocene, Eocene, and Cenomanian (90 Ma) suggest that tropical SSTs have not been significantly greater than present

for any of these time intervals (Dowsett et al., 1992; Zachos et al., 1994; Crowley and Kim, 1995). Although more data are still needed, these results place tighter constraints on the paleo record and suggest a significant disagreement with climate model predictions for the future tropical SST response to a CO_2 increase (Crowley, 1993). If this prediction is borne out by future studies, it has considerable implications about predictions of the regional response to the CO_2 increase.

13.4.4 *How have changes in solid-earth boundary conditions affected ocean heat transport in the past?* Some progress has been made in understanding aspects of this problem. Several modeling studies have now shown that increasing transport in an ocean model results in zonal SST patterns more similar to observations (e.g., Rind and Chandler, 1991; Barron et al., 1995). In particular, tropical SSTs tend to remain stable and high-latitude temperatures increase. The high-latitude increase is aided by meltback of sea ice, which causes a positive feedback (Rind and Chandler, 1991).

Despite the above progress, several key questions remain regarding the ocean heat transport problem:

(1) How specifically does the ocean transport the heat? Two logical pathways are via the western boundary currents (Chap. 2) or via formation of warm saline bottom water (Sec. 8.1.2). With respect to the first mechanism, we know of no observational evidence suggesting increased transport of, for example, the Gulf Stream during past warm time periods, and one observational and one modeling study suggesting that it was less prior to the latest Cenozoic (Kaneps, 1979; Maier-Reimer et al., 1990; cf. Fig. 10.25).

With respect to the problem of warm, saline bottom water (WSBW), although this continues to be a fascinating possibility, the actual evidence for its existence is rather slim. There are indications from isotopic records for WSBW occurring during some intervals of geologic time (e.g., Kennett and Stott, 1990; Barrera and Huber, 1993; Railsback et al., 1990). However, it is not at all clear that this mechanism dominated the thermohaline flow and was responsible for the very large-scale change in the zonal temperature gradients. For example, if WSBW was the sole mechanism responsible for altering the zonal temperature gradient, production rates of WSBW would be on the order of 60 Sv (Crowley, 1992)—four to five times the present rate of NADW formation. To our knowledge, no evidence exists for such a massive increase in transport. Furthermore, geochemical models suggest that flooding the ocean with warm, low oxygen water should result in anoxic conditions not seen in deep-sea sediments (Herbert and Sarmiento, 1991).

(2) Are the models properly simulating the increase in ocean heat transport? The way increased ocean heat transport is handled in current models is to force the model to transport more heat across latitude belts. This is akin to increasing diffusion in an energy balance model (cf. Fig. 1.6). The only problem with this approach involves the real way in which the ocean transports heat. In a geostrophic system, transport is along isopycnals (lines of equal density) rather than across isopycnals (e.g., from the equator to the pole). A classic example is the Gulf Stream. Increasing the transport in the Gulf Stream merely moves water more rapidly around the subtropical gyres, returning it to lower latitudes along the eastern boundary current. Thus, in the absence of a thermohaline cell (cf. discussion in Sec. 2.1.2), a spinup of the real ocean will not necessarily transport more heat to higher latitudes, and the model may therefore be prescribing a change that is unphysical.

(3) What caused the ocean heat transport to change? At the present stage of our understanding, the ocean heat transport is forced to change and, as stated above, it may be forced to change in an unphysical manner. But a further problem involves why it would change. Certainly, changes in ocean boundary conditions (e.g., gateways) may have had a significant effect on the ocean circulation (Maier-Reimer et al., 1990; Mikolajewicz et al., 1993), but these changes do not seem capable of generating SST changes as observed. Wind forcing could of course change, but there is no clear evidence that model-generated changes in surface wind forcing due to changes in tectonic boundary conditions generate increases in poleward flow as required by the concept. (Polar flow might increase in one or two basins, but the evidence suggests it is a more general problem than that).

One way to trigger an ocean response is to have CO_2 levels increase, with these levels then triggering a global shift in the ocean circulation. However, the visualized changes are not gener-

ated by our present generation of ocean models (Crowley, 1993). So it is not clear whether the models are wrong or that the proposed reconciliation is wrong. Until the reason for an ocean heat transport change is clarified, acceptance of this explanation must be held in abeyance.

13.4.5 *What is the origin of model-data discrepancies in the tropics at the glacial maximum?* The main progress in this area involves an increasing amount of geochemical data from land or coastal areas indicating that surface temperatures were close to 5°C cooler than present during the glacial maximum in the tropics (Stute et al., 1994; Guilderson et al., 1994). The data involve Sr/Ca measurements of corals and noble gas measurements of groundwater; observations suggest that these indices have strong correlations with temperature. As the geochemical data are consistent with pollen evidence for pronounced climate change on land (Sec. 3.1.4), a somewhat clearer picture is now emerging as to the significance of large-scale change in the tropics on land.

Several questions remain about full acceptance of the significance of these results. Open marine data still indicate little changes in tropical SSTs (CLIMAP, 1981). The small estimated changes simply reflect the fact that biotic changes were not large in the marine realm. It is possible that the planktonic data are somehow yielding a biased signal. For example, since the lifespan of a foraminifera is only about a month, seasonal changes in plankton abundances could be hampering interpretation of the record. We consider this unlikely, as some compensating changes in the opposite direction during other seasons would be required to generate a time series that had little variation. Furthermore, there is a relatively small seasonal variation of overturn in oceanic regions that have a permanent thermocline.

An alternate explanation for land–sea differences is that changes between land and open ocean were larger than often assumed. This possibility is not entirely unreasonable, for lower ice-age CO_2 levels would be expected to favor C4 plants (grasses, etc.) and transform the vegetation cover for reasons other than direct temperature/precipitation changes. Altered groundcover also increases the surface albedo, potentially causing greater cooling over land than over ocean. Whatever the reason, a fully satisfactory explanation of changes in the tropics requires reconciling the different data sets. An explanation is also needed for why conditions cooled so much in the tropics, for calculations suggest that such a large cooling can only result if the sensitivity of the climate system is rather high (Crowley, 1994).

13.4.6 *What is the origin of ice-age CO_2 fluctuations?* Relatively little progress has been made on this topic. Probably the most popular view involves changes in the vertical structure of the ocean, with more intermediate water production and less deep water production (Sec. 6.4.5). There is a striking correspondence between $\delta^{13}C$ changes in Atlantic intermediate waters and ice-age CO_2 variations (Oppo and Fairbanks, 1990), providing some support for the intermediate water model. However, other lines of evidence (Naqvi et al., 1994) suggest that ocean-wide changes in intermediate water production were less than earlier predicted; thus, the above correlation may be coincidental. Researchers are still pursuing alternate explanations for the ice-age CO_2 changes. For example, inferred changes in the ocean pH as estimated by boron isotopes (Sanyal et al., 1995) provides support for an alternate model linking CO_2 variations resulting from variations in the organic carbon/carbonate rain ratio (cf. Sec. 6.4.5).

Although less commented on, the phase relations between CO_2 and climate also present some puzzling dilemmas. Sowers et al. (1991) analyzed $\delta^{18}O$ variations of atmospheric oxygen in ice to directly correlate the marine and ice-core records, thereby establishing that at the time of the penultimate deglaciation ($\sim$130,000 years ago), CO_2 increased before temperature. However, CO_2 levels remained high even after local temperatures decreased and after global ice volume started to increase. Thus, there is not a consistent phase relationship between CO_2 and climate—a conclusion opposite to that originally inferred from proxy $\delta^{13}C$ measurements (Shackleton and Pisias, 1985). A satisfactory explanation for this response is lacking.

13.4.7 *What is the origin of the abrupt warming at $\sim$14,000 BP?* As discussed in the referenced section, this question is important for trying to understand factors responsible for glacial termination. Even though two ice cores have been drilled on Greenland since 1990, at this stage we do not know whether CO_2 increased at the same time as North Atlantic Deep Water increased (Charles and Fairbanks, 1992). Until the CO_2 rec-

ord is published, we cannot say for sure whether NADW is driving the deglaciation or perhaps responding to a CO_2 increase.

It is rather peculiar that we can say something more definitive about the penultimate deglaciation than the most recent deglaciation. Direct correlations of the Vostok ice-core record with the marine $\delta^{18}O$ record, using $\delta^{18}O$ variations in atmospheric O_2 in the ice, indicate that CO_2 increased before *local* warming on Antarctica (Sowers et al., 1991). Since the Vostok temperature record at 80°S has a high correlation with a marine record at 45°S ($r = 0.86$; Crowley, 1992), and that record leads global ice volume (Hays et al., 1976), for the last deglaciation CO_2 definitely leads ice volume and contributes to the deglaciation. Since NADW production is at best in phase with ice volume (or may even lags it) during the last interglacial (Crowley, 1992; Imbrie et al., 1993), this result implies that CO_2 is more important than NADW for driving the system toward melting during the penultimate deglaciation. One would think the same pattern might apply to the most recent deglaciation, but until the new Greenland CO_2 measurements are available, we will not know for sure.

13.4.8 *What factors are responsible for ice sheet inception during the early part of a glacial cycle?* Not a great deal of progress has been made on this topic. The cited paper in the discussion (Sec. 6.4.4) continues to present a dilemma that has not received an adequate answer. One explanation involves the possibility that some climate models may just be too warm in summer, preventing any buildup of permanent snowcover. Another explanation involves the possibility that the climate system sensitivity is greater than accounted for in our present generation of climate models. Thus, for a given orbital perturbation, a larger temperature (and snow) response may result than is presently produced in climate models. However, this explanation has implications for greenhouse warming projections, for it suggests that the future response to greenhouse warming may also be large.

A third explanation is that the climate sensitivity is variable. This is essentially the same as suggesting that the system is capable of rapid transitions in some areas of parameter space. In Chap. 1 we cite some examples of such snowline behavior in simple climate models (Sec. 1.3.3). A similar response has now been found in a GCM (Crowley et al., 1994), so perhaps this mechanism may apply to the cooling event after the last interglacial.

13.4.9 *What is the origin of the ice-age 100,000 year cycles?* Again, there has been little satisfactory progress on this topic. We still have several explanations but no rigorous testing of the different models to determine which gives the most satisfactory response.

13.4.10 *What is the origin of 100,000-year cycles in the pre-Pleistocene?* This question should not have been separated from the previous one in our original edition, as they are almost certainly linked. There has been a small amount of progress in this area. Energy-balance-model calculations (Crowley et al. 1992) suggest that the twice yearly passage of the sun across the equator could boost the temperature response at the 100,000 year period, especially at times when there was a large land mass in the tropics (e.g., Triassic). However, this response needs to be further tested with GCMs before it can be accepted.

13.4.11 *What is the origin of decadal-centennial scale climate fluctuations?* This subject has received a great deal of enhanced emphasis since original publication of our book. The motivation has been the recognition that understanding the origin of ''natural'' variability in the climate system is essential for predicting the future course of warming resulting from the anthropogenic greenhouse perturbation. In the area of observations, Bradley and Jones (1993) have composited a number of records of summer temperatures to produce an estimate of decadal-scale trends in temperatures for the northern hemisphere (summer, land) back to 1400. As unpublished results indicate that the instrumental record for the last 120 years for northern hemisphere summer (land) is not greatly different from northern hemisphere (mean annual), the Bradley and Jones compilation provides a very useful estimate of decadal-scale trends, suggesting perhaps a 0.6–0.7°C temperature fluctuation over the last few hundred years. This estimate is smaller than estimates from individual sites (Sec. 5.1.2). Unless the Bradley and Jones estimate is in error, the composite record suggests that estimates from individual sites may be biased.

Some progress has also been made on assessing mechanisms for decadal-centennial-scale climate

variability. Crowley et al. (1993) reanalyzed a record of volcanism by separating out the volcanic spikes from background variations that are probably of different origin (cf. Sec. 5.2.1). Results suggest that the correlations between decadal-scale volcanism and temperature change are much smaller than originally suggested ($r = -0.23$ rather than -0.52). This result suggests that, except for occasional clusters of volcanism, the latter mechanism may not be important for decadal-scale climate variability.

Despite the often contested conclusions about the role of solar variability in climate (Sec. 5.2.2), analysis of proxy records continues to find solar spectra in climate records (Crowley and Kim, 1993). In the opinion of the first author of this book, these results now suggest that solar variability plays a significant role in decadal-centennial-scale climate variability.

A third mechanism involves variations of the ocean-atmosphere system on decadal time scales. There is still strong reason for believing such variations have a significant effect on climate. The key question involves the amount of the effect. Although a great deal of work has been done on modeling variations in the North Atlantic thermohaline flow, and significant evidence exists for variations in this flow during the last glacial maximum (Sec. 3.1.8), there is less support for significant thermohaline changes on millennial scales in the last 10,000 years. For example, if the Cd/Ca record illustrated in Fig. 3.19 can be taken as a measure of NADW variations in the Holocene, translation of these changes into estimates of NADW variability suggests only about a 10–20% variation around the mean value for the last 10,000 years (Crowley and Kim, 1993).

Another mechanism for ocean-atmospheric interactions on decadal time scales involves fluctuations in the Pacific Basin. Detailed analyses of time series spanning the last 25 years indicate a significant decadal-shift in equatorial Pacific temperatures that resemble decadal-scale El Ninos (Graham, 1995). These fluctuations may significantly affect global temperatures. There has been very little modeling work done on understanding decadal-scale fluctuations in the Pacific realm. Because interannual fluctuations in these regions have a planetary scale influence, while changes in the North Atlantic thermohaline cell are restricted to the North Atlantic, Arctic, and northwestern

Europe (Delworth et al., 1993), a case can be made that in terms of the planetary scale influence, Pacific decadal fluctuations may explain more variance in global climates than Atlantic thermohaline fluctuations. For example, modeled decadal-scale variations in the Atlantic thermohaline circulation cause only about a 0.05°C change in global average temperature (T. Delworth, personal communication), while the recent decadal shift related to Pacific changes is on the order of 0.2–0.3°C (Graham, 1995; cf. Fig. 14.4). Thus, if we were to use global average temperatures as a gauge of relative importance of mechanisms, the "weight" of Pacific Basin changes is significantly greater than the more intensively studied Atlantic phenomenon.

New Trends

In addition to the above climate developments, a few new problems have emerged or become more important since this book was first published. The following section will briefly outline these results.

Abrupt transitions during the last glacial cycle (Sec. 6.3). New results have shed more light on this problem. Counting of annual layers in ice cores during the deglaciation indicate that some transitions could have occurred in five years (Alley et al., 1993; cf. Sec. 3.2.4). Charles et al. (1994) have argued that such changes may be more explicable in terms of rapid changes in the atmospheric circulation than in the thermohaline circulation. Although Bond et al. (1993) illustrate variations in the surface waters of the North Atlantic that are correlative with the ice-core fluctuations, it is not yet clear whether the thermohaline circulation is involved in such fluctuations.

The far-field extent of such rapid fluctuations is also a subject of considerable interest, but it is difficult to draw any definitive conclusions about results published to date. It is also not clear whether any such far-field changes reflect variations in the thermohaline circulation of concomitant changes in atmospheric CO_2 levels. We continue to maintain that thermohaline-driven changes produce responses of opposite signs in the northern and southern hemispheres due to processes associated with heat transfer between the North Atlantic and South Atlantic (see p. 28 and Crowley, 1992). So we find it difficult to accept that thermohaline changes alone can account for

any inferred global synchroneity of cooling or warming. CO_2 changes could produce such synchroneity, and preliminary results from Greenland support such rapid transitions (Fig. 6.7). But until such results are reproduced in the new Greenland ice cores, final judgment on the matter will have to be suspended.

An associated phenomenon involves the relation between the rapid fluctuations in Greenland and massive discharges of icebergs from North America (Bond et al., 1992)—the so-called "Heinrich Events" (Heinrich, 1988; Broecker, 1994). These discharges are associated with significant decreases in sea-surface temperature and salinity. Subsequent to each of the six major events of the last 100,000 years, the system flips to one of the warm modes illustrated in Fig. 6.6. There are also smaller discharges on the time scale of 2000–3000 years (Bond and Lotti, 1995).

The precise connection between the Heinrich Events and the subsequent switch to warm temperatures has not yet been determined. Nor have the reasons for the quasi-cyclical nature of the oscillations been clarified. One plausible explanation involves modeling studies suggesting inherently bistable conditions for ice sheets (Alley, 1990; MacAyeal, 1993). The mechanism involves a first phase in which the ice sheet is frozen to the bedrock and is associated with ice-sheet thickening. This thickening in turn leads to basal melting and lubrication of the discharge streams. Rapid discharge in ice streams and reduction in ice volume quickly result. Although the time scale for this mechanism could explain the 7000–10,000-year time scale of major iceberg discharges, the similarity of the 2000–3000-year time scale of smaller oscillations with the ^{14}C record of solar variability suggests that solar forcing may be playing a role in the higher frequency fluctuations.

A more contentious matter concerns evidence from one of the two new Greenland ice cores for instability during the last interglacial (GRIP Ice Core Project Members, 1993). However, this evidence is not detected in a companion core drilled within 30 km of the first core (Taylor et al., 1993), nor is it found in surface or deep waters from the North Atlantic (Keigwin et al., 1994; McManus et al., 1994). Although it is too early to conclude that there may be a disturbance in the first ice core, this possibility must be kept in mind. It is our opinion that the original conclusion concerning this matter requires substantially more verification before it can be accepted.

Concluding remarks: A disturbing trend in paleoclimatology. Although studies over the last five years have uncovered a vast amount of fascinating information concerning the history of past climates, there is a significant undercurrent running through some of this and previous research that we find distrubing. The problem involves one of "closure," that is, bringing a problem to completion in terms of developing a satisfactory answer to an observation. In several areas we find a lack of closure on long-pursued problems. For example, no real breakthrough has occurred on the origin of the 100,000-year cycle. We define breakthrough in terms of convincing evidence that reconciles one model in favor of others and offers solid testable predictions of the model. Similarly, the spate of early work on the ice-age CO_2 problem has failed to result in a satisfactory explanation, and it is our sense that there are so many possibilities now available that the fascination with the problem has been dimmed by lack of clinching support with respect to any one hypothesis. In fact, research enthusiasm on both the 100,000-year cycle and CO_2 problem seems to have waned, as investigators almost appear to be dismayed at the multiplicity of explanations and lack of clinching arguments for any one mechanism.

In addition to the above developments (or lack thereof), there are disturbing indications of difficult-to-understand disagreements in various types of paleodata. For example, cadmium/calcium and $\delta^{13}C$ records yield different indications of the history of deep-water circulation in the southern hemisphere (Oppo and Rosenthal, 1994). Similarly, there are very large differences between various proxy estimates of CO_2 during the Phanerozoic (see above discussion in Sec. 13.4.3). These developments lead us to wonder whether we are not encountering some fundamental limits to extracting sufficiently-precise environmental information from some types of proxy data. Such a development could make the problem of closure even more difficult than it already is.

Other disturbing developments involve misinterpretation of the far-field role of North Atlantic Deep Water (see discussion on abrupt transitions), invocation of the WSBW mechanism while failing to rigorously demonstrate its role (13.4.4), and

problems with application of the increased ocean heat transport argument (13.4.4). A final development involves exceedingly slow progress in some critical areas of paleoclimatology. For example, problems of "equable" climates and the history of tropical sea surface temperatures in the Cenozoic are getting relatively little attention, and a number of years will pass before enough data have been collected to clarify some of these issues.

So despite the significant advances that have taken place in the last five years, the future of paleoclimatology is clouded by the emergence of some troubling problems. These developments result in the peculiar situation where we continue to learn a great deal more about a subject but the ultimate goal of solving a particular problem seems no closer to resolution. It is almost as if the goal is receding at the same rate in which we are acquiring new knowledge. A similar "climate paradox" may be occurring with regard to the anthropogenic greenhouse effect.

Having identified some of these problems, it will be interesting to determine whether a revisitation of this subject in a few years will result in the persistence of these problems or gratifying clarifications. Perhaps some readers may be motivated to participate in these clarifications.

References

Alley, R. B., 1990. Multiple steady states in ice-water-till systems. *Annals of Glaciology* **14**:1–5.
Alley, R. B., et al., 1993. Abrupt increase in Greenland snow accumulation at the end of the Younger Dryas event. *Nature* **362**:527–529.
Arthur, M. A., K. Miller, and T. J. Crowley, 1991. Global episodes of moderate to extreme warmth. In: *Advisory Panel Report on Earth System History*. G. S. Mountain (Ed.). Joint Oceanographic Institutions Inc., Washington, D.C., pp. 51–74.
Barrera, E., and B. T. Huber, 1993. Eocene to Oligocene oceanography and temperatures in the Antarctic Indian Ocean. In: *The Antarctic Paleoenvironment: A Perspective on Global Change*. Antarctic Res. Ser. **60**, Am. Geophys. Union, Washington, D.C., pp. 49–65.
Barron, E., et al., 1995. A "simulation" of Mid-Cretaceous climate. *Paleoceanog.* (in press).
Berner, R. A., 1992. Palaeo-CO_2 and climate. *Nature* **358**:114.
Berner, R. A., 1994. GEOCARB II: A revised model of atmospheric CO_2 over Phanerozoic time. *Am. J. Sci.* **294**:56–91.

Bond, G. C., and R. Lotti, 1995. Iceberg discharges into the North Atlantic on millennial time scales during the last glaciation. *Science* **267**:1005–1008.

Bond, G., et al., 1992. Evidence for massive discharges of icebergs into the North Atlantic ocean during the last glacial period. *Nature* **360**:245–249.

Bond, G., et al., 1993. Correlations between climate records from North Atlantic sediments and Greenland ice. *Nature* **365**:143–147.

Bradley, R. S., and P. D. Jones, 1993. 'Little Ice Age' summer temperature variations: Their nature and relevance to recent global warming trends. *The Holocene* **3**:367–376.

Broecker, W. S., 1994. Massive iceberg discharges as triggers for global climate change. *Nature* **372**:421–424.

Cerling, T. E., 1991. Carbon dioxide in the atmosphere: Evidence from Cenozoic and Mesozoic paleosols. *Am. J. Sci.* **291**:377–400.

Charles, C. D., and R. G. Fairbanks, 1992. Evidence from Southern Ocean sediments for the effect of North Atlantic deep-water flux on climate. *Nature* **355**:416–419.

Charles, C. D., D. Rind, J. Jouzel, R. D. Koster, and R. G. Fairbanks, 1994. Glacial-interglacial changes in moisture sources for Greenland: Influences on the ice core record of climate. *Science* **263**:508–511.

Chen, J.-J., J. W. Farrell, D. W. Murray, and W. L. Prell, 1994. Atmospheric CO_2 variations over the last 3.6 my inferred from benthic and planktonic foraminfera $\delta^{13}C$ records, ODP Site 758. *Eos (Transactions, Am. Geophys. Union, Suppl. Spr. Meet.)* **74**:183.

Crowley, T. J., 1992. North Atlantic Deep Water cools the southern hemisphere. *Paleoceanog.* **7**:489–497.

Crowley, T. J., 1993. Geological assessment of the greenhouse effect. *Bull. Am. Meteorolog. Soc.* **74**:2363–2373.

Crowley, T. J., 1994. Pleistocene temperature changes. *Nature* **371**:664.

Crowley, T. J., and K.-Y. Kim, 1993. Towards development of a strategy for determining the origin of decadal-centennial scale climate variability. *Quat. Sci. Rev.* **12**:375–385.

Crowley, T. J., and K.-Y. Kim, 1995. Comparison of longterm greenhouse projections with the geologic record. *Geophys. Res. Lett.* **22**:933–936.

Crowley, T. J., and S. K. Baum, 1995. Reconciling Late Ordovician (440 Ma) glaciation with very high (14X) CO_2 levels. *J. Geophys. Res.* **100**:1093–1101.

Crowley, T. J., K.-Y. Kim, J. G. Mengel, and D. A. Short, 1994. Modeling 100,000 year climate fluctuations in pre-Pleistocene time series. *Science* **255**:705–707.

Crowley, T. J., T. A. Criste, and N. R. Smith, 1993. Reassessment of Crete (Greenland) ice core acidity/volcanism link to climate change. *Geophys. Res. Lett.* **20**:209–212.

Crowley, T. J., K.-J. Yip, and S. K. Baum, 1994. Snowline instability in a general circulation model: application to Carboniferous glaciation. *Clim. Dynam.* **10**:363–376.

Delworth, T., S. Manabe, and R. J. Stouffer, 1993. Interdecadal variability of the thermohaline circulation in a coupled ocean-atmosphere model. *J. Clim.* **6**:1993–2011.

Dowsett, H. J., et al., 1992. Micropaleontological evidence for increased meridional heat transport in the North Atlantic Ocean during the Pliocene. *Science* **258**:1133–1134.

Freeman, K. H., and J. M. Hayes, 1992. Fractionation of carbon isotopes by phytoplankton and estimates of ancient CO_2 levels. *Glob. Biogeochem. Cycles* **6**:185–198.

Graham, N. E., 1995. Simulation of recent global temperature trends. *Science* **267**:666–671.

GRIP Ice Core Project Members, 1993. Climate instability during the last interglacial period recorded in the GRIP ice core. *Nature* **364**:203–207.

Guilderson, T. P., R. G. Fairbanks, and J. L. Rubenstone, 1994. Tropical temperature variations since 20,000 years ago: Modulating interhemispheric climate change. *Science* **263**:663–665.

Heinrich, H., 1988. Origin and consequences of cyclic ice rafting in the northeast Atlantic Ocean during the past 130,000 years. *Quat. Res.* **29**:142–152.

Herbert, T. D., and J. L. Sarmiento, 1991. Ocean nutrient distribution and oxygenation: Limits of the formation of warm saline bottom water over the past 91 m.y. *Geol.* **19**:702–705.

Imbrie, J., et al., 1993. On the structure and origin of major glaciation cycles 2. The 100,000-year cycle. *Paleoceanog.* **8**:699–735.

Keigwin, L. D., W. B. Curry, S. J. Lehman, and S. Johnsen, 1994. The role of the deep ocean in North Atlantic climate change between 70 and 130 kyr ago. *Nature* **371**:323–326.

Kennett, J. P., and L. D. Stott, 1990. Proteus and Protooceanus: Ancestral Paleogene oceans as revealed from Antarctic stable isotopic results; ODP Leg 113. In: *Proceedings of the Ocean Drilling Program.* P. F. Barker, et al. (Eds.). Scientific Results, vol. 113, pp. 865–880.

MacAyeal, D. R., 1993. Binge/purge oscillations of the Laurentide Ice Sheet as a cause of the North Atlantic's Heinrich events. *Paleoceanog.* **8**:775–784.

Markwick, P. J., 1994. "Equability," continentality, and Tertiary "climate": The crocodilian perspective. *Geology* **22**:613–616.

McManus, J. F., et al., 1994. High-resolution climate records from the North Atlantic during the last interglacial. *Nature* **371**:326–329.

Mikolajewicz, U., E. Maier-Reimer, T. J. Crowley, and K.-Y. Kim, 1993. Effects of Drake and Panamanian gateways on the circulation of an ocean model. *Paleoceanog.* **8**:409–426.

Naqvi, W. A., C. D. Charles, and R. G. Fairbanks, 1994. Carbon and oxygen isotope records of benthic foraminifera from the Northeast Indain Ocean: Implications on glacial-interglacial atmospheric CO_2 changes. *Earth Plan. Sci. Lett.* **121**:99–110.

Oppo, D. W., and R. G. Fairbanks, 1990. Atlantic Ocean thermohaline circulation of the last 150,000 years: Relationship to climate and atmospheric CO_2 *Paleoceanog.* **5**:277–288.

Oppo, D. W., and Y. Rosenthal, 1994. Cd/Ca changes in a deep Cape Basin core over the past 730,000 years: Response of circumpolar deepwater variability to northern hemisphere ice sheet melting? *Paleoceanog.* **9**:661–675.

Raymo, M. E., and G. H. Rau, 1995. Mid-Pliocene warmth: Stronger greenhouse and stronger conveyor. *Mar. Micropaleontol.* (in press).

Rind, D., and M. Chandler, 1991. Increased ocean heat transports and warmer climate. *J. Geophys. Res.* **96**:7437–7461.

Sanyal, A., N. G. Hemming, G. H. Hanson, and W. S. Broecker, 1995. Evidence for a higher pH in the glacial ocean from boron isotopes in foraminifera. *Nature* **373**:234–236.

Sowers, T., et al., 1991. The $\delta^{18}O$ of atmospheric O_2 from air inclusions in the Vostok ice core: Timing of CO_2 and ice volume changes during the penultimate deglaciation. *Paleoceanog.* **6**:679–696.

Stute, M., et al., 1994. A continental paleotemperature record from equatorial Brazil derived from noble gases dissolved in goundwater. *Eos (Transactions, Am. Geophys. Union, Suppl. Fall Meet.)* **75**:381.

Taylor, E. L., T. N. Taylor, and R. Cúneo, 1992. The present is not the key to the past: A polar forest from the Permian of Antarctica. *Science* **257**:1675–1677.

Taylor, K. C., et al., 1993. Electrical conductivity measurements from the GISP2 and GRIP Greenland ice cores. *Nature* **366**:549–552.

Van der Burgh, J., H. Visscher, D. L. Dilcher, and W. M. Kürschner, 1993. Paleoatmospheric signatures in Neogene fossil leaves. *Science* **260**:1788–1790.

Wing, S. L., and D. R. Greenwood, 1993. Fossils and fossil climate: The case for equable continental interiors in the Eocene. *Phil. Trans. R. Soc. B* **41**:243–252.

Yapp, C. J., and H. Poths, 1992. Ancient atmospheric CO_2 pressures inferred from natural goethites. *Nature* **355**:342–344.

Yemane, K., 1993. Contribution of Late Permian palaeogeography in maintaining a temperate climate in Gondwana. *Nature* **361**:51–54.

Zachos, J. C., L. D. Stott, and K. C. Lohmann, 1994. Evolution of early Cenozoic marine temperatures. *Paleoceanog.* **9**:353–387.

INDEX